APPLICATIONS OF ARTIFICIAL INTELLIGENCE IN HEALTHCARE AND BIOMEDICINE

T0286294

Artificial Intelligence Applications in
Healthcare and Medicine

APPLICATIONS OF ARTIFICIAL INTELLIGENCE IN HEALTHCARE AND BIOMEDICINE

Edited by

ABDULHAMIT SUBASI

University of Turku, Turku, Finland
Effat University, Jeddah, Saudi Arabia

ELSEVIER

ACADEMIC PRESS
An imprint of Elsevier

Academic Press is an imprint of Elsevier
125 London Wall, London EC2Y 5AS, United Kingdom
525 B Street, Suite 1650, San Diego, CA 92101, United States
50 Hampshire Street, 5th Floor, Cambridge, MA 02139, United States

Notices
Knowledge and best practice in this field are constantly changing. As new research and experience broaden our understanding, changes in research methods, professional practices, or medical treatment may become necessary.

Practitioners and researchers must always rely on their own experience and knowledge in evaluating and using any information, methods, compounds, or experiments described herein. In using such information or methods they should be mindful of their own safety and the safety of others, including parties for whom they have a professional responsibility.

To the fullest extent of the law, neither the Publisher nor the authors, contributors, or editors, assume any liability for any injury and/or damage to persons or property as a matter of products liability, negligence or otherwise, or from any use or operation of any methods, products, instructions, or ideas contained in the material herein.

ISBN: 978-0-443-22308-2

For information on all Academic Press publications visit our website at https://www.elsevier.com/books-and-journals

Publisher: Stacy Masucci
Acquisitions Editor: Linda Buschman
Editorial Project Manager: Pat Gonzalez
Production Project Manager: Swapna Srinivasan
Cover Designer: Greg Harris

Typeset by TNQ Technologies

Working together to grow libraries in developing countries
www.elsevier.com • www.bookaid.org

Dedication

A huge thanks to my parents for always expecting me to do my best, and telling me I could accomplish anything, no matter what it was.

To my wife, Rahime, for her patience and support.

To my wonderful children, Seyma Nur, Tuba Nur, and Muhammed Enes. You are always in my heart and the joys in my life.

To those who read this book, and appreciate the work that goes into them, thank you. If you have any feedback, please let me know.

Abdulhamit Subasi

Contents

List of contributors

Wei Chen Center for Intelligent Medical Electronics (CIME), School of Information Science and Technology, Fudan University, China; Human Phenome Institute, Fudan University, Shanghai, China

Eman Hassanain Turku Brain and Mind Center, Faculty of Medicine, University of Turku, Turku, Finland

Safdar Wahid Inamdar Indian Institute of Technology, Indore, Madhya Pradesh, India

Muhammad Irfan Center for Intelligent Medical Electronics (CIME), School of Information Science and Technology, Fudan University, China; TIERS, University of Turku (UTU), Turku, Finland

Noman Mustafa Intac Optical Instruments Ltd, Suzhou, Jiangsu, China

Sohan Patnaik Department of Mechanical Engineering, Indian Institute of Technology Kharagpur, Kharagpur, West Bengal, India

Saeed Mian Qaisar LINEACT CESI, Lyon, France; College of Engineering, Effat University, Jeddah, Saudi Arabia

Tanisha Sahu Department of Computer Science and Engineering, Indian Institute of Technology, Indore, Madhya Pradesh, India

Rupal Shah Department of Electrical Engineering, Indian Institute of Technology, Indore, Madhya Pradesh, India

Muhammed Enes Subasi Faculty of Medicine, Izmir Katip Celebi University, Izmir, Turkey

Abdulhamit Subasi Institute of Biomedicine, Faculty of Medicine, University of Turku, Turku, Finland; College of Engineering, Effat University, Jeddah, Saudi Arabia; Faculty of Medicine, Institute of Biomedicine, University of Turku, Turku, Finland; Department of Computer Science, College of Engineering, Effat University, Jeddah, Saudi Arabia

Mirka Suominen Institute of Biomedicine, Faculty of Medicine, University of Turku, Turku, Finland

Fadime Tokmak Digital Health Technology, Faculty of Technology, University of Turku, Turku, Finland

Tomi Westerlund TIERS, University of Turku (UTU), Turku, Finland

Series preface

Artificial intelligence (AI) is a concept, which allows for the optimization of performance criteria utilizing a set of data and some experience. The learning process is essentially the execution of the model parameter optimization with a training dataset or past experience. Models can be predictive, for predicting the future, descriptive, for extracting knowledge from input data, or both. Two fundamental activities are accomplished in machine learning: (1) processing massive amounts of data and optimizing the model and (2) testing the model and efficiently displaying the solution.

The process of applying AI methods to a large dataset is called data mining. The logic behind data mining is when a large volume of raw data is processed, and efficient prediction model is constructed, with high predictive accuracy. AI applications are present in different areas: in finance, for credit scoring, fraud detection, or stock market prediction; in manufacturing, for optimization, control, and troubleshooting; in medicine, for efficient medical diagnosis; and in telecommunication, for network and quality of service optimization. Furthermore, AI enables algorithms to learn and adapt to changes in a variety of situations. Artificial neural networks, fuzzy logic, support vector machines, decision tree algorithms, and deep learning algorithms are all examples of AI techniques.

AI tools have been employed in different areas for several years. New AI methods such as deep learning have assisted uncovering information, which entirely altered the approach to different areas. AI has reached a certain maturity as an academic subject and there are many useful books related to this subject. Since AI is an interdisciplinary subject, it should be implemented in different ways depending on the application field. Nowadays, there is great interest in AI applications in several disciplines. This book series presents how AI and machine learning methods can be used in the analysis of different data. The application of various AI technologies spans across diverse domains such as Biomedical Engineering, Electrical Engineering, Computer Science, Information Technology, Medical Science, healthcare, Finance, and Economy. These fields leverage engineering methodologies to address complex challenges and advance their respective domains.

This book series will consist of numerous volumes, each of which will cover an application of AI techniques in a different field. This series will benefit a wide range of readers, including academicians, professionals, graduate students, and researchers from a variety of fields who are exploring artificial intelligence applications. This book series will provide an in-depth account of recent research in this emerging topic, and the principles discussed here will spark additional research in this multidisciplinary field.

The target audience is widespread since this series will include several areas of applications of AI, machine learning, and deep learning that are given below. Hence, the audience can be computer scientists, biomedical engineers and healthcare scientists, financial engineers, economists, researchers, and consultants in electrical, computer engineering and science, finance, economy, and security. Nowadays, AI is one of the hot topics that is used in several data analyses in the world. Therefore, this book series caters to a wide range of application areas and audiences. It will encompass emerging AI trends across various fields, with a particular focus on healthcare, cybersecurity, and finance, among others.

Preface

The rapid advancement of Artificial Intelligence (AI) solutions and their widespread adoption across healthcare industry has enabled the development of complex models to tackle real-world problems by learning from observed training data. The process of creating effective AI models and achieving reliable results demands a significant investment of time and effort. This involves understanding core project concepts, constructing robust data pipelines, and conducting data analysis and visualization, which includes feature extraction, selection, and modeling. Hence, there is a growing demand for a comprehensive AI framework that is not only well-suited for immersive AI modeling but also excels in preprocessing, visualization, system integration, and provides robust support for runtime deployment and maintenance settings. Python stands out as an innovative programming language with its versatility, ease of implementation, seamless integration, active developer community, and a continuously expanding AI ecosystem, making it a preferred choice for AI applications.

Artificial intelligence (AI) techniques have revolutionized healthcare and biomedicine, offering innovative solutions to complex challenges. This book provides an abstract of the literature on AI techniques applied in healthcare and biomedicine, highlighting their diverse applications and potential benefits. AI techniques offer tremendous potential in transforming healthcare and biomedicine. The applications of machine learning, deep learning, natural language processing, and computer vision in disease prediction, diagnosis, treatment, and drug discovery have shown promising results. Addressing ethical concerns and fostering interdisciplinary collaboration are key to unlocking the full potential of AI in improving patient care, advancing biomedical research, and shaping the future of healthcare.

Applications of Artificial Intelligence in Healthcare and Biomedicine is designed to cater to readers with varying skill levels and aims to equip them with the ability to develop practical AI solutions. This book serves as a guide for problem solvers, offering a systematic framework that encompasses principles, procedures, practical examples, and code snippets. It imparts the essential skills required for understanding and resolving a range of AI challenges. Moreover, this book also highlights the diverse applications of AI in healthcare and biomedicine. AI models have been developed for early disease detection, improving patient outcomes by identifying subtle patterns and biomarkers in patient data. AI-powered systems have been used for drug discovery, accelerating the identification of potential therapeutic compounds and reducing the time and cost of the development process. AI techniques also support remote patient monitoring, enabling continuous monitoring of vital signs and proactive intervention when necessary. Furthermore, this book addresses the challenges and considerations in implementing AI techniques in healthcare and biomedicine.

AI techniques encompass a variety of computational algorithms, including artificial neural networks, k-nearest neighbors, support vector machines, decision tree algorithms, and deep learning methods. The applications of AI have gained considerable attention in economics, security, healthcare, biomedicine, and biomedical engineering. This book explores various AI techniques employed, including machine learning, deep learning, natural language processing, and computer vision. Machine learning algorithms, such as decision trees, support vector machines, and random forests, have been used for disease prediction, diagnosis, and treatment outcome prediction. Deep learning models, such as convolutional neural networks and recurrent neural networks, have demonstrated exceptional performance in medical image analysis, genomics, and drug discovery. Natural language processing techniques enable the extraction of valuable information from clinical text, facilitating clinical decision support systems and pharmacovigilance. Computer vision algorithms enable automated analysis of medical images, assisting in early detection and diagnosis of diseases.

This book serves as an excellent resource for learning AI techniques through real-world case studies within the Python Machine Learning environment, enabling readers to become proficient practitioners. The book's focus is on establishing a foundational knowledge of Machine Learning for addressing real-world problems in healthcare and biomedicine including biomedical signal analysis, medical image analysis, image segmentation, DNA analysis, and disease diagnosis.

The primary goal of this book is to assist a wide range of readers in solving their own real-world problems, including IT professionals, analysts, developers, data scientists, and engineers. Furthermore, it is intended to serve as a textbook for postgraduate and research students working in the field of data science and machine learning. Additionally, it lays the groundwork

for researchers interested in the application of AI methods to healthcare and biomedical data analysis. This book aims to benefit a broad readership, including researchers, professionals, academics, and graduate students from diverse disciplines, who are just beginning to explore applications in biomedical signal analysis, healthcare data analysis, medical image analysis, disease diagnosis, and more. This book provides a comprehensive overview of recent advancements in this emerging field and seeks to inspire further research in this multidisciplinary domain by presenting the concepts outlined here.

To execute the code examples provided in this book, it is necessary to have Python 3.x or higher versions installed on macOS, Linux, or Microsoft Windows. The examples throughout this book frequently utilize essential Python libraries for scientific computing, including SciPy, NumPy, Scikit-learn, matplotlib, pandas, OpenCV, TensorFlow, and Keras.

This book comprises 16 chapters, each addressing the specific aspects of AI. The goal of Chapter 1, **AI techniques for Healthcare and Biomedicine**, is to assist scientists in selecting an appropriate AI model and then guiding them in determining the best strategy by utilizing healthcare data. Furthermore, to introduce readers to the fundamentals of AI before digging into tackling real-world issues with AI methodologies. Chapter 2, **Artificial Intelligence based Emotion Recognition using ECG Signals**, gives a succinct description and implementation of artificial intelligence (AI) algorithms for emotion recognition using electrocardiogram (ECG) signals. Chapter 3, **Artificial Intelligence based Depression Detection using EEG Signals**, presents different AI methods for depression detection using EEG signals. Chapter 4, **EMG signal Classification using Artificial Intelligence**, presents how AI methods are used in neuromuscular disease detection using EMG signals. Chapter 5, **An Evaluation of Pretrained Convolutional Neural Networks for Stroke Classification from Brain CT Images**, presents how to detect brain hemorrhage from CT images, with deep learning architectures. Chapter 6, **Brain Tumor Detection Using Deep Learning from Magnetic Resonance Images**, implements AI algorithms to detect the presence of brain tumors in MRI scans of the brain. Chapter 7, **Artificial Intelligence based Fatty Liver Disease Detection using Ultrasound Images**, presents the use of CNN models for fatty liver disease detection using ultrasound images. Chapter 8, **Deep learning approaches for Breast Cancer detection using Breast MRI**, presents early breast cancer detection from MRI using AI. In Chapter 9, **Automated Detection of Colon Cancer from Histopathological Images Using Deep Neural Networks**, various deep learning and feature extraction methods are investigated to classify histopathological images from colon cancer dataset. Chapter 10, **Optical Coherence Tomography Image Classification for Retinal Disease Detection using Artificial Intelligence**, presents deep learning-based approaches to get a proper and faster classification of OCT images. Chapter 11, **Heart Muscles Inflammation (Myocarditis) Detection using Artificial Intelligence**, focuses on cardiovascular disease (myocarditis) and uses relevant deep learning methods to realize automatic analysis and diagnosis of medical images and verify the feasibility of AI-assisted detection of heart muscles inflammation (myocarditis). Chapter 12, **Artificial intelligence for 3D Medical Image Analysis**, presents 3D medical image analysis using 3D CNN model. Chapter 13, **Medical Image Segmentation using Artificial Intelligence**, presents a biomedical image segmentation application of AI, emphasizing its potential to enhance precision, effectiveness, and therapeutic results. In Chapter 14, **DNA sequence classification using Artificial Intelligence**, typical application of AI in DNA sequence analysis is presented. Chapter 15, **Artificial intelligence in drug discovery and development**, discusses the roles of AI in facilitating drug development and discovery processes, making them more cost-effective or eliminating the need for clinical trials. Chapter 16, **Hospital Readmission forecasting using Artificial Intelligence**, presents an AI-based approach to forecast diabetes patients' hospital readmissions. The research on diabetes patients' readmission rates may be useful in clinical practice and offer useful suggestions to stakeholders for minimizing readmission and lowering public healthcare expenses.

Acknowledgments

First of all, I would like to thank my publisher Elsevier and its team of dedicated professionals who have made this book-writing journey very simple and effortless and many who have worked in the background to make this book a success.

I would like to thank Linda Buschman and Pat Gonzalez who provided great support and did a lot of work for this book. Also, I would like to thank the technical team for being patient in getting everything necessary completed for this book.

Abdulhamit Subasi

AI techniques for healthcare and biomedicine*

Abdulhamit Subasi[1,2]

[1]Institute of Biomedicine, Faculty of Medicine, University of Turku, Turku, Finland; [2]Department of
Computer Science, College of Engineering, Effat University, Jeddah, Saudi Arabia

1. Introduction

Artificial intelligence (AI) techniques have emerged as formidable tools with enormous promise to alter the world of healthcare and biomedicine. AI provides new possibilities for addressing complex challenges and enhancing patient care by employing powerful computational methods, machine learning, and deep learning models. This chapter introduces AI approaches in healthcare and biomedicine, highlighting their various applications and the benefits they bring to the sector.

AI approaches include a variety of technologies such as machine learning, deep learning, natural language processing, and computer vision. These strategies have found use in a variety of areas of healthcare and biomedicine, including disease diagnosis, customized treatment, drug development, and remote patient monitoring. AI models can process and analyze vast amounts of data, detect patterns and correlations, and offer actionable insights, allowing for more accurate and efficient decision-making in healthcare settings.

Decision trees, support vector machines, and random forests are examples of machine learning algorithms that have been effectively applied to disease prediction, risk stratification, and treatment outcome prediction. Deep learning models, such as convolutional neural networks and recurrent neural networks, have excelled in medical image processing, genomics, and drug development. These artificial intelligence algorithms enable the extraction of useful information from large datasets and provide valuable insights for clinical decision assistance.

Natural language processing techniques allow information to be extracted and analyzed from clinical material, such as electronic health records and medical literature. This allows for the development of clinical decision support systems, pharmacovigilance, and the detection of adverse events. Computer vision algorithms are

---

* This chapter is adapted from "Subasi, A. (2023). Introduction to artificial intelligence techniques for medical image analysis. In Applications of Artificial Intelligence in Medical Imaging (pp. 1−49). Academic Press."

important in medical imaging because they allow for automated image processing and interpretation, which aids in illness detection and diagnosis.

The use of AI techniques in healthcare and biomedicine has the potential to provide various benefits. Improved diagnostic accuracy, individualized treatment plans, quicker medication discovery, and greater remote patient monitoring are among the benefits. However, implementing AI in these sectors is fraught with difficulties. Data privacy, security, and algorithmic bias are all ethical concerns that must be addressed. Integrating AI technology into existing healthcare systems and workflows necessitates the development of infrastructure, interoperability standards, and collaboration between healthcare practitioners and AI experts.

The primary obstacle faced by researchers is to design a CAD framework that is both effective and robust in accurately classifying tissues, especially considering the overlapping nature of tissue representations. Moreover, the current level of accuracy falls short of commercial implementation of these diagnostic devices. Consequently, there is a pressing demand for the development of a diagnostic framework that can swiftly and accurately differentiate tissues, including tumors, cysts, stones, and normal tissues (Rawal et al., 2020). The building blocks of the CAD framework are shown in Fig. 1.1.

To summarize, artificial intelligence approaches have the potential to change healthcare and biomedicine by enabling more precise diagnosis, tailored therapies, and efficient healthcare delivery. Machine learning, deep learning, natural language processing, and computer vision are strong methods for extracting meaningful insights from large datasets and advancing patient care. Addressing ethical concerns and resolving implementation obstacles, on the other hand, are critical to realizing AI's full promise in healthcare.

2. Unsupervised learning (clustering)

Clustering is a widely used technique for analyzing experimental data in diverse fields, such as social sciences, computer science, and biology. Its primary purpose is to group data points into meaningful categories to gain

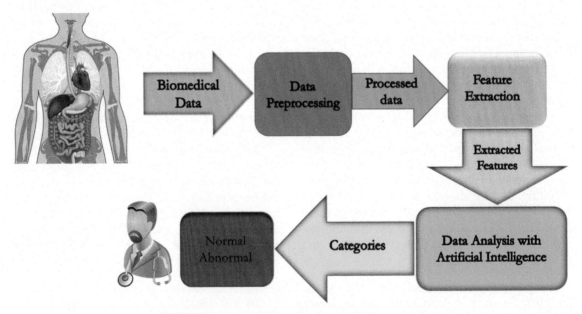

FIGURE 1.1   The building blocks of CAD framework.

insights into the underlying patterns and relationships. For example, businesses use clustering to segment customers based on similar profiles for targeted marketing, astronomers cluster stars based on proximity, and bioinformaticians cluster genes based on expression similarities. Clustering involves organizing objects in a way that similar items belong to the same group, while dissimilar ones are placed in separate groups. However, defining clustering precisely can be challenging due to conflicting goals. The notion of closeness or similarity among objects is not always transitive, while cluster membership is an equivalence relation, specifically transitive. This discrepancy can make clustering ambiguous and difficult to define clearly. As a result, clustering methods may prioritize retaining nearby points in a cluster rather than strictly adhering to similarity relationships (Shalev-Shwartz & Ben-David, 2014; Subasi, 2020).

Clustering faces a fundamental challenge of lacking a "ground truth," which is a common issue in unsupervised learning. Unlike supervised learning, where the goal is to train a classifier to accurately predict future sample labels using labeled training data, clustering is an unsupervised learning task where no labels are predicted. Instead, the objective is to organize the data in a meaningful way without predefined classes. As a result, evaluating the performance of clustering becomes complex, and there is no straightforward method to assess clustering results. Even with complete knowledge of the data distribution, determining the "correct" clustering or evaluating the quality of a proposed clustering remains ambiguous (Shalev-Shwartz & Ben-David, 2014; Subasi, 2020).

Clustering is an approach that groups similar items together, and there are various types of inputs that can be used. One approach, called similarity-based clustering, involves using a dissimilarity matrix or distance matrix D as the input. This method allows for the incorporation of domain-specific similarity or kernel functions, providing flexibility in handling different data types. Another type is feature-based clustering, which is suitable for handling "raw" data that may be potentially noisy. In terms of outputs, there are two possibilities: hierarchical clustering, which generates a nested partition tree, and flat clustering, also known as partition clustering, which divides objects into disjoint sets. Some methods specify that D represents a true distance matrix, while others do not. In cases where we have a similarity matrix S, it can be transformed into a dissimilarity matrix using a monotonically decreasing function. The dissimilarity between items is often described based on their properties, and common attribute dissimilarity functions include the hamming distance, city block distance, square (Euclidean) distance, and correlation coefficient (Murphy, 2012; Subasi, 2020).

Clustering is a straightforward technique utilized by humans to manage the vast amount of information they encounter daily. Instead of dealing with each piece of information separately, people tend to group similar things into clusters, each representing the specific characteristics of the entities within it. Similar to supervised learning, patterns are represented as l-dimensional feature vectors. However, the presence of a clustering tendency is essential in some cases. This involves various tests to determine whether the data exhibits a clustering pattern or not. For instance, if the dataset is entirely random, attempting to identify clusters would be fruitless. The choice of features and proximity measures can significantly influence the clustering results, leading to different clustering outcomes depending on the criteria and methods used (Subasi, 2020; Theodoridis et al., 2010).

2.1 Image segmentation with clustering

Images are widely acknowledged as a vital means of conveying information, and they serve various purposes, including robotic navigation and medical diagnostics, such as detecting cancerous tissues in body scans. An essential step in image recognition involves segmenting the images and identifying distinct objects within them. This can be achieved through various techniques, including frequency domain transformations and histogram plots, which help to identify and distinguish important features in the images (Subasi, 2020; Tatiraju & Mehta, 2008).

Image segmentation plays a crucial role in computer vision and image recognition as a necessary preprocessing step. It involves dividing an image into distinct and nonoverlapping relevant sections, each possessing similar qualities. This process is essential in digital image processing, and the accuracy of segmentation directly influences the effectiveness of subsequent tasks and activities (Zheng et al., 2018). Over the past few decades, various image segmentation algorithms have been developed to address the critical importance of image segmentation in numerous image processing applications. However, researchers continue to seek better solutions, as image segmentation remains a challenging task that significantly impacts the effectiveness of subsequent image processing stages. While clustering was not initially intended for image processing, the computer vision community has adopted it for image segmentation. For instance, the k-means clustering method requires prior knowledge of the number of clusters (k) to be formed. Each pixel in the image is then iteratively assigned to the cluster whose centroid is closest to it. The centroids of the clusters are determined based on the pixels assigned to each cluster, utilizing distance calculations, with the Euclidean distance being the most commonly used due to its simplicity. However, this use of Euclidean distance may introduce errors in the final image segmentation (Gaura et al., 2011; Subasi, 2020). A simple k-means clustering Python code for image segmentation is given below.

```python
import numpy as np # linear algebra
import pandas as pd # data processing, CSV file I/O (e.g. pd.read_csv)
from IPython.display import Image
import matplotlib.pyplot as plt
from mpl_toolkits.mplot3d import Axes3D
from sklearn.cluster import KMeans
import cv2

img = cv2.imread(filepath)
img = cv2.cvtColor(img, cv2.COLOR_BGR2RGB)
r, g, b = cv2.split(img)
r = r.flatten()
g = g.flatten()
b = b.flatten()

K=3
attempts=10
ret,label,center=cv2.kmeans(vectorized,K,None,criteria,attempts,cv2.KMEANS_RA
NDOM_CENTERS)
label = label.flatten()

center = np.uint8(center)
res = center[label.flatten()]
result_image = res.reshape((img.shape))

plt.imshow(result_image)
plt.show()
```

3. Supervised learning

3.1 k-nearest neighbor

The k-nearest-neighbor (KNN) approach initially faced challenges with large training sets and limited computing power, but as computing capabilities improved, it gained popularity in pattern recognition. KNN classifiers are based on analogy learning, comparing a test instance to similar training instances. The classifier identifies the k training instances that are closest to the unknown instance in the pattern space, using a distance metric such as Euclidean distance to determine proximity. In KNN classification, the unknown instance is assigned to the most frequent class among its k-nearest neighbors. For numeric prediction, the classifier provides the average of the real-valued labels associated with the k-nearest neighbors of the unknown instance. The k value is determined through error rate assessment using a test set, with the lowest error rate selected. While KNN classifiers can suffer from low accuracy when dealing with noisy or irrelevant attributes, improvements have been made, including attribute weighting and the pruning of noisy data instances. Various distance metrics, such as Manhattan distance, can be used in addition to Euclidean distance. One limitation of nearest-neighbor classifiers is their potentially slow classification process for test instances. Techniques like partial distance computations and data instance pruning have been employed to reduce classification time. The partial distance technique involves computing distances using a subset of the characteristics, while the editing approach eliminates unnecessary training instances to minimize storage and processing requirements (Han et al., 2011). A simple Python code for k-NN is given below.

3.1.1 Support vector machine

Support vector machines (SVM) is a classification technique suitable for linear and nonlinear data. It achieves this by transforming the original training data into a higher-dimensional space through a nonlinear mapping. In this new space, SVM searches for the optimal linear separation hyperplane. The algorithm relies on support vectors and margins to identify this hyperplane. Despite SVMs having relatively long training times, they exhibit high accuracy and can effectively predict complex nonlinear decision boundaries. Their advantage lies in their reduced risk of overfitting compared to other methods. Furthermore, the support vectors identified during training serve as a concise representation of the learned model. SVMs find application in various domains, including medical imaging, object recognition, handwritten digit recognition, and speaker identification (Han et al., 2011).

In the realm of classification, there are numerous ways to draw separating lines. Our goal is to identify the "best" line, one that minimizes classification errors when applied to new, unseen data. But how can we determine this optimal line, especially in higher-dimensional spaces? Regardless of the number of input features, we refer to this sought-after decision boundary as a "hyperplane." In essence, we seek the optimal hyperplane. SVMs tackle this challenge by searching for the maximum margin hyperplane. We expect the hyperplane with a larger margin to be more accurate in classifying future data instances compared to a hyperplane with a smaller margin. The margin represents the significant gap between the categories, enabling better discrimination between them (Han et al., 2011). A simple Python code for SVM is given below.

```
#Import k-NN Model
from sklearn.neighbors import KNeighborsClassifier
#Create the Model
clf = KNeighborsClassifier(n_neighbors = 5)
#Train the model with Training Dataset
clf.fit(Xtrain,ytrain)
#Test the model with Testset
ypred = clf.predict(Xtest)
```

```
#Import SVM Model
from sklearn import svm
C = 10.0 # SVM regularization parameter
#Create the Model
clf =svm.SVC(kernel = 'linear', C = C)
#Train the model with Training Dataset
clf.fit(Xtrain,ytrain)
#Test the model with Testset
ypred = clf.predict(Xtest)
```

3.2 Decision tree

The decision tree is a nonparametric method that constructs binary trees using data containing both discrete and continuous attributes. To evaluate split characteristics, impurity reduction criteria are applied, where impurity is typically represented by the Gini index (a measure of diversity). Alternatively, entropy or other impurity measures can be utilized in place of the Gini index. Breiman et al. (1984) also suggested twoing, a strategy for dealing with multiclass issues using two-class criteria. In contrast to analyzing the original classes, the twoing approach in-

for impurity criteria with two classes, which is compatible with the Gini index. It also allows for the consideration of misclassification costs when making decisions. To handle missing data values, surrogate splits are employed. If a data item lacks a value necessary for a tree node test, it is directed to another test that utilizes a different characteristic to create a split that closely resembles the original one. Multiple surrogate splits can be identified and used in the correct sequence for data with missing values (Grąbczewski, 2014). A simple Python code for decision tree is given below.

```
#Import Decision Tree Model
from sklearn import tree
#Create the Model
clf = tree.DecisionTreeClassifier()
#Train the Model with Training dataset
clf.fit(Xtrain,ytrain)
#Test the Model with Testing dataset
ypred = clf.predict(Xtest)
```

volves dividing the classes into two larger groups and performing a two-class analysis on these groupings. However, when dealing with a large number of classes, considering all possible groupings can lead to a combinatorial explosion of alternative configurations. Breiman et al. (1984) suggested an efficient approach for determining ideal superclasses for each potential split instead of examining all splits for all possible class groups. This method is designed

3.3 Random forests

Random forest (RF) constitutes a collection of classifiers, with each classifier being a decision tree. The split calculation in individual decision trees involves a random selection of features at each node. Technically, each tree's values are determined by a random vector, sampled independently and uniformly across all trees within the forest. During classification, each tree contributes to the final decision through voting. The construction of random forests can be achieved using bagging along with random attribute selection, proving particularly beneficial when dealing with a limited set of features as it reduces the correlation between the classifiers. Random forests exhibit similar accuracy to AdaBoost but offer greater resistance to errors and outliers. As the forest size increases, the generalization error converges, avoiding overfitting concerns. The accuracy of a random forest relies on the strength of individual classifiers and the measure of their independence, where the ideal scenario is to maintain strong classifiers with reduced correlation. The number of features considered at each split has no impact on random forests, making them efficient for large databases as they examine far fewer features for each split. Random forests also have the potential to be faster than both bagging and boosting methods. Additionally, they provide internal estimations of variable significance, offering valuable insights in various applications (Han et al., 2011). A simple Python code for random forest is given below.

created using the same machine-learning approach on different randomly selected training datasets from the problem area. While one might expect these trees to be almost identical and make the same predictions for each new test case, this assumption is often incorrect, particularly with small training datasets. Decision tree induction is an unstable process, and even small changes in the training data can lead to different features being selected at specific nodes, resulting in significant differences in the subtrees below those nodes. This instability raises concerns about the reliability of the overall ensemble. To address this issue, experts, represented by individual decision trees, can "vote" on each test case, and the predictions are combined accordingly. If one class receives more votes than others, that class is considered the final prediction. Generally, the more votes counted, the more reliable the predictions become, making voting an effective way to aggregate the decisions of multiple models (Witten et al., 2016).

Bagging aims to mitigate the instability of learning methods by reiterating the process

```
#Import Random Forest Ensemble Model
from sklearn.ensemble import RandomForestClassifier
#Create the Model
clf = RandomForestClassifier(n_estimators = 200)
#Train the model with Training set
clf.fit(Xtrain,ytrain)
#Test the model with Test set
ypred = clf.predict(Xtest)
```

3.4 Bagging

The combination of decisions from multiple models involves aggregating the various outputs into a single prediction. For classification tasks, a common method is to take a majority vote, while for numerical prediction, the average is often used. In bagging, several decision trees are

described earlier using a specific training set. Instead of selecting entirely new training datasets each time, bagging modifies the original training data by randomly selecting examples with replacements to form a new dataset of the same size. This resampling approach creates new datasets that contain duplicates of some instances while excluding others. The concept of bagging is closely related to the bootstrap technique, which is used to estimate the generalization error of a learning system. In fact, the term "bagging" stands for "bootstrap aggregating." For each of these artificially generated datasets, bagging applies a learning method, resulting in multiple classifiers that vote to predict the class. The difference between bagging and the idealized technique mentioned earlier lies in the way the training datasets are generated. Bagging does not produce entirely independent datasets from the problem domain; instead, it resamples the existing training data, making the resampled datasets distinct from each other but not independent because they are all based on the same original dataset (Witten et al., 2016). A simple Python code for Bagging is given below.

```
#Import Bagging Ensemble Model
from sklearn.ensemble import BaggingClassifier
#Create a Bagging Ensemble Classifier
bagging = BaggingClassifier(tree.DecisionTreeClassifier(),max_samples = 0.5,
max_features = 0.5)
#Train the model using the training set
bagging.fit(Xtrain,ytrain)
#Predict the response for test set
ypred = bagging.predict(Xtest)
```

3.5 Boosting

The boosting approach for combining multiple models leverages this knowledge by seeking models. Similar to bagging, boosting combines the outputs of individual models through voting for classification or averaging for numerical prediction. It works with models of the same type, such as decision trees, in a manner similar to bagging. However, boosting is an iterative process, unlike bagging. While individual models are generated independently in bagging, the performance of prior models in boosting influences the performance of subsequent models. Boosting achieves this by giving more weight to examples that were incorrectly handled by previous

models, encouraging subsequent models to specialize in those specific cases. Moreover, unlike bagging, where all models are treated equally, boosting assigns different weights to each model based on its confidence level. This weighting mechanism ensures that the contribution of each model is proportionate to its level of certainty in making predictions (Hall et al., 2011).

3.6 AdaBoost

Boosting has various variations, with one of the most well-known being the AdaBoost method, which is widely used for classification tasks. Similar to bagging, AdaBoost can be applied to any classification learning method.

For simplicity, let's assume that the learning algorithm is capable of handling weighted instances, where each instance is assigned a positive weight. When using instance weights, the classifier's error is computed as the sum of the weights of the misclassified instances divided by the total weight of all instances, rather than just the proportion of misclassified instances. By incorporating instance weights, the learning algorithm can be influenced to focus on specific groups of examples, especially those with high weights. These weighted instances become more significant as the algorithm is incentivized to correctly identify them due to their higher importance (Hall et al., 2011).

The boosting process commences by assigning equal weights to all samples in the training data. A classifier is then created for this data using the learning process, and each sample's weight is adjusted based on the classifier's performance. Correctly classified samples receive reduced weights, while incorrectly classified samples receive increased weights. This results in a set of "easy" instances with low weight and a set of "hard" instances with high weight. In the next iteration, a new classifier is constructed for the reweighted data, focusing on successfully categorizing the difficult instances. The weights of the instances are further adjusted based on the output of this new classifier. Consequently, some difficult instances may become even more challenging, while easier ones may become even easier. However, some difficult instances may turn out to be easier, and previously easy instances could become harder. The weights reflect how frequently the samples have been misclassified by the classifiers developed in each iteration. This process creates a sophisticated ensemble of experts that complement one another by retaining a measure of "hardness" with each instance. Boosting often yields classifiers that are significantly more accurate on new data compared to those generated by bagging. However, it is worth noting that unlike bagging, boosting may occasionally face challenges in practical applications (Hall et al., 2011). A simple Python code for AdaBoost is given below.

3.7 XGBoost

XGBoost (Chen & Guestrin, 2016, pp. 785–794) is an efficient gradient tree boosting technique that generates decision trees sequentially. XGBoost is widely known for its ability to perform rapid and relevant computations on various computer platforms. This makes it a popular choice for modeling new features and classifying labels. The algorithm has gained significant interest, especially in tabular and structured datasets. Originally, XGBoost was based on decision tree techniques, which create graphical representations of potential decision solutions based on specific conditions. The ensemble meta-algorithm called "bagging" was then introduced to combine predictions from multiple decision trees using majority voting. To further enhance the performance, the bagging method evolved into a decision tree forest, where attributes were randomly chosen to construct multiple trees, reducing errors associated with individual models. To improve the sequential models even further, the gradient descent approach was incorporated to minimize errors. This made XGBoost an effective method for optimizing the gradient boosting algorithm by handling missing data and mitigating overfitting concerns through parallel processing. As a result, XGBoost has become a popular and powerful algorithm for various tasks in machine learning (Bhattacharya et al., 2020). A simple Python code for XGBoost is given below.

```
#Import Adaboost ensemble model
from sklearn.ensemble import AdaBoostClassifier
#Create an Adaboost Ensemble Classifier
clf = AdaBoostClassifier(tree.DecisionTreeClassifier(),n_estimators = 10, alg
orithm= 'SAMME',learning_rate = 0.5)
#Train the model using the training set
clf.fit(Xtrain,ytrain)
#Predict the response for test set
ypred = clf.predict(Xtest)
```

```
% pip install xgboost
#Import XGBoost ensemble model
from xgboost import XGBClassifier
# Create XGB model
model = XGBClassifier()
#Train the model using the training set
model.fit(Xtrain, ytrain)
# make predictions for test data
ypred = model.predict(Xtest)
```

3.8 Artificial neural networks

Artificial neural networks (ANNs) are computational models inspired by the structure of biological neural networks found in the human brain. ANNs consist of interconnected artificial neurons, forming a network. Each connection between neurons allows the transmission of signals. These signals are processed by the receiving neuron, which can then pass them on to other neurons connected downstream. Neurons have states represented by

especially in challenging applications where conventional rule-based programming faces limitations. Originally, the neural network approach aimed to mimic problem-solving processes similar to the human brain. Over time, the focus shifted to capturing specific cognitive abilities, leading to the development of techniques like backpropagation to incorporate feedback information (Vasuki & Govindaraju, 2017). A simple Python code for ANN is given below.

```
#Import ANN model
from sklearn.neural_network import MLPClassifier
#Create the Model
mlp = MLPClassifier(hidden_layer_sizes = (100, ), learning_rate_init = 0.001,
alpha = 1, momentum = 0.9,max_iter = 1000)
#Train the Model with Training set
mlp.fit(Xtrain,ytrain)
#Test the Model with Test set
ypred = mlp.predict(Xtest)
```

real numbers between 0 and 1, and they also possess weights that adjust as they learn. These weights control the strength of the signals transmitted to other neurons in the network. Additionally, neurons can have a threshold, determining whether the output signal will be transmitted based on a certain level. Neurons are typically organized in layers, each layer performing different transformations on their inputs. Neural networks have gained popularity,

3.9 Deep learning

Deep neural networks have proven to be the most suitable choice for image-processing tasks. While conventional neural networks typically consist of one input layer, two or three hidden layers, and one output layer, deep neural networks go much deeper, incorporating multiple hidden layers between the input and output layers. The number of hidden layers determines the depth of the network. These layers are interconnected, with the output of one layer serving as the input for the next. The network's performance heavily relies on the weights assigned to its inputs and outputs, which are determined during the training process. Training deep networks requires significant computational power, processing capacity, access to extensive datasets, and the use of appropriate parallel processing software (Vasuki & Govindaraju, 2017).

Deep learning is a specialized field of artificial intelligence that centers around constructing extensive neural network models capable of making accurate decisions based on data. It is particularly effective in situations where the data is complex and large datasets are available. In the healthcare industry, deep learning is utilized to analyze medical images such as X-rays, MRI scans, and CT scans, aiding in the diagnosis of various health conditions (Kelleher, 2019). Machine learning algorithms derive their own logic from the provided data, eliminating the need for manual coding to address each specific problem. To effectively learn and classify, the algorithm requires a substantial volume of medical images. If the images are precategorized and used for training, they fall under supervised learning; otherwise, they fall under unsupervised learning. A common application is binary classification, such as discerning whether a medical image corresponds to tumor tissue or not (Vasuki & Govindaraju, 2017). A simple Python code for deep neural networks is given below.

3.10 The overfitting problem in neural network training

Although neural networks are renowned for their ability to approximate universal functions, they encounter significant challenges during training, such as overfitting. Overfitting occurs when a model is tailored to a specific training dataset, but it fails to perform well on unseen test data. This issue becomes more pronounced when dealing with complex models and limited training data, leading to a performance gap between the training and test datasets (Kelleher, 2019).

Applying our solution to unfamiliar test data is likely to yield unsatisfactory outcomes due to incorrectly inferred learned parameters that do not generalize well to new situations. This erroneous inference arises from a lack of sufficient training data, which permits random details to

```
from keras.models import Sequential
from keras.layers import Dense
# define the keras model
model = Sequential()
model.add(Dense(10, input_shape=(4,), activation='relu', name='fc1'))
model.add(Dense(10, activation='relu', name='fc2'))
model.add(Dense(3, activation='softmax', name='output'))
# compile the keras model
model.compile(loss='binary_crossentropy', optimizer='adam',
              metrics=['accuracy'])


# Train the model
model.fit(Xtrain, ytrain, verbose=2, batch_size=10, epochs=200)
# Test on unseen data
results = model.evaluate(Xtest, ytest)
# Pront the results
print('Final test set loss: {:4f}'.format(results[0]))
print('Final test set accuracy: {:4f}'.format(results[1]))
```

be embedded into the model. Consequently, the solution fails to effectively adapt to previously unseen test data. Increasing the number of training examples enhances the model's generalization ability, but augmenting its complexity diminishes this power. Similarly, in cases of abundant training data, a model that is too simplistic may struggle to capture intricate correlations between features and targets. A useful guideline is to ensure that the total amount of training data points is at least 2–3 times greater than the number of neural network parameters. In machine learning, overfitting is often comprehended through the balance between bias and variance. Even with substantial data, neural networks necessitate a well-considered strategy to counteract the adverse impacts of overfitting (Kelleher, 2019).

3.10.1 Regularization

Restricting the model to use fewer nonzero parameters is a clear solution to prevent overfitting since a higher number of parameters often leads to overfitting. Regularization becomes particularly crucial when there is a limited amount of available data. One biological explanation of regularization is related to progressive forgetting, where the model focuses on "less important" examples and discards them. In general, more complex models with regularization are preferred over simpler models without regularization (Kelleher, 2019).

3.11 Convolutional neural networks

Convolutional neural network (CNN) is a widely used deep learning network for image categorization. The CNN architecture, as shown in Fig. 1.2, consists of an input layer and multiple feature detection layers. These layers perform three main actions: convolution, pooling, and rectified linear unit (ReLU). Convolution is a process used for feature detection in CNNs. Pooling and ReLU are also integral parts of the CNN architecture (Vasuki & Govindaraju, 2017). CNNs are computer vision networks inspired by the biological visual system, designed to classify images and recognize objects. The key operation in the

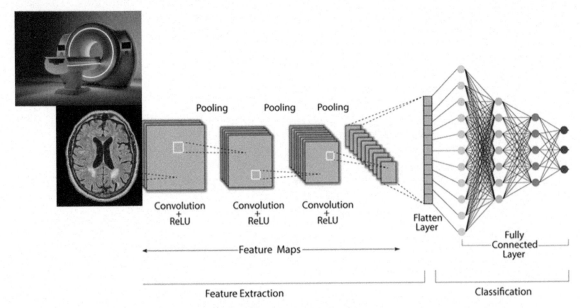

FIGURE 1.2   CNN architecture.

convolution layers involves using a filter to pass activations from one layer to the next. This filter is a three-dimensional weighted structure, matching the depth of the current layer but having a smaller spatial area. The value of the hidden state in the subsequent layer is determined by taking the dot product of the filter's weights with the spatial region in the layer. This interaction is performed at every possible location to generate the next layer, maintaining the spatial connections from the previous layer. Due to the limited region affecting each activation in a specific layer, CNNs have sparse connections. Apart from the last two or three layers, all layers retain their spatial structure, allowing for a physical visualization of the impact of picture elements on activations in a layer. Lower-level layers capture simple features like lines or basic shapes, while higher-level layers capture more intricate patterns. As a result, subsequent layers can combine these features to generate complex shapes. Additionally, CNNs incorporate subsampling layers that average data in local 2×2 regions to reduce the spatial size of the layers by a factor of 2. In the past, CNNs have proven to be the most effective among different types of neural networks. They find widespread applications in tasks such as image recognition, object detection, localization, and even natural language processing (Aggarwal, 2018).

In image processing, convolution involves using convolution filters to trigger specific features in an image. Pooling, on the other hand, utilizes nonlinear down sampling to reduce the data that needs to be processed. ReLU helps retain positive values while setting negative values to zero. Before the output layer, there is a classification layer in the CNN. This layer is fully connected and has an N-dimensional output, where N represents the number of categories for classification. It generates an N-dimensional vector, where each element represents the probability of the input image belonging to one of the N classes. The final output layer uses a SoftMax function to provide the categorized output.

CNNs are inspired by the biological structure of the visual cortex. In the visual cortex, simple and complex cells activate receptive field subregions of a visual field. Similarly, in CNNs, subregions of the preceding layer are connected to neurons in the subsequent layer, and other subregions have no impact on these neurons. Unlike traditional neural networks, CNNs allow subregions to overlap, providing spatially related results. This spatial relationship is a key distinction between CNNs and other neural networks (Vasuki & Govindaraju, 2017).

3.11.1 Functioning of CNN

In the CNN operation, an image containing one or more items to be classified is input into the network. The number of input values depends on the image size and pixel depth, representing objects of a particular type. CNN aims to mimic the human visual cortex, which has specialized neurons in tiny clusters that are sensitive to specific areas of the visual field, such as curves and edges. The first layer of the CNN is the convolution layer, which performs spatial convolution using a predefined mask with pixel values. This process creates a feature map at a higher level when the activation output of the first layer is fed into the second hidden layer, and the convolution process is repeated. This generates distinct activation maps for different image features, including complex features. At the end of the network, there is a fully connected layer that takes input from the previous layer and produces an N-dimensional output. For an N-class problem, it can return N alternative probability values, indicating the likelihood of the item belonging to each class. To achieve accurate classification, the CNN must be trained on a large number of images using backpropagation. The procedure in CNN is nonlinear due to the activation ReLU layer following each convolution layer. Negative activation values are set to zero, while positive activation values are retained. Pooling layers, such as the max pooling layer, downsample the activation inputs to reduce the computational burden. During the network's training phase, dropout layers are used to remove certain activation outputs, preventing overfitting. This ensures that the CNN generalizes well to new data (Vasuki & Govindaraju, 2017).

CNNs are structured in a way that maintains spatial correlations among grid cells in each layer. The conditions in every layer follow a spatial grid pattern, with each feature value depending on a small local geographic location in the previous layer. These spatial associations are crucial for the convolution operation and transitioning to the next layer, as they heavily influence the connections between layers. It is important to note that a layer's depth in a CNN is different from the network's overall depth. While CNN operates similarly to a standard feed-forward neural network, its layers are spatially structured, and the connections between layers are sparse and well-planned. Typically, CNNs consist of convolution, pooling, and ReLU layers. The ReLU activation functions work similarly to those in standard neural networks. Additionally, there are fully connected layers that translate to a set of output nodes in a specific application. The input data for a CNN is organized into a two-dimensional grid structure, where pixels represent values of individual grid points. Each pixel corresponds to a particular spatial position, and a multidimensional array is used to represent the specific hue of the pixel in the RGB color scheme, which includes intensities for the three primary colors: red, green, and blue (Aggarwal, 2018).

3.11.2 Padding

It is important to note that the convolution process in CNNs can lead to the shrinking of the $(q + 1)$th layer compared to the qth layer. This reduction in size is generally undesirable as it can cause the loss of information near the edges of the image. To address this issue, padding can be employed. Padding involves adding $(Fq - 1)/2$ "pixels" around the edges of the feature map to maintain the spatial footprint. In the case of padding hidden layers, these pixels are actual feature values. Whether applied to input or hidden layers, the value of each padded feature is set to 0. This allows the spatial height and breadth of the input volume to grow by $(Fq - 1)$, effectively compensating for the reduction caused by convolution. The padded areas, being set to 0, do not impact the final dot product. Padding enables the convolution operation to be performed with a portion of the filter extending beyond the layer's boundaries, ensuring that the dot product is carried out only over the region of the layer where the values are defined (Aggarwal, 2018).

3.11.3 Strides

Convolution can be employed in different ways to reduce the spatial footprint of an image. The conventional approach involves performing convolution at every point in the feature map's spatial location. However, it is not necessary to conduct convolution at every spatial location in the layer. Strides can be utilized to reduce the granularity of convolution. The most common stride used is one, but occasionally a stride of two is employed, and strides greater than two are rare in typical scenarios. Larger strides can be beneficial for managing memory constraints or reducing overfitting if the spatial resolution is excessively high. Moreover, a larger receptive field, which allows the detection of complex features in a larger spatial region of the image, can be achieved. The hierarchical feature engineering process of a CNN captures increasingly complex patterns in later layers. Additionally, another technique called max-pooling has historically been utilized to expand the receptive fields (Aggarwal, 2018).

3.11.4 The ReLU layer

The pooling and ReLU operations are integrated with the convolution operation in the CNN. The ReLU activation function behaves similarly to its usage in a standard neural network, serving as a one-to-one mapping of activation values without affecting the layer's dimensions. In traditional neural networks, the activation function is combined with a linear transformation using a weight matrix to generate the next layer's activations. In CNN designs, a ReLU layer is often not explicitly shown in graphical representations and is commonly placed after a convolution operation. It is worth noting that the ReLU activation function is a relatively new addition to neural network architecture, replacing saturating activation functions like tanh and sigmoid, which were commonly used in the past. The ReLU offers significant advantages in terms of both speed and accuracy compared to these older activation functions. The increased speed is directly related to improved accuracy because it allows for the utilization of more complex models and longer training times. Consequently, the ReLU activation function has gradually replaced other activation functions in CNN architectures, and it is now considered the default activation function (Aggarwal, 2018).

3.11.5 Pooling

The pooling process is distinct from filters and operates on small grid patches of size $Pq \times Pq$ in each layer, creating a new layer with the same depth. In this process, referred to as max-pooling, the maximum value is selected for each square region of size $Pq \times Pq$ in every dq activation map. Pooling significantly reduces the spatial dimensions of each activation map but does not affect the number of feature maps, occurring independently at the level of each activation map. Consequently, the depth of the pooled layer remains unchanged from the layer where the pooling operation was applied. Typically, a region size of 2×2, denoted as Pq, is commonly used for pooling, and a stride of 2 is employed, ensuring no overlap between the regions being pooled. However, some argue that introducing at least some overlap between the regions being pooled can be advantageous, as it helps reduce the risk of overfitting (Aggarwal, 2018).

3.11.6 Fully connected layers

Each hidden state in the first fully connected layer is linked to every feature in the last spatial layer, serving the same function as a standard feed-forward network. In many cases, more than one fully connected layer may be employed to increase computational capacity toward the end of the network. These connections between the layers follow the typical arrangement found in a traditional feed-forward network. Due to their dense connections, the fully connected layers have numerous parameters. Despite convolutional layers having more activations, the fully connected layers generally contain more connections. Depending on the application's requirements, various activation functions like logistic, softmax, or linear activation can be used in the fully connected layers. For instance, average pooling across the entire spatial region of the final set of activation maps is one approach to utilize fully connected layers to produce a single value (Aggarwal, 2018).

3.11.7 Training a convolutional network

The CNN is trained using the backpropagation technique and is made up of three types of layers: convolution, ReLU, and max pooling. Backpropagation through the ReLU layer is simple, similar to that of a normal neural network. When it comes to max pooling without overlapping pools, identifying the unit with the highest value within each pool is sufficient. The partial derivative of the loss is sent back to the unit with the greatest value during backpropagation through the max-pooled state, while all other entries in the grid, except the largest, are set to 0. The processes for ReLU and backpropagation via max pooling are similar to those used in traditional neural networks (Aggarwal, 2018).

3.11.8 Dropout

Dropout is a neural network technique that creates an ensemble by randomly sampling nodes instead of edges. When a node is dropped, all its connections to and from other nodes are removed as well. This dropout procedure is exclusively applied to the network's input and hidden layers, combining node sampling with weight sharing. During training, backpropagation updates the weights of the sampled network using a single example from the data. The primary effect of dropout is regularization, as it introduces noise by randomly dropping input and hidden units. This noise prevents feature coadaptation among hidden units, thereby enhancing the network's robustness. Dropout is efficient in regularization because it trains each subnetwork with a limited number of sampled examples, preventing overfitting. However, it may introduce redundancy among learned features at different hidden units due to the impact of masking noise on removing certain units. While dropout can be computationally efficient, requiring additional work only in sampling the hidden units, it restricts the network's expressive power. To fully benefit from dropout, larger models with more units are needed, which come with an added computational overhead. In scenarios where the initial training dataset is already large enough to reduce overfitting risk, the computational advantages of dropout might be relatively minor but still noticeable (Aggarwal, 2018).

3.11.9 Early stopping

Early stopping is a popular regularization method in which the gradient descent is stopped after a few iterations. A portion of the training data is kept and the model's error on the hold-out set is examined to decide the stopping point. The gradient-descent operation is terminated when the error on the hold-out set begins to grow. This effectively constricts the parameter space to a smaller region around the initial parameter values. As a result, by effectively reducing the parameter space, early stopping functions as a regularizer (Kelleher, 2019).

3.11.10 Batch normalization

Batch normalization is a recently developed technique designed to tackle issues like vanishing and exploding gradients, which lead to significant decreases or increases in activation gradients across consecutive layers. Additionally, a crucial problem in training deep networks is the internal covariate shift, where the activations of hidden variables change during training due to parameter adjustments. To mitigate these effects, batch normalization introduces "normalization layers" between hidden layers, ensuring consistent variance among features. The backpropagation algorithm accounts for this additional node, ensuring that the loss derivative of layers preceding the batch normalization layer compensates for the transformation introduced by these new nodes. An intriguing aspect of batch normalization is its role as a regularizer. Slight variations in adjustments may occur when different batches of the same data point are processed, introducing a form of noise to the updating process. This noise acts similarly to the regularization achieved by adding a small amount of noise to the training data. While there is some debate on this matter, empirical evidence indicates that once batch normalization is applied, regularization techniques like dropout do not offer significant performance improvements (Aggarwal, 2018). A simple Python code for CNN is given below.

```python
from tensorflow.keras.preprocessing.image import load_img ,img_to_array
import matplotlib.pyplot as plt
from tensorflow.keras.models import Sequential
from tensorflow.keras.layers import Conv2D
from tensorflow.keras.layers import MaxPooling2D,AveragePooling2D
from tensorflow.keras.layers import BatchNormalization
from tensorflow.keras.layers import Dense,Activation
from tensorflow.keras.layers import Flatten,Dropout,SpatialDropout2D
from tensorflow.keras.optimizers import SGD
from tensorflow.keras.optimizers import Adam
from tensorflow.keras.callbacks import EarlyStopping, ModelCheckpoint
early_stopping = EarlyStopping(monitor='val_loss', mode='min', patience=10,re
store_best_weights=True, verbose=1)

#Create the CNN Model
model = Sequential()
#1st Convolutional Layer
model.add(Conv2D(32, (3, 3), padding='valid', strides=(1, 1),input_shape=img_
shape))
model.add(BatchNormalization())
model.add(Activation('relu'))
model.add(MaxPooling2D(pool_size=(2,2), strides=(1,1), padding='same'))

#2nd Convolutional Layer
model.add(Conv2D(32, (3, 3), padding='same', strides=(1, 1)))
model.add(BatchNormalization())
model.add(Activation('relu'))
model.add(MaxPooling2D(pool_size=(2, 2),strides=(2,2), padding='same'))
model.add(Dropout(0.2))
```

```
#3rd Convolutional Layer
model.add(Conv2D(64, (3, 3), padding='same', strides=(1, 1)))
model.add(BatchNormalization())
model.add(Activation('relu'))
model.add(MaxPooling2D(pool_size=(2, 2),strides=2))

#4th Convolutional Layer
model.add(Conv2D(64, (3, 3), padding='same', strides=(1, 1)))
model.add(BatchNormalization())
model.add(Activation('relu'))
model.add(MaxPooling2D(pool_size=(2, 2),strides=2))

#Passing it to a Fully Connected layer
model.add(Flatten())
# 1st Fully Connected Layer
model.add(Dense(64, input_shape=(32,32,3,)))
model.add(BatchNormalization())
model.add(Activation('relu'))
# Add Dropout to prevent overfitting
model.add(Dropout(0.4))

#2nd Fully Connected Layer
model.add(Dense(32))
model.add(BatchNormalization())
model.add(Activation('relu'))
# Add Dropout to prevent overfitting
model.add(Dropout(0.4))
```

```
#Output Layer
model.add(Dense(3))
model.add(BatchNormalization())
model.add(Activation('softmax'))

model.compile(optimizer='adam',loss='categorical_crossentropy',metrics=['acc'
])

# Train the model
history = model.fit(
    Xtrain,
    ytrain,
    epochs=200,
    validation_split=0.25,
    batch_size=10,
    verbose=2,
    callbacks=[early_stopping])
# Test on unseen data
results = model.evaluate(Xtest, ytest)
# Pront the results
print('Final test set loss: {:4f}'.format(results[0]))
print('Final test set accuracy: {:4f}'.format(results[1]))

# plot training history
from matplotlib import pyplot
pyplot.plot(history.history['loss'], label='train')
pyplot.plot(history.history['val_loss'], label='validation')
pyplot.legend()
pyplot.show()
```

3.12 Recurrent neural networks

Every neural network is designed to handle multidimensional data where the features are generally considered independent of each other. However, certain data types, such as biological data, text, and time series, exhibit sequential relationships between their attributes. Some examples of such dependencies include the following:

1. In a time-series data collection, the values at successive timestamps are strongly correlated with each other. Treating these values as separate features can lead to the loss of crucial information about their connections. On the other hand, processing the values at different timestamps separately may result in information loss.

2. While text data is often treated as a collection of individual words (bag of words), considering the sequence of words can offer more meaningful semantic insights. Therefore, it is essential to develop models that take into account the sequential nature of the words. Recurrent neural networks are commonly used for text data in such scenarios.

3. Biological data often contains sequences, where the symbols represent nucleobases or amino acids, the fundamental building blocks of DNA.

Sequences can consist of either real-valued or symbolic values. Real-valued sequences are commonly referred to as time-series data. Recurrent neural networks (RNNs) can be applied to various types of data, including both real-valued and symbolic sequences. However, the vanishing and exploding gradient problem poses a significant challenge in this domain, especially in deep networks like RNNs. As a result, researchers have developed various RNN variants, such as the gated recurrent unit (GRU) and long short-term memory (LSTM), to address this issue. RNNs and their derivatives have found application in diverse fields, including image captioning, sequence-to-sequence learning, sentiment analysis, and machine translation, among others. These applications demonstrate the versatility and usefulness of RNNs in handling sequential data (Aggarwal, 2018). A simple Python code for RNN is given below.

```
from keras.layers import Dense
from keras.models import Sequential
from keras.layers import SimpleRNN
#Create Model
model = Sequential()
model.add(SimpleRNN(32))
model.add(Dense(1, activation = 'sigmoid'))
# Compile model
model.compile(optimizer = 'rmsprop', loss = 'binary_crossentropy', metrics
= ['acc'])
# Fit model
history = model.fit(Xtrain, ytrain,epochs = 100, batch_size = 20, validation_
split = 0.2)
# Evaluate the model
_, train_acc = model.evaluate(Xtrain, ytrain, verbose = 0)
_, test_acc = model.evaluate(Xtest, ytest, verbose = 0)
print('Training Accuracy: %.3f, Testing Accuracy: %.3f' % (train_acc,test_acc
))
```

3.13 Long short-term memory

RNNs face challenges related to the vanishing and exploding gradients problem. This is a common issue in neural network updates, where repeated matrix multiplication can result in either vanishing gradients during backpropagation or unstable, excessively large gradients. This instability arises from the repeated multiplication with the weight matrix at different time steps. A neural network that relies solely on multiplicative updates is well-suited for learning short sequences, exhibiting good short-term memory but poor long-term memory by default. To overcome this limitation, LSTM units are employed as a solution. LSTM modifies the recurrence equation for the hidden vector, enabling the network to effectively retain long-term memory. These LSTM operations are designed to offer precise control over the data stored in the long-term memory, enhancing the network's ability to remember and learn from longer sequences (Aggarwal, 2018). A simple Python code for LSTM is given below.

3.14 Transfer learning

Image analysis faces a challenge when labeled training data is not readily available for a specific purpose, such as image retrieval. In such cases, it is important to ensure that the extracted image features are semantically meaningful, even in the absence of labels. The ImageNet dataset, containing over a million images from 1000 different classes, has been curated to serve as a resource for deriving general-purpose image features. By passing the dataset through a pretrained CNN and extracting multidimensional features from the fully connected layers, one can obtain a new representation of the images that can be used for various tasks like clustering or retrieval. This approach of using pretrained CNNs for feature extraction is widely adopted, and it is considered a form of transfer learning. Instead of training CNNs from scratch, researchers utilize shared resources like ImageNet to extract features when specific training data is unavailable for addressing different problems. In cases where additional training data is available, fine-tuning can be applied to the deeper layers closer to the output, while keeping the early layers fixed. The rationale behind this is that early layers capture simple features like edges,

```
from keras.layers import Dense
from keras.models import Sequential
from keras.layers import LSTM
#Create Model
model = Sequential()
model.add(LSTM(32))
model.add(Dense(1, activation = 'sigmoid'))
# Compile model
model.compile(optimizer = 'rmsprop', loss = 'binary_crossentropy', metrics= [
'acc'])
# Fit model
history = model.fit(Xtrain, ytrain,epochs = 100,batch_size = 10, validation_s
plit = 0.2)
# Evaluate the model
_, train_acc = model.evaluate(Xtrain, ytrain, verbose = 0)
_, test_acc = model.evaluate(Xtest, ytest, verbose = 0)
print('Training Accuracy: %.3f, Testing Accuracy: %.3f' % (train_acc, test_ac
c))
```

which are generally applicable across tasks, while deeper layers capture more complex features that might be more specific to the task at hand. By fine-tuning only the deeper layers, one can adapt the network to the specific application without drastically altering the simple features captured by the early layers (Aggarwal, 2018).

3.14.1 VGG

VGG (Simonyan & Zisserman, 2014) also noted the rising trend of increased network depth. The networks under study were constructed with various topologies, having layer sizes ranging from 11 to 19, and the most effective ones had 16 or more layers. VGG's groundbreaking approach involved reducing filter sizes while increasing depth. It is crucial to note that using smaller filter sizes necessitates greater depth. Without sufficient depth, a small filter

can only capture limited information from the image. VGG consistently employs 3×3 spatial footprint filters and 2×2 pooling size. Convolution is performed with a stride of 1 and padding of 1, while pooling uses a stride of 2. Another notable aspect of VGG's architecture is that the number of filters is often doubled after each max-pooling. The objective is to maintain a balanced processing effort between layers by doubling the depth as the spatial footprint is reduced by half. This design principle has influenced subsequent topologies like ResNet. However, the use of deep configurations has the drawback of increasing sensitivity to initialization, leading to instability. Pretraining was employed to address this issue, where a shallow architecture was initially trained, and then more layers were added gradually (Aggarwal, 2018). A simple Python code for VGG16 is given below.

```
from tensorflow.keras.models import Sequential
from tensorflow.keras.layers import BatchNormalization
from tensorflow.keras.layers import Dense,Activation
from tensorflow.keras.layers import Flatten,Dropout
from tensorflow.keras.optimizers import Adam
from tensorflow.keras.callbacks import EarlyStopping, ModelCheckpoint
early_stopping = EarlyStopping(monitor='val_loss', mode='min', patience=10,re
store_best_weights=True, verbose=1)


base_model = tf.keras.applications.VGG16(
    include_top=False,
    weights="imagenet",
    input_tensor=None,
    input_shape=img_shape,
    pooling=None
)


for l in base_model.layers:
    l.trainable = False

model = Sequential()
model.add(base_model)

model.add(Flatten())
model.add(BatchNormalization())
model.add(Dense(512,activation='relu'))
model.add(Dropout(0.5))
model.add(Dense(256,activation='relu'))
```

```
model.add(Dropout(0.5))
model.add(Dense(64,activation='relu'))
model.add(Dropout(0.5))
model.add(Dense(3,activation='softmax'))

model.compile(optimizer='adam',loss='categorical_crossentropy',
              metrics=['acc'])

# Train the model
history = model.fit(
    Xtrain,
    ytrain,
    epochs=200,
    validation_split=0.25,
    batch_size=10,
    verbose=2,
    callbacks=[early_stopping])
# Test on unseen data
results = model.evaluate(Xtest, ytest)
# Pront the results
print('Final test set loss: {:4f}'.format(results[0]))
print('Final test set accuracy: {:4f}'.format(results[1]))

# plot training history
from matplotlib import pyplot
pyplot.plot(history.history['loss'], label='train')
pyplot.plot(history.history['val_loss'], label='test')
pyplot.legend()
pyplot.show()
```

3.14.2 ResNet

ResNet (He et al., 2016, pp. 770–778) included 152 layers that are nearly an order of size more than existing architectures. Training a 152-layered architecture often becomes unfeasible without significant advancements. The primary challenge lies in hindering the gradient flow between layers due to the extensive number of operations in deep layers, potentially leading to changes in gradient magnitude. This increased depth gives rise to issues like vanishing and exploding gradients. Many deep networks demonstrate high errors on both training and test data, indicating that the optimization process has not reached an optimal state. In neural network learning, layer-wise implementations assume that all image concepts should exist at the same level of abstraction, whereas in reality, a preferred approach is hierarchical feature engineering. Certain concepts require fine-grained connections, while others can be effectively learned using shallower networks (Aggarwal, 2018). A simple Python code for ResNet101 is given below.

```python
from tensorflow.keras import Model
from tensorflow.keras.models import Sequential
from tensorflow.keras.layers import Conv2D
from tensorflow.keras.layers import MaxPooling2D,AveragePooling2D,BatchNormal
ization
from tensorflow.keras.layers import Dense,Activation
from tensorflow.keras.layers import Flatten,Dropout,SpatialDropout2D,AverageP
ooling2D,GlobalAveragePooling2D
from tensorflow.keras.optimizers import SGD
from tensorflow.keras.optimizers import Adam
import tensorflow as tf
from tensorflow.keras.callbacks import EarlyStopping, ModelCheckpoint
early_stopping = EarlyStopping(monitor='val_loss', mode='min', patience=10,re
store_best_weights=True, verbose=1)

base_model = tf.keras.applications.ResNet101(
                include_top=False,
                weights="imagenet",
                input_tensor=None,
                input_shape=img_shape,
                pooling=None,
            )
for layer in base_model.layers:
    layer.trainable = False

model = Sequential()
model.add(base_model)

model.add(Flatten())
model.add(BatchNormalization())
model.add(Dense(128,activation='relu'))
model.add(Dropout(0.2))
model.add(Dense(64,activation='relu'))
model.add(Dropout(0.2))
model.add(Dense(32,activation='relu'))
model.add(Dropout(0.2))
```

```
model.add(Dense(3,activation='softmax'))

model.compile(optimizer='adam',loss='categorical_crossentropy',
            metrics=['acc'])


# Train the model
history = model.fit(
    Xtrain,
    ytrain,
    epochs=200,
    validation_split=0.25,
    batch_size=10,
    verbose=2,
    callbacks=[early_stopping])
# Test on unseen data
results = model.evaluate(Xtest, ytest)
# Pront the results
print('Final test set loss: {:4f}'.format(results[0]))
print('Final test set accuracy: {:4f}'.format(results[1]))


# plot training history
from matplotlib import pyplot
pyplot.plot(history.history['loss'], label='train')
pyplot.plot(history.history['val_loss'], label='test')
pyplot.legend()
pyplot.show()
```

3.14.3 MobileNet architecture

The MobileNet model is built upon the concept of depthwise separable convolutions, a type of factorized convolution that breaks down a regular convolution into two distinct parts: depthwise convolution and pointwise convolution (1×1 convolution). In the depthwise convolution, each input channel is filtered with a single filter, and the outputs of this step are then combined using the pointwise convolution. Unlike ordinary convolutions that perform filtering and combining in a single step, depthwise separable convolutions split these processes into two layers, resulting in improved computational efficiency. The depthwise convolution filters input channels without merging them, while the pointwise convolution linearly combines the outputs from the depthwise convolution to create new features. MobileNets leverage batch norm and ReLU nonlinearities for both depthwise and pointwise convolutions. Downsampling is accomplished through strided convolution in the depthwise convolutions and the first layer. The spatial resolution is subsequently reduced to 1 through average pooling before reaching the fully connected layer. Overall, MobileNet consists of 28 layers when considering depthwise and pointwise convolutions as separate entities (Howard et al., 2017). A simple Python code for MobileNet is given below.

```python
from tensorflow.keras.models import Sequential
from tensorflow.keras.layers import  BatchNormalization
from tensorflow.keras.layers import Dense,Activation
from tensorflow.keras.layers import Flatten,Dropout,SpatialDropout2D
from tensorflow.keras.optimizers import SGD
from tensorflow.keras.optimizers import Adam
import tensorflow as tf
from tensorflow.keras.callbacks import EarlyStopping, ModelCheckpoint
early_stopping = EarlyStopping(monitor='val_loss', mode='min', patience=10,
                              restore_best_weights=True, verbose=1)

base_model = tf.keras.applications.MobileNet(
             alpha = 0.75,
             include_top=False,
             weights="imagenet",
             input_tensor=None,
             input_shape=img_shape,
             pooling=None,
         )
for layer in base_model.layers:
    layer.trainable = False

model = Sequential()
model.add(base_model)

model.add(Flatten())
model.add(BatchNormalization())
model.add(Dense(256,activation='relu'))
model.add(Dropout(0.2))
model.add(Dense(128,activation='relu'))
model.add(Dropout(0.2))
```

```
model.add(Dropout(0.2))
model.add(Dense(3,activation='softmax'))

model.compile(optimizer='adam',loss='categorical_crossentropy',
              metrics=['acc'])

# Train the model
history = model.fit(
    Xtrain,
    ytrain,
    epochs=200,
    validation_split=0.25,
    batch_size=10,
    verbose=2,
    callbacks=[early_stopping])
# Test on unseen data
results = model.evaluate(Xtest, ytest)
# Pront the results
print('Final test set loss: {:4f}'.format(results[0]))
print('Final test set accuracy: {:4f}'.format(results[1]))

# plot training history
from matplotlib import pyplot
pyplot.plot(history.history['loss'], label='train')
pyplot.plot(history.history['val_loss'], label='test')
pyplot.legend()
pyplot.show()
```

3.14.4 Inception-v4 and Inception-ResNet

The Inception architecture offers remarkable flexibility, allowing easy adjustments to the number of filters in different layers without compromising the fully trained network's quality. Layer sizes are carefully tuned to distribute computation efficiently among various subnetworks, leading to improved training speed. In Inception-v4, consistent selections are made for Inception blocks across all grid sizes to eliminate unnecessary overhead. Cheaper Inception blocks are utilized for the residual versions compared to the original Inception networks. Each Inception block is followed by a filter-expansion layer, which increases the dimensionality of the filter bank to match the input requirements, compensating for dimensionality reduction caused by the Inception block.

The residual versions of Inception, including "Inception-ResNet-v1" and "Inception-ResNet-v2," have been subject to various variations. Inception-ResNet-v1 bears computational similarity to Inception-v3, while Inception-ResNet-v2 aligns with the computational cost of the recently proposed Inception-v4 network. One technical distinction between the residual and nonresidual Inception versions lies in the use of batch-normalization. In Inception-ResNet, batch normalization is applied only to standard layers and not to summations. This decision ensures each model replica remains trainable on a single GPU, despite the possibility that extensive batch normalization could be beneficial. By eliminating batch normalization on certain layers, the overall number of Inception blocks can be significantly increased, reducing the need for resource trade-offs (Szegedy et al., 2017). A simple Python code for InceptionRes-NetV2 is given below.

```python
from tensorflow.keras.models import Sequential
from tensorflow.keras.layers import BatchNormalization
from tensorflow.keras.layers import Dense,Activation
from tensorflow.keras.layers import Flatten,Dropout,SpatialDropout2D
from tensorflow.keras.optimizers import Adam
import tensorflow as tf
from tensorflow.keras.callbacks import EarlyStopping, ModelCheckpoint
early_stopping = EarlyStopping(monitor='val_loss', mode='min', patience=10,
                              restore_best_weights=True, verbose=1)

base_model = tf.keras.applications.InceptionResNetV2(
               include_top=False,
               weights="imagenet",
               input_tensor=None,
               input_shape=img_shape,
               pooling=None,
          )
for layer in base_model.layers:
    layer.trainable = False

model = Sequential()
model.add(base_model)

model.add(Flatten())
model.add(BatchNormalization())
model.add(Dense(256,activation='relu'))
model.add(Dropout(0.2))
model.add(Dense(128,activation='relu'))
model.add(Dropout(02))
model.add(Dense(64,activation='relu'))
model.add(Dropout(0.2))
```

```
model.add(Dense(32,activation='relu'))
model.add(Dropout(0.2))
model.add(Dense(3,activation='softmax'))

model.compile(optimizer='adam',loss='categorical_crossentropy',
            metrics=['acc'])

# Train the model
history = model.fit(
    Xtrain,
    ytrain,
    epochs=200,
    validation_split=0.25,
    batch_size=10,
    verbose=2,
    callbacks=[early_stopping])
# Test on unseen data
results = model.evaluate(Xtest, ytest)
# Pront the results
print('Final test set loss: {:4f}'.format(results[0]))
print('Final test set accuracy: {:4f}'.format(results[1]))

# plot training history
from matplotlib import pyplot
pyplot.plot(history.history['loss'], label='train')
pyplot.plot(history.history['val_loss'], label='test')
pyplot.legend()
pyplot.show()
```

3.14.5 Xception

Chollet, (2017) suggested a depthwise separable CNN model based purely on convolution layers. The Xception architecture is founded on the concept that cross-channel and spatial correlations within CNN feature maps can be completely separated. It represents an enhanced version of the Inception design, aptly named Xception, short for "Extreme Inception." In this novel approach, an "extreme" Inception module utilizes a 1×1 convolution to map cross-channel correlations and then individually maps spatial correlations for each output channel. This extreme Inception module closely resembles a depthwise separable convolution, a popular neural network-building technique adopted in frameworks like TensorFlow. Comprising 36 convolutional layers, the Xception architecture forms the foundation for feature extraction in the network. For experimental assessments, a suggested convolutional base is followed by a logistic regression layer, with the option to add further fully connected layers if needed. The 36 convolutional layers are organized into 14 modules, each having linear residual connections, except for the first and last modules. Essentially, Xception is a linear stack of depthwise separable convolution layers with residual links, rendering it highly adaptable and easy to define and modify using high-level libraries like Keras or TensorFlow-Slim. This simplicity in definition sets it apart from more complex architectures like Inception V2 or V3, while still achieving remarkable performance in image classification tasks (Chollet, 2017). A simple Python code for Xception is given below.

```python
from tensorflow.keras.models import Sequential
from tensorflow.keras.layers import BatchNormalization
from tensorflow.keras.layers import Dense,Activation
from tensorflow.keras.layers import Flatten,Dropout,SpatialDropout2D
from tensorflow.keras.optimizers import Adam
import tensorflow as tf
from tensorflow.keras.callbacks import EarlyStopping, ModelCheckpoint
early_stopping = EarlyStopping(monitor='val_loss', mode='min', patience=10,
                               restore_best_weights=True, verbose=1)

base_model = tf.keras.applications.Xception(
            include_top=False,
            weights="imagenet",
            input_tensor=None,
            input_shape=img_shape,
            pooling=None,
        )
for layer in base_model.layers:
    layer.trainable = False

model = Sequential()
model.add(base_model)

model.add(Flatten())
model.add(BatchNormalization())
model.add(Dense(256,activation='relu'))
model.add(Dropout(0.2))
model.add(Dense(128,activation='relu'))
model.add(Dropout(02))
model.add(Dense(64,activation='relu'))
model.add(Dropout(0.2))
model.add(Dense(32,activation='relu'))
```

```
model.add(Dropout(0.2))
model.add(Dense(3,activation='softmax'))

model.compile(optimizer='adam',loss='categorical_crossentropy',
              metrics=['acc'])

# Train the model
history = model.fit(
    Xtrain,
    ytrain,
    epochs=200,
    validation_split=0.25,
    batch_size=10,
    verbose=2,
    callbacks=[early_stopping])
# Test on unseen data
results = model.evaluate(Xtest, ytest)
# Pront the results
print('Final test set loss: {:4f}'.format(results[0]))
print('Final test set accuracy: {:4f}'.format(results[1]))

# plot training history
from matplotlib import pyplot
pyplot.plot(history.history['loss'], label='train')
pyplot.plot(history.history['val_loss'], label='test')
pyplot.legend()
pyplot.show()
```

3.14.6 Densely connected convolutional networks (DenseNets)

ResNets offer the advantage of allowing the gradient to flow directly through the identity function from later layers to earlier layers. However, the merging of the identity function and the output of H′ through summation can limit information flow within the network. To address this, a new connectivity pattern called dense convolutional network (DenseNet) is proposed, which introduces direct connections from any layer to all subsequent layers, further enhancing information flow between layers. The term "DenseNet" is used because of its dense connectivity.

In conventional convolutional networks, down-sampling layers are important for adjusting the size of feature maps. In DenseNet, the network is divided into multiple densely linked dense blocks to aid in down-sampling. The layers between these dense blocks are called transition layers as they perform convolution and pooling operations. During testing, batch normalization layers are used with 1×1 convolutional layers and 2×2 average pooling layers as transition layers. One unique characteristic of DenseNet is that it can have very compact layers, setting it apart from traditional network topologies (Huang et al., 2017, pp. 4700–4708). A simple Python code for DenseNet121 is given below.

```python
from tensorflow.keras.models import Sequential
from tensorflow.keras.layers import BatchNormalization
from tensorflow.keras.layers import Dense,Activation
from tensorflow.keras.layers import Flatten,Dropout,SpatialDropout2D
from tensorflow.keras.optimizers import Adam
import tensorflow as tf
from tensorflow.keras.callbacks import EarlyStopping, ModelCheckpoint
early_stopping = EarlyStopping(monitor='val_loss', mode='min', patience=10,
                              restore_best_weights=True, verbose=1)


base_model = tf.keras.applications.DenseNet121(
                include_top=False,
                weights="imagenet",
                input_tensor=None,
                input_shape=img_shape,
                pooling=None,
            )
for layer in base_model.layers:
    layer.trainable = False

model = Sequential()
model.add(base_model)

model.add(Flatten())
model.add(BatchNormalization())
model.add(Dense(256,activation='relu'))
model.add(Dropout(0.2))
model.add(Dense(128,activation='relu'))
model.add(Dropout(02))
model.add(Dense(64,activation='relu'))
model.add(Dropout(0.2))
```

```
model.add(Dense(32,activation='relu'))
model.add(Dropout(0.2))
model.add(Dense(3,activation='softmax'))

model.compile(optimizer='adam',loss='categorical_crossentropy',
              metrics=['acc'])

# Train the model
history = model.fit(
    Xtrain,
    ytrain,
    epochs=200,
    validation_split=0.25,
    batch_size=10,
    verbose=2,
    callbacks=[early_stopping])
# Test on unseen data
results = model.evaluate(Xtest, ytest)
# Pront the results
print('Final test set loss: {:4f}'.format(results[0]))
print('Final test set accuracy: {:4f}'.format(results[1]))

# plot training history
from matplotlib import pyplot
pyplot.plot(history.history['loss'], label='train')
pyplot.plot(history.history['val_loss'], label='test')
pyplot.legend()
pyplot.show()
```

3.14.7 *Feature extraction with pretrained models*

Pretrained CNNs sourced from publicly available repositories like ImageNet are widely employed across various applications and datasets. The common approach is to retain most of the pretrained weights in the neural network, excluding the final classification layer. The weights of this classification layer are adjusted to align with the specific dataset at hand because the class labels may differ from those in ImageNet. However, the weights in the earlier layers of the pretrained network still hold value as they have learned to recognize diverse shapes in images, proving beneficial for nearly any type of classification task. Additionally, the activations in the penultimate layer can be leveraged for unsupervised applications. Given the extensive use of pretrained convolutional networks, initiating the training process from scratch has become an uncommon practice (Aggarwal, 2018).

In more extensive feed-forward architectures, multiple layers are employed, with each layer transforming the inputs from the preceding layer to generate increasingly sophisticated data representations. The output layer utilizes these well-transformed feature interpretations to make more straightforward predictions.

This level of complexity is achieved through the use of nonlinear activations in intermediate layers. In the past, tanh and sigmoid activation functions were common in hidden layers, but recently, the ReLU activation has become popular due to its ability to address vanishing and exploding gradient issues. The division of work between hidden layers and the final prediction layer can be understood as follows: the early layers provide a feature representation tailored to the specific task, which is then utilized by the final layer for prediction. These learned features in the hidden layers are generally transferable to different datasets and problems within the same domain, enabling easy reuse of pretrained networks. To do so, one can simply replace the output node(s) with a new application-specific output layer for the particular problem and dataset. Each hidden layer generates a transformed feature representation of the data, with the dimensionality of these representations determined by the number of units in each layer. This approach can be viewed as a form of hierarchical feature engineering, where early layers capture basic data qualities, and later layers capture more complex attributes that hold semantic meaning related to the class labels. The concept of transfer learning describes the practice of using pretrained models, leveraging existing network weights to solve new tasks or datasets (Aggarwal, 2018). A DenseNet121 deep feature extraction example is given below. For the details, check the machinelearningmastery.com web page (https://machinelearningmastery.com/how-to-use-transfer-learning-when-developing-convolutional-neural-network-models/).

```python
# Example of using the DenseNet121 model as a feature extraction model (Adapt
ed from machinelearningmastery.com)
from keras.preprocessing.image import load_img
from keras.preprocessing.image import img_to_array
from keras.applications.densenet import preprocess_input
from keras.applications.densenet import decode_predictions
from keras.applications.densenet import DenseNet121
from keras.models import Model
from pickle import dump
# load an image from file
image = load_img('BrainMRI.jpg', target_size=(224, 224))
# convert the image pixels to a numpy array
image = img_to_array(image)
# reshape data for the model
image = image.reshape((1, image.shape[0], image.shape[1], image.shape[2]))
# prepare the image for the DenseNet121 model
image = preprocess_input(image)
# load model
model = DenseNet121()
# remove the output layer
model = Model(inputs=model.inputs, outputs=model.layers[-2].output)
# get extracted features
features = model.predict(image)
print(features.shape)
```

References

Aggarwal, C. C. (2018). *Neural networks and deep learning.* Springer.

Bhattacharya, S., Maddikunta, P. K. R., Kaluri, R., Singh, S., Gadekallu, T. R., Alazab, M., & Tariq, U. (2020). A novel PCA-firefly based XGBoost classification model for intrusion detection in networks using GPU. *Electronics, 9*(2), 219.

Breiman, L., Friedman, J. H., Olshen, R. A., & Stone, C. J. (1984). *Classification and regression trees.* Hall/CRC.

Chen, T., & Guestrin, C. (2016). *Xgboost: A scalable tree boosting system.*

Chollet, F. (2017). *Xception: Deep learning with depthwise separable convolutions* (pp. 1251–1258).

Gaura, J., Sojka, E., & Krumnikl, M. (2011). *Image segmentation based on k-means clustering and energy-transfer proximity* (pp. 567–577).

Grąbczewski, K. (2014). *Meta-learning in decision tree induction* (Vol. 1). Springer.

Hall, M., Witten, I., & Frank, E. (2011). *Data mining: Practical machine learning tools and techniques.* Burlington: Kaufmann.

Han, J., Pei, J., & Kamber, M. (2011). *Data mining: Concepts and techniques.* Elsevier.

He, K., Zhang, X., Ren, S., & Sun, J. (2016). *Deep residual learning for image recognition.*

Howard, A. G., Zhu, M., Chen, B., Kalenichenko, D., Wang, W., Weyand, T., Andreetto, M., & Adam, H. (2017). Mobilenets: Efficient convolutional neural networks for mobile vision applications. *ArXiv Preprint ArXiv:1704.04861.*

Huang, G., Liu, Z., Van Der Maaten, L., & Weinberger, K. Q. (2017). *Densely connected convolutional networks.*

Kelleher, J. D. (2019). *Deep learning.* MIT Press.

Murphy, K. P. (2012). *Machine learning: A probabilistic perspective.* MIT Press.

Rawal, K., Sethi, G., & Ghai, D. (2020). Medical imaging in healthcare applications. In *Artificial intelligence and machine learning in 2D/3D medical image processing* (pp. 97–106). CRC Press.

Shalev-Shwartz, S., & Ben-David, S. (2014). *Understanding machine learning: From theory to algorithms.* Cambridge University Press.

Simonyan, K., & Zisserman, A. (2014). Very deep convolutional networks for large-scale image recognition. *ArXiv Preprint ArXiv:1409.1556.*

Subasi, A. (2020). *Practical machine learning for data analysis using Python.* Academic Press.

Szegedy, C., Ioffe, S., Vanhoucke, V., & Alemi, A. (2017). Inception-v4, inception-resnet and the impact of residual connections on learning. *31*(1).

Tatiraju, S., & Mehta, A. (2008). *Image segmentation using k-means clustering, EM and normalized cuts* (Vol. 1, pp. 1–7). Department of EECS.

Theodoridis, S., Pikrakis, A., Koutroumbas, K., & Cavouras, D. (2010). *Introduction to pattern recognition: A matlab approach.* Academic Press.

Vasuki, A., & Govindaraju, S. (2017). Deep neural networks for image classification. In *Deep learning for image processing applications* (Vol 31, p. 27). IOS Press.

Witten, I. H., Frank, E., Hall, M. A., & Pal, C. J. (2016). *Data Mining: Practical machine learning tools and techniques.* Morgan Kaufmann.

Zheng, X., Lei, Q., Yao, R., Gong, Y., & Yin, Q. (2018). Image segmentation based on adaptive K-means algorithm. *EURASIP Journal on Image and Video Processing, 2018*(1), 68.

Artificial intelligence-based emotion recognition using ECG signals

Fadime Tokmak[1], Abdulhamit Subasi[2,3], Saeed Mian Qaisar[4,5]

[1]Digital Health Technology, Faculty of Technology, University of Turku, Turku, Finland; [2]Institute of Biomedicine, Faculty of Medicine, University of Turku, Turku, Finland; [3]Department of Computer Science, College of Engineering, Effat University, Jeddah, Saudi Arabia; [4]LINEACT CESI, Lyon, France; [5]College of Engineering, Effat University, Jeddah, Saudi Arabia

1. Introduction

The combination of artificial intelligence (AI) with physiological signal analysis has opened up new paths in the field of emotion recognition in recent years. Emotions are complex psychological states that have a significant impact on human interactions, decision-making, and general well-being. The potential applications of developing accurate and efficient systems that discern emotions from physiological signals in many domains such as human—computer interaction, healthcare, and affective computing have sparked increased attention. The Electrocardiogram (ECG) is one physiological signal that has gained popularity in emotion recognition studies. The ECG, which records the electrical activity of the heart over time, has traditionally been used to diagnose cardiac problems. Recent research has discovered that emotions cause minor changes in the electrical rhythms of the heart, making ECG a promising tool for emotion

detection. The intrinsic advantage of ECG is its noninvasiveness, low cost, and widespread availability, which makes it a feasible candidate for real-time emotion recognition applications.

Traditional emotion recognition methods rely heavily on facial expressions, voice, and other exterior clues. While these strategies have had a lot of success, they are vulnerable to ambient influences and individual variances, which limits their reliability in different situations. The combination of AI techniques, particularly machine learning algorithms, with physiological signal analysis provides a viable option for overcoming these constraints. AI enables computers to discover deep patterns and relationships in physiological data, allowing for accurate and context-independent emotion recognition. This chapter aims to add to the developing field of emotion recognition by offering a novel strategy that makes use of ECG signals and AI methodologies. The suggested approach employs AI algorithms to decipher the subtle physiological

Applications of Artificial Intelligence in Healthcare and Biomedicine
https://doi.org/10.1016/B978-0-443-22308-2.00002-0

37

changes in ECG waveforms associated with various emotional states. Machine learning and deep learning models are used to effectively determine emotions by leveraging a large dataset containing varied emotional experiences. The system's effectiveness is examined by a thorough evaluation method incorporating controlled emotional stimuli, revealing light on its potential efficacy in real-world circumstances.

The following sections of this chapter go into detail on previous studies, the methodology used, the dataset creation process, the AI models used, and the experimental results gained. This study's findings not only advance the science of emotion recognition, but they also have implications for applications such as human–computer interaction and mental health monitoring. This discovery represents a big step forward in unlocking the potential of ECG-based emotion identification systems in an era marked by the confluence of AI and biomedical signal analysis. Typically, the process of classifying biomedical signals can be broken down into four main phases: (1) data collection and segmentation, (2) data preprocessing, (3) feature extraction, and (4) recognition and classification. As depicted in Fig. 2.1, the ECG signal is initially recorded from the human body and then subjected to preprocessing. Data preprocessing serves the purpose of converting the raw data into a more useful and efficient format. This is essential because the data may contain noise, irrelevant information, and missing elements that need to be removed. Subsequently, features are extracted from the processed biomedical data and converted into a feature vector. The feature vector encapsulates the relevant structure within the raw data. Following this, dimension reduction techniques are applied to eliminate any redundant information from the feature vector, resulting in a reduced feature vector. Finally, a classifier is employed to categorize and classify the reduced feature vector (Subasi, 2019).

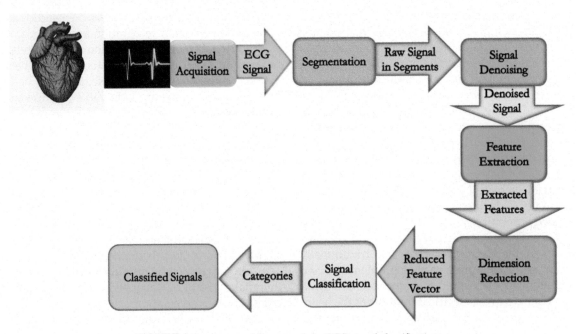

FIGURE 2.1 A general framework for ECG signal classification.

2. Literature review

In the recent years, the areas of psychology, neuroscience, and human—machine interaction have gained interest in the use of ECG for emotion identification (Bulagang et al., 2020; Nita et al., 2022). Researchers have looked into the possibility of using ECG to comprehend and identify emotional states, even though it has traditionally been used to monitor heart health and diagnose cardiac diseases (Hsu et al., 2017; Sarkar & Etemad, 2020).

The recent advances in ECG technology, signal processing methods, and machine learning have made it possible to identify human emotions by exploring ECG signals. Beginning with images, videos, and audible cues, researchers have been examining how ECG signals respond to different emotional stimuli (Katsigiannis & Ramzan, 2017). The common emotional states targeted in these studies include happiness, sadness, anger, fear, and neutral emotions (Sharma & Dhall, 2021; Wang et al., 2022).

ECG-based emotion recognition has found use in a variety of applications such as virtual reality, mental health monitoring, and human—machine interaction (Ismail et al., 2023; Patil & Pawar, 2022, pp. 1—6). Systems that can modify the content or user experience based on the user's emotional state have been developed using it. ECG-based emotion detection in the context of mental health may help in the monitoring of people with illnesses like anxiety and depression (Ismail et al., 2023; Patil & Pawar, 2022, pp. 1—6).

The requirement for vast and diverse datasets, the impact of confounding variables, and the generalization of models across the intended population are difficulties in ECG-based emotion identification. Research is now being done to increase the precision and dependability of emotion identification systems and investigate possible practical applications.

The autonomic nerve system, which affects heart rate and heart rate variability, alters as a result of emotions. Therefore, using the ECG to identify emotions has a number of benefits that make it a desirable method for researching and comprehending human emotions. The objective and nonintrusive approach and the potential for continuous plus nonobtrusive monitoring are some of the major benefits.

The ECG provides a number of benefits for identifying emotions, but it also has certain drawbacks and restrictions. Subject dependence, the impact of artifacts and physiological factors, privacy concerns, gathering data, accurate labeling of emotions, generalization of emotion detection models, and real-time processing are some of the main obstacles in the realization of effective ECG-based emotion recognition systems (Mian Qaisar, 2020; Mian Qaisar & Subasi, 2020). To cope with these challenges the usage of ECG for emotion recognition is an emerging research area with promising potential. In recent years a variety of studies have been conducted in this field (Bulagang et al., 2020; Hsu et al., 2017; Nita et al., 2022; Sarkar & Etemad, 2020).

Nita et al. (2022) suggested a novel data-augmentation convolutional neural network (CNN) for ECG-based human emotion identification. They proposed an approach of adding a considerable number of representative ECG samples to the considered ECG dataset using a procedure called randomize, concatenate, and resample. To recognize human emotions in terms of valence, arousal, and dominance levels, a suggested CNN model was used. The functionality was tested using the DREAMER database. The proposed model attained accuracy scores of 95.16% for valence, 85.56% for arousal, and 77.54% for dominance.

Hsu et al. (2017) employed a musical induction technique to elicit participants' true emotional states and gather their ECG signals. The features were extracted from the ECG signals using the time-frequency and nonlinear analysis. Then, to efficiently select significant ECG features they used a "sequential forward floating selection-kernel-based class separability" (SFFS-KBCS) "generalized discriminant analysis" (GDA) based feature selection algorithms. The selected features set is processed using the "least square support vector machine" (LS-SVM) classifier. The model secured accuracy scores of 82.78%, 72.91%, and 61.52%, respectively, for the valence, arousal, and four considered categories of emotions.

Sarkar and Etemad (2020) used a self-supervised deep multitask learning framework to recognize emotions based on the ECG signals. The suggested approach involves two phases: firstly, learning the ECG representations; secondly, learning to categorize emotions. A signal transformation and recognition network is trained to recognize ECG representations. From unlabeled ECG data, the network learns high-level abstract representations. The ECG data are subjected to six various signal transformations, and transformation recognition is carried out as pretext tasks. The network learns spatiotemporal representations that generalize effectively across diverse datasets and different emotion categories by training the model on pretext tasks. The model achieved 85.9% and 85% accuracy scores, respectively, for arousal and valance while processing the DREAMER dataset.

Nisa'Minhad et al. (2017) recorded the "skin conductance response" (SCR) and ECG signals to study the sympathetic reactions of human emotions. The SCR and ECG signals are filtered digitally for noise diminishing. Onward, the statistical features are extracted from the preprocessed signals. The prepared features set is classified using the "support vector machine" (SVM) classifier. This model attained accuracy scores between 82.6% and 95.7% for the classification of four considered emotions namely happy, sad, anger, disgust, and fear.

Goshvarpour et al. (2017) recorded the SCR and ECG signals of 11 subjects while they were listening to emotional music clips. They applied wavelet packets and discrete cosine transform-based dictionaries with statistical analysis for features extraction. The dimension reduction performance of the "principal component analysis" (PCA), "linear discriminant analysis," and Kernel PCA was compared. The selected features set was processed using the "probabilistic neural network" for automated categorization of emotions. This model secured an accuracy score of 100% while the identification of five considered emotions namely happiness, peacefulness, sadness, scary, and neutral.

Ferdinando et al. (2017) extracted features, from the ECG signal, using the bivariate empirical mode decomposition. Onward, the dimension reduction performance is compared among the linear discriminant analysis (LDA), neighborhood components analysis (NCA), and maximally collapsing metric learning (MCML). The categorization of emotions, valence, and arousal, is carried out using the k-nearest eighbor (k-NN) algorithm. The applicability is tested using the Mahnob-HCI database. This model secures accuracy scores of 64.1% and 66.1%, respectively, for valence and arousal.

Wei et al. (2018) proposed a decision-level weight fusion technique for physiological inputs from many channels to recognize emotions. They used four different physiological signals namely the ECG, respiratory amplitude (RA), and galvanic skin response (GSR). Additionally, the extraction of physiological emotion components was made. The extracted features set was processed using the SVM classifier. The model achieved an accuracy score of 84.6% while solving the classification problem for five emotions namely sadness, happiness, disgust, neutral, and fear.

Sepúlveda et al. (2021) employed the wavelet transform to analyze ECG signals. The ECG signal's attributes were extracted using a wavelet scattering approach, which makes it possible to acquire the signal's features at various time scales. These features were then utilized as inputs for several classifiers to assess each one's performance. Using the AMIGOS database, the suggested model's performance was evaluated. According to the findings, the suggested method for extracting features and categorizing the signals achieves an accuracy of 88.8% for valence, 90.2% for arousal, and 95.3% for 2-class emotions categorization.

Hasnul et al. (2023) developed a method for enhancing ECG signals with several filters. The sample size of the ECG data is expanded. In this work, three ECG datasets, A2ES, AMIGOS, and DREAMER, that were labeled with emotion states were employed. 66 statistical features were mined from each ECG segment. The classification was carried out using the k-NN. The model secures the highest classification accuracy score of 90%.

Ismail et al. (2022) made a comparison between the ECG and Photoplethysmogram (PPG) in categorizing emotions. They collected the dataset from 47 participants and also used two public datasets namely the DREAMER and DEEP. The result, from the collected dataset, showed that the ECG was superior at recognizing arousal emotion with an accuracy score of up to 68.75%. Whereas the PPG was superior at recognizing valence with an accuracy score of 64.94%. The features from ECG signals were extracted using the "Augsburg biosignal toolbox" (AUBT). The "Toolbox for Extracting Emotional FeAtures from Physiological Signals" (TEAP) was used for mining the features from PPG signals. The SVM was used for classification purpose.

He et al. (2021, pp. 1001–1005) proposed an online cross-subject "emotion recognition" (ER) approach from ECG signals via "unsupervised domain adaptation" (UDA). First, they trained a classifier on a common subspace with a smaller intersubject discrepancy during the training phase. The "online data adaptation" (ODA) approach is then employed in an online recognition stage to modify time-varying ECG by lowering the intrasubject disagreement, and online recognition results may then be achieved by the trained classifier. The method was tested using the DREAMER and AMOGOS datasets. The results for valence and arousal emotions showed that the suggested strategy is effective for the time-varying ECG in online settings. The model secured an increased accuracy score, roughly 12% high, compared to the baseline methods.

Singson et al. (2021, pp. 15–18) used ECG and facial expressions for emotion recognition. The intended dataset comprised five emotions namely, happy, sad, neutral, fear, and anger. The facial images and ECG signals were processed using the CNN using ResNet architecture for an automated categorization. The proposed model achieved an accuracy score of 68.42%.

Ye et al. (2022, pp. 23–29) suggested the usage of "hypergraph-based online transfer learning" (HOTL) to implement a unique online cross-subject ECG-based emotion identification approach. The identification model was trained on source subjects. It was shown that such a model can be more successfully adapted to target subjects because the suggested hypergraph structure can learn the high-order correlation among data. The time-frequency analysis is performed for extracting the features from the ECG signal. The model is tested using the AMIGOS dataset. The method secured accuracy scores of 89.4% and 82.7%, respectively, for the arousal and valance.

Vazquez-Rodriguez et al. (2022, pp. 2605–2612) suggested processing the ECG using a transformer-based model to identify emotions. Contextualized representations for a signal were built using the transformer's attention processes, giving pertinent components additional weight. A fully linked network was then used to process these representations to forecast emotions. The model was tested using the AMIGOS dataset. This research demonstrated that pretraining and transformers were effective techniques for identifying emotions from the ECG signals.

Alam et al. (2023) proposed a unimodal emotion classifier system in which the ECG signal has been used to identify human emotions. The ECG signal was acquired using a capacitive sensor-based noncontact ECG belt system. The considered dataset is composed of seven emotions namely, anger, disgust, fear, happy, neutral, sad, and surprise. The ECG signals are denoised and statistical features are extracted from the considered ECG segments. The "random forest" (RF) classifier is used for the categorization of emotions. The model secured the highest accuracy score of 98%.

3. Atrificial intelligence based emotion recognotion

3.1 Data set information

This dataset[1] is designed for emotion recognition in young adults, particularly beneficial in human—computer interactions for individuals dealing with cognitive and physical challenges. The data is collected using wearable shimmer3 sensors, capturing both ECG and GSR from a total of 25 participants. This dataset comprises three main components: raw data, extracted ECG features, and self-annotation labels. The raw data encompasses multimodal and single-modal data in.mat format. The multimodal data includes ECG and GSR readings from 12 participants who watched 21 stimulus videos divided into three sessions. Similarly, the single modal data contains ECG data from 13 other participants. These physiological signals correspond to the recognition of six primary emotions (surprise, anger, fear, happiness, sadness, disgust), as well as a neutral emotional state. Each of these seven emotional states is further categorized into five intensity levels, resulting in a total of 35 emotional states. Participants self-annotated these emotional states and provided scores for valence, arousal, and dominance. Additionally, we extracted 20 emotion-related features from each ECG sample, which

are included as supplementary material for further analysis. This dataset significantly enhances the resources available for physiological signal-based emotion recognition, offering a broader range of emotions and focusing on young adults.

3.2 Emotion recognition from ECG signals

ECG records are segmented into eight nonoverlapping segments, each lasting 11 s. To process these segments, we applied the continuous wavelet transform (CWT) to individual ECG signal channels, which transforms them into 2D representations. The CWT is a mathematical tool that decomposes signals and images into localized waveforms called wavelets, offering a multiresolution representation capturing both frequency and time information. Commonly used wavelets in literature include Haar, Daubechies, and Morlet (Daubechies, 1988; Grossmann & Morlet, 1984; Haar, 1910). In our case, we used the Morlet wavelet to perform a 128-level wavelet transform on each channel record within a segment. The resulting 2D transformations of ECG channels were then combined into a single 2D array, representing the entire segment's transformation. These 2D images were resized to 224 × 224 dimensions and used in the training process (Fig. 2.2).

For training, we employed two different learning approaches. Firstly, we created new CNN models with varying numbers of CNN layers, ranging from 2 to 8, and trained them from the ground up. Secondly, we experimented with transfer learning, utilizing established classification models to evaluate their performance. We transferred the weights of the CNN layers pretrained on the Image Net dataset without alterations. However, we added and trained new

---

[1] https://data.mendeley.com/datasets/g2p7vwxyn2/4

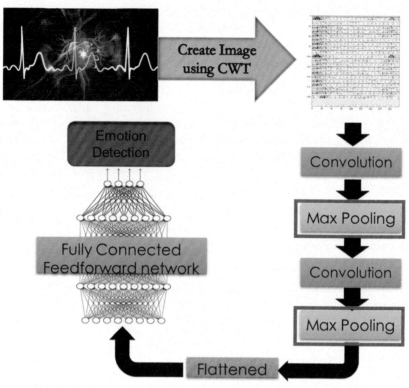

FIGURE 2.2 Artificial intelligence-based emotion recognition from ECG signals.

fully connected layers after the transferred CNN layer. Throughout the training processes, we utilized the Adam optimizer with binary cross-entropy loss as hyperparameters (Kingma & Ba, 2014).

3.3 CNN models

CNNs have emerged as a game-changing deep learning architecture, transforming biomedical image analysis and having a substantial impact on a variety of other domains. CNNs are specifically built to excel at image and spatial data analysis jobs. Their success can be ascribed to their ability to learn hierarchical data representations automatically, beginning with simple features like edges and textures and evolving to increasingly sophisticated and abstract concepts

as network layers increase. Convolutional layers are used to enable hierarchical feature learning by applying a series of learnable filters to the input data, successfully capturing patterns and features at multiple spatial scales. Pooling layers, such as max pooling or average pooling, are also widely implemented into CNN systems, lowering feature map spatial dimensions while keeping crucial information. Following these layers are fully linked layers, which allow the network to make high-level decisions based on the learned features. CNNs have been used in a variety of applications, such as image classification, object detection, face recognition, medical image analysis, and even natural language processing tasks where the input may be represented as an image-like structure, such as text or speech spectrograms.

One of CNN models' primary benefits is their capacity to learn and generalize from vast datasets. CNNs can capture a wide range of visual patterns by training on large and diverse datasets, making them very adaptable and relevant to a wide range of fields. In this chapter, we designed several CNN configurations with varying quantities of CNN layers, spanning from 2 to 8, to assess how the depth of the model influences its performance. With each increase in the number of layers, we incorporated extra convolutional layers, along with max pooling and dropout layers, to capture more intricate image characteristics. CNNs have actually transformed machine learning and artificial intelligence, making them a crucial tool in today's data-driven world.

```python
import os
from os import path
import glob
import pandas as pd
import numpy as np
import shutil
from sklearn.preprocessing import StandardScaler
from tensorflow.keras.utils import to_categorical
from sklearn.preprocessing import LabelEncoder
from tensorflow.keras.models import Sequential
from tensorflow.keras.layers import Dense, Dropout, Activation, Flatten,
Conv1D, BatchNormalization
from tensorflow.keras.optimizers import Adam
from sklearn import metrics
import matplotlib.pyplot as plt
from sklearn.model_selection import train_test_split
from tensorflow.keras.callbacks import ModelCheckpoint
from datetime import datetime
```

```python
import tensorflow as tf
# signal processing
from scipy import signal
from scipy.ndimage import label as sci_label
from scipy.stats import zscore
from scipy.interpolate import interp1d
from scipy.integrate import trapz
# misc
import warnings
```

3.3.1 *Data reading and preparation*

The sole difference between the multimodal and single-modal folders is that the single-modal folder contains only ECG signals. The multimodal folder, on the other hand, has both ECG and GSR recordings recorded at the same time. In this research, the user classified only the ECG recordings and did not view the GSR data.

```python
from google.colab import drive
drive.mount('/content/gdrive')
stimulus_desc_file =
pd.read_excel('/content/gdrive/MyDrive/ECG_GSR_Emotions/Stimulus_Descripti
on.xlsx')
stimulus_desc_file.to_csv('Stimulus_Description.csv', index = None,
header=True)
stimulus_desc = pd.read_csv('Stimulus_Description.csv')
self_annotation_multimodal_file =
pd.read_excel('/content/gdrive/MyDrive/ECG_GSR_Emotions/Self-Annotation
Labels/Self-annotation Multimodal_Use.xlsx')
self_annotation_multimodal_file.to_csv('Self-annotation
Multimodal_Use.csv', index = None, header=True)
self_annotation_multimodal = pd.read_csv('Self-annotation
Multimodal_Use.csv')
self_annotation_multimodal['annotation'] = 'M'
self_annotation_multimodal.rename(columns = {'V_Label':'Valence',
'A_Label':'Arousal', 'Four_Labels':'Four_Label'}, inplace = True)
self_annotation_singlemodal_file =
pd.read_excel('/content/gdrive/MyDrive/ECG_GSR_Emotions/Self-Annotation
Labels/Self-annotation Single Modal_Use.xlsx')
```

```python
self_annotation_singlemodal_file.to_csv('Self-annotation Single
Modal_Use.csv', index = None, header=True)
self_annotation_singlemodal = pd.read_csv('Self-annotation Single
Modal_Use.csv')
self_annotation_singlemodal['annotation'] = 'S'
self_annotation_singlemodal.rename(columns = {'Male':'Gender', 'Session
Id':'Session ID', 'Video Id':'Video ID'}, inplace = True)
self_annotation_frames = [self_annotation_multimodal,
self_annotation_singlemodal]
merged_dataframe = pd.concat(self_annotation_frames)
table_frame = merged_dataframe.copy()
cols = ['Raw Data', "GSR"]
for col in merged_dataframe.columns:
    cols.append(col)
arr_shape = 1000
raw_data_arr = []
def form_data(data_array = [], annotation = '', data_path = ''):
    for filename in os.listdir(data path): #for each record
```

```
          data = np.loadtxt(os.path.join(data_path, filename),
delimiter=',') #get the signal as array
          data = data ## get first 5000 element of array
          try:
            gsr = np.loadtxt(os.path.join(data_path[:-4]+"GSR/",
"GSR"+filename[3:]), delimiter=',')
          except:
            continue
          ### split the file name to get session, participant and video
name#
          filenames = filename.split('ECGdata_')[1]
          filenames = filenames.split('.dat')[0]
          filenames = filenames.lower()
          s = filenames.split('s')[1].split('p');
          p = s[1].split('v')
          s = s[0]
          v = p[1]
          p = p[0]




          ###########################################################
          data_row = merged_dataframe.loc[(merged_dataframe['Session
ID'] == int(s)) & #get the row of data_frame that has corresponding
features
                          (merged_dataframe['Participant Id'] ==
int(p)) &
                          (merged_dataframe['Video ID'] == int(v)) &
                          (merged_dataframe['annotation'] ==
annotation)]
          stim_row = stimulus_desc.loc[(stimulus_desc['Session ID'] ==
int(s)) &  #get the row of stimulus_desc that has corresponding features
                          (stimulus_desc['Video ID'] == int(v))]
          for index, row in data_row.iterrows(): # combine all the
information into one row and add this to the new array
            data_array.append([data, gsr,
            row['Participant Id'], row['Session ID'], row['Video ID'],
            row['Name'], row['Age'], row['Gender'], row['Valence
level'],
            row['Arousal level'], row['Dominance level'], row['Happy'],
            row['Sad'], row['Fear'], row['Anger'], row['Neutral'],
            row['Disgust'], row['Surprised'], row['Familiarity Score'],
            row['Emotion'], row['Valence'], row['Arousal'],
row['Four_Label'],
```

```
                        row['annotation'],  stim_row['Target Emotion'].iat[0]
                        ])
        return data_array


raw_data_arr1 =  form_data(data_array = raw_data_arr, annotation = 'M',
data_path = "/content/gdrive/MyDrive/ECG_GSR_Emotions/Raw
Data/Multimodal/ECG/") #get only multimodal
#raw_data_arr2 =  form_data(data_array = raw_data_arr, annotation = 'S',
data_path = "/content/gdrive/MyDrive/ECG_GSR_Emotions/Raw Data/Single
Modal/ECG/") #get only single modal
cols.append('Target Emotion')
raw_dataframe = pd.DataFrame(raw_data_arr1, columns = cols)

raw_dataframe.rename(columns = {'Participant Id':'Participant ID',
'annotation':'Modal', 'Four_Label':'Four label'}, inplace = True)
raw_dataframe['Familiarity Score'] = raw_dataframe['Familiarity
Score'].fillna('Never watched')
raw_dataframe = raw_dataframe.replace(np.nan, '', regex=True)
```

3.3.1.1 Data plotting

```
plot_frame = raw_dataframe.copy()
plot_frame = plot_frame.drop(['Participant ID', 'Session ID', 'Familiarity
Score', 'Age', 'Gender', 'Name'], axis = 1)
sad_data = plot_frame.loc[(plot_frame['Emotion'] == 'Sad') &
(plot_frame['Target Emotion'] == 'sad')]
fear_data = plot_frame.loc[(plot_frame['Emotion'] == 'Fear')  &
(plot_frame['Target Emotion'] == 'fear')]
happy_data = plot_frame.loc[(plot_frame['Emotion'] == 'Happy') &
(plot_frame['Target Emotion'] == 'happy')]
anger_data = plot_frame.loc[(plot_frame['Emotion'] == 'Anger') &
(plot_frame['Target Emotion'] == 'anger')]
neutral_data = plot_frame.loc[(plot_frame['Emotion'] == 'Neutral') &
(plot_frame['Target Emotion'] == 'neutral')]
mixed_data = plot_frame.loc[(plot_frame['Emotion'] == 'Mixed') &
(plot_frame['Target Emotion'] == 'neutral')]
disgust_data = plot_frame.loc[(plot_frame['Emotion'] == 'Disgust') &
(plot_frame['Target Emotion'] == 'disgust')]
```

```python
surprised_data = plot_frame.loc[(plot_frame['Emotion'] == 'Surprise') &
(plot_frame['Target Emotion'] == 'surprise')]

def plot_signals(data_arr, title = ''):
    plt.clf()
    plt.figure(figsize=(12, 4))
    for index, row in data_arr.iterrows():
        y = row['Raw Data']
        plt.plot(y)
        #x = np.arange(y.size)
        #plt.plot(x, y)
    plt.tight_layout()
    plt.title(title)
    plt.show()
```

3.3.1.2 Training

```python
from google.colab import drive
import numpy as np
import pandas as pd
import scipy
from tensorflow.keras.applications import *
from tensorflow.keras.optimizers import *
from tensorflow.keras.losses import *
from tensorflow.keras.layers import *
from tensorflow.keras.models import *
from tensorflow.keras.callbacks import *
from tensorflow.keras.preprocessing.image import *
from tensorflow.keras.utils import *
import matplotlib.pyplot as plt
from sklearn.model_selection import *
from sklearn.metrics import f1_score, roc_auc_score, cohen_kappa_score,
precision_score, recall_score, accuracy_score, confusion_matrix
import tensorflow as tf
```

```
import os
import cv2
import seaborn as sns
import pywt
print("All modules have been imported")

def get_accuracy_metrics(model_, X_train, y_train, X_val, y_val,X_test,
y_test ):
    train_pred= model_.predict(X_train)
    val_pred= model_.predict(X_val)
    test_pred = model_.predict(X_test)
    y_train = np.argmax(y_train, axis = 1)
    y_val = np.argmax(y_val, axis = 1)
    y_test = np.argmax(y_test, axis = 1)
    train_pred = np.argmax(train_pred, axis = 1)
    val_pred = np.argmax(val_pred, axis = 1)
    test_pred = np.argmax(test_pred, axis = 1)
    print("Train accuracy Score------------>")
    print ("{0:.3f}".format(accuracy_score(y_train, train_pred)*100), "%")




    print("Val accuracy Score--------->")
    print("{0:.3f}".format(accuracy_score(y_val, val_pred)*100), "%")
    print("Test accuracy Score--------->")
    print("{0:.3f}".format(accuracy_score(y_test, test_pred)*100), "%")
    cf_matrix_test = confusion_matrix(y_test, test_pred)
    cf_matrix_val = confusion_matrix(y_val, val_pred)
    plt.figure(figsize = (12, 6))
    plt.subplot(121)
    sns.heatmap(cf_matrix_val, annot=True, cmap='Blues')
    plt.title("Val Confusion matrix")
    plt.subplot(122)
    sns.heatmap(cf_matrix_test, annot=True, cmap='Blues')
    plt.title("Test Confusion matrix")
    plt.show()
train_d_frame = plot_frame.copy().drop(['Video ID', 'Happy', 'Sad',
'Fear','Anger', 'Neutral', 'Disgust', 'Surprised', 'Four label', 'Modal'
                                       ], axis = 1)
```

```
train_d_frame.rename(columns = {'Raw Data':'feature','Emotion':'emotion'},
inplace = True)
#train_d_frame.head()
x = np.array([train_d_frame['feature'].tolist(),
train_d_frame['GSR'].tolist()])
y = np.array(train_d_frame['emotion'].tolist())
sc = StandardScaler()
labelencoder = LabelEncoder()
y = to_categorical(labelencoder.fit_transform(y))
x_3d = []
y_2 = []
for i in range(x[0].shape[0]):
    for j in range(0,x[0][i].shape[0]-2560, 500):
        a = pywt.cwt(x[0][i][j:j+5120],range(1,64),"morl",)[0]
        a_ = pywt.cwt(x[1][i][j:j+5120],range(1,64),"morl",)[0]
        a = np.concatenate([a, a_], axis = 0)
        x_3d.append(cv2.resize(np.array([a,a,a]).T,(128, 128)))
        y_2.append(y[i])
y_2 = np.array(y_2)
```

```
x_train, x_test, y_train, y_test = train_test_split(np.array(x_3d),
y_2.astype(int), test_size = 0.2, random_state = 0, stratify =
y_2.astype(int))
x_train, x_valid, y_train, y_valid = train_test_split(np.array(x_train),
y_train.astype(int), test_size = 0.125, random_state = 0, stratify =
y_train.astype(int))
plt.imshow(x_train[0])
```

3.3.1.3 5 layer CNN model for emotion detection

```
inp = Input(shape = (128, 128, 3))
model = BatchNormalization()(inp)
model = Conv2D(filters = 128, kernel_size = (3, 3), padding = 'same',
activation='relu')(model)
model = MaxPooling2D()(model)
model = Dropout(0.2)(model)
model = Conv2D(filters = 128, kernel_size = (3, 3), padding = 'same',
activation='relu')(model)
model = MaxPooling2D()(model)
model = Dropout(0.2)(model)
model = Conv2D(filters = 64, kernel_size = (3, 3), padding = 'same',
activation='relu')(model)
model = MaxPooling2D()(model)
model = Dropout(0.2)(model)
model = Conv2D(filters = 32, kernel_size = (3, 3), padding = 'same',
activation='relu')(model)
model = MaxPooling2D()(model)
model = Dropout(0.2)(model)
```

```
model = Conv2D(filters = 32, kernel_size = (3, 3), padding = 'same',
activation='relu')(model)
model = Flatten()(model)
output = Dense(units = 128, activation = 'relu')(model)
output = Dense(units = 64, activation = 'relu')(model)
output = Dense(units = 32, activation = 'relu')(model)
output = Dense(units = 8, activation = 'relu')(model)
output = Dense(units = 8, activation = 'softmax')(model)
model = Model(inputs=inp, outputs=output)
model.compile(loss = 'categorical_crossentropy', metrics = ['accuracy'],
optimizer =tf.keras.optimizers.Adam(learning_rate=1e-5) )
num_epochs = 500
num_batch_size = 2
checkpointer = ModelCheckpoint(filepath = './Raw
Data/ecg_emotion_recognizer.hdf5',
                              verbose = 1, save_best_only = True)
start = datetime.now()
```

```
model.fit(x_train, y_train, batch_size = num_batch_size, epochs =
num_epochs,
        validation_data = (x_valid, y_valid), callbacks =
[checkpointer], verbose = 1)
duration = datetime.now() - start
print("Training completed in time: ", duration)
get_accuracy_metrics(model, x_train, y_train, x_valid, y_valid,  x_test,
y_test)
```

3.4 Transfer learning

Transfer learning is a machine learning technique that has grown in popularity in recent years, altering the way models are trained and deployed across several domains. Transfer learning, at its heart, applies knowledge obtained from one activity or dataset to a different but related task or dataset. This method has various advantages, most notably when getting a big and tagged dataset for a certain task is difficult or unfeasible. Transfer learning allows practitioners to start with a pretrained model on a source job, frequently including a large and diverse dataset, and fine-tune it for a target task rather than training a model from scratch, which can be computationally expensive and data-intensive. As a result, the model can acquire valuable features, representations, and patterns from the source task, accelerating convergence and enhancing performance on the target goal. Transfer learning has proved especially disruptive in computer vision, natural language processing, and speech recognition, where pretrained models like CNNs and transformer models have served as the foundation for developing cutting-edge applications. This method saves time and computational resources while often improving model performance. CNN designs are growing in tandem with the area of deep learning, with variations such as residual networks (ResNets) and densely connected networks (DenseNets) significantly improving model depth and performance.

Furthermore, transfer learning makes it easier to create more robust and generalized models. Because pretrained models have often learned generic features that are broadly applicable, they frequently outperform models trained from scratch, particularly when the target dataset is limited or lacks diversity. Transfer learning also encourages knowledge and expertise sharing across areas, boosting collaboration and innovation. With the availability of pretrained models and open-source libraries, practitioners from a variety of domains can gain access to and profit from cutting-edge solutions without requiring considerable machine-learning skills. Overall, transfer learning has opened a new era of efficiency and efficacy in AI model development, enabling improvements in a wide range of applications and democratizing access to AI technology.

```
import os
from google.colab import drive
import numpy as np
import pandas as pd
import scipy
from tensorflow.keras.applications import *
from tensorflow.keras.optimizers import *
from tensorflow.keras.losses import *
from tensorflow.keras.layers import *
from tensorflow.keras.models import *
from tensorflow.keras.callbacks import *
from tensorflow.keras.preprocessing.image import *
from tensorflow.keras.utils import *
import matplotlib.pyplot as plt
from sklearn.model_selection import *
from sklearn.metrics import f1_score, roc_auc_score, cohen_kappa_score,
precision_score, recall_score, accuracy_score, confusion_matrix
```

```
from sklearn.preprocessing import StandardScaler,LabelEncoder
import tensorflow as tf
import cv2
import seaborn as sns
import pywt
print("All modules have been imported")
drive.mount('/content/gdrive')
datasheet_multi =
pd.read_excel('/content/gdrive/MyDrive/ECG_GSR_Emotions/Self-Annotation
Labels/Self-annotation Multimodal_Use.xlsx')
datasheet_multi = datasheet_multi[['Participant Id','Session ID','Video
ID','Emotion']]
datasheet_multi['annotation'] = 'M'
datasheet_single =
pd.read_excel('/content/gdrive/MyDrive/ECG_GSR_Emotions/Self-Annotation
Labels/Self-annotation Single Modal_Use.xlsx')
datasheet_single = datasheet_single[['Participant Id','Session Id','Video
Id','Emotion']]
datasheet_single['annotation'] = 'S'
datasheet_single = datasheet_single.rename(columns={"Session Id": "Session
ID", "Video Id": "Video ID"})
cols = ['ECG']
```

```
data_arr = []
def form_data(data_array = [], annotation = '', datasheet = [], data_path
= ''):
    for filename in os.listdir(data_path):
            data = np.loadtxt(os.path.join(data_path, filename),
delimiter=',')

            filenames = filename.split('ECGdata_')[1]
            filenames = filenames.split('.dat')[0]
            filenames = filenames.lower()
            session = filenames.split('s')[1].split('p');
            participant = session[1].split('v')
            session = session[0]
            video = participant[1]
            participant = participant[0]
            data_row = datasheet.loc[(datasheet['Session ID'] ==
int(session)) &
                            (datasheet['Participant Id'] ==
int(participant)) &
                            (datasheet['Video ID'] == int(video)) &
                            (datasheet['annotation'] == annotation)]
            for index, row in data_row.iterrows():



                data_array.append([data,row['Emotion']])
    return data_array
data_arr = form_data(data_array = data_arr, annotation = 'M', datasheet =
datasheet_multi,  data_path =
"/content/gdrive/MyDrive/ECG_GSR_Emotions/Raw Data/Multimodal/ECG/")
data_arr = form_data(data_array = data_arr, annotation = 'S', datasheet =
datasheet_single,  data_path =
"/content/gdrive/MyDrive/ECG_GSR_Emotions/Raw Data/Single Modal/ECG/")
cols.append('Emotion')
dataframe = pd.DataFrame(data_arr, columns = cols)
def get_accuracy_metrics(model_, X_train, y_train, X_val, y_val,X_test,
y_test ):
    train_pred= model_.predict(X_train)
    val_pred= model_.predict(X_val)
    test_pred = model_.predict(X_test)
    y_train = np.argmax(y_train, axis = 1)
    y_val = np.argmax(y_val, axis = 1)
    y_test = np.argmax(y_test, axis = 1)
    train_pred = np.argmax(train_pred, axis = 1)
```

```
val_pred = np.argmax(val_pred, axis = 1)
test_pred = np.argmax(test_pred, axis = 1)
print("Train accuracy Score------------>")
print ("{0:.3f}".format(accuracy_score(y_train, train_pred)*100), "%")
print("Val accuracy Score--------->")
print("{0:.3f}".format(accuracy_score(y_val, val_pred)*100), "%")
print("Test accuracy Score--------->")
print("{0:.3f}".format(accuracy_score(y_test, test_pred)*100), "%")
cf_matrix_test = confusion_matrix(y_test, test_pred)
cf_matrix_val = confusion_matrix(y_val, val_pred)
plt.figure(figsize = (12, 6))
plt.subplot(121)
sns.heatmap(cf_matrix_val, annot=True, cmap='Blues')
plt.title("Val Confusion matrix")
plt.subplot(122)
sns.heatmap(cf_matrix_test, annot=True, cmap='Blues')
plt.title("Test Confusion matrix")
plt.show()
```

```
x = np.array(dataframe['ECG'].tolist())
y = np.array(dataframe['Emotion'].tolist())
labelencoder = LabelEncoder()
y = to_categorical(labelencoder.fit_transform(y))
images = []
labels = []
for i in range(x.shape[0]):
    for j in range(0,x[i].shape[0]-1280, 1280):
      ecg_cwt = pywt.cwt(x[i][j:j+1280],range(1,128),"morl",)[0]
      images.append(cv2.resize(np.array([ecg_cwt,ecg_cwt,ecg_cwt]).T,(224,
224)))
      labels.append(y[i])
labels = np.array(labels)

x_train, X_test, y_train, y_test = train_test_split(np.array(images),
labels.astype(int), test_size = 0.2, random_state = 0, stratify =
labels.astype(int))
X_train, X_val, y_train, y_val = train_test_split(np.array(x_train),
y_train.astype(int), test_size = 0.125, random_state = 0, stratify =
y_train.astype(int))
plt.imshow(X_train[0])
```

3.4.1 VGG16 model for emotion detection

```
base_Neural_Net= VGG16(include_top=False, input_shape=(128, 128, 3),
weights='imagenet')
model=Sequential()
model.add(base_Neural_Net)
model.add(BatchNormalization())
model.add(Flatten())
###fifth layer
model.add(Dense(64))
model.add(Activation('relu'))
model.add(BatchNormalization())
model.add(Dense(128))
model.add(Activation('relu'))
model.add(BatchNormalization())
model.add(Dense(64))
model.add(Activation('relu'))
model.add(BatchNormalization())
model.add(Dense(32))
model.add(Activation('relu'))
model.add(BatchNormalization())
model.add(Dense(8))
```

```
model.add(Activation('relu'))
model.add(Dense(8))
model.add(Activation('softmax'))
for layer in base_Neural_Net.layers:
    layer.trainable = False
model.compile(loss = 'categorical_crossentropy', metrics = ['accuracy'],
optimizer =tf.keras.optimizers.Adam(learning_rate=1e-5) )
num_epochs = 1000
num_batch_size = 2
checkpointer = ModelCheckpoint(filepath = './Raw
Data/ecg_emotion_recognizer.hdf5',
                                verbose = 1, save_best_only = True)
start = datetime.now()
model.fit(x_train, y_train, batch_size = num_batch_size, epochs =
num_epochs,
        validation_data = (x_valid, y_valid), callbacks =
[checkpointer], verbose = 1)
duration = datetime.now() - start
print("Training completed in time: ", duration)
get_accuracy_metrics(model, x_train, y_train, x_valid, y_valid,  x_test,
y_test)
```

3.4.2 VGG19 model for emotion detection

```
base_Neural_Net= VGG19(include_top=False, input_shape=(128, 128, 3),
weights='imagenet')
model=Sequential()
model.add(base_Neural_Net)
model.add(BatchNormalization())
model.add(Flatten())
###fifth layer
model.add(Dense(64))
model.add(Activation('relu'))
model.add(BatchNormalization())
model.add(Dense(128))
model.add(Activation('relu'))
model.add(BatchNormalization())
model.add(Dense(64))
model.add(Activation('relu'))
model.add(BatchNormalization())
model.add(Dense(32))
model.add(Activation('relu'))
model.add(BatchNormalization())
```

```
model.add(Dense(8))
model.add(Activation('relu'))
model.add(Dense(8))
model.add(Activation('softmax'))
for layer in base_Neural_Net.layers:
    layer.trainable = False
model.compile(loss = 'categorical_crossentropy', metrics = ['accuracy'],
optimizer =tf.keras.optimizers.Adam(learning_rate=1e-5) )
num_epochs = 1000
num_batch_size = 2
checkpointer = ModelCheckpoint(filepath = './Raw
Data/ecg_emotion_recognizer.hdf5',
                               verbose = 1, save_best_only = True)
start = datetime.now()
model.fit(x_train, y_train, batch_size = num_batch_size, epochs =
num_epochs,
          validation_data = (x_valid, y_valid), callbacks =
[checkpointer], verbose = 1)
duration = datetime.now() - start
print("Training completed in time: ", duration)
get_accuracy_metrics(model, x_train, y_train, x_valid, y_valid,  x_test,
y_test)
```

3.4.3 InceptionV3 model for emotion detection

```
# Inception V3
base_Neural_Net = InceptionV3(include_top=False, input_shape=(224, 224,
3), weights='imagenet')
model=Sequential()
model.add(base_Neural_Net)
model.add(Flatten())
model.add(BatchNormalization())
model.add(Dense(256,kernel_initializer='he_uniform', activation='relu'))
model.add(BatchNormalization())
model.add(Dense(64,kernel_initializer='he_uniform', activation='relu'))
model.add(BatchNormalization())
model.add(Activation('relu'))
model.add(Dropout(0.5))
model.add(Dense(8,activation='relu'))
model.add(Dense(8,activation='softmax'))
for layer in base_Neural_Net.layers:
    layer.trainable = False
es = EarlyStopping(monitor='val_loss', mode='min', verbose=1, patience=5)
```

```
model.compile(
    loss='categorical_crossentropy',
    optimizer=tf.keras.optimizers.Adam(learning_rate=5*1e-7),
    metrics=['accuracy' , 'AUC'])
print(model.summary())
EPOCHS = 5
model.fit(X_train, y_train, validation_data=(X_val, y_val), epochs=1000,
verbose=1)
get_accuracy_metrics(model, X_train, y_train, X_val, y_val,  X_test,
y_test)
```

3.4.4 MobileNet model for emotion detection

```
# MobileNet
base_Neural_Net = MobileNet(include_top=False, input_shape=(224, 224, 3),
weights='imagenet')
model=Sequential()
model.add(base_Neural_Net)
model.add(Flatten())
model.add(BatchNormalization())
model.add(Dense(256,kernel_initializer='he_uniform', activation='relu'))
model.add(BatchNormalization())
model.add(Dense(64,kernel_initializer='he_uniform', activation='relu'))
model.add(BatchNormalization())
model.add(Activation('relu'))
model.add(Dropout(0.5))
model.add(Dense(8,activation='relu'))
model.add(Dense(8,activation='softmax'))
for layer in base_Neural_Net.layers:
    layer.trainable = False
es = EarlyStopping(monitor='val_loss', mode='min', verbose=1, patience=5)
```

```
model.compile(loss='categorical_crossentropy',
    optimizer=tf.keras.optimizers.Adam(learning_rate=5*1e-7),
    metrics=['accuracy' , 'AUC'])
print(model.summary())
EPOCHS = 5
model.fit(X_train, y_train, validation_data=(X_val, y_val), epochs=1000,
verbose=1)
get_accuracy_metrics(model, X_train, y_train, X_val, y_val,  X_test,
y_test)
```

3.4.5 DenseNet 169 model for emotion detection

```
# DenseNet 169
base_Neural_Net = DenseNet169(include_top=False, input_shape=(224, 224,
3), weights='imagenet')
model=Sequential()
model.add(base_Neural_Net)
model.add(Flatten())
model.add(BatchNormalization())
model.add(Dense(256,kernel_initializer='he_uniform', activation='relu'))
model.add(BatchNormalization())
model.add(Dense(64,kernel_initializer='he_uniform', activation='relu'))
model.add(BatchNormalization())
model.add(Activation('relu'))
model.add(Dropout(0.5))
model.add(Dense(8,activation='relu'))
model.add(Dense(8,activation='softmax'))
for layer in base_Neural_Net.layers:
    layer.trainable = False
es = EarlyStopping(monitor='val_loss', mode='min', verbose=1, patience=5)
```

```
model.compile(loss='categorical_crossentropy',
    optimizer=tf.keras.optimizers.Adam(learning_rate=5*1e-7),
    metrics=['accuracy' , 'AUC'])
print(model.summary())
EPOCHS = 5
model.fit(X_train, y_train, validation_data=(X_val, y_val), epochs=1000,
verbose=1)
get_accuracy_metrics(model, X_train, y_train, X_val, y_val,  X_test,
y_test)
```

3.4.6 DenseNet 121 model for emotion detection

```python
# DenseNet 121
base_Neural_Net = DenseNet121(include_top=False, input_shape=(224, 224,
3), weights='imagenet')
model=Sequential()
model.add(base_Neural_Net)
model.add(Flatten())
model.add(BatchNormalization())
model.add(Dense(256,kernel_initializer='he_uniform', activation='relu'))
model.add(BatchNormalization())
model.add(Dense(64,kernel_initializer='he_uniform', activation='relu'))
model.add(BatchNormalization())
model.add(Activation('relu'))
model.add(Dropout(0.5))
model.add(Dense(8,activation='relu'))
model.add(Dense(8,activation='softmax'))
for layer in base_Neural_Net.layers:
    layer.trainable = False
es = EarlyStopping(monitor='val_loss', mode='min', verbose=1, patience=5)
```

```python
model.compile(loss='categorical_crossentropy',
    optimizer=tf.keras.optimizers.Adam(learning_rate=5*1e-7),
    metrics=['accuracy' , 'AUC'])
print(model.summary())
EPOCHS = 5
model.fit(X_train, y_train, validation_data=(X_val, y_val), epochs=1000,
verbose=1)
get_accuracy_metrics(model, X_train, y_train, X_val, y_val,  X_test,
y_test)
```

3.4.7 InceptionResNet V2 model for emotion detection

```python
# InceptionResNet V2
base_Neural_Net = InceptionResNetV2(include_top=False, input_shape=(224,
224, 3), weights='imagenet')
model=Sequential()
model.add(base_Neural_Net)
model.add(Flatten())
model.add(BatchNormalization())
model.add(Dense(256,kernel_initializer='he_uniform', activation='relu'))
model.add(BatchNormalization())
model.add(Dense(64,kernel_initializer='he_uniform', activation='relu'))
model.add(BatchNormalization())
model.add(Activation('relu'))
model.add(Dropout(0.5))
model.add(Dense(8,activation='relu'))
model.add(Dense(8,activation='softmax'))
for layer in base_Neural_Net.layers:
    layer.trainable = False
es = EarlyStopping(monitor='val_loss', mode='min', verbose=1, patience=5)
```

```python
model.compile(loss='categorical_crossentropy',
    optimizer=tf.keras.optimizers.Adam(learning_rate=5*1e-7),
    metrics=['accuracy' , 'AUC'])
print(model.summary())
EPOCHS = 5
model.fit(X_train, y_train, validation_data=(X_val, y_val), epochs=1000,
verbose=1)
get_accuracy_metrics(model, X_train, y_train, X_val, y_val,  X_test,
y_test)
```

3.4.8 MobileNetV2 model for emotion detection

```
# MobileNet V2
base_Neural_Net = MobileNetV2(include_top=False, input_shape=(224, 224,
3), weights='imagenet')
model=Sequential()
model.add(base_Neural_Net)
model.add(Flatten())
model.add(BatchNormalization())
model.add(Dense(256,kernel_initializer='he_uniform', activation='relu'))
model.add(BatchNormalization())
model.add(Dense(64,kernel_initializer='he_uniform', activation='relu'))
model.add(BatchNormalization())
model.add(Activation('relu'))
model.add(Dropout(0.5))
model.add(Dense(8,activation='relu'))
model.add(Dense(8,activation='softmax'))
for layer in base_Neural_Net.layers:
    layer.trainable = False
es = EarlyStopping(monitor='val_loss', mode='min', verbose=1, patience=5)
```

```
model.compile(loss='categorical_crossentropy',
    optimizer=tf.keras.optimizers.Adam(learning_rate=5*1e-7),
    metrics=['accuracy' , 'AUC'])
print(model.summary())
EPOCHS = 5
model.fit(X_train, y_train, validation_data=(X_val, y_val), epochs=1000,
verbose=1)
get_accuracy_metrics(model, X_train, y_train, X_val, y_val,  X_test,
y_test)
```

3.4.9 ResNet 101 model for emotion detection

```
# ResNet 101
base_Neural_Net = ResNet101(include_top=False, input_shape=(224, 224, 3),
weights='imagenet')
model=Sequential()
model.add(base_Neural_Net)
model.add(Flatten())
model.add(BatchNormalization())
model.add(Dense(256,kernel_initializer='he_uniform', activation='relu'))
model.add(BatchNormalization())
model.add(Dense(64,kernel_initializer='he_uniform', activation='relu'))
model.add(BatchNormalization())
model.add(Activation('relu'))
model.add(Dropout(0.5))
model.add(Dense(8,activation='relu'))
model.add(Dense(8,activation='softmax'))
for layer in base_Neural_Net.layers:
    layer.trainable = False
es = EarlyStopping(monitor='val_loss', mode='min', verbose=1, patience=5)
```

```
model.compile(loss='categorical_crossentropy',
    optimizer=tf.keras.optimizers.Adam(learning_rate=5*1e-7),
    metrics=['accuracy' , 'AUC'])
print(model.summary())
EPOCHS = 5
model.fit(X_train, y_train, validation_data=(X_val, y_val), epochs=1000,
verbose=1)
get_accuracy_metrics(model, X_train, y_train, X_val, y_val,  X_test,
y_test)
```

3.5 Discussion

The current study sought to investigate the feasibility and efficacy of an AI-based emotion recognition system based on ECG signals. This chapter contributes to the booming field of emotion recognition and prepares the way for creative applications in a variety of domains by harnessing the power of AI algorithms, notably CNNs and transfer learning approaches. The observed results highlight the utility of using ECG data for emotion identification. AI approaches enable the extraction of small changes in ECG rhythms linked with different emotional states. The proposed approach's ability to accurately identify emotions supports the idea that emotional experiences have a discernible impact on physiological signals. This finding is consistent with earlier research indicating the link between emotions and cardiac activity, hence supporting ECG as a tool for emotion assessment.

One important finding from this study is the critical role that AI models play in improving the accuracy of emotion recognition. CNNs and transfer learning models both performed admirably, demonstrating these algorithms' ability to learn detailed patterns from complex physiological data. CNNs' ability to learn relevant characteristics from ECG waveforms, paired with transfer learning models' ability to capture temporal dependencies, resulted in a robust and accurate emotion classification system. The use of AI techniques alleviated the limits of traditional systems that rely on external cues, making the suggested system more versatile and trustworthy in a variety of settings.

Another significant asset of this research is the generation of a large and diversified dataset. The model's generalizability and adaptability were enhanced by the incorporation of varied emotional experiences and controlled inputs. The dataset's richness aided not just the training of AI models but also the construction of a system capable of recognizing a wide range of emotional states. The availability of this dataset and its potential for growth provide researchers and practitioners with a significant resource for future studies in emotion identification.

Despite the positive results, various restrictions should be considered. The created system's generalizability across varied demographics, cultural backgrounds, and real-world contexts warrants additional exploration. Furthermore, because emotional experiences are complex and multidimensional, emotion extraction just from ECG signals may not reflect the fullness of human feelings. Integrating several modalities, such as facial expressions and voice analysis, has the potential to improve the accuracy and resilience of the system.

Finally, this study emphasizes the transformational potential of AI-based emotion identification systems based on ECG signals. This study provides a road to more accurate and context-independent emotion recognition by merging AI algorithms, physiological signal analysis, and a large dataset. The findings have implications for human–computer interface applications, mental health monitoring, and personalized healthcare therapies. This discovery represents a big step forward in utilizing ECG-based emotion identification to improve human–machine interactions and individual well-being as the disciplines of AI and physiological signal analysis continue to evolve.

4. Conclusion

This chapter reveals the intriguing potential of an AI-based emotion recognition system based on ECG signals. This chapter contributes to the growing field of emotion recognition by leveraging the capabilities of AI algorithms, providing insights that cross academic borders and open up new paths for application. The ability of this study to bridge the gap between physiological signals and emotional experiences is what makes it successful. The incorporation of artificial intelligence algorithms aided in the accurate extraction of emotional states from minor fluctuations in ECG waveforms. This accomplishment verifies ECG's usefulness as an emotion evaluation method, increasing the potential for real-time and unobtrusive emotion detection.

The ability of AI models to discriminate between emotions strengthens their ability to learn detailed patterns from physiological data. The adaptability and versatility of this technology across numerous settings have enormous promise for improving human–computer interactions, individualized healthcare interventions, and breakthroughs in affective computing. The creation of a broad and extensive dataset increased the system's dependability and generalizability. The dataset allows the AI models to reliably classify emotions across a range of scenarios by combining a wide range of emotional experiences and controlled stimuli. This dataset, in conjunction with the AI algorithms, lays the groundwork for future research and innovation in emotion recognition.

Limitations are obvious, as with any pioneering research. The system's applicability across a wide range of demographics and cultural contexts merits additional investigation. Integrating several modalities, such as facial expressions and voice analysis, could improve the system's accuracy and ability to capture the complexities of human emotions. Finally, this study demonstrates the synergy between AI and physiological signal processing, highlighting the transformational potential of ECG-based emotion recognition. This study extends the terrain of human—computer interaction, mental health monitoring, and healthcare interventions by decoding emotional states from ECG rhythms. As artificial intelligence continues to transform the technological landscape, the findings of this study lay the groundwork for building meaningful connections between humans and machines, ultimately leading to increased well-being and quality of life.

References

Alam, A., Urooj, S., & Ansari, A. Q. (2023). Design and development of a non-contact ECG-based human emotion recognition system using SVM and RF classifiers. *Diagnostics, 13*(12), 2097.

Bulagang, A. F., Weng, N. G., Mountstephens, J., & Teo, J. (2020). A review of recent approaches for emotion classification using electrocardiography and electrodermography signals. *Informatics in Medicine Unlocked, 20*, 100363.

Daubechies, I. (1988). Orthonormal bases of compactly supported wavelets. *Communications on Pure and Applied Mathematics, 41*(7), 909—996.

Ferdinando, H., Seppänen, T., & Alasaarela, E. (2017). Enhancing emotion recognition from ECG signals using supervised dimensionality reduction. In *6th international conference on pattern recognition applications and methods (ICPRAM)*.

Goshvarpour, A., Abbasi, A., & Goshvarpour, A. (2017). An accurate emotion recognition system using ECG and GSR signals and matching pursuit method. *Biomedical Journal, 40*(6), 355—368.

Grossmann, A., & Morlet, J. (1984). Decomposition of Hardy functions into square integrable wavelets of constant shape. *SIAM Journal on Mathematical Analysis, 15*(4), 723—736. https://doi.org/10.1137/0515056

Haar, A. (1910). Zur Theorie der orthogonalen Funktionensysteme. *Mathematische Annalen, 69*(3), 331—371. https://doi.org/10.1007/BF01456326

Hasnul, M. A., Ab. Aziz, N. A., & Abd. Aziz, A. (2023). Augmenting ECG data with multiple filters for a better emotion recognition system. *Arabian Journal for Science and Engineering*, 1—22.

He, W., Ye, Y., Li, Y., Pan, T., & Lu, L. (2021). *Online cross-subject emotion recognition from ECG via unsupervised domain adaptation*.

Hsu, Y.-L., Wang, J.-S., Chiang, W.-C., & Hung, C.-H. (2017). Automatic ECG-based emotion recognition in music listening. *IEEE Transactions on Affective Computing, 11*(1), 85—99.

Ismail, S. N. M. S., Aziz, N. A. A., & Ibrahim, S. Z. (2022). A comparison of emotion recognition system using electrocardiogram (ECG) and photoplethysmogram (PPG). *Journal of King Saud University-Computer and Information Sciences, 34*(6), 3539—3558.

Ismail, S. N. M. S., Aziz, N. A. A., Ibrahim, S. Z., & Mohamad, M. S. (2023). *A systematic review of emotion recognition using cardio-based signals*. ICT Express.

Katsigiannis, S., & Ramzan, N. (2017). Dreamer: A database for emotion recognition through EEG and ECG signals from wireless low-cost off-the-shelf devices. *IEEE Journal of Biomedical and Health Informatics, 22*(1), 98—107.

Kingma, D. P., & Ba, J. (2014). Adam: A method for stochastic optimization. *arXiv Preprint arXiv:1412.6980*.

Mian Qaisar, S. (2020). Baseline wander and power-line interference elimination of ECG signals using efficient signal-piloted filtering. *Healthcare Technology Letters, 7*(4), 114—118.

Mian Qaisar, S., & Subasi, A. (2020). Cloud-based ECG monitoring using event-driven ECG acquisition and machine learning techniques. *Physical and Engineering Sciences in Medicine, 43*, 623—634.

Nisa'Minhad, K., Ali, S. H. M., & Reaz, M. B. I. (2017). A design framework for human emotion recognition using electrocardiogram and skin conductance response signals. *Journal of Engineering Science & Technology, 12*(11), 3102—3119.

Nita, S., Bitam, S., Heidet, M., & Mellouk, A. (2022). A new data augmentation convolutional neural network for human emotion recognition based on ECG signals. *Biomedical Signal Processing and Control, 75*, 103580.

Patil, V. K., & Pawar, V. R. (2022). *How can emotions be classified with ECG sensors, AI techniques and IOT setup?*.

Sarkar, P., & Etemad, A. (2020). Self-supervised ECG representation learning for emotion recognition. *IEEE Transactions on Affective Computing, 13*(3), 1541–1554.

Sepúlveda, A., Castillo, F., Palma, C., & Rodriguez-Fernandez, M. (2021). Emotion recognition from ECG signals using wavelet scattering and machine learning. *Applied Sciences, 11*(11), 4945.

Sharma, G., & Dhall, A. (2021). A survey on automatic multimodal emotion recognition in the wild. *Advances in Data Science: Methodologies and Applications*, 35–64.

Singson, L. N. B., Sanchez, M. T. U. R., & Villaverde, J. F. (2021). *Emotion recognition using short-term analysis of heart rate variability and ResNet architecture*.

Subasi, A. (2019). *Practical guide for biomedical signals analysis using machine learning techniques, a MATLAB based approach* (1st ed.). Academic Press.

Vazquez-Rodriguez, J., Lefebvre, G., Cumin, J., & Crowley, J. L. (2022). *Transformer-based self-supervised learning for emotion recognition*.

Wang, Y., Song, W., Tao, W., Liotta, A., Yang, D., Li, X., Gao, S., Sun, Y., Ge, W., & Zhang, W. (2022). A systematic review on affective computing: Emotion models, databases, and recent advances. *Information Fusion, 83*, 19–52.

Wei, W., Jia, Q., Feng, Y., & Chen, G. (2018). Emotion recognition based on weighted fusion strategy of multichannel physiological signals. *Computational Intelligence and Neuroscience, 2018*.

Ye, Y., Pan, T., Meng, Q., Li, J., & Lu, L. (2022). *Online ECG emotion recognition for unknown subjects via hypergraph-based transfer learning*.

3

Artificial intelligence—based depression detection using EEG signals

Fadime Tokmak[1], Abdulhamit Subasi[2,3]

[1]Digital Health Technology, Faculty of Technology, University of Turku, Turku, Finland; [2]Institute of Biomedicine, Faculty of Medicine, University of Turku, Turku, Finland; [3]Department of Computer Science, College of Engineering, Effat University, Jeddah, Saudi Arabia

1. Introduction

Depression or depressive disorder is a phenomenon that manifests with psychological symptoms as well as physical symptoms, which may include fatigue, changes in appetite and sleep patterns, aches and pains in the body, and problems in the digestive system. It is a common disorder that affects numerous people every year and causes much harm, such as suicides. Therefore, it is important to diagnose and apply treatment to the patients before it causes any harm. In a clinical setting, there are scales used for diagnosing depression. The Montgomery Asberg Depression Rating Scale (MADRS) (Montgomery & Asberg, 1979) and the Hamilton Rating Scale for Depression (HAM-D) (Hamilton, 1960) are two of those scales used for diagnosing depression. In addition to diagnosis, these tests also evaluate the severity of depression at different levels: "mild," "moderate," "moderate to severe," and "severe."

A significant number of patients suffering from depression visit hospitals and describe the symptoms they recognize. However, somatization is a significantly common pattern in patients (American Psychiatric Association, 1997, p. 317; Simon et al., 1999; World Health Organization, 1992). If the patient reports only the physical symptoms, the depression case can easily be confused with other physical diseases because the physical symptoms of depression are not only associated with depression but also with various other diseases. This may cause delays in the diagnosis and treatment process of the patient. Additionally, this can even result in mistreatment.

Physical symptoms become severe depending on whether the patient is experiencing the disease mildly or severely. Also, the existence of physical symptoms is related to the presence of depression and the expected continuation of the current depression. In the literature, it is shown that the duration of depression is on average 6 months longer for patients who report chronic pain (Ohayon & Schatzberg, 2003). This direct relationship between depression and aches is associated with their neurological background (Trivedi, 2004). Furthermore, it is proven that serotonin has a significant effect on mood (Meltzer, 1990). Similarly, studies have shown that serotonin is associated with pain threshold (Martin et al., 2017) and this common background causes the relationship between depression and pain to form. Additionally, physical symptoms are related not only to the level of depression but also to the next onset time of depression (Judd et al., 1998; Kanai et al., 2003; Paykel et al., 1995). Furthermore, patients who are suffering from physical symptoms after their psychological symptoms have a higher frequency of relapse. This highlights the need for evaluating physical effects as more important and crucial. It is clear that physical effects should be carefully considered, as well as physiological effects, and treated to complete the treatment process.

Although pain is an important physical symptom of depression, the physical effects of depression are not limited to pain alone. Depression has severe silent effects on the human brain that are not noticeable by patients and health professionals in the clinical environment. These effects involve a decrease in the size of neurons in the prefrontal cortex and hippocampus (Rajkowska et al., 1999; Stockmeier et al., 2004). Additionally, depression also results in synaptic dysfunction in the mood circuitry of the brain (Duman & Aghajanian, 2012). Depression is a crucial health problem with serious effects on neurobiological implications on the brain, alterations in neural circuitry, neurotransmitter imbalances, and structural changes. Thus, it has an impact beyond emotional distress and physical pain.

Fortunately, with the help of brain imaging techniques, we now know crucial brain regions where depression is visible, that is, the dorsal anterior cingulate cortex and the bilateral dorsal posterior insula. These common grounds are promising, as they suggest that we can find neurological signs and measurements that can be used to detect or even measure the level of depression. In fact, in the literature, numerous

studies analyze the markers of depression in electroencephalography (EEG) signals (Mumtaz et al., 2018; Mumtaz, Xia, Ali, et al., 2017; Mumtaz, Xia, Mohd Yasin, et al., 2017). These analyses include channel spectral analysis of the EEG signal, asymmetry analysis, and other feature extraction techniques. These studies and techniques will be discussed in the literature review part. In this chapter, we will introduce a technique that creates 2D images from EEG signals and detects and evaluates depression using deep learning.

2. Background/literature review

Detection of depression is an important topic, and EEG is a highly promising biosignal for the detection of depression. Fortunately, in the literature, there are lots of studies that have discovered some EEG biomarkers for depression. These biomarkers can be classified into different groups according to their calculation and the underlying physiology that they try to represent. For analyzing EEG signals, it is a very common and widely used idea to decompose them into different frequency components. In the literature, there is a widely accepted classification that separates the EEG signal into five frequency bands: alpha, beta, theta, delta, and gamma bands. These bands are mainly associated with the brain's activity state (The Normal EEG Throughout Life: The EEG of the Waking Adult, 1976), and these frequency components have been acquired and analyzed with spectral analysis techniques.

Delta wave has the lowest frequency range in this classification and includes a frequency range between 0.5 and 4 Hz. It is highly related to sleep (especially deep sleep) (Freeman & Quiroga, 2013). It has also been shown that delta activity is associated with attention when performing mental tasks in the brain (Harmony et al., 1996). Theta waves are considered as the frequency components between 4 and 8 Hz, and they are related to emotions (Aftanas et al., 2002; Aftanas & Golocheikine, 2001). Moreover, there are studies that show a relationship between memory and theta and alpha waves (Klimesch et al., 1994). The 8—14 Hz band is named as alpha waves, and they are dominant frequency bands, especially in an awake but relaxed subject (Freeman & Quiroga, 2013). In addition, they are affected by visual and emotional processing (Barry et al., 2007; Freeman & Quiroga, 2013). Beta waves are the fourth class of the EEG bands, and they are the waves between 14 and 30 Hz, and they are considered to be linked with anxiety and stress (Abhang et al., 2016; Jena, 2015). Lastly, gamma waves are waves over 30 Hz, and they are considered to be linked with attention, sensory systems, mood, and depression (Fitzgerald & Watson, 2018; Freeman & Quiroga, 2013).

Each of the frequency bands serves as an important reflector of different parts of the physiological and cognitive states of the brain. This highly informative structure of the EEG signals is used in the trials to predict depression. Hosseinifard et al. (2013) extracted nonlinear features and band powers of each of the band classes, and they achieved 90% accuracy in predicting depression by the logistic regression algorithm. They also applied a feature selection method to investigate the expressive power of the bands, and they showed that alpha and theta bands are the most informative bands for predicting depression. In a different study with 100 participants, high alpha and beta power from the C3 channel are shown to be linked with depression by using logistic regression (Lee et al., 2018). Cai et al. (2016, 2018) conducted research using 178 subjects and collecting EEG from only three channels: Fp1, Fp2, and Fpz, and they reported that the absolute power of beta waves was the most informative feature during their tasks. Additionally, Grin-Yatsenko et al. (2010) reported a difference in the power of theta and alpha waves, especially in parietal and occipital regions that can be used as a biomarker to detect depression. The study included 637 subjects with 526 healthy subjects and used independent component analysis as a method.

There are several studies that have different results about the most informative component by analyzing the band powers. Other than band powers, a different kind of complex feature that combines the information from different bands is proposed in the literature. Rather than examining each band individually, these features express the relative activity of the bands according to each other. Firstly, relative gamma power is used by Bachmann et al. (Bachmann et al., 2018) and is defined as the ratio between the power of the gamma band and the power of all components between 3 and 46 Hz. In addition, they used alpha band power variability (APV) because they claimed that variable alpha power can be an indicator of problems in fundamental brain function. For that aim, they divided the signal into 2-second-long windows and calculated the alpha band power for each window. The APV is defined as the ratio between the standard deviation of the powers of the windows and the power of the first window. As another example of complex features, it is introduced by Hinrikus et al. (2009). The spectral asymmetry index represents the balance between high and low-frequency components in the power spectrum (Bachmann et al., 2018; Mahato & Paul, 2019).

In addition to power-related features, it is possible to define some features on the predictability or regularity of the signal. For instance, Hosseinifard et al. (2013) used maximum Lyapunov exponent as a feature in their study. They represented the predictability (or randomness) of the signal with the MLE feature inspired by a study on epilepsy (Mormann et al., 2005). For representing randomness, another study used the C0-complexity feature, which was proposed by Fang et al. (Cai et al., 2016, 2018; Fang et al., 1998). The correlation dimension is another example of the features of randomness (Shen et al., 2017).

The features that are discussed earlier are extracted from only one channel of the EEG, and they are calculated for each channel independently. Although they are good enough to provide good results, they lack the capability to show the spatial relationships and differences of different brain parts. For this reason, features that represent brain laterality and asymmetry were proposed. Brain laterality means the specialization of the brain parts for performing specific functions. Moreover, it is known that EEG asymmetry in the frontal cortex represents emotional processing, disorders, and changes (Coan & Allen, 2004), and alpha asymmetry is a widely used feature.

Alpha asymmetry is a metric that shows the activity difference of the alpha waves between the hemispheres of the brain (de Aguiar Neto & Rosa, 2019). Firstly, a study showed that for females, depression is linked with alpha asymmetry, but they reported that they could not find any significant results for males (Jesulola et al., 2017). In the same year, another group reported that alpha asymmetry is not useless for the classification task (Cai et al., 2016, pp. 1239–1246). However, in the following years, two different studies showed that regardless of gender, there was a significant difference in the alpha asymmetry between depressive and healthy subjects (Koo et al., 2019; Smith et al., 2018). Nevertheless, alpha variability is also shown to be linked with anxiety (Nusslock et al., 2018), and because of the inconsistency between the studies, the diagnostic power of alpha asymmetry is not clearly known.

In addition to studies that search for biomarkers and apply machine learning techniques, there are also plenty of studies that apply deep models for the detection of depression. Especially, convolutional neural networks (CNN) structures allow models to learn useful features, and also the transfer learning approach is applied to this problem. As the CNN and transfer learning structures require 2D signals, different methods to create 2D signals from

EEG are also proposed. These structures, methods, and studies will be explained in the next sections.

3. Artificial intelligence for depression detection

3.1 Machine learning techniques

Machine learning is a term that is mainly used to refer to a group of algorithms and their applications to existing data to predict future outcomes of the dependent variable of interest in the task. Machine learning algorithms assume a mathematical model that the given data could follow and attempt to find the best parameter settings to fit this mathematical model to the given data. As an example, in the basic linear regression algorithm, the underlying assumption is the existence of a linear model between independent variables (input variables) and the dependent variable (output variable), and the algorithm tries to find coefficients of the model that will result in a minimum error given the data and assumption. There are different machine learning algorithms and their variants.

In biomedical signal processing applications, it is not appropriate to give the raw signal to the algorithm. This will generally result in unsuccessful results because the signal itself has many samples to fit, many dependencies between nearby samples, and noise. Additionally, the magnitude of the samples is not meaningful independently due to intersubject variability, possible noise, and baselines. For this reason, rather than giving the raw signal, we need to extract features to explain the behavior of the signal and represent the underlying physiology. As an example of features, the alpha asymmetry feature aims to represent the underlying physiological difference between the right and left hemispheres of the brain as discussed in the literature review section. Note that the performance of the machine learning model highly depends on these features, and mostly, this is an important bottleneck in the success of the machine learning models. In the literature, different machine-learning techniques are applied to various features of EEG signals. Hosseinfard, Bachmann, and Lee (Bachmann et al., 2018; Hosseinifard et al., 2013; Lee et al., 2018) preferred to apply a basic logistic regression model to their proposed features.

3.1.1 Deep learning

Deep learning is an algorithm that aims to discover the relationship between input and the desired outcome. Deep learning algorithms include lots of interconnected neurons that are constructed in a layered structure. Each neuron has weight information, bias, and an activation function that determines the behavior of the neuron. The power of deep learning models comes from their common layered and connected architecture, and this nature allows them to be universal with enough data and computational resources. As machine learning models assume a mathematical model, deep learning models are preferred when there is enough data. For deep learning models, the amount of data is important because the universality of the deep learning algorithm can result in overfitting, that is, the deep learning model can fit the training data and learn it by heart rather than generalizing it. The structure of the deep learning model is crucial for the potential capabilities of the model. For instance, when more hidden layers are added to an artificial neural network (ANN), the model gains more learning capacity. However, for a basic ANN, it is still challenging to learn directly from the raw signal due to the resulting number of parameters and computational resource limitations, and it still requires some feature engineering. Fortunately, there are sophisticated neural network structures that can take raw signals as input.

3.1.2 Convolutional neural networks

CNNs are special neural networks that include convolutional and pooling units within the networks to classify images. Thanks to these units, the neural network structure gains feature extraction and feature selection capabilities. In CNNs, each convolutional unit has its own convolutional kernel, and during the training of the network, the coefficients in the kernels are adjusted. At the end of the training, the model has its own feature extractor that utilizes the convolution operation. In addition to this, the pooling layers operate as fixed feature reduction units. The fully connected structure following the convolutional architecture works similarly to ANN models and learns the relationship between the extracted features by the convolutional layers and the desired outcome.

Compared with an ANN that is complex enough for the same problem, CNN networks have fewer parameters by allowing the sharing of parameters across all parts of the image. Each CNN kernel has a constant number of parameters independent of the image's size. These kernels are not region-specific, and they are applied to the entire image by sliding. Additionally, CNN kernels are useful for images since they preserve spatial relationships between pixels. Despite the advantage of parameter sharing, CNN structures are still extensive. Hence, they have thousands of parameters to tune and require a lot of data for training. Despite their great capability to learn complex problems, they may not be suitable for some applications due to their data requirements. For the depression detection problem with EEG signals, there are different trials of CNN in the literature. For instance, Seal et al. introduced a CNN network called DeprNet and achieved a subject-wise accuracy of 0.914 (Seal et al., 2021). Additionally, Kang et al. trained another CNN model and reported an accuracy of 0.969 in 10-fold cross-validation (Kang et al., 2020).

```python
# Import required packages
from google.colab import drive
import mne
import numpy as np
import pandas as pd
import scipy
from tensorflow.keras.applications import *
from tensorflow.keras.optimizers import *
from tensorflow.keras.losses import *
from tensorflow.keras.layers import *
from tensorflow.keras.models import *
from tensorflow.keras.callbacks import *
from tensorflow.keras.preprocessing.image import *
from tensorflow.keras.utils import *
import matplotlib.pyplot as plt
from sklearn.model_selection import *
from sklearn.metrics import f1_score, roc_auc_score, cohen_kappa_score, precision_score,
recall_score, accuracy_score, confusion_matrix
import tensorflow as tf


import os
import cv2
import seaborn as sns
import pywt

# Read the data
def read_file(file_name):
    file = file_name
    data = mne.io.read_raw_edf(file)
    raw_data = data.get_data()
    info = data.info
    channels = data.ch_names
    return (raw_data, info, channels)


def read_folder():
    MDD = []
    H = []
    for i in os.listdir("/content/gdrive/MyDrive/data"):
        if ("edf" in i) and ("TASK" in i) :
            if "H" in i:
                H.append( read_file( "/content/gdrive/MyDrive/data/" + i))
            else:
                MDD.append(read_file( "/content/gdrive/MyDrive/data/" +i))
    return MDD, H
```

```python
MDD, H = read_folder()

# Define the mother wavelet for CWT
wavelet = 'morl'
# Set the scale range and number of scales for CWT
scales = range(1, 129)

# Define a function to convert the EEG signals to CWT or Spectrogram images
def create_images(signal, labels):
    images  = []
    for i in range(0,20000, 2816):
        channels = []
        for label in labels:
            signal_part =  signal[labels.index( label)][i:i+2816]
            channels.append( pywt.cwt(signal_part, scales, wavelet,)[0])
        image = np.concatenate(channels)
        images.append(image)
    return images

# Prepare the dataset
```

```python
subject_photo_dict = {}
count = 0
subjects = []
labels = []
for i in H:
    signals = i[0]
    channel_names = i[2]
    images = create_images(signals, channel_names)
    for image in images:
        image = cv2.resize(np.array([image,image,image]).T,(224, 224))
        if count not in subject_photo_dict:
            subject_photo_dict[count] = [image]
        else:
            subject_photo_dict[count].append(image)
    subjects.append(count)
    count+=1
    labels.append(0)
for i in MDD:
    signals = i[0]
    channel_names = i[2]
    images = create_images(signals, channel_names)
    for image in images:
        image = cv2.resize(np.array([image,image,image]).T,(224, 224))
        if count not in subject_photo_dict:
            subject_photo_dict[count] =  [image]
        else:
            subject_photo_dict[count].append(image)
    subjects.append(count)
    count+=1
    labels.append(1)
```

```
# Split the dataset to train, validation, and test sets
train_subjects, remaining_subjects,train_y, remaining_y = train_test_split(subjects,
labels, random_state =1,train_size = 0.7, stratify = labels)
valid_subjects, test_subjects,valid_y, test_y =train_test_split(remaining_subjects,
remaining_y, train_size = 0.3,random_state = 1, stratify = remaining_y)

X_train = []
y_train = []
X_val = []
y_val = []
X_test = []
y_test = []
for i, y in zip(train_subjects, train_y):
    for image in subject_photo_dict[i]:


        X_train.append(image)
        y_train.append(y)
for i, y in zip(valid_subjects, valid_y):
    for image in subject_photo_dict[i]:
        X_val.append(image)
        y_val.append(y)
for i, y in zip(test_subjects, test_y):
    for image in subject_photo_dict[i]:
        X_test.append(image)
        y_test.append(y)

X_train = np.array(X_train)
X_val = np.array(X_val)
X_test = np.array(X_test)
y_train = np.array(y_train).astype("float32")
y_val = np.array(y_val).astype("float32")
y_test = np.array(y_test).astype("float32")
```

```python
# Define a function to print accuracy metrics in the specified format
def get_accuracy_metrics(model_, X_train, y_train, X_val, y_val,X_test, y_test ):
    train_pred= model_.predict(X_train)
    val_pred= model_.predict(X_val)
    test_pred = model_.predict(X_test)
    train_pred = [1 if x>0.5 else 0 for x in train_pred]
    val_pred = [1 if x>0.5 else 0 for x in val_pred]
    test_pred = [1 if x>0.5 else 0 for x in test_pred]
    print("Train accuracy Score------------>")
    print ("{0:.3f}".format(accuracy_score(y_train, train_pred)*100), "%")
    print("Val accuracy Score--------->")
    print("{0:.3f}".format(accuracy_score(y_val, val_pred)*100), "%")
    print("Test accuracy Score--------->")
    print("{0:.3f}".format(accuracy_score(y_test, test_pred)*100), "%")
    print("F1 Score--------------->")
    print("{0:.3f}".format(f1_score(y_test, test_pred, average = 'weighted')*100), "%")
    print("Cohen Kappa Score------------->")
    print("{0:.3f}".format(cohen_kappa_score(y_test, test_pred)*100), "%")
    print("ROC AUC Score------------->")
        print("{0:.3f}".format(roc_auc_score(y_test,    test_pred   ,multi_class='ovo',
average='weighted')*100), "%")
    print("Recall------------->")
     print("{0:.3f}".format(recall_score(y_test, test_pred, average = 'weighted')*100),
"%")
    print("Precision------------->")
     print("{0:.3f}".format(precision_score(y_test, test_pred, average = 'weighted')*100),
"%")


    cf_matrix_test = confusion_matrix(y_test, test_pred)
    cf_matrix_val = confusion_matrix(y_val, val_pred)
    plt.figure(figsize = (12, 6))
    plt.subplot(121)
    sns.heatmap(cf_matrix_val, annot=True, cmap='Blues')
    plt.title("Val Confusion matrix")
    plt.subplot(122)
    sns.heatmap(cf_matrix_test, annot=True, cmap='Blues')
    plt.title("Test Confusion matrix")

    plt.show()
```

```
# Construct the CNN architecture
inp = Input(shape = (224, 224, 3))
model = BatchNormalization()(inp)
model = Conv2D(filters = 64, kernel_size = (3, 3), padding = 'same',
activation='relu')(model)
model = BatchNormalization()(model)
model = Conv2D(filters = 128, kernel_size = (3, 3), padding = 'same',
activation='relu')(model)
model = MaxPooling2D()(model)
model = Dropout(0.2)(model)
model = Conv2D(filters = 64, kernel_size = (3, 3), padding = 'same',
activation='relu')(model)
model = Flatten()(model)
model = Dense(units = 64, activation = 'relu')(model)
output = Dense(units = 1, activation = 'sigmoid')(model)
model = Model(inputs=inp, outputs=output)
model.compile(
    loss='binary_crossentropy',
    optimizer=tf.keras.optimizers.Adam(learning_rate=1e-7),
    metrics=['accuracy' , 'AUC']
)
# Fit the model
model.fit(X_train, y_train, validation_data=(X_val, y_val), epochs=1000, verbose=1)
# Get the accuracy metrics
get_accuracy_metrics(model, X_train, y_train, X_val, y_val,  X_test, y_test)
```

3.1.3 Transfer learning

As discussed in the previous part, CNNs require a lot of data to train, and transfer learning is a technique that presents a solution to this problem. In biomedical applications, the datasets are not large enough to train a CNN. However, there are already massive databases available for other problems like ImageNet, and there are models that have been trained on these datasets. Additionally, these models have great capabilities to classify hundreds of different objects, which means that they have a high ability to extract sophisticated features. With transfer learning, we can utilize their feature extraction capabilities for our problems without training a new CNN.

In transfer learning, a pretrained CNN model is selected, and without altering its convolutional layers, the fully connected layers are retrained to adapt it to the specific problem of interest. This way, it is possible to leverage the sophisticated feature extraction capabilities of the chosen model and apply it to different problems. While the convolutional layers remain fixed during training, the fully connected layers adjust their weights, and in some applications, they can be removed, or their structure can be modified. As it requires less data, transfer learning is a popular technique in the biomedical image analysis field, and in the literature, there are numerous successful applications of transfer learning methods. In the depression detection problem, Uyulan et al. applied transfer learning and achieved an accuracy of 0.926 using the MobileNet architecture (Uyulan et al., 2021).

```python
# Import required packages
from google.colab import drive
import mne
import numpy as np
import pandas as pd
import scipy
from tensorflow.keras.applications import *
from tensorflow.keras.optimizers import *
from tensorflow.keras.losses import *
from tensorflow.keras.layers import *
from tensorflow.keras.models import *
from tensorflow.keras.callbacks import *
from tensorflow.keras.preprocessing.image import *
from tensorflow.keras.utils import *
import matplotlib.pyplot as plt
from sklearn.model_selection import *
from sklearn.metrics import f1_score, roc_auc_score, cohen_kappa_score, precision_score,
recall_score, accuracy_score, confusion_matrix
import tensorflow as tf
import os
import cv2
import seaborn as sns
import pywt

# Read the data
def read_file(file_name):
    file = file_name
    data = mne.io.read_raw_edf(file)
    raw_data = data.get_data()
    info = data.info
    channels = data.ch_names
    return (raw_data, info, channels)
```

```
def read_folder():
    MDD = []
    H = []
    for i in os.listdir("/content/gdrive/MyDrive/data"):
        if ("edf" in i) and ("TASK" in i) :
            if "H" in i:
                H.append( read_file( "/content/gdrive/MyDrive/data/" + i))
            else:
                MDD.append(read_file( "/content/gdrive/MyDrive/data/" +i))
    return MDD, H


MDD, H = read_folder()
# Define the mother wavelet for CWT
wavelet = 'morl'
# Set the scale range and number of scales for CWT
scales = range(1, 129)




# Define a function to convert the EEG signals to CWT or Spectrogram images
def create_images(signal, labels):
    images  = []
    for i in range(0,20000, 2816):
        channels = []
        for label in labels:
            signal_part =  signal[labels.index( label)][i:i+2816]
            channels.append( pywt.cwt(signal_part, scales, wavelet,)[0])
        image = np.concatenate(channels)
        images.append(image)
    return images

# Prepare the dataset
subject_photo_dict = {}
count = 0
subjects = []
labels = []
for i in H:
    signals = i[0]
    channel_names = i[2]
    images = create_images(signals, channel_names)
    for image in images:
        image = cv2.resize(np.array([image,image,image]).T,(224, 224))
        if count not in subject_photo_dict:
            subject_photo_dict[count] = [image]
        else:
            subject_photo_dict[count].append(image)
```

```python
        subjects.append(count)
        count+=1
        print(count)
        labels.append(0)
for i in MDD:
    signals = i[0]
    channel_names = i[2]
    images = create_images(signals, channel_names)
    for image in images:
        image = cv2.resize(np.array([image,image,image]).T,(224, 224))
        if count not in subject_photo_dict:
            subject_photo_dict[count] =  [image]
        else:
            subject_photo_dict[count].append(image)
    subjects.append(count)
    count+=1
    print(count)
    labels.append(1)
```

```python
# Split the dataset to train, validation, and test sets
train_subjects, remaining_subjects,train_y, remaining_y  = train_test_split(subjects,
labels, random_state =1,train_size = 0.7, stratify = labels)
valid_subjects,  test_subjects,valid_y,  test_y  =train_test_split(remaining_subjects,
remaining_y, train_size = 0.3,random_state = 1, stratify = remaining_y)


X_train = []
y_train = []
X_val = []
y_val = []
X_test = []
y_test = []
for i, y in zip(train_subjects, train_y):
    for image in subject_photo_dict[i]:
        X_train.append(image)
        y_train.append(y)
for i, y in zip(valid_subjects, valid_y):
    for image in subject_photo_dict[i]:
        X_val.append(image)
        y_val.append(y)
for i, y in zip(test_subjects, test_y):
    for image in subject_photo_dict[i]:
        X_test.append(image)
        y_test.append(y)


X_train = np.array(X_train)
```

```python
X_val = np.array(X_val)
X_test = np.array(X_test)
y_train = np.array(y_train).astype("float32")
y_val = np.array(y_val).astype("float32")
y_test = np.array(y_test).astype("float32")


# Define a function to print accuracy metrics in the specified format
def get_accuracy_metrics(model_, X_train, y_train, X_val, y_val,X_test, y_test ):
    train_pred= model_.predict(X_train)
    val_pred= model_.predict(X_val)
    test_pred = model_.predict(X_test)
    train_pred = [1 if x>0.5 else 0 for x in train_pred]
    val_pred = [1 if x>0.5 else 0 for x in val_pred]
    test_pred = [1 if x>0.5 else 0 for x in test_pred]
    print("Train accuracy Score------------>")
    print ("{0:.3f}".format(accuracy_score(y_train, train_pred)*100), "%")
    print("Val accuracy Score--------->")
    print("{0:.3f}".format(accuracy_score(y_val, val_pred)*100), "%")
    print("Test accuracy Score--------->")
    print("{0:.3f}".format(accuracy_score(y_test, test_pred)*100), "%")
    print("F1 Score--------------->")
    print("{0:.3f}".format(f1_score(y_test, test_pred, average = 'weighted')*100), "%")
    print("Cohen Kappa Score------------->")
    print("{0:.3f}".format(cohen_kappa_score(y_test, test_pred)*100), "%")
    print("ROC AUC Score------------->")
        print("{0:.3f}".format(roc_auc_score(y_test,      test_pred    ,multi_class='ovo',
average='weighted')*100), "%")
    print("Recall------------->")
     print("{0:.3f}".format(recall_score(y_test, test_pred, average = 'weighted')*100),
"%")
    print("Precision--------------->")
     print("{0:.3f}".format(precision_score(y_test, test_pred, average = 'weighted')*100),
"%")



    cf_matrix_test = confusion_matrix(y_test, test_pred)
    cf_matrix_val = confusion_matrix(y_val, val_pred)
    plt.figure(figsize = (12, 6))
    plt.subplot(121)
    sns.heatmap(cf_matrix_val, annot=True, cmap='Blues')
    plt.title("Val Confusion matrix")
    plt.subplot(122)
    sns.heatmap(cf_matrix_test, annot=True, cmap='Blues')
    plt.title("Test Confusion matrix")

    plt.show()
```

```
Construct the Transfer Learning model
base_Neural_Net= VGG19(include_top=False, input_shape=(224, 224, 3), weights='imagenet')
model=Sequential()
model.add(base_Neural_Net)
model.add(Flatten())
model.add(BatchNormalization())
model.add(Dense(256,kernel_initializer='he_uniform', activation='relu'))
model.add(BatchNormalization())
model.add(Dense(64,kernel_initializer='he_uniform', activation='relu'))
model.add(BatchNormalization())
model.add(Activation('relu'))
model.add(Dropout(0.5))
model.add(Dense(1,activation='sigmoid'))

for layer in base_Neural_Net.layers:
    layer.trainable = False

model.compile(
    loss='binary_crossentropy',
    optimizer=tf.keras.optimizers.Adam(learning_rate=1e-7),
    metrics=['accuracy' , 'AUC']
)
 # Fit the model
model.fit(X_train, y_train, validation_data=(X_val, y_val), epochs=1000, verbose=1)
 # Get the accuracy metrics
get_accuracy_metrics(model, X_train, y_train, X_val, y_val,  X_test, y_test)
```

3.2 Image Generation from EEG signals

EEG signals are in the form of 1D discrete signals, while popular CNN networks take 2D images as input. To train a model with EEG signals, these signals need to be represented as images. In literature, various methods have been applied to similar problems. For instance, Seal et al. (2021) created a 2D matrix by concatenating all channels of the EEG and used this matrix as input for their DeprNet. It is also possible to employ different techniques to represent the signal or underlying physiology, as seen in the study by Uyulan et al. (2021). Spectrogram extraction and continuous wavelet transform (CWT) are other alternative methods that are suitable for this problem. CWT of EEG signals is shown in Fig. 3.1.

3.3 Proposed method

We utilized the recordings during the mental task state in this chapter. Eight nonoverlapping segments of 11 s were created from each EEG record. For each segment, the individual channels of the EEG signal were transformed into 2D representations by applying the continuous wavelet transform (CWT). The CWT, a mathematical tool

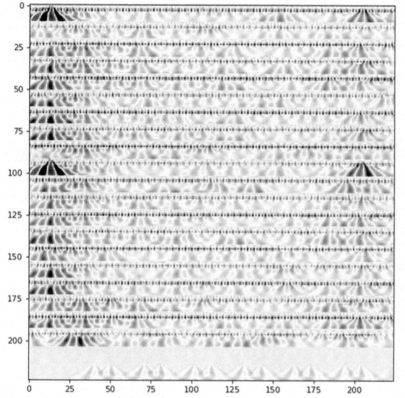

FIGURE 3.1 Example image representation of one EEG segment.

for analyzing signals and images, decomposes them into localized waveforms known as wavelets, providing a multiresolution representation that captures frequency and time information. The most popular wavelets used in the literature are Haar, Daubechies, and Morlet (Daubechies, 1988; Grossmann & Morlet, 1984; Haar, 1910). Each channel record within a segment underwent a 128-level wavelet transform using the Morlet wavelet. The resulting 2D transformations of the EEG channels were then concatenated and combined into a single 2D array, representing the transformation of the entire segment. The resulting 2D images were then resized into 224 × 224 shapes and used during the training process.

During the training process, we employed two different learning methods. Firstly, we experimented with transfer learning using well-known classification models to analyze their performance. We transferred the weights of the CNN layers that were pretrained on the Image-Net dataset without any modifications. However, we appended and trained new fully connected layers after the transferred CNN layer. Secondly, we constructed new CNN models with varying numbers of CNN layers, ranging from 2 to 8, and trained them from scratch. For the training processes, we employed the Adam optimizer along with binary cross-entropy loss as hyperparameters (Kingma & Ba, 2014).

3.3.1 CNN models

In this study, we constructed various CNN structures with different numbers of CNN layers, ranging from 2 to 8, to analyze the impact of model depth on performance. As we increased the number of layers, additional convolutional layers were added with max pooling and drop-out layers to represent more complex image features. The architecture of each model is shown in Fig. 3.2.

3.3.2 Transfer learning

We imported the CNN blocks of the well-known models without changing their weights learned on ImageNet dataset. We added fully connected layers after pretrained CNN block with the following layers:

- Batch normalization layer
- Dense layer with 256 neuron unit (ReLU)
- Batch normalization layer
- Dense layer with 64 neuron unit (ReLU)
- Batch normalization layer
- ReLU activation layer
- Drop-out layer (0.5)
- Dense layer with one neuron unit (Sigmoid).

The models we utilized included ResNet50, VGG16, VGG19, InceptionV3, MobileNet, DenseNet169, DenseNet121, InceptionResNetV2, MobileNetV2, and ResNet101.

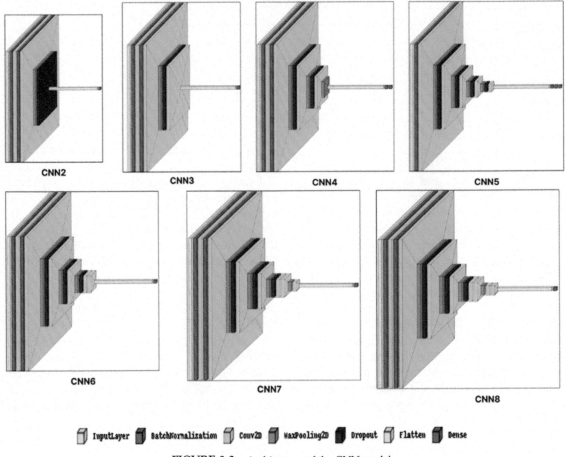

FIGURE 3.2   Architecture of the CNN models.

3.3.2.1 ResNet-50 and ResNet-101

ResNet-50 and ResNet-101 are pretrained network architectures introduced by Microsoft Research in 2015 (He et al., 2016, pp. 770–778). They consist of 50 and 101 convolutional layers, respectively, divided into several blocks with different configurations. These are Residual Network structures, which are deeply layered and utilize residual connections. These residual connections, also known as skip connections, bypass one or more layers, contrasting with standard deep models that only receive inputs from the preceding layer. These residual connections aid the model in learning residual mappings, facilitating effective gradient flow through earlier layers. The gradients can directly flow through the shortcuts, helping them reach earlier layers effectively in deep models. ResNet-50 and ResNet-101 also include pooling and fully connected layers and are trained for classification tasks.

3.3.2.2 VGG16 and VGG19

VGG16 and VGG19 are two pretrained CNN models proposed by the visual geometry group (VGG) at the University of Oxford. The models are named based on the number of layers they employ: VGG16 has 16 layers, and VGG19 has 19 layers (Simonyan & Zisserman, 2014). These models employ a series of convolutional layers followed by max pooling layers. These convolutional layers use small 3×3 size convolutional kernels, stacked one after another to create a deep network.

3.3.2.3 DenseNet121 and DenseNet169

DenseNet121 and DenseNet169 are two pretrained CNN models that utilize dense connections between layers (Huang et al., 2017, pp. 4700–4708). In the DenseNet structure, the model comprises dense blocks, and each layer of these dense blocks receives input from all previous layers in the same block. Similar to the ResNet structure, this architecture allows gradients to flow efficiently. These dense blocks are connected with transition layers, consisting of a batch normalization layer, convolutional layer, and 2×2 average pooling layer. DenseNet121 has 121 layers, and DenseNet169 has 169 layers.

3.3.2.4 MobileNet and MobileNetV2

MobileNet and MobileNetV2 are lightweight pretrained CNN architectures designed for mobile systems. They offer resource efficiency, easier deployment, and lower computational complexity while still providing reasonable accuracy (Howard et al., 2017; Sandler et al., 2018, pp. 4510–4520). To achieve this, they employ depthwise separable convolutions, reducing the number of parameters and required computations. MobileNetV2 is an improved version of MobileNet with inverted residual blocks and linear bottlenecks. It introduces bottleneck and expansion layers and incorporates linear activations in bottleneck layers to decrease information loss.

3.3.2.5 InceptionV3 and InceptionResNetV2

InceptionV3 is a pretrained CNN architecture introduced by Google Research. The key idea behind InceptionV3 is the use of "Inception modules," which combine different filter sizes and pooling operations within the same layer (Szegedy et al., 2016, pp. 2818–2826). This allows the network to extract information at various levels of abstraction. InceptionV3 consists of multiple stacked Inception modules, each containing parallel convolutional layers of different filter sizes ($1 \times 1, 3 \times 3, 5 \times 5$) and pooling operations. While both InceptionV3 and InceptionResNetV2 employ the Inception module, InceptionResNetV2 further enhances the architecture by introducing residual connections from the ResNet model, resulting in improved performance and better training of very deep networks (Szegedy et al., 2017).

4. Result and discussion

4.1 Experimental data

The dataset utilized in this chapter was taken from Mumtaz et al. (2018) and consists of publicly available EEG data recordings from 34 individuals diagnosed with major depression (17 females and 17 males, 40.33 ± 12.86 years), as well as 30 healthy individuals (9 females and 21 males, 38.23 ± 15.64 years) (Wajid Mumtaz, 2016). The study took place at Concordia University of Medical Sciences Malaysia (HUSM) with HUSM Ethics Committee approval. EEG data were collected using the international 10—20 system (Jasper, 1958), including frontal (Fp1, F3, F7, Fz, Fp2, F4, and F8), central (C3, C4, and Cz), parietal (P3, Pz, and P4), occipital (O1 and O2), and temporal (T3, T4, T5, and T6) regions with three reference EEG channels. The data was recorded during periods of eyes closed, eyes open, and with a specific visual stimulus with a 256 Hz sampling rate. The diagnosis of depression was done by the DSM IV criteria, and all participants were informed about the procedure and provided their consent by signing a consent form.

4.2 Performance evaluation measures

After the image creation process, the patient set was divided into train, validation, and test sets, with each set including 70%, 10%, and 20% of the patients respectively. The train, validation, and test datasets were constructed using images of the patients belonging to their corresponding patient set. After the training operation, the model's performance metrics on each set were calculated. The performance metrics used include accuracy, precision, recall, F1 score, and Cohen's kappa score (Cohen, 1960).

Accuracy represents the ratio of correctly predicted instances to the total number of instances, reflecting the overall correctness of the trained model. However, accuracy is highly sensitive to unbalanced data and should be complemented by different metrics for a more comprehensive evaluation. Precision is another metric that can be employed to assess classification performance. It evaluates the accuracy of positive predictions and is calculated as the proportion of correctly predicted positive instances out of the total instances predicted as positive. Recall focuses on the detection of positive cases and measures the proportion of correctly predicted positive instances out of the actual positive instances. On the other hand, the F1 score is the harmonic means of precision and recall, offering a balanced measure of performance. The formulas for accuracy, precision, recall, and F1 score are as follows:

$$Accuracy = \frac{True\ Positive + True\ Negative}{All\ samples}$$

$$Precision = \frac{True\ Positive}{True\ Positive + False\ Positive}$$

$$Recall = \frac{True\ Positive}{True\ Positive + False\ Negative}$$

$$F1\ Score = \frac{2 * precision * recall}{precision + recall}$$

Cohen's kappa score is a metric used to assess agreement between true labels and predicted labels. It gives a measure between 1 and −1 considering also the possibility of the agreement occurring by chance. While Cohen's kappa score of 1 means total agreement between the labels, 0 means that the agreement is equivalent to chance. In binary classification case, Cohen's kappa score is calculated as below:

$$Cohen's\ kappa\ score = \frac{2 * (TP * TN - FN * FP)}{(TP + FP) * (FP + TN) * (TP + FN) * (FN + TN)}$$

where TP = True Positive, TN = True Negative, FP = False Positive, and FN = False Negative.

4.3 Experimental results

4.3.1 *Experimental results for CNNs*

In Table 3.1, the classification performance of different CNN structures is given for our problem. CNN with three convolution layers resulted in the best performance (%95.536) than other CNNs and the transfer learning models. CNN with three convolutional layers was followed by 6-layer, 2-layer, and 8-layer CNN structures, respectively.

4.3.2 *Experimental results for transfer learning*

This section presents the results obtained from various transfer learning models. These models were trained for 1000 epochs using a learning rate of 10^{-7} and the Adam optimizer. As seen in Table 3.2, various models resulted with distinct results, while VGG19 outperformed the remaining models.

4.4 Discussion

In the selected pretrained model set, the VGG models have basic CNN architectures without utilizing complex deep learning tools, and

TABLE 3.1 Results for CNN models.

Classifier	Training accuracy	Validation accuracy	Test accuracy	Precision	Recall	F1 measure	Kappa
CNN 2 layer	87.202	97.500	93.750	94.102	93.750	93.738	87.500
CNN 3 layer	**99.405**	**100.000**	**95.536**	**95.550**	**95.536**	**95.535**	**91.071**
CNN 4 layer	99.702	100.000	88.393	88.701	88.393	88.370	76.786
CNN 5 layer	94.048	100.000	84.821	85.374	84.821	84.762	69.643
CNN 6 layer	91.071	100.000	94.643	94.643	94.643	94.643	89.286
CNN 7 layer	89.881	100.000	91.964	92.085	91.964	91.959	83.929
CNN 8 layer	88.393	100.000	92.857	93.077	92.857	92.848	85.714

The bold values are the maximum values.

TABLE 3.2 Results for transfer learning models.

Classifier	Training accuracy	Validation accuracy	Test accuracy	Precision	Recall	F1 measure	Kappa
ResNet50	94.643	100.000	92.857	92.857	92.857	92.857	85.714
VGG16	91.964	100.000	91.071	91.071	91.124	91.069	82.143
VGG19	**90.774**	**100.000**	**94.643**	**94.872**	**94.643**	**94.636**	**89.286**
InceptionV3	94.940	100.000	86.607	86.901	86.607	86.580	73.214
MobileNet	74.405	77.500	73.214	74.332	73.214	72.903	46.429
DenseNet169	91.667	100.000	87.500	87.500	87.500	87.500	75.000
DenseNet121	91.667	100.000	91.964	91.978	91.964	91.964	83.929
InceptionResNetV2	80.060	57.500	77.679	79.818	77.679	77.271	55.357
MobileNetV2	82.440	62.500	76.786	76.923	76.786	76.756	53.571

The bold values are the maximum values.

VGG19 reached the highest performance in our problem. Additionally, our CNN model with only three convolutional layers achieved an accuracy of 95.536%, surpassing the models with more convolutional layers. Therefore, the models with a plain and simple CNN architecture performed the best in our problem settings. This demonstrates that depression has a severe and easily recognizable effect on the EEG, and even with a basic CNN model, it may be detectable with high accuracy.

On the other hand, the good performance achieved by the basic models may be related to the representative power of the CWT. The CWT provides us with a 2D multiresolution representation that includes both time and frequency information. Additionally, by concatenating all channels, we gather all the information from different brain parts. Ultimately, we created a very compact representation that allows the model to extract all the useful features discussed in the background section, such as band powers, relative band powers, randomness, and spatial features.

We divided our patient set into three distinct subsets, verifying that each patient's images were included in only one of the training, validation, or test sets. This division was important because EEG segments from the same individual tend to show similarities, and as a result, our trained model has a higher probability of performing well on test samples that belong to the same person as any training instance. This approach helps to reduce optimistic bias in our performance evaluation. Taking into account our results and the aforementioned strategy, it is evident that both the 3-layer CNN and VGG19 models show strong performance not only on the data used for training but also on unseen patient data. This highlights the robustness and generalization capabilities of these models, indicating their potential for accurate predictions across various individuals.

We used 11-second segments for prediction, and this allows five predictions within a single 1-minute EEG recording. This was an important advantage of our model, and in clinical settings, this model may be used to get multiple predictions and combine predictions with a majority voting strategy to reduce the error rate even more. According to our results, a segment duration of 11 seconds was long enough to reflect the behavior of the brain and also short enough to allow feasible multiple checks in the clinical settings. Therefore, this duration strikes a balance, allowing for accurate predictions without sacrificing the feasibility of multiple assessments.

5. Conclusion

Deep neural networks possess a remarkable ability to learn from data without requiring feature engineering. Additionally, CNNs make the learning process even simpler by preserving spatial relationships, sharing parameters, and learning feature extraction filters. The applications of deep learning have great power in solving various classification problems, and they are particularly popular in healthcare. In healthcare, they find utility in tasks such as medical image classification, segmentation, and the processing of physiological signals. The usage of AI in medical applications has improved the healthcare industry by providing valuable insights, improving diagnostic accuracy, and enabling personalized treatment plans. AI algorithms can analyze a large amount of medical data, such as imaging scans and patient records, to help in early detection, disease diagnosis, and treatment decision-making, enhancing patient care and outcomes.

This chapter focuses on the application of CNN models to analyze 1D physiological signals, specifically using EEG signals to detect depression. We explain different physiological markers of depression and mention several studies that have utilized artificial intelligence to address this problem. To create a 2D representation of EEG records, we employed CWT and found that CNN structures with a simple architecture outperformed others on the dataset. Based on our results, we thoroughly discuss various aspects of our algorithm and reach the conclusion that a CNN model with three convolutional layers achieved the highest performance. Our findings reveal the potential of deep learning models in detecting depression using EEG signals, as demonstrated by the impressive 95% accuracy achieved on signals from unseen subjects.

To further enhance this project, we can delve deeper into comparing different parameter settings, datasets, performance evaluation methods, or alternative CNN architectures. Within the context of this chapter, we employed multiple convolutional neural networks to detect depression using EEG signals. Depression is a phenomenon encompassing both physiological and psychological symptoms, and it can be easily misdiagnosed in clinical settings when patients fail to recognize the psychological symptoms. As a solution, our model provides a feasible, fast, and reliable approach for identifying depression in clinical settings.

References

Abhang, P. A., Gawali, B. W., & Mehrotra, S. C. (2016). Technical aspects of brain rhythms and speech parameters. In *Introduction to EEG- and speech-based emotion recognition* (pp. 51–79). Elsevier. https://doi.org/10.1016/B978-0-12-804490-2.00003-8

Aftanas, L. I., & Golocheikine, S. A. (2001). Human anterior and frontal midline theta and lower alpha reflect emotionally positive state and internalized attention: High-resolution EEG investigation of meditation. *Neuroscience Letters, 310*(1), 57–60. https://doi.org/10.1016/S0304-3940(01)02094-8

Aftanas, L. I., Varlamov, A. A., Pavlov, S. V., Makhnev, V. P., & Reva, N. V. (2002). Time-dependent cortical asymmetries induced by emotional arousal: EEG analysis of event-related synchronization and desynchronization in individually defined frequency bands. *International Journal of Psychophysiology, 44*(1), 67–82. https://doi.org/10.1016/S0167-8760(01)00194-5

American Psychiatric Association. (1997). *Diagnostic and statistical manual of mental disorders*. Washington: American Psychiatric Association.

Bachmann, M., Päeske, L., Kalev, K., Aarma, K., Lehtmets, A., Ööpik, P., Lass, J., & Hinrikus, H. (2018). Methods for classifying depression in single channel EEG using linear and nonlinear signal analysis. *Computer Methods and Programs in Biomedicine, 155*, 11–17. https://doi.org/10.1016/j.cmpb.2017.11.023

Barry, R. J., Clarke, A. R., Johnstone, S. J., Magee, C. A., & Rushby, J. A. (2007). EEG differences between eyes-closed and eyes-open resting conditions. *Clinical Neurophysiology, 118*(12), 2765–2773. https://doi.org/10.1016/j.clinph.2007.07.028

Cai, H., Chen, Y., Han, J., Zhang, X., & Hu, B. (2018). Study on feature selection methods for depression detection using three-electrode EEG data. *Interdisciplinary Sciences: Computational Life Sciences, 10*, 558–565.

Cai, H., Sha, X., Han, X., Wei, S., & Hu, B. (2016). *Pervasive EEG diagnosis of depression using Deep Belief Network with three-electrodes EEG collector*.

Coan, J. A., & Allen, J. J. B. (2004). Frontal EEG asymmetry as a moderator and mediator of emotion. *Frontal EEG Asymmetry, Emotion, and Psychopathology, 67*(1), 7–50. https://doi.org/10.1016/j.biopsycho.2004.03.002

Cohen, J. (1960). A coefficient of agreement for nominal scales. *Educational and Psychological Measurement, 20*(1), Article 1.

Daubechies, I. (1988). Orthonormal bases of compactly supported wavelets. *Communications on Pure and Applied Mathematics, 41*(7), 909–996.

de Aguiar Neto, F. S., & Rosa, J. L. G. (2019). Depression biomarkers using non-invasive EEG: A review. *Neuroscience & Biobehavioral Reviews, 105*, 83–93. https://doi.org/10.1016/j.neubiorev.2019.07.021

Duman, R. S., & Aghajanian, G. K. (2012). Synaptic dysfunction in depression: Potential therapeutic targets. *Science (New York, N.Y.), 338*(6103), 68–72. https://doi.org/10.1126/science.1222939

Fang, C., Fanji, G., Jinghua, X., Zengrong, L., & Ren, L. (1998). A new measurement of complexity for studying EEG mutual information. *Shengwu Wuli Xuebao, 14*(3), 508–512.

Fitzgerald, P. J., & Watson, B. O. (2018). Gamma oscillations as a biomarker for major depression: An emerging topic. *Translational Psychiatry, 8*(1), 177. https://doi.org/10.1038/s41398-018-0239-y

Freeman, W. J., & Quiroga, R. Q. (2013). *Imaging brain function with EEG: Advanced temporal and spatial analysis of electroencephalographic signals*. New York: Springer. https://doi.org/10.1007/978-1-4614-4984-3

Grin-Yatsenko, V. A., Baas, I., Ponomarev, V. A., & Kropotov, J. D. (2010). Independent component approach to the analysis of EEG recordings at early stages of depressive disorders. *Clinical Neurophysiology, 121*(3), 281–289. https://doi.org/10.1016/j.clinph.2009.11.015

Grossmann, A., & Morlet, J. (1984). Decomposition of Hardy functions into square integrable wavelets of constant shape. *SIAM Journal on Mathematical Analysis, 15*(4), 723–736. https://doi.org/10.1137/0515056

Haar, A. (1910). Zur Theorie der orthogonalen Funktionensysteme. *Mathematische Annalen, 69*(3), 331–371. https://doi.org/10.1007/BF01456326

Hamilton, M. (1960). A rating scale for depression. *Journal of Neurology, Neurosurgery, and Psychiatry, 23*(1), 56–62.

Harmony, T., Fernández, T., Silva, J., Bernal, J., Díaz-Comas, L., Reyes, A., Marosi, E., Rodríguez, M., & Rodríguez, M. (1996). EEG delta activity: An indicator of attention to internal processing during performance of mental tasks. *New Advances in EEG and Cognition, 24*(1), 161–171. https://doi.org/10.1016/S0167-8760(96)00053-0

He, K., Zhang, X., Ren, S., & Sun, J. (2016). *Deep residual learning for image recognition.*

Hinrikus, H., Suhhova, A., Bachmann, M., Aadamsoo, K., Võhma, Ü., Lass, J., & Tuulik, V. (2009). Electroencephalographic spectral asymmetry index for detection of depression. *Medical, & Biological Engineering & Computing, 47*(12), 1291–1299. https://doi.org/10.1007/s11517-009-0554-9

Hosseinifard, B., Moradi, M. H., & Rostami, R. (2013). Classifying depression patients and normal subjects using machine learning techniques and nonlinear features from EEG signal. *Computer Methods and Programs in Biomedicine, 109*(3), 339–345. https://doi.org/10.1016/j.cmpb.2012.10.008

Howard, A. G., Zhu, M., Chen, B., Kalenichenko, D., Wang, W., Weyand, T., Andreetto, M., & Adam, H. (2017). Mobilenets: Efficient convolutional neural networks for mobile vision applications. *ArXiv Preprint ArXiv:1704.04861.*

Huang, G., Liu, Z., Van Der Maaten, L., & Weinberger, K. Q. (2017). *Densely connected convolutional networks.*

Jasper, H. H. (1958). Ten-twenty electrode system of the international federation. *Electroencephalography and Clinical Neurophysiology, 10*, 371–375.

Jena, S. K. (2015). Examination stress and its effect on EEG. *International Journal of Medical Science and Public Health, 11*(4), 1493–1497.

Jesulola, E., Sharpley, C. F., & Agnew, L. L. (2017). The effects of gender and depression severity on the association between alpha asymmetry and depression across four brain regions. *Behavioural Brain Research, 321*, 232–239. https://doi.org/10.1016/j.bbr.2016.12.035

Judd, L. L., Akiskal, H. S., Maser, J. D., Zeller, P. J., Endicott, J., Coryell, W., Paulus, M. P., Kunovac, J. L., Leon, A. C., Mueller, T. I., Rice, J. A., & Keller, M. B. (1998). Major depressive disorder: A prospective study of residual sub-threshold depressive symptoms as predictor of rapid relapse. *Journal of Affective Disorders, 50*(2–3), 97–108. https://doi.org/10.1016/S0165-0327(98)00138-4

Kanai, T., Takeuchi, H., Furukawa, T. A., Yoshimura, R., Imaizumi, T., Kitamura, T., & Takahashi, K. (2003). Time to recurrence after recovery from major depressive episodes and its predictors. *Psychological Medicine, 33*(5), 839–845. https://doi.org/10.1017/S0033291703007827. Cambridge Core.

Kang, M., Park, J., Kang, S., & Lee, Y. (2020). Low channel electroencephalogram based deep learning method to pre-screening depression. In *2020 international conference on information and communication technology convergence (ICTC)* (pp. 449–451). https://doi.org/10.1109/ICTC49870.2020.9289308

Kingma, D. P., & Ba, J. (2014). Adam: A method for stochastic optimization. *ArXiv Preprint ArXiv:1412.6980.*

Klimesch, W., Schimke, H., & Schwaiger, J. (1994). Episodic and semantic memory: An analysis in the EEG theta and alpha band. *Electroencephalography and Clinical Neurophysiology, 91*(6), 428–441. https://doi.org/10.1016/0013-4694(94)90164-3

Koo, P. C., Berger, C., Kronenberg, G., Bartz, J., Wybitul, P., Reis, O., & Hoeppner, J. (2019). Combined cognitive, psychomotor and electrophysiological biomarkers in major depressive disorder. *European Archives of Psychiatry and Clinical Neuroscience, 269*(7), 823–832. https://doi.org/10.1007/s00406-018-0952-9

Lee, P. F., Kan, D. P. X., Croarkin, P., Phang, C. K., & Doruk, D. (2018). Neurophysiological correlates of depressive symptoms in young adults: A quantitative EEG study. *Journal of Clinical Neuroscience, 47*, 315–322. https://doi.org/10.1016/j.jocn.2017.09.030

Mahato, S., & Paul, S. (2019). Electroencephalogram (EEG) signal analysis for diagnosis of major depressive disorder (MDD): A review. In V. Nath, & J. K. Mandal (Eds.), *Nanoelectronics, circuits and communication systems* (Vol 511, pp. 323–335). Springer Singapore. https://doi.org/10.1007/978-981-13-0776-8_30

Martin, S. L., Power, A., Boyle, Y., Anderson, I. M., Silverdale, M. A., & Jones, A. K. P. (2017). 5-HT modulation of pain perception in humans. *Psychopharmacology, 234*(19), 2929–2939. https://doi.org/10.1007/s00213-017-4686-6

Meltzer, H. Y. (1990). Role of serotonin in depression. *Annals of the New York Academy of Sciences, 600*, 486–500. https://doi.org/10.1111/j.1749-6632.1990.tb16904.x

Montgomery, S. A., & Åsberg, M. (1979). A new depression scale designed to be sensitive to change. *British Journal of Psychiatry, 134*(4), 382–389. https://doi.org/10.1192/bjp.134.4.382

Mormann, F., Kreuz, T., Rieke, C., Andrzejak, R. G., Kraskov, A., David, P., Elger, C. E., & Lehnertz, K. (2005). On the predictability of epileptic seizures. *Clinical*

Neurophysiology, *116*(3), 569–587. https://doi.org/10.1016/j.clinph.2004.08.025

Mumtaz, W. (2016). *MDD Patients and healthy controls EEG data (new) [dataset].* https://doi.org/10.6084/m9.figshare.4244171.v2

Mumtaz, W., Ali, S. S. A., Yasin, M. A. M., & Malik, A. S. (2018). A machine learning framework involving EEG-based functional connectivity to diagnose major depressive disorder (MDD). *Medical, & Biological Engineering & Computing,* *56*(2), 233–246. https://doi.org/10.1007/s11517-017-1685-z

Mumtaz, W., Xia, L., Ali, S. S. A., Yasin, M. A. M., Hussain, M., & Malik, A. S. (2017). Electroencephalogram (EEG)-based computer-aided technique to diagnose major depressive disorder (MDD). *Biomedical Signal Processing and Control,* *31*, 108–115. https://doi.org/10.1016/j.bspc.2016.07.006

Mumtaz, W., Xia, L., Mohd Yasin, M. A., Azhar Ali, S. S., & Malik, A. S. (2017). A wavelet-based technique to predict treatment outcome for Major Depressive Disorder. *PLoS One,* *12*(2), e0171409. https://doi.org/10.1371/journal.pone.0171409

Nusslock, R., Shackman, A. J., McMenamin, B. W., Greischar, L. L., Davidson, R. J., & Kovacs, M. (2018). Comorbid anxiety moderates the relationship between depression history and prefrontal EEG asymmetry. *Psychophysiology,* *55*(1), e12953. https://doi.org/10.1111/psyp.12953

Ohayon, M. M., & Schatzberg, A. F. (2003). Using chronic pain to predict depressive morbidity in the general population. *Archives of General Psychiatry,* *60*(1), 39. https://doi.org/10.1001/archpsyc.60.1.39

Paykel, E. S., Ramana, R., Cooper, Z., Hayhurst, H., Kerr, J., & Barocka, A. (1995). Residual symptoms after partial remission: An important outcome in depression. *Psychological Medicine,* *25*(6), 1171–1180. https://doi.org/10.1017/s0033291700033146

Rajkowska, G., Miguel-Hidalgo, J. J., Wei, J., Dilley, G., Pittman, S. D., Meltzer, H. Y., Overholser, J. C., Roth, B. L., & Stockmeier, C. A. (1999). Morphometric evidence for neuronal and glial prefrontal cell pathology in major depression. *Biological Psychiatry,* *45*(9), 1085–1098.

Sandler, M., Howard, A., Zhu, M., Zhmoginov, A., & Chen, L.-C. (2018). *Mobilenetv2: Inverted residuals and linear bottlenecks.*

Seal, A., Bajpai, R., Agnihotri, J., Yazidi, A., Herrera-Viedma, E., & Krejcar, O. (2021). DeprNet: A deep convolution neural network framework for detecting depression using EEG. *IEEE Transactions on Instrumentation and Measurement,* *70*, 1–13. https://doi.org/10.1109/TIM.2021.3053999

Shen, J., Zhao, S., Yao, Y., Wang, Y., & Feng, L. (2017). A novel depression detection method based on pervasive EEG and EEG splitting criterion. In *2017 IEEE international conference on bioinformatics and biomedicine (BIBM)* (pp. 1879–1886). https://doi.org/10.1109/BIBM.2017.8217946

Simon, G. E., VonKorff, M., Piccinelli, M., Fullerton, C., & Ormel, J. (1999). An international study of the relation between somatic symptoms and depression. *New England Journal of Medicine,* *341*(18), 1329–1335. https://doi.org/10.1056/NEJM199910283411801

Simonyan, K., & Zisserman, A. (2014). Very deep convolutional networks for large-scale image recognition. *ArXiv Preprint ArXiv:1409.1556.*

Smith, E. E., Cavanagh, J. F., & Allen, J. J. B. (2018). Intracranial source activity (eLORETA) related to scalp-level asymmetry scores and depression status. *Psychophysiology,* *55*(1), e13019. https://doi.org/10.1111/psyp.13019

Stockmeier, C. A., Mahajan, G. J., Konick, L. C., Overholser, J. C., Jurjus, G. J., Meltzer, H. Y., Uylings, H. B., Friedman, L., & Rajkowska, G. (2004). Cellular changes in the postmortem hippocampus in major depression. *Biological Psychiatry,* *56*(9), 640–650.

Szegedy, C., Ioffe, S., Vanhoucke, V., & Alemi, A. (2017). Inception-v4, inception-resnet and the impact of residual connections on learning. In *Proceedings of the AAAI Conference on Artificial Intelligence, 31*(1).

Szegedy, C., Vanhoucke, V., Ioffe, S., Shlens, J., & Wojna, Z. (2016). *Rethinking the inception architecture for computer vision.*

The Normal EEG Throughout Life. (1976). *The EEG of the waking adult (Issue pt. 1).* Elsevier. https://books.google.fi/books?id=g0y4zQEACAAJ.

Trivedi, M. H. (2004). The link between depression and physical symptoms. *Primary Care Companion to the Journal of Clinical Psychiatry,* *6*(Suppl. 1), 12–16.

Uyulan, C., Ergüzel, T. T., Unubol, H., Cebi, M., Sayar, G. H., Nezhad Asad, M., & Tarhan, N. (2021). Major depressive disorder classification based on different convolutional neural network models: Deep learning approach. *Clinical EEG and Neuroscience,* *52*(1), 38–51. https://doi.org/10.1177/1550059420916634

World Health Organization (Ed.). (1992). *The ICD-10 classification of mental and behavioural disorders: Clinical descriptions and diagnostic guidelines.* World Health Organization.

Electromyography signal classification using artificial intelligence

Abdulhamit Subasi[1,2]

[1]Institute of Biomedicine, Faculty of Medicine, University of Turku, Turku, Finland; [2]Department of Computer Science, College of Engineering, Effat University, Jeddah, Saudi Arabia

1. Introduction

Neuromuscular disorders are a broad category of ailments that affect the neuromuscular system and cause abnormalities in muscle control, movement, and coordination. It is critical for successful treatment and management to recognize these diseases accurately and on time. Traditional diagnostic approaches frequently entail intrusive procedures, subjective assessments, and dependence on expert interpretation, all of which can be time consuming, expensive, and vulnerable to interobserver variability. Artificial intelligence (AI) approaches, notably machine learning and deep learning algorithms, have shown promise in improving neuromuscular problem detection utilizing electromyography (EMG) signals in recent years.

EMG is a diagnostic procedure that detects muscle electrical activity. Electrodes are placed on the skin or directly into the muscles to capture the electrical impulses generated during muscle contractions. EMG signals provide unique insights into the neuromuscular system's functioning, making them an important resource for diagnosing anomalies associated with numerous neuromuscular illnesses. Myopathy, amyotrophic lateral sclerosis (ALS), and myasthenia gravis are examples of these illnesses.

AI approaches have transformed the processing and interpretation of EMG signals in the context of neuromuscular problem detection. AI methods including support vector machines (SVMs), random forests, and neural networks have been used to evaluate enormous amounts of EMG data and discover complicated patterns that are symptomatic of certain illnesses. These techniques extract meaningful information from EMG data and train models to categorize them as normal or diseased. Convolutional neural networks (CNNs) and recurrent neural networks (RNNs) excel at capturing spatial and temporal relationships in EMG signals, allowing for more accurate and robust identification of neuromuscular diseases.

There are various advantages to incorporating AI approaches into the detection of neuromuscular illnesses. For starters, it allows for the automation of the diagnostic process, which reduces reliance on subjective judgments and increases efficiency. Traditionally, clinicians manually scrutinize and visually assess EMG signals, a time consuming and subject to interobserver variability process. The examination of EMG signals can be automated using AI approaches, allowing for more objective and consistent judgments. Large datasets of labeled EMG signals can be used to train AI models, capturing complicated correlations between signal patterns and related neuromuscular disorders.

Furthermore, AI approaches have the ability to detect neuromuscular disorders early and accurately. Appropriate interventions and therapies, which can have a major impact on patient outcomes, must be initiated as soon as possible. AI models can detect small patterns and deviations in EMG signals that are not always visible to the naked eye. This skill for early diagnosis may lead to earlier therapies and more effective management of neuromuscular diseases, thereby improving patient prognosis.

However, implementing AI approaches for neuromuscular problem identification utilizing EMG signals successfully presents its own set of challenges. The variety and complexity of EMG data are a major concern. Electrode placement, signal artifacts, and individual variances in muscle activity can all have an effect on EMG signals. These causes of variability must be accounted for in AI models, which must also adapt to different patient groups and recording settings. Furthermore, the interpretability of AI models is critical, especially in clinical settings where decisions have serious consequences for patient care. It is critical to develop explainable AI models that can provide insights into decision-making processes to acquire the trust and acceptance of healthcare practitioners.

In this chapter, we will show how to use AI approaches to diagnose neuromuscular disorders using EMG signals. We will look at the many machine learning approaches used in this domain, noting their advantages, disadvantages, and claimed outcomes. We will also examine the obstacles and potential solutions for adopting AI-based systems for neuromuscular condition identification, such as data variability, model interpretability, and integration into clinical processes. We intend to contribute to the evolution of AI-driven techniques for improved neuromuscular disorder detection and management by evaluating the current state of the field and identifying future research possibilities.

Neuromuscular diseases (NMDs) are a broad category of disorders that affect the peripheral nerves, neuromuscular junctions, and muscles. These disorders can cause major motor impairments, functional restrictions, and decreased quality of life in those who suffer from them. It is critical to detect NMDs early and accurately to provide prompt treatments, proper management, and improved patient outcomes. In recent years, there has been a surge of interest in harnessing the capabilities of AI and EMG data to create better NMD diagnosis tools. This study discusses the problems, improvements, and future possibilities in the field of AI for NMD detection utilizing EMG signals.

The diagnosis of NMDs traditionally relies on a combination of clinical examinations, patient history, and diagnostic tests, including EMG. EMG is a noninvasive technique that measures the electrical activity of muscles, providing valuable information about muscle function and detecting abnormal patterns indicative of NMDs. However, the interpretation of EMG signals can be complex and subjective, requiring specialized expertise. This is where AI techniques can play a significant role by automating the analysis of EMG signals, extracting relevant features, and developing predictive models for NMD detection.

Despite encouraging advances, various hurdles must be overcome before AI in NMD detection using EMG signals can be widely adopted. The availability of high-quality and well-annotated EMG datasets is a significant barrier, as vast and diverse datasets are required for training accurate and stable AI models. Data gathering standards, standardization, and sharing activities can aid in overcoming these obstacles and facilitating collaborative research efforts in this area. Another issue is the interpretability and explainability of AI models, as providing physicians with transparent and intelligible information is crucial for obtaining confidence and adoption.

Furthermore, incorporating AI models into clinical practice necessitates addressing regulatory and ethical concerns, preserving patient privacy, and adhering to relevant healthcare rules. Transparency in model construction, validation, and performance assessment is critical for establishing the dependability and trustworthiness of AI-based diagnostic tools. Collaboration among physicians, AI researchers, and regulatory agencies is critical for defining rules and frameworks for the responsible and ethical use of AI in NMD diagnosis.

Finally, the use of AI approaches for NMD diagnosis using EMG signals offers a lot of potential in terms of boosting diagnostic accuracy, allowing early therapies, and improving patient outcomes. The incorporation of machine learning and deep learning algorithms can help to automate and standardize NMD diagnosis, minimizing subjectivity and unpredictability. To ensure the reliable and responsible application of AI in clinical practice, however, difficulties relating to data availability, model interpretability, and ethical considerations must be addressed. Collaboration between healthcare practitioners, AI researchers, and regulatory agencies is critical for moving this field forward and attaining AI's full potential in NMD detection.

2. Literature review

Because of their various symptoms and complex underlying mechanisms, neuromuscular disorders pose major diagnostic and treatment hurdles. In recent years, there has been an increase in interest in using AI techniques, including machine learning and deep learning, to evaluate EMG signals for the detection and diagnosis of neuromuscular illnesses. The purpose of this literature review is to provide an overview of present research on this topic, highlighting accomplishments, problems, and future directions.

Gokgoz and Subasi (2015) focused on the classification of biomedical signals, specifically EMG signals, using decision tree algorithms in combination with denoising and feature extraction techniques. The researchers introduced a framework that incorporates multiscale principal component analysis (MSPCA) for denoising, discrete wavelet transform (DWT) for feature extraction, and decision tree algorithms (CART, C4.5, and random forest) for classification. The framework automatically categorizes EMG signals as myopathic, ALS, or normal. Performance evaluation metrics such as sensitivity, specificity, accuracy, F-measure, and area under the receiver operating characteristic curve (AUC) are used to assess the framework's effectiveness. The combination of DWT and the random forest algorithm achieves the highest performance, with a total classification accuracy of 96.67% using k-fold cross-validation. The results indicate that the proposed approach demonstrates significant accuracy in classifying EMG signals, making it a valuable tool for clinicians in diagnosing neuromuscular disorders.

In another study, Subasi (2015) focused on optimizing the performance of SVMs in diagnosing neuromuscular disorders. SVM classifiers' accuracy is affected by kernel parameter selection, but SVM lacks an inherent mechanism for determining these parameters effectively. To address this, the study proposed a novel approach called evolutionary support vector machine (ESVM), which uses a genetic algorithm for automatic parameter tuning. The ESVM method is applied to classify EMG signals into normal, myopathic, and neurogenic datasets after decomposing the signals using DWT and extracting statistical features from the subbands. The experimental results show that ESVM achieves a high accuracy of 97% on the EMG datasets using tenfold cross-validation.

Various approaches have been employed to quantitatively analyze EMG signals. This research introduces the impact of the MSPCA denoising method on the classification of EMG signals. Gokgoz and Subasi (2014) examined the effects of MSPCA denoising on EMG signal classification, as well as the application and comparison of the multiple signal classification (MUSIC) feature extraction method. The classification task involves distinguishing between normal, ALS, and myopathic EMG signals. Additionally, the study discusses the overall accuracy of classifiers such as k-nearest neighbor (k-NN), artificial neural network (ANN), and SVMs. Remarkable outcomes are observed with the utilization of the MSPCA denoising method. The comparison among the developed classifiers is based on various performance metrics, including sensitivity, specificity, accuracy, F-measure, and area under the Receiver operating characteristic (ROC) curve (AUC). The findings demonstrate that MSPCA denoising substantially enhances the accuracy compared to EMG data without MSPCA denoising.

Neuromuscular disorders are analyzed using EMG signals, and machine learning algorithms have been applied as a decision support system for detecting these disorders. However, EMG signals often contain noise originating from various sources, including electrical and electronic instruments, as well as movement artifacts. Subasi (2019a) presented the utilization of MSPCA to effectively eliminate impulsive noise from EMG signals. Subsequently, the dual-tree complex wavelet transform (DT-CWT) is employed to extract features, while the recognition of EMG signals is carried out using the rotation forest ensemble classifier. Additionally, the performance of several classifiers in combination with the rotation forest is investigated. Through the efficient integration of DT-CWT and rotation forest, promising results are obtained using tenfold cross-validation, particularly in terms of total classification accuracy. The rotation forest approach achieves an impressive accuracy of 99.7% when applied to clinical EMG signals using a support vector machine and a notable accuracy of 96.6% when applied to simulated EMG signals using the ANN.

Samanta et al. (2020, pp. 694−697) presented a deep learning framework for diagnosing neuromuscular disorders by detecting and classifying EMG signals. The framework utilizes cross-wavelet transform, a modified version of the continuous wavelet transform, to analyze nonstationary signals in time scale and time-frequency domains. EMG signals from healthy individuals, myopathy, and ALS patients were used in the study. A healthy EMG signal was selected as the reference, and cross-wavelet transform was applied to the other EMG signals using the reference signal. The resulting cross-wavelet spectrum images were subjected to deep feature extraction using a CNN. Significant features were identified through feature ranking with a one-way analysis of variance (ANOVA) test, and these features were then inputted into machine learning classifiers for binary classification of EMG signals. The results demonstrate that the proposed method achieves a high mean classification accuracy of 100% using the statistically significant extracted deep features.

Subasi et al. (2018) utilized a bagging ensemble classifier for the automated classification of EMG signals, which combines multiple classifier models to improve performance. Previous research has shown the effectiveness of ensemble classifiers in various applications, but their use in diagnosing neuromuscular disorders has been less explored. The proposed method involves applying DWT to extract features from each type of EMG signal, computing statistical values of the DWT to represent wavelet coefficients' distribution. The resulting feature set is then inputted into the bagging ensemble classifier for diagnosing neuromuscular disorders.

Experimental results demonstrate the feasibility of the bagging ensemble classifier, achieving an impressive 99% accuracy when combined with SVM for diagnosing neuromuscular disorders.

The diagnosis of neuromuscular diseases relies on the analysis of EMG signals. Automatic classification systems are used to support specialists and improve the diagnostic process. Vallejo et al., (2018) utilized time-frequency analysis, fuzzy entropy, and neural networks to identify specific disorder characteristics in EMG signals, including myopathy and ALS. The study extracts and selects features from EMG signals using DWT and Fuzzy Entropy, and recognition results are generated using ANNs. The research employs a publicly available database from EMGLAB, achieving an accuracy of approximately 98% in identifying EMG signals from three classes: healthy individuals, patients with myopathy, and those displaying evidence of ALS.

Chatterjee et al. (2020) introduced a novel approach using multifractal detrended fluctuation analysis (DFA) to distinguish between myopathy, ALS, and healthy EMGs. EMGs are complex electrical signals capturing muscle and nerve cell activities in the human body. The study analyzes the nonlinear and dynamic characteristics of EMGs using fractal geometry and extracts five novel feature parameters from the multifractal spectrum of the signals. Support vector machine and k-NN classifiers are used for classification. The proposed method demonstrates satisfactory performance in distinguishing between different categories of EMGs, outperforming or comparable to existing methods applied to the same dataset.

Yaman and Subasi (2019) compared bagging and boosting ensemble learning methods for classifying electromyographic (EMG) signals to diagnose neuromuscular disorders. Although ensemble classifiers have shown effectiveness in various applications, their feasibility for EMG-based diagnosis is not well explored. The study aims to assess the potential of these classifiers for diagnosing neuromuscular disorders using EMG signals. The methodology involves calculating wavelet packed coefficients (WPCs) for each EMG signal, computing statistical values of WPC to represent wavelet coefficients' distribution, and utilizing an ensemble classifier for diagnosis. Experimental results indicate that ensemble classifiers outperform other methods in diagnosing neuromuscular disorders. The AdaBoost with random forest ensemble method achieves impressive accuracy (99.08%), F-measure (0.99), AUC (1), and kappa statistic (0.99).

Torres-Castillo et al. (2022) presented a machine learning strategy for classifying EMG signals to detect NMD. The traditional manual interpretation of EMG signals by clinicians is enhanced by developing a comprehensive technique for analysis. The proposed method decomposes each signal into amplitude or frequency-modulated subbands and extracts time-frequency features using the Hilbert Transform. Feature selection and dimensionality reduction are performed using statistical analysis and uncorrelated linear discriminant analysis (ULDA). Three machine learning techniques—linear discriminant analysis (LDA), decision tree (TREE), and k-NNs—are employed for classification using five decomposition methods: empirical mode decomposition (EMD), ensemble EMD (EEMD), complementary EEMD (CEEMD), empirical wavelet transform (EWT), and variational mode decomposition (VMD). The best results are achieved using EEMD with KNN, with an accuracy of 99.5% and high sensitivity and specificity rates for neuropathic and myopathic signals.

Although the application of AI techniques in neuromuscular disease detection using EMG signals shows promising results, several challenges need to be addressed. One critical challenge lies in the acquisition and preprocessing of high-quality EMG data. Factors such as electrode placement, signal artifacts, and noise can impact the accuracy of the results. Standardized protocols for data collection and rigorous quality control measures are necessary to ensure reliable and consistent EMG data for analysis. Furthermore, the integration of multimodal approaches holds the potential to improve the accuracy of neuromuscular disease detection. The combination of EMG data with other diagnostic modalities, such as genetic testing, clinical assessments, and imaging techniques, can provide a more comprehensive understanding of the underlying pathologies. The fusion of data from multiple sources can enable a more accurate and specific diagnosis of neuromuscular diseases.

3. Artificial intelligence for diagnosis of neuromuscular disorders

The primary function of the human skeletal-muscular system is to provide the forces required to perform a variety of actions. The neuromuscular system is comprised of two subsystems: the neurological system and the muscular system (Begg et al., 2008). Electrical signals that travel back and forth between muscles and the peripheral and central neural systems affect movement and limb configuration (Bronzino, 1999). Nerves can be thought of as electrical current-carrying wires, with nerve heads (nuclei) beginning in the spinal column and their long axonal bodies enlarging distant and deep, stimulating single motor units in diverse muscles. The skeletal-muscular system is made up of muscle groups that are linked to bones via tendons, and movement occurs when nerve signals cause muscle contractions and re-laxations, which either attract or repel the bone (Begg et al., 2008; Subasi, 2019b).

The EMG signal indicates the electrical condition of skeletal muscles and contains data connected with muscle anatomy, which causes a range of body parts to activate. The EMG signal carries information about the central and peripheral neural systems' control of the muscles. Naturally, the EMG signal provides an exceedingly valuable description of the neuromuscular system since variations in the signal features reveal numerous pathogenic measures, whether they originate in the neurological system or the muscle. In the previous few decades, the quality of EMG signal interpretation has significantly increased, along with the advent of novel signal processing techniques to understand muscle electrophysiology (Sörnmo & Laguna, 2005). In characterizing the muscle fiber types involved in an anomaly, EMG analysis is more accurate than clinical examinations. Even if the patient has good motor-nerve conduction, EMG investigation can reveal abnormal sensory nerve physical phenomena. Furthermore, EMG analysis may assist a physician in identifying without the necessity for a muscle biopsy (Begg et al., 2008; Subasi, 2019b).

Neuromuscular disorders are characterized by abnormalities in the nervous system, neuromuscular junctions, and muscle fibers. The severity of these abnormalities can range from mild muscle damage to severe cases leading to amputation due to neuron or muscle death, such as in ALS. Early and accurate diagnosis is essential for better patient prognosis and increased chances of successful rehabilitation. Clinical testing alone is often insufficient to identify and prevent the spread of the disorder as various abnormalities can lead to similar symptoms. Therefore, precise diagnosis is crucial to enable more effective treatment. Currently, electrodiagnostic studies, including nerve conduction studies and EMG, are used to evaluate and diagnose patients with neuromuscular disorders. EMG, which measures electrical signals during muscle activity, provides valuable information about the location, etiology, and type of abnormality. For instance, the characteristics of the EMG waveform, such as its voltage range (0.01–10 mV) and frequency (10–2000 Hz on average), can indicate the position and metabolic status of the muscle fiber network. For instance, the time interval of an EMG signal can indicate the position and metabolic status of the muscle fiber network, while irregular spikes may indicate myopathy (Emly et al., 1992).

However, while electrodiagnostic methods aid doctors in diagnosing diseases, they often cannot conclusively prove the diagnosis, requiring more invasive tools like muscle biopsies or advanced imaging methods like ultrasound or MRI in challenging cases. Currently, EMG analysis is typically performed by trained neurologists who interpret EMG waveforms and may use additional methods like needle conduction studies and muscle acoustics. However, a shortage of specialists can create challenges in meeting the demand for patient diagnosis, necessitating the development of automated diagnostic systems based on EMG readings. This presents an opportunity to utilize machine learning techniques for the automated detection and classification of neuromuscular disorders based on EMG processing. Intelligent systems can assist doctors in identifying anomalies in the neuromuscular system. The goal of these intelligent diagnostic systems is to preprocess raw EMG signals and extract characteristic features, including time and frequency domain data, Fourier coefficients, autoregressive coefficients, wavelet coefficients, and other signal processing—derived quantities. These features can then serve as input data for classifiers like decision trees and SVMs, which can effectively classify various neuromuscular diseases. Neuromuscular disorders typically involve abnormalities in the peripheral nervous system and can be categorized based on their location and underlying cause, with two main categories being Neuropathy and Myopathy (Begg et al., 2008; Subasi, 2019b).

Neuropathy is a term used to describe a nerve condition that causes pain and incapacity. Injury, infection, diabetes, alcoholism, and cancer chemotherapy are all potential causes of neuropathic diseases. Myopathy is a skeletal muscle disorder that is usually caused by a muscle group injury or a genetic mutation. Myopathy inhibits affected muscles from performing their regular functions. As a result, patients suffering from myopathic illnesses have weak muscles and, depending on the degree of the disorder, have difficulty completing daily tasks or making any movement with the impacted muscles (Begg et al., 2008; Subasi, 2019b).

The needle EMG is a common clinical recording method used for diagnosing neuromuscular conditions. For instance, when a patient presents with muscle weakness, the needle EMG is performed during muscle contractions of specific muscles. These data can help identify irregular activity in cases such as muscle pain, nerve injuries in the arms and legs, pinched nerves, and muscular dystrophy. Moreover, the needle EMG is used in conjunction with nerve injury assessments to determine if the injury is recovering and returning to normal with complete muscle reactivity. This is achieved by analyzing changes in motor unit activity over a specific period. The diagnostic EMG involves the examination of unexpected motor action, even during muscle relaxation. Under normal circumstances, the muscle remains electrically quiet during relaxation. However, irregular spontaneous waveforms and waveform patterns can occur, which are associated with spontaneous muscular activities (Sörnmo & Laguna, 2005; Subasi, 2019b).

Dataset information: EMG data are obtained from the EMGLAB website (http://www.emglab.net/). For MUAP analysis, clinical EMG data were collected under normal conditions. The EMG signals are recorded with a typical concentric needle electrode at a low voluntary and continuous level of contraction. The EMG data were filtered between 2 Hz and 10 kHz and included a control group, ALS patients, and myopathy patients. The control group consisted of 10 normal people (4 females and 6 males) ranging in age from 21 to 37 years. In the ALS group, there were 8 patients (4 females and 4 men) aged 35—67. In the myopathy group, there were 7 patients (19—63 years old) (2 females and 5 males) (Nikolic, 2001). You can download the data from the following website: http://www.emglab.net/emglab/Signals/signals.php.

```python
"""
Created on Thu May  9 12:18:30 2019
@author: asubasi
"""
#==============================================================================
# Feature Extraction Using the Statistical Values of Stationary Wavelet Transform
#==============================================================================
import scipy.io as sio
# descriptive statistics
import scipy as sp
import pywt
import matplotlib.pyplot as plt
import numpy as np
import scipy.stats as stats
waveletname='db1'
level=6 #Decomposition Level
#Load mat file
mat_contents = sio.loadmat('EMGDAT.mat')
sorted(mat_contents.keys())


CONTROL=mat_contents['CON']
ALS=mat_contents['ALS']
MYOPATHIC=mat_contents['MYO']

Labels = [] #Empty List For Labels
Length = 8192;    # Length of signal
Nofsignal=200; #Number of Signal
NofClasses=3; #Number of Classes
numrows =83    #Number of features extracted from Wavelet Packet Decomposition
#Create Empty Array For Features
Extracted_Features=np.ndarray(shape=(NofClasses*Nofsignal,numrows), dtype=float,
order='F')
#==============================================================================
```

```
# Define Utility Functions for feature extraction
#===========================================================================
def SWT_Feature_Extraction(signal, i, wname, level):
    coeffs = pywt.swt(signal, wname, level=level)
    #Mean Values of each subbands
    Extracted_Features[i,0]=sp.mean(abs(coeffs[0][0]))
    Extracted_Features[i,1]=sp.mean(abs(coeffs[1][0]))
    Extracted_Features[i,2]=sp.mean(abs(coeffs[2][0]))
    Extracted_Features[i,3]=sp.mean(abs(coeffs[3][0]))
    Extracted_Features[i,4]=sp.mean(abs(coeffs[4][0]))
    Extracted_Features[i,5]=sp.mean(abs(coeffs[5][0]))
    Extracted_Features[i,6]=sp.mean(abs(coeffs[0][1]))
    Extracted_Features[i,7]=sp.mean(abs(coeffs[1][1]))
    Extracted_Features[i,8]=sp.mean(abs(coeffs[2][1]))
    Extracted_Features[i,9]=sp.mean(abs(coeffs[3][1]))
    Extracted_Features[i,10]=sp.mean(abs(coeffs[4][1]))
    Extracted_Features[i,11]=sp.mean(abs(coeffs[5][1]))


    #Standart Deviation of each subbands
    Extracted_Features[i,12]=sp.std(coeffs[0][0])
    Extracted_Features[i,13]=sp.std(coeffs[1][0])
    Extracted_Features[i,14]=sp.std(coeffs[2][0])
    Extracted_Features[i,15]=sp.std(coeffs[3][0])
    Extracted_Features[i,16]=sp.std(coeffs[4][0])
    Extracted_Features[i,17]=sp.std(coeffs[5][0])
    Extracted_Features[i,18]=sp.std(coeffs[0][1])
    Extracted_Features[i,19]=sp.std(coeffs[1][1])
    Extracted_Features[i,20]=sp.std(coeffs[2][1])
    Extracted_Features[i,21]=sp.std(coeffs[3][1])
    Extracted_Features[i,22]=sp.std(coeffs[4][1])
    Extracted_Features[i,23]=sp.std(coeffs[5][1])
    #Median Values of each subbands
    Extracted_Features[i,24]=sp.median(coeffs[0][0])
    Extracted_Features[i,25]=sp.median(coeffs[1][0])
    Extracted_Features[i,26]=sp.median(coeffs[2][0])
    Extracted_Features[i,27]=sp.median(coeffs[3][0])
    Extracted_Features[i,28]=sp.median(coeffs[4][0])
    Extracted_Features[i,29]=sp.median(coeffs[5][0])
    Extracted_Features[i,30]=sp.median(coeffs[0][1])
    Extracted_Features[i,31]=sp.median(coeffs[1][1])
    Extracted_Features[i,32]=sp.median(coeffs[2][1])
    Extracted_Features[i,33]=sp.median(coeffs[3][1])
    Extracted_Features[i,34]=sp.median(coeffs[4][1])
    Extracted_Features[i,35]=sp.median(coeffs[5][1])
    #Skewness of each subbands
    Extracted_Features[i,36]=stats.skew(coeffs[0][0])
```

```python
Extracted_Features[i,37]=stats.skew(coeffs[1][0])
Extracted_Features[i,38]=stats.skew(cocffs[2][0])
Extracted_Features[i,39]=stats.skew(coeffs[3][0])
Extracted_Features[i,40]=stats.skew(coeffs[4][0])
Extracted_Features[i,41]=stats.skew(coeffs[5][0])
Extracted_Features[i,42]=stats.skew(coeffs[0][1])
Extracted_Features[i,43]=stats.skew(coeffs[1][1])
Extracted_Features[i,44]=stats.skew(coeffs[2][1])
Extracted_Features[i,45]=stats.skew(coeffs[3][1])
Extracted_Features[i,46]=stats.skew(coeffs[4][1])
Extracted_Features[i,47]=stats.skew(coeffs[5][1])
#Kurtosis of each subbands
Extracted_Features[i,48]=stats.kurtosis(coeffs[0][0])
Extracted_Features[i,49]=stats.kurtosis(coeffs[1][0])
Extracted_Features[i,50]=stats.kurtosis(coeffs[2][0])
Extracted_Features[i,51]=stats.kurtosis(coeffs[3][0])
Extracted_Features[i,52]=stats.kurtosis(coeffs[4][0])
Extracted_Features[i,53]=stats.kurtosis(coeffs[5][0])
Extracted_Features[i,54]=stats.kurtosis(coeffs[0][1])
Extracted_Features[i,55]=stats.kurtosis(coeffs[1][1])
Extracted_Features[i,56]=stats.kurtosis(coeffs[2][1])
Extracted_Features[i,57]=stats.kurtosis(coeffs[3][1])
Extracted_Features[i,58]=stats.kurtosis(coeffs[4][1])
Extracted_Features[i,59]=stats.kurtosis(coeffs[5][1])


#RMS Values of each subbands
Extracted_Features[i,60]=np.sqrt(np.mean(coeffs[0][0]**2))
Extracted_Features[i,61]=np.sqrt(np.mean(coeffs[1][0]**2))
Extracted_Features[i,62]=np.sqrt(np.mean(coeffs[2][0]**2))
Extracted_Features[i,63]=np.sqrt(np.mean(coeffs[3][0]**2))
Extracted_Features[i,64]=np.sqrt(np.mean(coeffs[4][0]**2))
Extracted_Features[i,65]=np.sqrt(np.mean(coeffs[5][0]**2))
Extracted_Features[i,66]=np.sqrt(np.mean(coeffs[0][1]**2))
Extracted_Features[i,67]=np.sqrt(np.mean(coeffs[1][1]**2))
Extracted_Features[i,68]=np.sqrt(np.mean(coeffs[2][1]**2))
Extracted_Features[i,69]=np.sqrt(np.mean(coeffs[3][1]**2))
Extracted_Features[i,70]=np.sqrt(np.mean(coeffs[4][1]**2))
Extracted_Features[i,71]=np.sqrt(np.mean(coeffs[5][1]**2))
#Ratio of subbands
Extracted_Features[i,72]=sp.mean(abs(coeffs[0][0]))/sp.mean(abs(coeffs[1][0]))
Extracted_Features[i,73]=sp.mean(abs(coeffs[1][0]))/sp.mean(abs(coeffs[2][0]))
Extracted_Features[i,74]=sp.mean(abs(coeffs[2][0]))/sp.mean(abs(coeffs[3][0]))
Extracted_Features[i,75]=sp.mean(abs(coeffs[3][0]))/sp.mean(abs(coeffs[4][0]))
Extracted_Features[i,76]=sp.mean(abs(coeffs[4][0]))/sp.mean(abs(coeffs[5][0]))
Extracted_Features[i,77]=sp.mean(abs(coeffs[5][0]))/sp.mean(abs(coeffs[0][1]))
Extracted_Features[i,78]=sp.mean(abs(coeffs[0][1]))/sp.mean(abs(coeffs[1][1]))
```

```python
        Extracted_Features[i,79]=sp.mean(abs(coeffs[1][1]))/sp.mean(abs(coeffs[2][1]))
        Extracted_Features[i,80]=sp.mean(abs(coeffs[2][1]))/sp.mean(abs(coeffs[3][1]))
        Extracted_Features[i,81]=sp.mean(abs(coeffs[3][1]))/sp.mean(abs(coeffs[4][1]))
        Extracted_Features[i,82]=sp.mean(abs(coeffs[4][1]))/sp.mean(abs(coeffs[5][1]))
#%%
#=============================================================================
# FEATURE EXTRACTION FROM CONTROL EMG SIGNAL
#=============================================================================
for i in range(Nofsignal):
    SWT_Feature_Extraction(CONTROL[:,i], i, waveletname, level)
    Labels.append("CONTROL")
#=============================================================================
# FEATURE EXTRACTION FROM ALS EMG SIGNAL
#=============================================================================
for i in range(Nofsignal, 2*Nofsignal):
    SWT_Feature_Extraction(ALS[:,i-Nofsignal], i, waveletname, level)
    Labels.append("ALS")
# =============================================================================
# FEATURE EXTRACTION FROM MYOPATHIC EMG SIGNAL
#=============================================================================
for i in range(2*Nofsignal, 3*Nofsignal):
    SWT_Feature_Extraction(MYOPATHIC[:,i-2*Nofsignal], i, waveletname, level)
    Labels.append("MYOPATHIC")
#%%


# =============================================================================
# CLASSIFICATION
#=============================================================================
X = Extracted_Features
y = Labels
# Import train_test_split function
from sklearn.model_selection import train_test_split
# Split dataset into training set and test set
# 70% training and 30% test
Xtrain, Xtest, ytrain, ytest = train_test_split(X, y,test_size=0.3, random_state=1)
#%%
#=============================================================================
# SVM Example with Training and Test Set
#=============================================================================
from sklearn import svm
""" The parameters and kernels of SVM classifierr can be changed as follows
C = 10.0  # SVM regularization parameter
svm.SVC(kernel='linear', C=C)
svm.LinearSVC(C=C, max_iter=10000)
svm.SVC(kernel='rbf', gamma=0.7, C=C)
svm.SVC(kernel='poly', degree=3, gamma='auto', C=C)
```

```python
"""
C = 10.0  # SVM regularization parameter
#Create the Model
clf =svm.SVC(kernel='linear', C=C)
#Train the model with Training Dataset
clf.fit(Xtrain,ytrain)
#Test the model with Testset
ypred = clf.predict(Xtest)
#Evaluate the Model and Print Performance Metrics
from sklearn import metrics
print('Accuracy:', np.round(metrics.accuracy_score(ytest,ypred),4))
print('Precision:', np.round(metrics.precision_score(ytest,
                          ypred,average='weighted'),4))
print('Recall:', np.round(metrics.recall_score(ytest,ypred,
                                        average='weighted'),4))
print('F1 Score:', np.round(metrics.f1_score(ytest,ypred,
                                        average='weighted'),4))
print('Cohen Kappa Score:', np.round(metrics.cohen_kappa_score(ytest, ypred),4))
print('Matthews Corrcoef:', np.round(metrics.matthews_corrcoef(ytest, ypred),4))
print('\t\tClassification Report:\n', metrics.classification_report(ypred, ytest))
#Plot Confusion Matrix
from sklearn.metrics import confusion_matrix
print("Confusion Matrix:\n",confusion_matrix(ytest, ypred))
#%%


#=============================================================================
# ROC curves for the multiclass problem
#=============================================================================
import numpy as np
import matplotlib.pyplot as plt
from itertools import cycle
from sklearn.metrics import roc_curve, auc
from sklearn.model_selection import train_test_split
from sklearn.preprocessing import label_binarize
from sklearn.multiclass import OneVsRestClassifier
from scipy import interp
# Binarize the output
y = label_binarize(y, classes=['CONTROL','ALS','MYOPATHIC' ])
n_classes = y.shape[1]
# shuffle and split training and test sets
Xtrain, Xtest, ytrain, ytest = train_test_split(X, y,test_size=0.3, random_state=1)
# Learn to predict each class against the other
classifier = OneVsRestClassifier(svm.SVC(kernel='linear', C=C))
yscore = classifier.fit(Xtrain, ytrain).decision_function(Xtest)
# Compute ROC curve and ROC area for each class
fpr = dict()
```

```
tpr = dict()
roc_auc = dict()
for i in range(n_classes):
    fpr[i], tpr[i], _ = roc_curve(ytest[:, i], yscore[:, i])
    roc_auc[i] = auc(fpr[i], tpr[i])
# Compute micro-average ROC curve and ROC area
fpr["micro"], tpr["micro"], _ = roc_curve(ytest.ravel(), yscore.ravel())
roc_auc["micro"] = auc(fpr["micro"], tpr["micro"])
##############################################################################
# Plot of a ROC curve for a specific class
plt.figure()
lw = 2
plt.plot(fpr[2], tpr[2], color='darkorange',
         lw=lw, label='ROC curve (area = %0.2f)' % roc_auc[2])
plt.plot([0, 1], [0, 1], color='navy', lw=lw, linestyle='--')
plt.xlim([0.0, 1.0])
plt.ylim([0.0, 1.05])
plt.xlabel('False Positive Rate')
plt.ylabel('True Positive Rate')
plt.title('Receiver operating characteristic')
plt.legend(loc="lower right")
plt.show()
##############################################################################
# Plot ROC curves for the multiclass problem
# Compute macro-average ROC curve and ROC area
# First aggregate all false positive rates
all_fpr = np.unique(np.concatenate([fpr[i] for i in range(n_classes)]))


# Then interpolate all ROC curves at this points
mean_tpr = np.zeros_like(all_fpr)
for i in range(n_classes):
    mean_tpr += interp(all_fpr, fpr[i], tpr[i])
# Finally average it and compute AUC
mean_tpr /= n_classes
fpr["macro"] = all_fpr
tpr["macro"] = mean_tpr
roc_auc["macro"] = auc(fpr["macro"], tpr["macro"])
# Plot all ROC curves
plt.figure()
plt.plot(fpr["micro"], tpr["micro"],
         label='micro-average ROC curve (area = {0:0.2f})'
               ''.format(roc_auc["micro"]),
         color='deeppink', linestyle=':', linewidth=4)
plt.plot(fpr["macro"], tpr["macro"],
         label='macro-average ROC curve (area = {0:0.2f})'
               ''.format(roc_auc["macro"]),
```

```
            color='navy', linestyle=':', linewidth=4)
colors = cycle(['aqua', 'darkorange', 'cornflowerblue'])
for i, color in zip(range(n_classes), colors):
    plt.plot(fpr[i], tpr[i], color=color, lw=lw,
            label='ROC curve of class {0} (area = {1:0.2f})'
            ''.format(i, roc_auc[i]))
plt.plot([0, 1], [0, 1], 'k--', lw=lw)
plt.xlim([0.0, 1.0])
plt.ylim([0.0, 1.05])
plt.xlabel('False Positive Rate')
plt.ylabel('True Positive Rate')
plt.title('Detailed Receiver operating characteristic')
plt.legend(loc="lower right")
plt.show()
```

4. Discussion

The use of AI approaches in the identification of neuromuscular diseases utilizing EMG signals has yielded encouraging results in terms of performance and accuracy. Several research works have shown that machine learning algorithms and deep learning models are effective in classifying EMG signals and accurately identifying various neuromuscular diseases. The interpretability and explainability of the models are some of the hurdles in the implementation of AI approaches in neuromuscular illness detection. It is critical for practitioners in the medical profession to comprehend the decision-making process of AI algorithms and to trust their outcomes. Deep learning models, such as CNNs and RNNs, frequently operate as black boxes, making interpretation difficult. Researchers are hard at work constructing explainable AI models that provide insights into the features and patterns that the models learn. By incorporating explainability methodologies, physicians can have a better understanding of the AI model's conclusion and make more educated decisions in neuromuscular disease diagnosis and therapy.

The availability of high-quality and well-curated datasets is another key difficulty in the field of neuromuscular illness identification utilizing AI and EMG signals. Due to ethical constraints and patient recruitment limits, collecting a varied variety of EMG data from individuals with various neuromuscular illnesses can be difficult. Furthermore, the variation in EMG signal-gathering techniques and equipment complicates the dataset collection procedure. To solve this difficulty, researchers, doctors, and data scientists must work together to compile large-scale datasets that span a wide spectrum of neuromuscular illnesses and patient demographics.

Although EMG signals give essential information for neuromuscular illness identification, incorporating multimodal data can improve disease classification accuracy and specificity. When EMG data are combined with other diagnostic modalities, such as genetic testing, clinical assessments, or imaging techniques, a comprehensive picture of the patient's state can be obtained. For example, genetic information can aid in the identification of specific gene mutations linked to certain neuromuscular illnesses. AI models can use complimentary information from diverse modalities by combining many data sources, resulting in more robust and accurate disease identification and diagnosis.

Real-time monitoring and the utilization of wearable sensors are the future of neuromuscular illness identification. Technology advancements have made it possible to install AI models on portable and wearable devices, allowing for continuous monitoring of patients' muscle activity. This real-time monitoring can provide crucial information about disease development and therapy efficacy. Wearable devices with EMG sensors can gather real-time muscle activity data, allowing for the early diagnosis of anomalies or changes in neuromuscular function. This has the potential to provide rapid interventions, individualized treatment plans, and remote patient monitoring, particularly in impoverished areas or where access to specialized healthcare is limited.

Finally, the use of AI approaches in the diagnosis of neuromuscular diseases utilizing EMG signals offers considerable promise for enhancing diagnostic accuracy and facilitating timely therapies. AI models' performance and accuracy have been proved in several research works, demonstrating their potential in differentiating neuromuscular illnesses based on EMG signals. However, problems relating to model interpretability, data availability, and multimodal data integration must be overcome before the technology can be widely adopted and seamlessly integrated into clinical practice. Future research should concentrate on developing explainable AI models, curating large-scale and diverse datasets, and investigating the real-time monitoring capabilities of wearable devices in the field of neuromuscular disease diagnosis to improve patient treatment.

5. Summary

The combination of AI approaches and EMG signals has shown considerable promise in the identification of neuromuscular diseases.

Machine learning and deep learning algorithms have increased the accuracy and efficiency of classifying EMG signals and identifying specific neuromuscular illnesses. These developments have the potential to transform the diagnosis and management of neuromuscular illnesses, resulting in earlier interventions, tailored treatment programs, and better patient outcomes.

The use of machine learning algorithms, such as SVMs, decision trees, random forests, and deep learning models like CNNs and RNNs, has demonstrated excellent performance in accurately classifying EMG signals. These AI models can effectively learn and recognize patterns and features within the EMG signals, enabling the identification of specific neuromuscular disorders with high accuracy. This provides clinicians with valuable insights for making informed decisions regarding patient diagnosis, treatment planning, and monitoring disease progression.

Future research in this sector should concentrate on overcoming these obstacles and constraints. Efforts should be made to develop robust and interpretable AI models, provide access to diverse and well-curated datasets, and investigate the integration of multimodal data for comprehensive illness identification. Furthermore, the use of AI models in real-time monitoring via wearable devices has the potential to revolutionize neuromuscular illness management by allowing for continuous assessment of muscle function and promoting early intervention.

Finally, combining AI approaches with EMG signals holds considerable promise for enhancing neuromuscular illness identification and diagnosis. The application of machine learning algorithms and deep learning methodologies has resulted in tremendous progress in reliably classifying EMG signals and detecting specific neuromuscular illnesses. These AI-powered technologies have the potential to improve disease detection accuracy and efficiency, allowing for early intervention, individualized treatment regimens, and improved patient outcomes.

References

Begg, R., Lai, D. T., & Palaniswami, M. (2008). *Computational intelligence in biomedical engineering*. CRC Press.

Bronzino, J. D. (1999). *Biomedical engineering handbook* (Vol. 2). CRC press.

Chatterjee, S., Singha Roy, S., Bose, R., & Pratiher, S. (2020). Feature extraction from multifractal spectrum of electromyograms for diagnosis of neuromuscular disorders. *IET Science, Measurement & Technology, 14*(7), 817–824.

Emly, M., Gilmore, L. D., & Roy, S. H. (1992). Electromyography. *IEEE Potentials, 11*(2), 25–28.

Gokgoz, E., & Subasi, A. (2014). Effect of multiscale PCA denoising on EMG signal classification for diagnosis of neuromuscular disorders. *Journal of Medical Systems, 38*(4), Article 4.

Gokgoz, E., & Subasi, A. (2015). Comparison of decision tree algorithms for EMG signal classification using DWT. *Biomedical Signal Processing and Control, 18*, 138–144.

Nikolic, M. (2001). *Findings and firing pattern analysis in controls and patients with myopathy and amytrophic lateral sclerosis*. Copenhagen: Faculty of Health Science, University of Copenhagen.

Samanta, K., Roy, S. S., Modak, S., Chatterjee, S., & Bose, R. (2020). *Neuromuscular disease detection employing deep feature extraction from cross spectrum images of electromyography signals.*

Sörnmo, L., & Laguna, P. (2005). *Bioelectrical signal processing in cardiac and neurological applications* (Vol. 8). Academic Press.

Subasi, A. (2015). A decision support system for diagnosis of neuromuscular disorders using DWT and evolutionary support vector machines. *Signal, Image and Video Processing, 9*(2), Article 2.

Subasi, A. (2019a). Diagnosis of neuromuscular disorders using DT-CWT and rotation forest ensemble classifier. *IEEE Transactions on Instrumentation and Measurement, 69*(5), 1940–1947.

Subasi, A. (2019b). *Practical guide for biomedical signals analysis using machine learning techniques: A MATLAB based approach* (1st ed.). Academic Press.

Subasi, A., Yaman, E., Somaily, Y., Alynabawi, H. A., Alobaidi, F., & Altheibani, S. (2018). Automated EMG signal classification for diagnosis of neuromuscular disorders using DWT and bagging. *Procedia Computer Science, 140*, 230–237.

Torres-Castillo, J. R., Lopez-Lopez, C. O., & Padilla-Castaneda, M. A. (2022). Neuromuscular disorders detection through time-frequency analysis and classification of multi-muscular EMG signals using Hilbert-Huang transform. *Biomedical Signal Processing and Control, 71*, 103037.

Vallejo, M., Gallego, C. J., Duque-Muñoz, L., & Delgado-Trejos, E. (2018). Neuromuscular disease detection by neural networks and fuzzy entropy on time-frequency analysis of electromyography signals. *Expert Systems, 35*(4), e12274.

Yaman, E., & Subasi, A. (2019). Comparison of bagging and boosting ensemble machine learning methods for automated EMG signal classification. *BioMed Research International, 2019*.

Further reading

Subasi, A. (2012). Medical decision support system for diagnosis of neuromuscular disorders using DWT and fuzzy support vector machines. *Computers in Biology and Medicine, 42*(8), Article 8.

Subasi, A. (2013). Classification of EMG signals using PSO optimized SVM for diagnosis of neuromuscular disorders. *Computers in Biology and Medicine, 43*(5), Article 5.

An evaluation of pretrained convolutional neural networks for stroke classification from brain CT images

Muhammad Irfan[1,2], Abdulhamit Subasi[3,4], Noman Mustafa[5], Tomi Westerlund[2], Wei Chen[1,6]

[1]Center for Intelligent Medical Electronics (CIME), School of Information Science and Technology, Fudan University, China; [2]TIERS, University of Turku (UTU), Turku, Finland; [3]Institute of Biomedicine, Faculty of Medicine, University of Turku, Turku, Finland; [4]Department of Computer Science, College of Engineering, Effat University, Jeddah, Saudi Arabia; [5]Intac Optical Instruments Ltd, Suzhou, Jiangsu, China; [6]Human Phenome Institute, Fudan University, Shanghai, China

1. Introduction

There is a growing global concern related to cerebral vascular accidents, commonly referred to as strokes. The World Health Organization reports that strokes affect 15 million people annually, resulting in approximately 6 million deaths and 5 million disablement (WHO, 2023). In light of the high incidence of strokes in the United States, where one occurs every 40 seconds, such figures highlight the severity of this condition (CDC, 2023).

A patient's return to health can be improved by rapid and accurate stroke detection, primarily by computer tomography (CT). According to Kakkar et al. (2021), these scans are essential in the preliminary assessment of suspected stroke victims, enabling physicians to differentiate between ischemic and hemorrhagic stroke variants or any other diseases, which differ in treatment pathways.

Machine learning (ML) capability to detect different types of strokes or whether CT scans show strokes or not can have a profound effect on patients' recovery expectations. Therefore, this study optimizes three DL algorithms for stroke identification and classification from CT images. Using cutting-edge technology, clinicians can speed up diagnosis and implement effective interventions, reducing stroke-related disability and mortality.

In strokes or cerebrovascular accidents, blood flow to the brain is disrupted or constrained, resulting in the death of brain cells. There are two types of strokes: ischemic strokes and hemorrhagic strokes. An ischemic stroke takes place when a blood clot blocks a blood vessel leading to the brain, responsible for approximately 87% of all strokes (Hopkins, 2023). Hemorrhagic strokes result when a compromised blood vessel ruptures, causing brain bleeding. Either disorder can lead to paralysis, speech impediments, or cognitive deficits (Hopkins, 2023).

Several factors increase stroke risk, including hypertension, smoking, diabetes, obesity, hypercholesterolemia, and sedentary lifestyles (WHO, 2023). Additionally, there is a strong correlation between age and stroke history in families, with older individuals and those with hereditary predispositions at higher risk. To prevent strokes, it is, therefore, crucial to recognize these risk factors and take preventative measures.

2. Related work

The use of artificial intelligence and ML to detect and classification of stroke has gained widespread acceptance in the scientific and engineering communities. Al-Qazzaz et al. (2021) developed a system, based on ML, that monitors rehabilitation progress through brain–computer interfaces (BCIs). It classified EEG signals from stroke patients with upper limb hemiparesis using random forest, support vector machine, and k-nearest neighbor (k-NN) classifiers.

Simultaneously, Sung et al. (2021) developed an ML algorithm adept at diagnosing possible stroke cases during emergency department triage. The innovative algorithm, which could be integrated into an electronic triage system, activated code strokes. Several classifiers, including SVM, RF, k-NN, C4.5, CART, and logistic regression (LR), were evaluated in their study. The rate of accuracy for these algorithms ranged from 80.6% to 82.6%.

Soltanpour et al. (2021) developed a model to accurately identify stroke lesions in CT perfusion images. This method, validated using the ISLES 2018 dataset, yielded a sensitivity and positive predictive value of 67% and 68%, respectively. A novel approach was put forth by Peixoto and Rebouças Filho (2018), in which primary frequencies of CT brain scans were employed to differentiate between hemorrhagic and ischemic strokes. The approach, which focuses exclusively on image parameters, displayed commendable stroke classification performance with specificity, sensitivity, and accuracy of 99.1%, 97%, and 98%, respectively.

The advancement of computer science has led to significant improvements in stroke detection, with DL and ML algorithms being used more and more to classify CT and MR images. Bento et al. (2019) proposed an SVM classifier-based model for detecting individuals with carotid artery atherosclerosis from MR brain images, exhibiting impressive accuracy, sensitivity, and specificity.

On another front, Reboucas Filho et al. (2017) outlined a technique for brain tissue density analysis to extract features from radiological brain density models. This strategy was applied to CT images to detect and classify stroke occurrences, and all classifiers achieved average accuracy values above 95%.

Vargas et al. (2019) developed a model to predict perfusion deficits in acute ischemic stroke patients. The model was designed as a convolutional neural network (CNN) stacked atop a long short-term memory layer, attaining an 85.8% accuracy score on validation data.

Gautam and Raman (2021) proposed a computational framework that combines image fusion and DL to identify cerebral palsy, with classification accuracy fluctuating between 98.33% and 98.77%.

Kanchana and Menaka (2020) developed a histogram-based algorithm that could detect ischemic stroke and implemented a variety of classification methods, with accuracy rates ranging from 88.77% to 99.79%. Raghavendra et al. (2021) put forth a probabilistic NN model for early detection of intracerebral hemorrhage on CT images, achieving a 97.37% accuracy rate.

Herzog et al. (2020) developed a Bayesian-integrated CNN model for detecting ischemic stroke patients based on MR images, achieving 95.33% accuracy, a 2% improvement over its non-Bayesian counterpart. Badriyah et al. (2020) employed a range of techniques such as RF, Naive Bayes, k-NN, LR, decision tree, SVM, multilayer perceptron neural network, and DL for classifying stroke into its two subtypes (ischemic and hemorrhagic).

In another interesting study, Diker et al. (2023) developed several DL models, including VGG 19 and VGG16, for binary classification of stroke images. Remarkably, the VGG 19 model showed a remarkable accuracy score of 97%.

3. Methodology

3.1 Preprocessing

Reducing the size and cropping of CT images of strokes can help streamline data handling and analysis, resulting in better diagnostic results. As illustrated in Fig. 5.1, it reduces data size, enhances signal by removing noise from irrelevant regions, and highlights critical diagnostic features. This standardizes image sizes for easier analysis and more efficient processing, especially for ML applications. Moreover, it promotes interoperability, allowing integration with other biomedical data. The process, however, must be carefully monitored to avoid losing vital information.

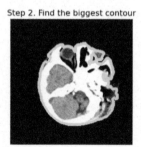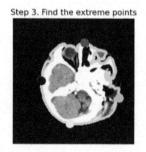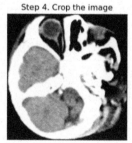

FIGURE 5.1   Image cropping.

```
1  !apt-get install tree
2  clear_output()
3  # create new folders
4  !mkdir TRAIN VAL TRAIN/YES TRAIN/NO VAL/YES VAL/NO
5  !tree -d
6  IMG_PATH = "/kaggle/input/brain-stroke-updated/Brain_Stroke/"
7  ignored = {"pred"}
8  # split the data by train/val
9  for CLASS in os.listdir(IMG_PATH):
10     if CLASS not in ignored:
11         if not CLASS.startswith('.'):
12             IMG_NUM = len(os.listdir(IMG_PATH +"/"+ CLASS))
13             for (n, FILE_NAME) in enumerate(os.listdir(IMG_PATH +"/"+ CLASS)):
14                 img = IMG_PATH+ '/' +  CLASS + '/' + FILE_NAME
15                 if n < 0.80*IMG_NUM:
16                     shutil.copy(img, 'TRAIN/'+ CLASS.upper() + '/' + FILE_NAME)
17                 else:
18                     shutil.copy(img, 'VAL/'+ CLASS.upper() + '/' + FILE_NAME)
19 def load_data(dir_path, img_size=(100,100)):
20     X = []
21     y = []
22     i = 0
23     labels = dict()
24     for path in tqdm(sorted(os.listdir(dir_path))):
25         if not path.startswith('.'):
26             labels[i] = path
27             for file in os.listdir(dir_path + path):
28                 if not file.startswith('.'):
29                     img = cv2.imread(dir_path + path + '/' + file)
30                     X.append(img)
31                     y.append(i)
32             i += 1
33     X = np.array(X)
34     y = np.array(y)
35     print(f'{len(X)} images loaded from {dir_path} directory.')
36     return X, y, labels
37 def plot_confusion_matrix(cm, classes,
38                           normalize=False,
39                           title='Confusion matrix',
40                           cmap=plt.cm.Blues):
```

```
41      plt.figure(figsize = (6,6))
42      plt.imshow(cm, interpolation='nearest', cmap=cmap)
43      plt.title(title)
44      plt.colorbar()
45      tick_marks = np.arange(len(classes))
46      plt.xticks(tick_marks, classes, rotation=90)
47      plt.yticks(tick_marks, classes)
48      if normalize:
49          cm = cm.astype('float') / cm.sum(axis=1)[:, np.newaxis]
50      thresh = cm.max() / 2.
51      cm = np.round(cm,2)
52      for i, j in itertools.product(range(cm.shape[0]), range(cm.shape[1])):
53          plt.text(j, i, cm[i, j],
54                  horizontalalignment="center",
55                  color="white" if cm[i, j] > thresh else "black")
56      plt.tight_layout()
57      plt.ylabel('True label')
58      plt.xlabel('Predicted label')
59      plt.show()
60  TRAIN_DIR = 'TRAIN/'
61  VAL_DIR = 'VAL/'
62  IMG_SIZE = (224,224)
63  X_train, y_train, labels = load_data(TRAIN_DIR, IMG_SIZE)
64  X_val, y_val, _ = load_data(VAL_DIR, IMG_SIZE)
65  y = dict()
66  y[0] = []
67  y[1] = []
68  for set_name in (y_train, y_val):
69      y[0].append(np.sum(set_name == 0))
70      y[1].append(np.sum(set_name == 1))
71  trace0 = go.Bar(
72      x=['Train Set', 'Validation Set'],
73      y=y[0],
74      name='No',
75      marker=dict(color='#33cc33'),
76      opacity=0.7
77  )
78  trace1 = go.Bar(
79      x=['Train Set', 'Validation Set'],
80      y=y[1],
```

```python
81      name ='Yes ',
82      marker= dict( color='# ff3300 '),
83      opacity =0.7
84 )
85 data = [trace0 , trace1]
86 layout = go.Layout(
87      title ='Count of classes in each set',
88      xaxis ={' title ': 'Set'},
89      yaxis ={' title ': 'Count'}
90 )
91 fig = go.Figure (data , layout)
92 iplot( fig)
93 def plot_samples (X, y, labels_dict , n=50):
94      for index in range (len(labels_dict )):
95          imgs = X[np.argwhere (y == index )][:n]
96          j = 10
97          i = int( n/j)
98
99          plt. figure ( figsize =(15 ,6))
100         c = 1
101         for img in imgs:
102             plt. subplot(i,j,c)
103             plt. imshow ( img [0])
104
105             plt. xticks ([])
106             plt. yticks ([])
107             c += 1
108         plt. suptitle (' Tumor: {}'. format( labels_dict [index ]))
109         plt. show ()
110 plot_samples (X_train , y_train , labels , 30)
111     set_new = []
112     for img in set_name :
113         gray = cv2.cvtColor( img, cv2.COLOR_RGB2GRAY)
114         gray = cv2.GaussianBlur (gray , (5, 5), 0)
115         thresh = cv2.threshold (gray , 45, 255, cv2.THRESH_BINARY )[1]
116         thresh = cv2.erode (thresh , None, iterations =2)
117         thresh = cv2.dilate (thresh , None, iterations =2)
118         # find contours in thresholded image , then grab the largest one
119         cnts = cv2.findContours (thresh.copy (), cv2.RETR_EXTERNAL , cv2.
        CHAIN_APPROX_SIMPLE )
120         cnts = imutils.grab_contours (cnts)
```

```
121          c = max(cnts, key=cv2.contourArea)
122          # find the extreme points
123          extLeft = tuple(c[c[:, :, 0].argmin()][0])
124          extRight = tuple(c[c[:, :, 0].argmax()][0])
125          extTop = tuple(c[c[:, :, 1].argmin()][0])
126          extBot = tuple(c[c[:, :, 1].argmax()][0])
127          ADD_PIXELS = add_pixels_value
128          new_img = img[extTop[1]-ADD_PIXELS:extBot[1]+ADD_PIXELS, extLeft[0]-
        ADD_PIXELS:extRight[0]+ADD_PIXELS].copy()
129          set_new.append(new_img)
130      return np.array(set_new)
131 import imutils
132 img = cv2.imread('/kaggle/input/brain-stroke-updated/Brain_Stroke/no/n1000.jpg')
133 img = cv2.resize(
134          img,
135          dsize=IMG_SIZE,
136          interpolation=cv2.INTER_CUBIC
137          )
138 gray = cv2.cvtColor(img, cv2.COLOR_RGB2GRAY)
139 gray = cv2.GaussianBlur(gray, (5, 5), 0)
140 thresh = cv2.threshold(gray, 45, 255, cv2.THRESH_BINARY)[1]
141 thresh = cv2.erode(thresh, None, iterations=2)
142 thresh = cv2.dilate(thresh, None, iterations=2)
143 cnts = cv2.findContours(thresh.copy(), cv2.RETR_EXTERNAL, cv2.CHAIN_APPROX_SIMPLE)
144 cnts = imutils.grab_contours(cnts)
145 c = max(cnts, key=cv2.contourArea)
146 # find the extreme points
147 extLeft = tuple(c[c[:, :, 0].argmin()][0])
148 extRight = tuple(c[c[:, :, 0].argmax()][0])
149 extTop = tuple(c[c[:, :, 1].argmin()][0])
150 extBot = tuple(c[c[:, :, 1].argmax()][0])
151 # add contour on the image
152 img_cnt = cv2.drawContours(img.copy(), [c], -1, (0, 255, 255), 4)
153 # add extreme points
154 img_pnt = cv2.circle(img_cnt.copy(), extLeft, 8, (0, 0, 255), -1)
155 img_pnt = cv2.circle(img_pnt, extRight, 8, (0, 255, 0), -1)
156 img_pnt = cv2.circle(img_pnt, extTop, 8, (255, 0, 0), -1)
157 img_pnt = cv2.circle(img_pnt, extBot, 8, (255, 255, 0), -1)
158 # crop
159 ADD_PIXELS = 0
```

```python
160 new_img = img[ extTop [1]- ADD_PIXELS : extBot [1]+ ADD_PIXELS , extLeft [0]- ADD_PIXELS :
        extRight [0]+ ADD_PIXELS ].copy ()
161 plt. figure (figsize =(15 ,6))
162 plt. subplot (141)
163 plt. imshow (img)
164 plt. xticks ([])
165 plt. yticks ([])
166 plt. title ('Step  1. Get  the  original  image')
167 plt. subplot (142)
168 plt. imshow (img_cnt)
169 plt. xticks ([])
170 plt. yticks ([])
171 plt. title ('Step  2. Find  the  biggest  contour')
172 plt. subplot (143)
173 plt. imshow (img_pnt)
174 plt. xticks ([])
175 plt. yticks ([])
176 plt. title ('Step  3. Find  the  extreme  points')
177 plt. subplot (144)
178  plt. imshow (new_img )
179 plt. xticks ([])
180 plt. yticks ([])
181 plt. title ('Step  4. Crop  the  image')
```

```
182  plt.show()
183  X_train_crop = crop_imgs(set_name=X_train)
184  X_val_crop = crop_imgs(set_name=X_val)
185  def save_new_images(x_set, y_set, folder_name):
186      i = 0
187      for (img, imclass) in zip(x_set, y_set):
188          if imclass == 0:
189              cv2.imwrite(folder_name+'NO/'+str(i)+'.jpg', img)
190          else:
191              cv2.imwrite(folder_name+'YES/'+str(i)+'.jpg', img)
192          i += 1
193  # saving new images to the folder
194  !mkdir TRAIN_CROP VAL_CROP TRAIN_CROP/YES TRAIN_CROP/NO VAL_CROP/YES VAL_CROP/N
195  def preprocess_imgs(set_name, img_size):
196      set_new = []
197      for img in set_name:
198          img = cv2.resize(
199              img,
200              dsize=img_size,
201              interpolation=cv2.INTER_CUBIC
202          )
```

```
203        set_new.append(preprocess_input(img))
204    return np.array(set_new)
205 X_train_prep = preprocess_imgs(set_name=X_train_crop, img_size=IMG_SIZE)
206 X_val_prep  = preprocess_imgs(set_name=X_val_crop, img_size=IMG_SIZE)
207 plot_samples(X_train_prep, y_train, labels, 30)
208 TRAIN_DIR = 'TRAIN_CROP/'
209 VAL_DIR = 'VAL_CROP/'
210 RANDOM_SEED = 42
211 train_datagen = ImageDataGenerator(
212     featurewise_center=False,
213     samplewise_center=False,
214     featurewise_std_normalization=False,
215     samplewise_std_normalization=False,
216     zca_whitening=False,
217     zca_epsilon=1e-06,
218     rotation_range=0,
219     width_shift_range=0.0,
220     height_shift_range=0.0,
221     brightness_range=None,
222     shear_range=0.0,
223     zoom_range=0.0,
224     channel_shift_range=0.0,

225     fill_mode='nearest',
226     cval=0.0,
227     horizontal_flip=False,
228     vertical_flip=False,
229     rescale=None,
230     # preprocessing_function=None,
231     data_format=None,
```

```
232     validation_split =0.0,
233     dtype =None,
234     preprocessing_function = preprocess_input
235 )
236 test_datagen  =  ImageDataGenerator(
237     featurewise_center = False,
238     samplewise_center = False,
239     featurewise_std_normalization = False,
240     samplewise_std_normalization = False,
241     zca_whitening = False,
242     zca_epsilon =1e-06,
243     rotation_range =0,
244     width_shift_range =0.0,
245     height_shift_range =0.0,
246     brightness_range =None,
247     shear_range =0.0,
248     zoom_range =0.0,
249     channel_shift_range =0.0,
250     fill_mode  =' nearest ',
251     cval   =0.0   ,
252     horizontal_flip = False,
253     vertical_flip = False,
254     rescale       =None        ,
255     # preprocessing_function =None ,
256     data_format =None,
257     validation_split  =0.0 ,
258     dtype =None,
259     preprocessing_function = preprocess_input
260 )
261 train_generator = train_datagen.flow_from_directory(
262     TRAIN_DIR,
263     color_mode ='rgb',
264     target_size =IMG_SIZE,
265     batch_size =20,
266     class_mode ='binary',
267     seed =RANDOM_SEED
268 )
269 validation_generator = test_datagen.flow_from_directory(
270     VAL_DIR,
271     color_mode ='rgb',
```

```
272     target_size = IMG_SIZE ,
273     batch_size =16 ,
274     class_mode ='binary',
275     seed = RANDOM_SEED
```

3.2 Transfer learning models

In our research, we evaluated various ML, DL, and transfer learning models, though we have only featured the top three models based on given dataset. Fig. 5.2 illustrates a schematic illustration of the proposed architecture for brain stroke detection.

3.3 MobileNetV2

MobileNetV2 is a DL architecture for efficient on-device computer vision tasks. It is a follow-up version of the original MobileNetV1 and was introduced by Google researchers in the paper titled "MobileNetV2: Inverted Residuals and Linear Bottlenecks" in 2018 (Sandler et al., 2018). The key objective behind MobileNetV2 was to build upon the efficiency and speed of Mo- bileNetV1 while sustaining or improving the accuracy. MobileNetV1 relied on depth-wise separable convolutions, which considerably reduced the computational complexity of the model. However, it sometimes led to a decrease in accuracy. MobileNetV2 resolves these issues by incorporating two major upgrades:

- Inverted residuals: MobileNetV2 uses a type of block called "Inverted Residuals" to store and propagate information more efficiently through the network. In these blocks, the input is first routed through a lightweight expansion layer (1 × 1 convolution) to boost the number of channels. Afterward, a depth-wise separable convolution reduces the spatial resolution and calculates the spatial dependencies. For the final step, a 1 × 1 convolution is used to return the desired number of output channels. Despite being computationally efficient, this structure maintains accuracy.
- Linear bottlenecks: MobileNetV2 includes linear bottlenecks, that use 1 × 1 convolutions with a linear activation function in the middle of the inverted residual blocks. As a result, better information flows through the network and accuracy is improved.

With these improvements, MobileNetV2 is more efficient and accurate than MobileNetV1, making it suitable for a variety of on-device applications, such as real-time object detection and image classification. Many mobile and embedded devices use it because of its limited computational resources (Fig. 5.3).

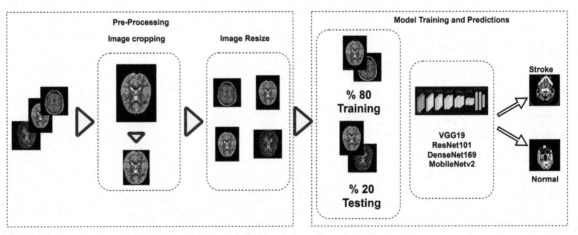

FIGURE 5.2   Block illustration of brain stroke detecting scheme.

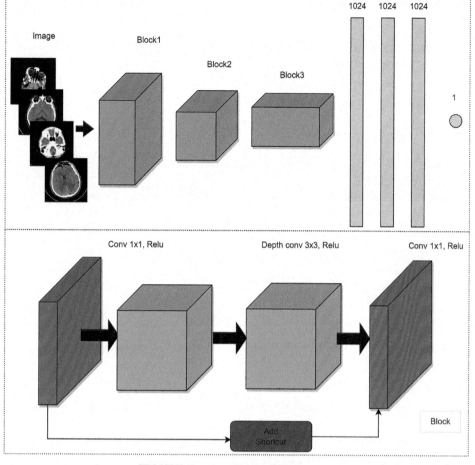

FIGURE 5.3 MobileNetV2 architecture.

```
1 base_model = MobileNet(weights='imagenet', include_top=False, layers=tf.keras.
      layers, input_shape=(224, 224,3))
2 x = base_model.output
3 x = GlobalAveragePooling2D()(x)
4 x = Dense(1024, activation='relu')(x)
5 x = Dropout(0.35)(x)
6 x = Dense(1024, activation='relu')(x)
7 x = Dropout(0.35)(x)
8 x = Dense(1024, activation='relu')(x)
9 x = Dropout(0.35)(x)
10 preds = Dense(1, activation='sigmoid')(x)
11 model = Model(inputs=base_model.input, outputs=preds)
12 opt = tf.keras.optimizers.Adam(learning_rate=0.0001)
13 model.compile(loss='binary_crossentropy',
14                optimizer=opt,
15                metrics=['accuracy', AUC()])
```

```python
17  model.compile(loss='binary_crossentropy',
18                optimizer=opt,
19                metrics =['accuracy', 'AUC'])
20  v = tf.keras.callbacks.TensorBoard(
21      log_dir="logs",
22      histogram_freq =0,
23      write_graph =True,
24      write_images =False,
25      write_steps_per_second = False,
26      update_freq =" epoch ",
27      profile_batch =0,
28      embeddings_freq =0,
29      embeddings_metadata = None
30      #** kwargs
31  )
32  import tensorflow
33  import tensorflow.keras.backend as K
34  early_stopping_patience  =  50
35  early_stopping = tensorflow.keras.callbacks.EarlyStopping(
36      monitor="val_loss", patience =early_stopping_patience, restore_best_weights =True
37  )
38  tb_cb = tensorflow.keras.callbacks.ModelCheckpoint('savedmodels /mobilenetv2 -epoch ={
        epoch :02 d}-val.loss ={val_loss :.2 f}.h5', monitor='val_loss',save_best_only =True,
        period =1, mode ='min')
39  tb_cb1 = tensorflow.keras.callbacks.ModelCheckpoint('savedmodels /mobilenetv2 -epoch
        ={epoch :02 d}-val.acc ={val_acc :.3 f}.h5', monitor='val_accuracy', save_best_only =
        True,period =1,mode ='max')
40  v = tf.keras.callbacks.TensorBoard(log_dir='./logs'),
41  history =model.fit(
42          train_generator,
43          steps_per_epoch = train_generator .n// train_generator .batch_size,
44          epochs =200, initial_epoch =0,
45          validation_data = validation_generator,
46          validation_steps = validation_generator .n// validation_generator .batch_size,
        callbacks =[early_stopping ,tb_cb, v])
47  model.save('savedmodels /hmobilenetv2_last_epoch.h5')
48  #We can fit here again with 200 epochs
```

```
50 predictions = model.predict(X_val_prep)
51 predictions = [1 if x>0.5 else 0 for x in predictions]
52
53 accuracy = accuracy_score(y_val, predictions)
54 print('Pred Accuracy = %.4f' % accuracy)
55
56 from sklearn import metrics
57 print(' Accuracy:', np.round(metrics.accuracy_score(y_val, predictions),4))


58 print('Precision:', np.round(metrics.precision_score(y_val, predictions,average='
       weighted'),4))
59 print('Recall:', np.round(metrics.recall_score(y_val, predictions,
60                                             average='weighted'),4))
61 print('F1 Score:', np.round(metrics.f1_score(y_val, predictions,
62                                             average=' weighted ') ,4))
63 print('Cohen Kappa Score:', np.round(metrics.cohen_kappa_score(y_val, predictions)
       ,4))
64 print('Matthews Corrcoef:', np.round(metrics.matthews_corrcoef(y_val, predictions)
       ,4))
65 print('\t\tClassification Report:\n', metrics.classification_report(y_val,
       predictions))
66 confusion_mtx = confusion_matrix(y_val, predictions)
67 cm = plot_confusion_matrix(confusion_mtx, classes = list(labels.items()), normalize
       =False)
```

3.4 ResNet101

ResNet101 (also known as Residual Network-101) is a DL architecture that belongs to the ResNet family. ResNet was introduced by He et al. (2016). The "101" in ResNet101 refers to the number of layers in the network, indicating that it has 101 layers. The main contribution of ResNet is the use of residual blocks, which are used to overcome the problem of gradient disappearance during the training of very deep neural networks. The vanishing gradient issue occurs when gradients become too small as they are propagated backward through many layers, making it challenging for the algorithm to learn effectively. The residual blocks enable the system to learn residual mappings, which makes it possible to optimize the deeper layers.

A residual block in ResNet consists of two or more convolutional layers with shortcut connections (also called skip connections). The skip connections bypass one or more convolutional layers and directly connect the input of the block to its output. The network learns the residual mapping, that is, the difference between the input and the output of the block. The addition of the residual mapping to the original input is then passed through an activation function (usually a rectified linear unit, or ReLU) to obtain the final output of the block. The ResNet101 model has 101 layers and is deeper than the original ResNet-50. A deeper architecture can learn more complex features and patterns but comes with a greater computational cost and a greater risk of overfitting. On benchmark datasets like ImageNet and COCO, ResNet101 and its variants have achieved state-of-the-art performance for image classification, object detection, and segmentation. The ResNet101 architecture for brain stroke classification is given in Fig. 5.4.

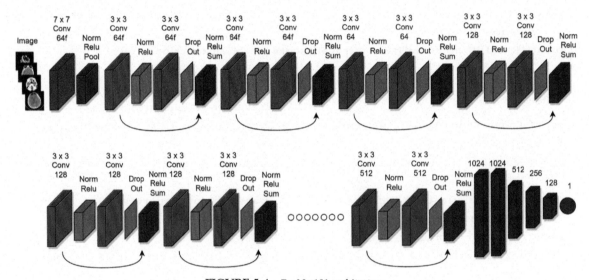

FIGURE 5.4 ResNet101 architecture.

```
1  from tensorflow.keras.applications import ResNet101
2  from tensorflow.keras.models import Sequential
3  from tensorflow.keras.layers import Flatten, BatchNormalization, Dense, Dropout
```

```
4  base_model = ResNet101 (input_shape =(224, 224, 3), weights='imagenet', include_top =
       False )
5  x=       base_model    .    output
6  x= GlobalAverage Pooling 2 D ()( x)
7  x=Dense (1024, activation ='relu')( x)
8  x       =         Dropout   (0.2)( x)
9  x=Dense (1024, activation ='relu')( x)
10 x = Dropout (0.2)( x)
11 x=Dense (512, activation ='relu')( x) #dense layer 2
12 x               =               Dropout   (0.2)( x)
13 x=Dense (256, activation ='relu')( x) #dense layer 2
14 x               =               Dropout   (0.2)( x)
15 x=Dense (128, activation ='relu')( x) #dense layer 2
16 x = Dropout (0.2)( x)
17 preds= Dense (1, activation ='sigmoid')( x) #final layer with sigmoid activation
18 model=   Model(  inputs=  base_model  .  input   ,  outputs=  preds)
19 opt=tf.  keras.  optimizers.  Adam  (  learning_rate   = 0 .0001 )
20 model . compile (loss='binary_crossentropy',
21              optimizer=opt           ,
22              metrics =['accuracy' , 'AUC'])
23 v = tf. keras. callbacks. TensorBoard (
24     log_dir=" logs",
25     histogram_freq =0,
26     write_graph =True,
27     write_images =False,
28     write_steps_per_second =False,
29     update_freq =" epoch",
30     profile_batch =0,
31     embeddings_freq =0,
32     embeddings_metadata = None
33 )
34 #Fitting of this model is similar to MobileNet
35 # Also, the confusion matrix
```

3.5 DenseNet-169

DenseNet-169 is a deep CNN architecture that contributes to the DenseNet family of models. DenseNet was developed by Huang et al. (2017). In DenseNet, dense connections between layers are used to optimize information flow through the network. The information from one layer of a CNN is only passed onto the next layer in traditional convolutional neural networks. Each layer of DenseNet receives input from all previous layers in a feed-forward fashion. With dense connectivity, features can be reused, parameters can be reduced, and the vanishing gradient problem can be alleviated, allowing very deep networks to be trained. DenseNet-169 is an advanced variant of the DenseNet architecture with 169 layers. It is deeper than DenseNet-121 but

shallower than DenseNet-201. There are 169 layers in the network, as indicated by the "169" in the name. There are four dense blocks in DenseNet-169, each containing several convolutional layers with densely connected feature maps. Transition layers with pooling are applied between dense blocks to minimize the spatial dimensions and the number of channels, contributing to the control of the model size and computational cost.

A large-scale dataset, such as ImageNet, has been used to train DenseNet-169 for various computer vision tasks, including image classification, object detection, and segmentation. Researchers and industry alike use the model for its effectiveness and efficiency in tackling these tasks. The architecture for the brain stroke classification is given in Fig. 5.5.

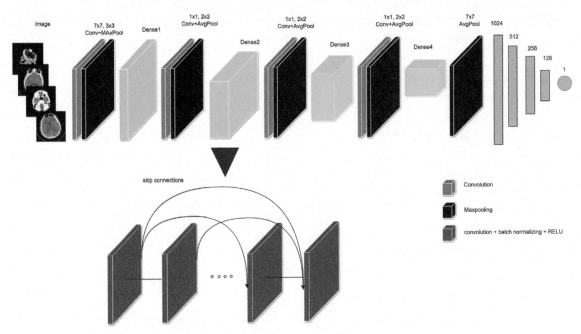

FIGURE 5.5   DenseNet 169 architecture.

```
1  base_Neural_Net = DenseNet169(input_shape=(224,224,3), weights='imagenet',
       include_top=False)
2  x=  base_model  .  output
3  x= GlobalAveragePooling2D()(x)
4  x=Dense(1024,activation='relu')(x) #we add dense layers so that the model can learn
       more complex functions and classify for better results.
5  x               =            Dropout   (0.2)(  x)
6  x=Dense(512,activation='relu')(x) #dense layer 2
7  x               =            Dropout   (0.2)(  x)
8  x=Dense(256,activation='relu')(x) #dense layer 2
9  x               =            Dropout   (0.2)(  x)
10 x=Dense(128,activation='relu')(x) #dense layer 2
11 x  =  Dropout(0.2)(x)
12 preds=Dense(1,activation='sigmoid')(x) #final layer with sigmoid activation
13 model=  Model(  inputs=  base_model . input  ,  outputs=  preds)
14 opt=tf.keras.optimizers.Adam(learning_rate=0.0001)
15 #opt=optimizers.Adam(lr=0.0001)



16 model.compile(loss='binary_crossentropy',
17              optimizer=opt        ,
18              metrics=['accuracy', 'AUC']))
19 #Fitting of this model is similar to MobileNet
20 # Also, the confusion matrix
```

4. Dataset

The Brain Stroke CT Image Dataset (Rahman, 2023) includes images from stroke-diagnosed and healthy individuals. The images in the dataset have a resolution of 650×650 pixels and are stored as JPEGs. Among the total 2501 images, 1551 belong to healthy individuals while the remainder represent stroke patients. Fig. 5.6 illustrates these images where the top row depicts healthy brain images and the subsequent row displays stroke-related images.

5. Experimental results

On Kaggle, experiments were conducted with a GPU100. The proposed model was evaluated based on a confusion matrix and a set of standard performance metrics. Accuracy (ACC), precision F-scores, kappa, and receiver operating

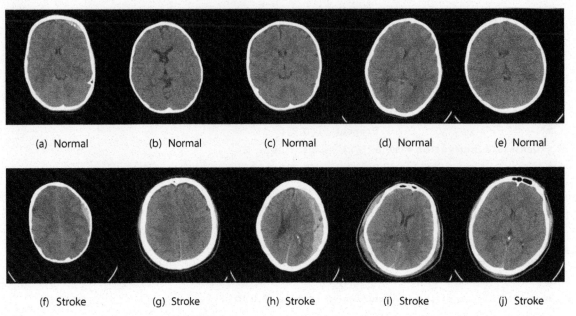

(a) Normal	(b) Normal	(c) Normal	(d) Normal	(e) Normal
(f) Stroke	(g) Stroke	(h) Stroke	(i) Stroke	(j) Stroke

FIGURE 5.6   CT images: normal versus stroke.

characteristics (ROCs) are the metrics that are used in this study. There are four elements in the confusion matrix: true positives (TPs), true negatives (TNs), false positives (FPs), and false negatives (FNs). The equations for accuracy, precision, F_1-score, and kappa are given in Eqs. (5.1), (5.2), (5.3), and (5.4).

$$\text{Accuracy(ACC)} = \frac{TP + TN}{TP + FP + TN + FN} \quad (5.1)$$

$$\text{Precision(PRE)} = \frac{TP}{TP + FP} \quad (5.2)$$

$$F_1 - \text{score}(F_1) = 2 \times \frac{Precision \cdot recall}{Precision + recall} \quad (5.3)$$

$$\text{Cohen's Kappa(KAP)} = \frac{(P_o - P_e)}{1 - P_e} \quad (5.4)$$

In Cohen's kappa: P_o is the relative observed agreement among raters (same as accuracy in binary classification). P_e is the hypothetical probability of chance agreement. The precise calculation of P_e depends on the counts of each category in your data.

In a practical exploration, we investigated the efficiency of various CNN architectures for processing brain stroke images. To validate and train the pretrained models, we divided the dataset, with 80% being used for training and 20% for validation. Figs. 5.7, 5.9 and 5.10 illustrate the training losses and corresponding accuracies for the various pretrained CNN models. In addition, we show the confusion matrices for these models in Figs. 5.11, 5.12, and 5.13. Finally, the ROCs of the pretrained CNN models are presented in Figs. 5.8, 5.14, and 5.15. This experimental approach provided a comprehensive analysis of the performance of pretrained models for classifying stroke images (Table 5.1).

5.1 Comparison with previous work

Table 5.2 provides a comparison of various research articles using different approaches

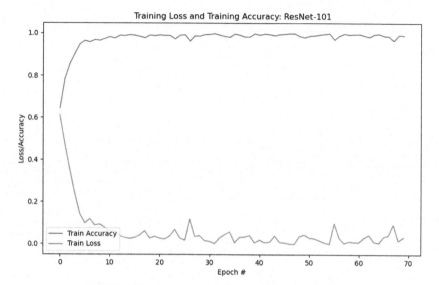

FIGURE 5.7 Training loss and accuracy for ResNet101.

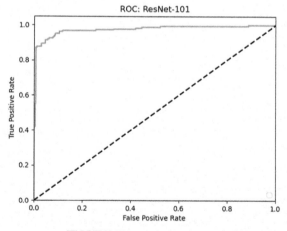

FIGURE 5.8 ROC: ResNet101.

and datasets for stroke detection and their achieved accuracy. Chin et al. (2017) used DL with CNN on a CT image dataset and obtained an accuracy of 90%. Karthik et al. (2019) utilized a fully convolutional network (FCN) on an MR image dataset with 4284 images and achieved a 70% accuracy rate. Liu et al. (2019) applied support vector machines (SVMs) to a CT-scan image dataset of 1157 samples, achieving an accuracy of 83.3%.

Gaidhani et al. (2019) employed DL models, in particular LeNet and SegNet, on an MR images dataset with 400 samples, and obtained a high accuracy range between 96% and 97%. Similar to this, Reddy et al. (2022) applied a DL model to a CT image dataset and reached an accuracy rate of 90%.

Contrary to that, we used three DL models, namely MobileNetV2, DenseNet169, and Rest-Net101, on a CT image dataset of 2501 samples. The models yielded different accuracy rates, with MobileNetV2 and DenseNet169 showing an almost identical accuracy of 95.80%, while RestNet101 achieved a slightly lower accuracy of 94.60%. The accuracy range with MobileNet reached as high as 96.62%, demonstrating the effectiveness of the models used.

FIGURE 5.9 Training loss and accuracy for DensNet169.

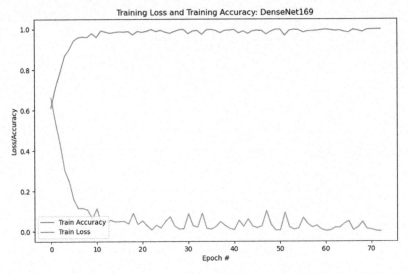

FIGURE 5.10 Training loss and accuracy for MobileNet.

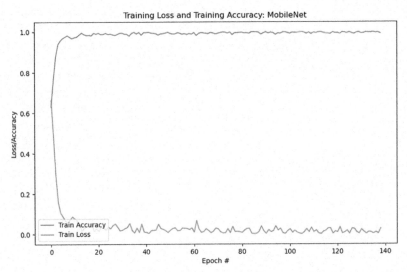

6. Conclusion

Stroke is a complex condition with 150 distinct causes, which requires early diagnosis to prevent long-term effects. CNNs, a type of deep learning model, provide clinical operators with a powerful adjunct to assist with image analysis. This study investigates three CNN models: MobileNetV2, DenseNet169, and RestNet101 for stroke classification from CT images. The leading model based on our tests was MobileNet, which yielded an accuracy, precision, F1-score, and

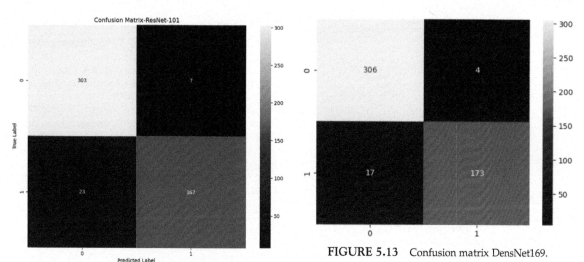

FIGURE 5.13   Confusion matrix DensNet169.

FIGURE 5.11   Confusion matrix ResNet101.

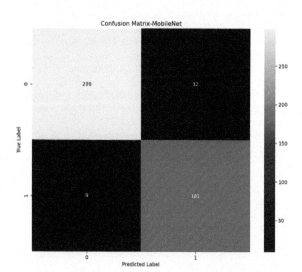

FIGURE 5.12   Confusion matrix MobileNet.

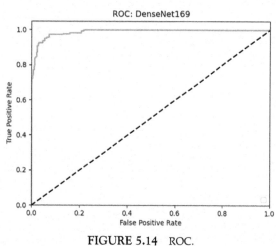

FIGURE 5.14   ROC.

Cohen's kappa of 95.80%, 95.82%, 95.81%, and 91.11%, respectively. MobileNet computational efficiency, coupled with a compact architecture, makes it ideal for real-time processing in resource-constrained environments, such as mobile and Internet of Things (IoT) devices, underscoring its utility in prompt stroke detection. DenseNet169 and RestNet101 also produced commendable results. These results highlight the potential of CNNs in classifying stroke cases, thus facilitating early and accurate diagnosis, yet also highlight the necessity for continued model optimization for specific applications like stroke classification.

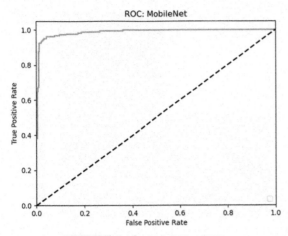

FIGURE 5.15   MobileNet: ROC.

TABLE 5.1   Stroke classification performance evaluation results.

CNN models	ACC (%)	Precision (%)	F1-score (%)	Kappa (%)
MobileNet	95.80	95.82	95.81	91.11
DenseNet169	95.80	95.88	95.77	90.97
RestNet101	94.00	94.10	93.94	87.06

TABLE 5.2   Comparison to previous studies.

References	Methods	Image datasets	Accuracy
Chin et al. (2017)	DL (CNN)	CT, image	90%
Karthik et al. (2019)	Fully convolutional network (FCN)	MR, 4284 samples	70%
Liu et al. (2019)	SVM	CT, 1157 samples	83.3%
Gaidhani et al. (2019)	DL (LeNet and Seg-Net)	MR, 400 samples	96%–97%
Reddy et al. (2022)	DL	CT, images	90%
This study	MobileNetV2	CT, 2501 samples	95.80%–96.62%
This study	DenseNet169	CT, 2501 samples	95.80%
This study	RestNet101	CT, 2501 samples	94.60%

References

Al-Qazzaz, N. K., Alyasseri, Z. A. A., Abdulkareem, K. H., Ali, N. S., Al-Mhiqani, M. N., & Guger, C. (2021). EEG feature fusion for motor imagery: A new robust framework towards stroke patients rehabilitation. *Computers in Biology and Medicine, 137*, 104799.

Badriyah, T., Sakinah, N., Syarif, I., & Syarif, D. R. (2020). Machine learning algorithm for stroke disease classification. In *2020 international conference on electrical, communication, and computer engineering (ICECCE)* (pp. 1–5).

Bento, M., Souza, R., Salluzzi, M., Rittner, L., Zhang, Y., & Frayne, R. (2019). Automatic identification of atherosclerosis subjects in a heterogeneous MR brain imaging data set. *Magnetic Resonance Imaging, 62*, 18–27.

CDC. (2023). *Preventing stroke deaths.* https://www.cdc.gov/vitalsigns/stroke/index.html. (Accessed 1 July 2023).

Chin, C.-L., Lin, B.-J., Wu, G.-R., Weng, T.-C., Yang, C.-S., Su, R.-C., & Pan, Y.-J. (2017). An automated early ischemic stroke detection system using CNN deep learning algorithm. In *2017 IEEE 8th International conference on awareness science and technology (iCAST)* (pp. 368–372).

Diker, A., Elen, A., & Subasi, A. (2023). Brain stroke detection from computed tomography images using deep learning algorithms. In *Applications of artificial intelligence in medical imaging* (pp. 207–222). Elsevier.

Gaidhani, B. R., Rajamenakshi, R., & Sonavane, S. (2019). Brain stroke detection using convolutional neural network and deep learning models. In *2019 2nd International conference on intelligent communication and computational techniques* (pp. 242–249). ICCT.

Gautam, A., & Raman, B. (2021). Towards effective classification of brain hemorrhagic and ischemic stroke using CNN. *Biomedical Signal Processing and Control, 63*, 102178.

He, K., Zhang, X., Ren, S., & Sun, J. (2016). Deep residual learning for image recognition. In *Proceedings of the IEEE conference on computer vision and pattern recognition* (pp. 770–778).

Herzog, L., Murina, E., Dürr, O., Wegener, S., & Sick, B. (2020). Integrating uncertainty in deep neural networks for MRI based stroke analysis. *Medical Image Analysis, 65*, 101790.

Hopkins, J. (2023). *Types of stroke.* https://www.hopkinsmedicine.org/health/conditions-and-diseases/stroke/types-of-stroke.html. (Accessed 1 July 2023).

Huang, G., Liu, Z., Van Der Maaten, L., & Weinberger, K. Q. (2017). Densely connected convolutional networks. In *Proceedings of the IEEE conference on computer vision and pattern recognition* (pp. 4700–4708).

Kakkar, P., Kakkar, T., Patankar, T., & Saha, S. (2021). Current approaches and advances in the imaging of stroke. *Disease Models & Mechanisms, 14*(12), dmm048785.

Kanchana, R., & Menaka, R. (2020). Ischemic stroke lesion detection, characterization and classification in CT images with optimal features selection. *Biomedical Engineering Letters, 10*, 333–344.

Karthik, R., Gupta, U., Jha, A., Rajalakshmi, R., & Menaka, R. (2019). A deep supervised approach for ischemic lesion segmentation from multimodal MRI using fully convolutional network. *Applied Soft Computing, 84,* 105685.

Liu, J., Xu, H., Chen, Q., Zhang, T., Sheng, W., Huang, Q., Song, J., Huang, D., Lan, L., Li, Y., & Chen, W. (2019). Prediction of hematoma expansion in spontaneous intracerebral hemorrhage using support vector machine. *EBioMedicine, 43,* 454–459.

Peixoto, S. A., & Rebouças Filho, P. P. (2018). Neurologist-level classification of stroke using a structural co-occurrence matrix based on the frequency domain. *Computers & Electrical Engineering, 71,* 398–407.

Raghavendra, U., Pham, T.-H., Gudigar, A., Vidhya, V., Rao, B. N., Sabut, S., Wei, J. K. E., Ciaccio, E. J., & Acharya, U. R. (2021). Novel and accurate non-linear index for the automated detection of haemorrhagic brain stroke using CT images. *Complex & Intelligent Systems, 7,* 929–940.

Rahman, A. (2023). *Brain stroke CT image dataset.* https://www.kaggle.com/datasets/afridirahman/brain-stroke-ct-image-dataset. (Accessed 29 July 2023).

Reboucas Filho, P. P., Sarmento, R. M., Holanda, G. B., & de Alencar Lima, D. (2017). New approach to detect and classify stroke in skull CT images via analysis of brain tissue densities. *Computer Methods and Programs in Biomedicine, 148,* 27–43.

Reddy, M. K., Kovuri, K., Avanija, J., Sakthivel, M., & Kaleru, S. (2022). Brain stroke prediction using deep learning: A CNN approach. In *2022 4th international conference on inventive research in computing applications (ICIRCA)* (pp. 775–780). https://doi.org/10.1109/ICIRCA54612.2022.9985596

Sandler, M., Howard, A., Zhu, M., Zhmoginov, A., & Chen, L.-C. (2018). Mobilenetv2: Inverted residuals and linear bottlenecks. In *Proceedings of the IEEE conference on computer vision and pattern recognition* (pp. 4510–4520).

Soltanpour, M., Greiner, R., Boulanger, P., & Buck, B. (2021). Improvement of automatic ischemic stroke lesion segmentation in CT perfusion maps using a learned deep neural network. *Computers in Biology and Medicine, 137,* 104849.

Sung, S.-F., Hung, L.-C., & Hu, Y.-H. (2021). Developing a stroke alert trigger for clinical decision support at emergency triage using machine learning. *International Journal of Medical Informatics, 152,* 104505.

Vargas, J., Spiotta, A., & Chatterjee, A. R. (2019). Initial experiences with artificial neural networks in the detection of computed tomography perfusion deficits. *World Neurosurgery, 124,* e10–e16.

WHO. (2023). *Stroke, cerebrovascular accident.* https://www.emro.who.int/health-topics/stroke-cerebrovascular-accident/index.html. (Accessed 1 July 2023).

Brain tumor detection using deep learning from magnetic resonance images

Eman Hassanain[1], Abdulhamit Subasi[2,3]

[1]Turku Brain and Mind Center, Faculty of Medicine, University of Turku, Turku, Finland; [2]Institute of Biomedicine, Faculty of Medicine, University of Turku, Turku, Finland; [3]Department of Computer Science, College of Engineering, Effat University, Jeddah, Saudi Arabia

1. Introduction

Brain tumors significantly cause mortality and morbidity in children and adults worldwide. They are one of the most complex and challenging medical conditions to treat due to their heterogeneity and origin in the brain (Lapointe et al., 2018). Generally, they refer to the growth of abnormal cells that form a mass or lump within the brain tissue. They can be either malignant or benign, and the symptoms usually vary based on the tumor's type, size, and location in the brain (Lapointe et al., 2018; Zeinalkhani et al., 2018). However, common symptoms include headaches, seizures, vision problems, speech difficulties, memory loss, cognitive impairments, and difficulty with coordination or balance (Alther et al., 2020). Brain tumors are commonly prevalent among children, and they are considered the leading tumor-type cause of death in females under 20 and males under 40 (Abolanle AA et al., 2020; Miller et al., 2021). Their prevalence was around 14.5 per 100,000 people worldwide in 2020. Additionally, approximately 83,000 new cases were diagnosed in the United States in 2021, 29% of malignant origin, and about 18,200 people died from brain tumors (Schaff & Mellinghoff, 2023).

The World Health Organization (WHO) reported that brain tumors have four grades, ranging from I to IV. Grade I tumors are the least malignant, while grade IV tumors are the most aggressive and have the poorest prognosis and a survival rate of fewer than 15 months with current treatments (Mabray et al., 2015). Brain tumors can also be classified as primary or secondary based on their origin. Primary brain tumors originate in the brain tissue. In contrast, secondary/metastatic brain tumors are formed as the cancer cells break away from their primary location, such as the lungs or the breasts, and propagate to other body parts, including the brain (Dhanwani & Bartere, 2014; Patil et al., 2017; Sravanthi Peddinti et al., 2021).

Although primary brain tumors have many types, the most common ones are astrocytoma, meningioma, schwannoma, and oligodendroglioma (Sravanthi Peddinti et al., 2021).

Astrocytoma is the most prevalent type that arises from astrocytes, accounting for about two-thirds of all cases. It is typically slow growing and can be classified into four grades, with grade IV, glioblastoma multiforme (GBM), being the most aggressive (Kabitha et al., 2013). Meningioma arises from the meninges that protect the brain and spinal cord. It has a good prognosis, with many patients being cured by surgically removing the tumor (Lapointe et al., 2018). Schwannoma arises from Schwann cells, and it is usually benign and slow growing, with acoustic neuroma being the most common type of schwannoma (Greene & Al-Dhahir, 2023). Finally, oligodendroglioma arises from the oligodendrocytes, and they are a slow-growing type that typically occurs in adults, usually between the ages of 35 and 50. They have a better prognosis than GBM, with patients surviving an average of 5–10 years (Kabitha et al., 2013; Liu et al., 2019). Medulloblastoma is the most common in children, accounting for about 20% of all primary brain tumors. This type is highly malignant and can spread rapidly to other parts of the brain or spinal cord (Schaff & Mellinghoff, 2023). Astrocytoma can also occur in children in grades I and II (Kabitha et al., 2013). Furthermore, infants are also subject to brain tumors with primitive neuroectodermal tumor (PNET) as the common one that arises from the embryonic cells in the brain (Simone et al., 2021).

The etiology of brain tumors is not comprehensively elucidated. However, several risk factors for the tumors have been identified. These risk factors are age, gender, a weakened immune system, certain genetic disorders, a family history of brain tumors, and environmental factors. Environmental factors include exposure to chemicals, ionizing radiation, infections like HIV/AIDS and the Epstein–Barr virus, alcohol, smoking, and high-fat diets (Kabitha et al., 2013; Kalan Farmanfarma et al., 2019; Ostrom et al., 2021). Moreover, it was found that hormones may play a role in the development of tumors,

as many researchers have observed an increased risk among women exposed to higher estrogen levels, particularly during hormone replacement therapy (Kalan Farmanfarma et al., 2019).

Despite the fast-paced advancements in brain cancer research, choosing the most effective treatment is still daunting due to its intricate and diverse nature. Treatment options usually involve the standard approaches, including surgery, radiation, and chemotherapy, alone or in combination (Kabitha et al., 2013; Lapointe et al., 2018). On the other hand, innovative treatments such as immunotherapy and targeted molecular therapy are being explored to improve patient outcomes (Lapointe et al., 2018). In addition, it should be noted that the brain tumors' location poses a significant challenge for treatment as they can be in critical areas of the brain that control essential bodily functions. Early detection is, thus, vital for better management of the disease, and the diagnosis process should be performed meticulously. As such, the diagnosis usually depends on the use of multiple imaging techniques, such as computed tomography (CT), magnetic resonance imaging (MRI), and positron emission tomography (PET), of which MRI is widely used (Abolanle AA et al., 2020; Kabitha et al., 2013; Schaff & Mellinghoff, 2023).

With the emergence of artificial intelligence (AI), detecting brain tumors has improved, making it an indispensable tool for neurosurgeons and neuro-oncologists. AI can analyze the brain's MR images and identify the tumor's presence (Jia & Chen, 2020). This analysis is done through machine learning and deep learning, where the models are trained on a vast dataset of MR images consisting of healthy brains and brains with tumors. The AI system then uses this information to recognize patterns and identify tumors in new MR images (Ostrom et al., 2019). One of the advantages of AI is its capability to analyze vast amounts of data accurately. This advantage is significant in time-sensitive cases where actions are required in shorter times, such as emergencies or where the tumor proliferates (Jiang et al., 2017). AI can also identify the type of tumor and its location to help make informed decisions about treatment options. For example, the tumor's location may affect the approach used for surgery, and the type of tumor may determine the course of treatment, such as radiation or chemotherapy. AI can also monitor the progression of tumors by analyzing changes in MR images, which can assess the effectiveness of treatment and any changes in the size or location of the tumor in the brain (Jiang et al., 2017; Nadeem et al., 2020; Patil et al., 2017).

Deep learning is a subset of machine learning, and it has been applied to medical image analysis, including detecting brain tumors from MR images. It uses neural networks to learn and make predictions from data that could indicate the presence of the tumor (Sravanthi Peddinti et al., 2021). Deep learning techniques, namely convolutional neural networks (CNNs), recurrent neural networks (RNNs), deep neural networks (DNNs), and artificial neural networks (ANNs), have been used for the prediction, detection, and segmentation of brain tumors (Jayanthi & Iyyanki, 2020). These techniques comprise classification networks that categorize the MR images into two distinct categories, one depicting the presence of a tumor and the other its absence. Following this categorizing, transfer learning is employed to retrain the models to detect brain tumors (Sangeetha et al., 2022).

This chapter aims to analyze the uses and effectiveness of deep learning models in detecting brain tumors using MRI. The chapter will critically evaluate the performance of various deep learning models and review recent studies that have attempted to enhance the accuracy of detecting brain tumors using deep learning. Finally, the chapter will provide insight into the deep learning models' potential in detecting and diagnosing brain tumors.

2. Background/literature review

Brain tumors are complicated and multifaceted tumors requiring extensive care and treatment. These tumors can manifest in various forms and severity levels, each requiring individualized attention from multiple medical professionals. In addition, detecting brain tumors requires effective diagnosis, which involves identifying their presence and understanding their specific location in the brain and associated symptoms or other complications. Recently, deep learning techniques have shown promising results in automating the detection and classification of brain tumors from MR images (Dhanwani & Bartere, 2014; Nadeem et al., 2020).

Several studies have demonstrated the potential uses of deep learning in brain tumor detection. Havaei et al. (2017) presented a Brain Tumor Segmentation (BRATS) framework to accurately segment distinct types of brain tumors extracted from MR images. The BRATS utilized a CNN to learn features from MR images and generate segmentation maps for tumor detection. CNN models successfully segmented the whole brain rapidly in a period ranging from 25 s to 3 min. Furthermore, Jia and Chen (2020) proposed a novel approach named Fully Automatic Heterogenous Segmentation using Support Vector Machine (FAHS-SVM) for detecting and segmenting brain tumors. The experimental findings exhibit a remarkable accuracy rate of 98.51% in identifying normal and abnormal tissues from MR images. Finally, Mohamed Shakeel et al. (2019) presented a classification system with the Machine Learning-Based Back Propagation Neural Network (MLBPNN), which could improve the identification of the location of the tumors. They employed infrared sensor imaging technology to examine the method. Specifically, a wireless infrared imaging sensor has been utilized to facilitate health monitoring and optimize ultrasound measurements, especially for elderly patients.

MobileNetV2 has exhibited remarkable accuracy rates in brain tumor classification by implementing transfer learning techniques. For example, Lu et al. (2020) used MobileNetV2 with transfer learning to accurately classify brain tumors, achieving an accuracy of 96%. Similarly, Maqsood et al. (2022) innovative approach for multimodal MRI-based brain tumor segmentation and classification using MobileNetV2 demonstrated superior levels of efficacy with accuracy rates of 97.47% and 98.92%. Tazin et al. (2021) also reached an accuracy rate of 92% using a novel method combining CNN and MobileNetV2.

Rasool et al. (2022) presented a hybrid deep learning model to differentiate between three types of brain tumors from MRI scans: glioma, meningioma, and pituitary tumors. The proposed method combined unsupervised SVM classification and SqueezeNet, a pretrained CNN model. The experimental results denoted a special overall accuracy rate of 96.5%. On the other hand, when utilizing the SqueezeNet method as a feature extractor while applying SVM classifier, the accuracy rate increased to 98.7% with 98.5% F1-score, 98.3% recall, and 98.7% precision rates.

Mehrotra et al. (2020) compared several deep learning models' efficiency in classifying brain tumors as malignant or benign. The data comprised 696 images, of which 472 were malignant and 224 were benign. Experiment results showed that AlexNet achieved the highest accuracy with a rate of 99.04% compared to SqueezeNet (98.56%), GoogleNet (96.65%), ResNet101 (94.74%), and ResNet50 (90.91%) using the stochastic gradient descent with momentum (SGDM) optimizer. Zeinalkhani et al. (2018) combined the K-means clustering algorithm and genetic algorithms for brain tumor

segmentation to help diagnose brain tumors. The dataset consisted of 10 images and achieved an accuracy rate of 96.99%.

ResNet is another deep learning model used for brain tumor detection and classification. Zahid et al. (2022) used ResNet101 with transfer learning to classify between T1, T2, FLAIR, and T1CE images with 130,200 images. Their methodology included the following: a preprocessing stage to normalize the images and convert them into multichannel; the RestNET was then fine-tuned for classification using transfer learning techniques; using differential evolution and particle swarm intelligence to calculate two feature vectors, which were combined to get a fused feature vector; and finally, PCA was applied to get the final optimized vector used as an input to be classified. The proposed method achieved accuracies of 92.6%, 93.9%, 94.4%, and 95.4% on Fine tree, narrow neural network, medium neural network, and wide neural network classifiers, respectively. Ghosal et al. (2019) proposed an improved deep learning method for brain tumor classification using 3064 T1-weighted images, where ResNet101 was used based on CNN, achieving an overall 93.83% accuracy, 97.15% specificity, and 93.84% sensitivity. Deshpande et al. (2021), Li et al. (2021), and Sahaai et al. (2022, p. 020014) proposed a ResNet50 model for brain tumors classification using a convolutional neural network, achieving an overall accuracy of 98.14%, 98.59%, and 95.3%, respectively.

Other studies have utilized the VGG method to assess its uses in detecting and classifying brain tumors. For example, Majib et al. (2021) proposed a deep learning-based framework to accelerate the time for detecting and segmenting brain tumors for radiologists. They used hybrid CNN models, in which VGG-SCNet achieved the highest performance with 99.2% precision, 99.1% recall, and 99.2% F1 score. When classifying the four grades of brain tumors, Agus et al. (2022) used VGG-16 on 520 MR images and achieved an accuracy rate of 100%. Similarly, Gupta et al. (2014) utilized VGG-16 to predict brain tumors using binary classification (whether or not there is a tumor). The model achieved an accuracy rate of 90% on the test data and 86% on the validation data. The model also achieved remarkable precision and recall rates of about 93% and 90%, respectively.

Furthermore, deep learning has been used to develop computer-aided diagnosis (CAD) systems. As proposed by several studies, CAD has been proven to detect and predict brain tumors effectively and accurately (e.g., Ahuja et al., 2021; El-Dahshan et al., 2014; Kadhim et al., 2022; Naseer et al., 2021). CAD was developed to assist medical professionals in decision-making by providing the necessary information to help diagnose tumors and detect their types. This system has achieved high accuracy of 99.61% (Kadhim et al., 2022), 98.8% (Naseer et al., 2021), 99% (El-Dahshan et al., 2014), and 98.72% (Ahuja et al., 2021). Additionally, Sarhan (2020) proposed a novel CAD system based on CNN and Discrete Wavelet Transform (DWT) to classify diverse types of brain tumors from MR images. The proposed CAD system achieved a high accuracy of 99.3% in tumor classification and could assist radiologists in diagnosis and treatment planning. Arunkumar et al. (2020) also presented a CAD system to detect brain tumors. To achieve high accuracy, they performed a Fourier transform technique for the preprocessing stage, and their algorithm employed pixel-level features for regions segmentation, and histogram-of-oriented gradients (HOGs).

Another recent development in deep learning is generative adversarial networks (GANs). GANs are algorithms consisting of a generator and discriminator to generate new data that closely resemble the original data (Goodfellow et al., 2014). For example, Li et al. (2020) proposed a GAN-based framework named TumorGAN for brain tumor segmentation from MR images. This method generated synthetic MR images used later to train a CNN for tumor segmentation, and it achieved high accuracy in tumor segmentation. It could improve the performance of deep learning models for brain tumor detection.

3. Artificial intelligence for brain tumors detection

3.1 Machine learning techniques

The use of machine learning in brain tumor detection is in its early stages, and much research is needed. However, the potential benefits of this technology are clear. Machine learning algorithms can analyze larger datasets, enabling faster and more accurate diagnoses. This can lead to earlier treatment and better patient outcomes (Jiang et al., 2017; Koh et al., 2022). Machine learning involves developing models to learn from data and then predicting new, unseen examples. The process begins with collecting and preparing the data, which consists of selecting the relevant features and cleaning the data to remove any noise or errors. Once the data are ready, it is divided into a training set to train the data and a test set to evaluate its performance (Koh et al., 2022; Linardatos et al., 2020; Sarker, 2021b).

In detail, machine learning involves several steps, starting with data preprocessing, then feature extraction, model training, and model evaluation. Data preprocessing refers to the step that includes cleaning and preparing data for machine learning model training. This includes removing irrelevant or duplicated data, handling missing values, and normalizing the data to ensure all features have a similar scale (Al-jabery et al., 2020). Once the data are preprocessed, the next step is to extract relevant features from it, through which machine learning models make predictions. This step involves selecting essential features, transforming them into a suitable format, and reducing the dimensionality of the data to make it more manageable (Gite et al., 2023). The next step is model training, in which the machine learning model is trained using a subset of the preprocessed data. Next, the model is built to identify patterns and relationships in the data and learn from them. This involves selecting an appropriate algorithm, setting its parameters, and training the model using an iterative process until it can accurately predict outcomes on new data. After training the model, it is tested using a separate data set. This step is critical to ensure the model is generalizable and not just overfitting the training data. Finally, the model's performance is evaluated based on various metrics, such as accuracy, F1 score, precision, and recall. Once the model is trained and assessed, it can predict new data. This involves feeding them into the model and using the trained model to make predictions based on the features of the new data (Hicks et al., 2022; Koh et al., 2022; Sarker, 2021b).

There are several machine learning algorithms, each with strengths and weaknesses. For example, supervised learning algorithms, such as linear regression, logistic regression, decision trees, random forests, and neural networks, are used with labeled datasets. Each input is associated with a known target or output. The goal of the model is to learn to predict the target for new data. On the other hand, unsupervised learning algorithms are used with unlabeled datasets with no unclear targets or outputs. The model aims to learn the underlying structure or specific patterns that could be found in the images (Sarker, 2021b). Clustering algorithms are a typical example of unsupervised learning, where the goal is to group similar models based on their features (Talabis et al., 2015).

Additionally, reinforcement learning algorithms, such as game playing, robotics, and autonomous driving, are used when the model interacts with an environment and receives feedback on its actions (Mahmud et al., 2018). Finally, hybrid approaches combine different types of machine learning algorithms. For example, semisupervised learning uses some labeled and many unlabeled data. Then, the model can leverage the labeled data to predict the unlabeled data and learn from the feedback (Chen et al., 2021).

3.2 Deep learning

Deep learning algorithms are expected to help detect brain tumors with high accuracy and speed, providing physicians with a powerful tool to aid in diagnosing this critical disease. These algorithms use ANNs to learn patterns in the data and identify features indicative of a brain tumor (Hashemzehi et al., 2020). Deep learning is based on the principle of backpropagation, a technique for adjusting the neural network weights, and it is used to minimize the error between the predicted and actual output (Jayanthi & Iyyanki, 2020; Mahmud et al., 2018). The mechanism of deep learning is based on using ANNs, which are composed of layers of interconnected nodes (Hashemzehi et al., 2020; Sarker, 2021a).

The process of deep learning involves several key steps. Firstly, data preparation is required, which involves cleaning and formatting the data to ensure that it is consistent and can be adequately interpreted by the algorithm. Next, the model architecture is designed, selecting the number of layers, nodes in each layer, and the activation function to be used. Once the neural network's architecture has been created, the model is trained on the dataset using backpropagation to adjust the neural network weights and lower the errors (Jia & Chen, 2020; Sarker, 2021a,b; Taye, 2023b). After the model has been trained, it is tested on a separate dataset to evaluate its performance. This helps ensure that the model can be used with other data. Finally, after testing and validating the model, it can be deployed in a real-world application. This step involves integrating the model into a more extensive system and using it to make predictions or classifications on new data (Dhanwani & Bartere, 2014).

ANNs are algorithms modeled and trained after the human brain's structure and function. They consist of layers of interconnected nodes; these nodes can learn to recognize patterns and perform the predictions from the input data (Chaplot et al., 2006). Nevertheless, DNNs have multiple hidden layers between the input and output layers. As a result, they can learn increasingly complex representations of the input data, making them well suited for image recognition tasks such as brain tumor detection. By stacking multiple layers of neurons, DNNs can learn to identify subtle patterns and features within the image that may indicate a tumor (Taye, 2023b).

3.3 Convolutional neural networks

CNNs are algorithms widely used in medical analysis. They have revolutionized the field of image analysis by providing state-of-the-art results on multiple image recognition tasks, including object detection, image classification, and image segmentation, by learning an image's hierarchical representations and handling variations in the input data (Rasool et al., 2022). The critical feature of CNNs is that they can identify local patterns in an image, such as edges, corners, and textures, and use them to make predictions about the entire image. They achieve that through convolutional layers, which apply filters to the input image to generate a feature map. Each filter is a small matrix of values convolved with the input image to have a new set of values highlighting specific features in the image (Agus et al., 2022).

CNNs typically consist of multiple layers, such as convolutional, pooling, and fully connected layers. Convolutional layers are the backbone of the network, and their outputs are usually passed through pooling layers, which downsample the feature maps to reduce the computational load of the network. Finally, fully connected layers are used at the network's end to produce the final output, such as a probability distribution over the different classes in an image classification task (Agus et al., 2022; Singh et al., 2021). During training, the weights of the filters in the convolutional layers are adjusted, which is done using a variant of stochastic gradient descent, where the weights are updated in small batches according to the gradients of the loss function (Taye, 2023a).

One of the critical advantages of CNNs is that they can learn hierarchical representations of an image. This means that the lower layers of the network learn to identify simple features, such as edges and textures, while the higher layers learn to identify more complex features, such as shapes and objects. In addition, by stacking multiple convolutional and pooling layers, CNNs can learn increasingly abstract representations of the input image to make accurate predictions about the image contents (Singh et al., 2021; Taye, 2023a). Another advantage of CNNs is their ability to handle variations in the input image, such as changes in lighting, scale, and orientation. This is achieved through data augmentation techniques, such as flipping, rotating, and scaling the input images during training. By exposing the network to a wide range of variations in the input data, the network can generalize better to new images (Alzubaidi et al., 2021). An example of the 8-layer CNN model used in the chapter can be found below.

8-layer CNN model code.

```
import numpy as np
import pandas as pd
import cv2
from PIL import Image
import scipy
import os
import shutil
import itertools
import imutils
import tensorflow as tf
from tensorflow.keras.applications import *
from tensorflow.keras.optimizers import *
from tensorflow.keras.losses import *
from tensorflow.keras.layers import *
from tensorflow.keras.models import *
from tensorflow.keras.callbacks import *
```

```
from tensorflow.keras.preprocessing.image import *
from tensorflow.keras.utils import *
# import pydot

from sklearn.metrics import *
from sklearn.model_selection import *
import tensorflow.keras.backend as K

from tqdm import tqdm, tqdm_notebook
from colorama import Fore
import json
import matplotlib.pyplot as plt
import seaborn as sns
from glob import glob
from skimage.io import *
%config Completer.use_jedi = False
import time
from sklearn.decomposition import PCA
from sklearn.svm import LinearSVC
from sklearn.linear_model import LogisticRegression
from sklearn.metrics import accuracy_score
import lightgbm as lgb
import xgboost as xgb
import matplotlib.pyplot as plt
from sklearn.preprocessing import LabelBinarizer
from sklearn.model_selection import train_test_split
from sklearn.metrics import confusion_matrix
import plotly.graph_objs as go
from plotly.offline import init_notebook_mode, iplot
from plotly import tools
from keras.preprocessing.image import ImageDataGenerator
from keras.applications.vgg16 import VGG16, preprocess_input
from keras import layers
from keras.models import Model, Sequential
from keras.optimizers import Adam, RMSprop
from keras.callbacks import EarlyStopping
from keras.utils.vis_utils import plot_model
```

```
init_notebook_mode(connected=True)
RANDOM_SEED = 123


IMG_PATH = "../input/brain-tumor-detection-mri/Brain_Tumor_Detection"


TRAIN_DIR = 'TRAIN/'
TEST_DIR = 'TEST/'
VAL_DIR = 'VAL/'
IMG_SIZE = (224,224)


X_train, y_train, labels = load_data(TRAIN_DIR, IMG_SIZE)
X_test, y_test, _ = load_data(TEST_DIR, IMG_SIZE)
X_val, y_val, _ = load_data(VAL_DIR, IMG_SIZE)


#Crop images
def crop_imgs(set_name, add_pixels_value=0):
    """
    Finds the extreme points on the image and crops the rectangular out of
them
    """
    set_new = []
    for img in set_name:
        gray = cv2.cvtColor(img, cv2.COLOR_RGB2GRAY)
        gray = cv2.GaussianBlur(gray, (5, 5), 0)

        # threshold the image, then perform a series of erosions +
        # dilations to remove any small regions of noise
        thresh = cv2.threshold(gray, 45, 255, cv2.THRESH_BINARY)[1]
        thresh = cv2.erode(thresh, None, iterations=2)
        thresh = cv2.dilate(thresh, None, iterations=2)

        # find contours in thresholded image, then grab the largest one
        cnts      =      cv2.findContours(thresh.copy(),      cv2.RETR_EXTERNAL,
cv2.CHAIN_APPROX_SIMPLE)
        cnts = imutils.grab_contours(cnts)
        c = max(cnts, key=cv2.contourArea)

        # find the extreme points
```

```
        extLeft = tuple(c[c[:, :, 0].argmin()][0])
        extRight = tuple(c[c[:, :, 0].argmax()][0])
        extTop = tuple(c[c[:, :, 1].argmin()][0])
        extBot = tuple(c[c[:, :, 1].argmax()][0])

        ADD_PIXELS = add_pixels_value
        new_img = img[extTop[1]-ADD_PIXELS:extBot[1]+ADD_PIXELS, extLeft[0]-
ADD_PIXELS:extRight[0]+ADD_PIXELS].copy()
        set_new.append(new_img)

    return np.array(set_new)


X_train_crop = crop_imgs(set_name=X_train)
X_val_crop = crop_imgs(set_name=X_val)
X_test_crop = crop_imgs(set_name=X_test)


#resize images
def preprocess_imgs(set_name, img_size):
    set_new = []
    for img in set_name:
        img = cv2.resize(
            img,
            dsize=img_size,
            interpolation=cv2.INTER_CUBIC
        )
        set_new.append(preprocess_input(img))
    return np.array(set_new)


X_train_prep = preprocess_imgs(set_name=X_train_crop, img_size=IMG_SIZE)
X_test_prep = preprocess_imgs(set_name=X_test_crop, img_size=IMG_SIZE)
X_val_prep = preprocess_imgs(set_name=X_val_crop, img_size=IMG_SIZE)


#Create 8-layer CNN model
from keras.layers import Dense, Dropout, Flatten, Conv2D, MaxPooling2D
model_CNN8 = Sequential()
model_CNN8.add(Conv2D(16,(3,3),padding='same',input_shape
=X_train_prep[0].shape))
model_CNN8.add(BatchNormalization())
```

```
model_CNN8.add(Activation('relu'))
model_CNN8.add(MaxPooling2D(pool_size=(2,2),strides=2,padding = 'same'))


model_CNN8.add(Conv2D(32,(3,3),padding='same'))
model_CNN8.add(BatchNormalization())
model_CNN8.add(Activation('relu'))
model_CNN8.add(MaxPooling2D(pool_size=(2,2),strides=2,padding = 'same'))


model_CNN8.add(Conv2D(32,(3,3),padding='same'))
model_CNN8.add(BatchNormalization())
model_CNN8.add(Activation('relu'))
model_CNN8.add(MaxPooling2D(pool_size=(2,2),strides=2,padding = 'same'))


model_CNN8.add(Conv2D(64,(3,3),padding='same'))
model_CNN8.add(BatchNormalization())
model_CNN8.add(Activation('relu'))
model_CNN8.add(MaxPooling2D(pool_size=(2,2),strides=2,padding = 'same'))


model_CNN8.add(Conv2D(64,(3,3),padding='same'))
model_CNN8.add(BatchNormalization())
model_CNN8.add(Activation('relu'))
model_CNN8.add(MaxPooling2D(pool_size=(2,2),strides=2,padding = 'same'))


model_CNN8.add(Conv2D(128,(3,3),padding='same'))
model_CNN8.add(BatchNormalization())
model_CNN8.add(Activation('relu'))
model_CNN8.add(MaxPooling2D(pool_size=(2,2),strides=2,padding = 'same'))


model_CNN8.add(Conv2D(128,(3,3),padding='same'))
model_CNN8.add(BatchNormalization())
model_CNN8.add(Activation('relu'))
model_CNN8.add(MaxPooling2D(pool_size=(2,2),strides=2,padding = 'same'))


model_CNN8.add(Conv2D(256,(3,3),padding='same'))
model_CNN8.add(BatchNormalization())
model_CNN8.add(Activation('relu'))
model_CNN8.add(MaxPooling2D(pool_size=(2,2),strides=2,padding = 'same'))
```

```
model_CNN8.add(Flatten())
model_CNN8.add(Dense(1,activation='sigmoid'))


model_CNN8.summary()


model_CNN8.compile(optimizer='adam',loss='binary_crossentropy',
metrics=['accuracy','AUC'])
#training
EPOCHS = 30
es = EarlyStopping(
    monitor='val_acc',
    mode='max',
    patience=6
)
history = model_CNN8.fit(
    train_generator,
    steps_per_epoch=50,
    epochs=EPOCHS,
    validation_data=validation_generator,
    validation_steps=25,
    callbacks=[es]
)


#get the train, test and validation accuracies
train_predictions = model_CNN8.predict(X_train_prep)
val_predictions = model_CNN8.predict(X_val_prep)
test_predictions = model_CNN8.predict(X_test_prep)


#Binarizing all the outputs by setting the threshold to 0.5
train_predictions = [1 if x>0.5 else 0 for x in train_predictions]
val_predictions = [1 if x>0.5 else 0 for x in val_predictions]
test_predictions = [1 if x>0.5 else 0 for x in test_predictions]


#Calculating the accuracy scores
train_accuracy = accuracy_score(y_train, train_predictions)
val_accuracy = accuracy_score(y_val, val_predictions)
test_accuracy = accuracy_score(y_test, test_predictions)
```

```
print('Train Accuracy = %.4f' % train_accuracy)
print('Val Accuracy = %.4f' % val_accuracy)
print('Test Accuracy = %.4f' % test_accuracy)


#Plotting the confusion matrix on the test data
confusion_mtx = confusion_matrix(y_train, train_predictions)
cm = plot_confusion_matrix(confusion_mtx, classes = list(labels.items()),
normalize=False)


confusion_mtx = confusion_matrix(y_val, val_predictions)
cm = plot_confusion_matrix(confusion_mtx, classes = list(labels.items()),
normalize=False)


confusion_mtx = confusion_matrix(y_test, test_predictions)
cm = plot_confusion_matrix(confusion_mtx, classes = list(labels.items()),
normalize=False)


test_prob_pred = model_CNN8.predict(X_test_prep)


from sklearn import metrics
print('Accuracy     score     is     :',     metrics.accuracy_score(y_test,
test_predictions))
print('Precision     score     is     :',     metrics.precision_score(y_test,
test_predictions, average='weighted'))
print('Recall score is :', metrics.recall_score(y_test, test_predictions,
average='weighted'))
print('F1    Score    is    :',    metrics.f1_score(y_test,    test_predictions,
average='weighted'))
print('ROC     AUC     Score     is     :',     metrics.roc_auc_score(y_test,
test_prob_pred,multi_class='ovo', average='weighted'))
print('Cohen     Kappa     Score:',     metrics.cohen_kappa_score(y_test,
test_predictions))


print('\t\tClassification Report:\n', metrics.classification_report(y_test,
test_predictions))
```

3.4 Transfer learning techniques

Transfer learning, a powerful technique, involves leveraging pretrained models to solve new tasks. It involves reusing a model already trained on an extensive dataset and adapting it to a new, related problem. This technique is widely used in various settings, including image classification, natural language processing (NLP), and speech recognition. It can be a valuable tool for reducing the time and effort required to train models (Singh et al., 2021).

Several types of transfer learning techniques can be used in machine learning. Feature extraction is one of the most straightforward transfer learning techniques, using the pretrained model as a feature extractor. This approach removes the last and adds a new classifier to the extracted features. The classifier is then trained on the new dataset. For example, a pretrained model like VGG-16, originally trained on the ImageNet dataset, can be used as a feature extractor for a new dataset of images. The pretrained model extracts the input image's features, which are then fed to a new classifier for the new task, such as object detection or image classification (Agus et al., 2022; Yamashita et al., 2018). Multitask learning is a transfer learning technique in which a model is trained to perform multiple tasks concurrently. This approach uses a shared feature extractor, the pretrained model, and each task is associated with a separate output layer. This model is then fine-tuned on all the tasks (Zhang et al., 2017). For example, a pretrained model like ResNet can be used for object detection, image segmentation, and image classification. Training the model on multiple tasks can improve the model's overall performance and reduce the required training data (Alzubaidi et al., 2021).

VGG-16 sample code.

```
import numpy as np
import pandas as pd
import cv2
from PIL import Image
import scipy
import os
import shutil
import itertools
import imutils
import tensorflow as tf
from tensorflow.keras.applications import *
from tensorflow.keras.optimizers import *
from tensorflow.keras.losses import *
from tensorflow.keras.layers import *
from tensorflow.keras.models import *
from tensorflow.keras.callbacks import *
from tensorflow.keras.preprocessing.image import *
from tensorflow.keras.utils import *
# import pydot

from sklearn.metrics import *
```

```
from sklearn.model_selection import *
import tensorflow.keras.backend as K


from tqdm import tqdm, tqdm_notebook
from colorama import Fore
import json
import matplotlib.pyplot as plt
import seaborn as sns
from glob import glob
from skimage.io import *
%config Completer.use_jedi = False
import time
from sklearn.decomposition import PCA
from sklearn.svm import LinearSVC
from sklearn.linear_model import LogisticRegression
from sklearn.metrics import accuracy_score
import lightgbm as lgb
import xgboost as xgb
import matplotlib.pyplot as plt
from sklearn.preprocessing import LabelBinarizer
from sklearn.model_selection import train_test_split
from sklearn.metrics import confusion_matrix
import plotly.graph_objs as go
from plotly.offline import init_notebook_mode, iplot
from plotly import tools
from keras.preprocessing.image import ImageDataGenerator
from keras.applications.vgg16 import VGG16, preprocess_input
from keras import layers
from keras.models import Model, Sequential
from keras.optimizers import Adam, RMSprop
from keras.callbacks import EarlyStopping
from keras.utils.vis_utils import plot_model
init_notebook_mode(connected=True)
RANDOM_SEED = 123


IMG_PATH = "../input/brain-tumor-detection-mri/Brain_Tumor_Detection"
```

```
TRAIN_DIR = 'TRAIN/'
TEST_DIR = 'TEST/'
VAL_DIR = 'VAL/'
IMG_SIZE = (224,224)


X_train, y_train, labels = load_data(TRAIN_DIR, IMG_SIZE)
X_test, y_test, _ = load_data(TEST_DIR, IMG_SIZE)
X_val, y_val, _ = load_data(VAL_DIR, IMG_SIZE)


#Crop images
def crop_imgs(set_name, add_pixels_value=0):
    """
    Finds the extreme points on the image and crops the rectangular out of
them
    """
    set_new = []
    for img in set_name:
        gray = cv2.cvtColor(img, cv2.COLOR_RGB2GRAY)
        gray = cv2.GaussianBlur(gray, (5, 5), 0)

        # threshold the image, then perform a series of erosions +
        # dilations to remove any small regions of noise
        thresh = cv2.threshold(gray, 45, 255, cv2.THRESH_BINARY)[1]
        thresh = cv2.erode(thresh, None, iterations=2)
        thresh = cv2.dilate(thresh, None, iterations=2)

        # find contours in thresholded image, then grab the largest one
        cnts     =     cv2.findContours(thresh.copy(),     cv2.RETR_EXTERNAL,
cv2.CHAIN_APPROX_SIMPLE)
        cnts = imutils.grab_contours(cnts)
        c = max(cnts, key=cv2.contourArea)

        # find the extreme points
        extLeft = tuple(c[c[:, :, 0].argmin()][0])
        extRight = tuple(c[c[:, :, 0].argmax()][0])
        extTop = tuple(c[c[:, :, 1].argmin()][0])
        extBot = tuple(c[c[:, :, 1].argmax()][0])
```

```
        ADD_PIXELS = add_pixels_value
        new_img = img[extTop[1]-ADD_PIXELS:extBot[1]+ADD_PIXELS, extLeft[0]-
ADD_PIXELS:extRight[0]+ADD_PIXELS].copy()
        set_new.append(new_img)

    return np.array(set_new)


X_train_crop = crop_imgs(set_name=X_train)
X_val_crop = crop_imgs(set_name=X_val)
X_test_crop = crop_imgs(set_name=X_test)


#resize images
def preprocess_imgs(set_name, img_size):
    set_new = []
    for img in set_name:
        img = cv2.resize(
            img,
            dsize=img_size,
            interpolation=cv2.INTER_CUBIC
        )
        set_new.append(preprocess_input(img))
    return np.array(set_new)


X_train_prep = preprocess_imgs(set_name=X_train_crop, img_size=IMG_SIZE)
X_test_prep = preprocess_imgs(set_name=X_test_crop, img_size=IMG_SIZE)
X_val_prep = preprocess_imgs(set_name=X_val_crop, img_size=IMG_SIZE)


#create VGG-16 model
from keras.applications import VGG16
from sklearn import metrics

base_model = VGG16(input_shape=(224,224,3), include_top=False)
model=Sequential()
model.add(base_model)
model.add(Flatten())
model.add(BatchNormalization())
model.add(Dense(256,kernel_initializer='he_uniform'))
model.add(BatchNormalization())
model.add(Activation('relu'))
model.add(Dropout(0.5))
model.add(Dense(1,activation='sigmoid'))
```

```
for layer in base_model.layers:
    layer.trainable = False


model.compile(
    loss='binary_crossentropy',
    optimizer='adam',
    metrics=['accuracy' , 'AUC']
)

model.summary()


#training

EPOCHS = 30
es = EarlyStopping(
    monitor='val_acc',
    mode='max',
    patience=6
)
history = model.fit(
    train_generator,
    epochs=EPOCHS,
    validation_data=validation_generator,
    callbacks=[es]
)

# validate on train set

predictions = model.predict(X_train_prep)
predictions = [1 if x>0.5 else 0 for x in predictions]

accuracy = accuracy_score(y_train, predictions)
print('Train Accuracy = %.4f' % accuracy)

confusion_mtx = confusion_matrix(y_train, predictions)
cm = plot_confusion_matrix(confusion_mtx, classes = list(labels.items()),
normalize=False)

# validate on val set

predictions = model.predict(X_val_prep)
predictions = [1 if x>0.5 else 0 for x in predictions]

accuracy = accuracy_score(y_val, predictions)
print('Val Accuracy = %.4f' % accuracy)

confusion_mtx = confusion_matrix(y_val, predictions)
cm = plot_confusion_matrix(confusion_mtx, classes = list(labels.items()),
normalize=False)

# validate on test set

predictions = model.predict(X_test_prep)
predictions = [1 if x>0.5 else 0 for x in predictions]
```

```
accuracy = accuracy_score(y_test, predictions)
print('Test Accuracy = %.4f' % accuracy)

confusion_mtx = confusion_matrix(y_test, predictions)
cm = plot_confusion_matrix(confusion_mtx, classes = list(labels.items()),
normalize=False)

prob_pred = model.predict(X_test_prep)

print('Accuracy score is :', metrics.accuracy_score(y_test, predictions))
print('Precision score is :', metrics.precision_score(y_test, predictions,
average='weighted'))
print('Recall score is :', metrics.recall_score(y_test, predictions,
average='weighted'))
print('F1 Score is :', metrics.f1_score(y_test, predictions,
average='weighted'))
print('ROC AUC Score is :', metrics.roc_auc_score(y_test,
prob_pred,multi_class='ovo', average='weighted'))
print('Cohen Kappa Score:', metrics.cohen_kappa_score(y_test, predictions))

print('\t\tClassification Report:\n', metrics.classification_report(y_test,
predictions))
```

3.5 Deep feature extraction

Deep feature extraction involves extracting high-level features from raw data. This approach transforms input data, such as images or speech signals, into a feature vector representation that a machine learning algorithm can quickly process. The technique involves using DNNs with multiple layers, each performing operations on the input data, gradually transforming it into a more abstract and higher-level representation. The final layer output of the neural network is a feature vector representation of the input data. This representation captures the input data's essential features relevant to the task. The feature vector can then be fed into an algorithm for further processing, such as classification or clustering (Sangeetha et al., 2022; Sethuram Rao & Vydeki, 2018).

Deep feature extraction has several advantages over traditional feature extraction techniques. First, deep learning models can learn more complex and abstract representations of the input data, improving accuracy on tasks like image classification and object recognition. Second, these models can be trained, allowing the feature extraction and algorithms to be optimized together for better performance. Finally, deep feature extraction is computationally efficient, allowing it to be used on large datasets and in real-time applications (Hashemzehi et al., 2020; Sethuram Rao & Vydeki, 2018). In addition, it has been successfully applied in several tasks, such as object detection, facial recognition, and image segmentation. For example, in object detection, deep feature extraction is used to extract features from the input image and then fed into an object detection algorithm to identify the location, type, and/or size of the objects in the image. Similarly, in facial recognition, features are extracted from images of faces, which are then compared to a database of known faces to identify the person (Nadeem et al., 2020; Taşcı, 2023). The used Python code for different deep feature extraction models is given below.

Classification models for Deep Feature extraction.

```python
from sklearn.pipeline import make_pipeline
from sklearn.pipeline import Pipeline
from sklearn.neural_network import MLPClassifier
from sklearn import metrics
from xgboost import XGBClassifier
from sklearn.ensemble import AdaBoostClassifier,RandomForestClassifier,
BaggingClassifier
from sklearn.neighbors import KNeighborsClassifier
from sklearn.svm import SVC


names = [
        "K Nearest Neighbour Classifier",
        'SVM',
        "Random Forest Classifier",
        "AdaBoost Classifier",
        "XGB Classifier",
        "Bagging",
        "ANN Classifier"
         ]
classifiers = [
    KNeighborsClassifier(),
    SVC(probability = True),
    RandomForestClassifier(),
    AdaBoostClassifier(),
    XGBClassifier(),
    BaggingClassifier(),
    MLPClassifier()
        ]
zipped_clf = zip(names,classifiers)
def     classifier_summary(pipeline,     X_train,     y_train,     X_val,
y_val,X_test,y_test):
    sentiment_fit = pipeline.fit(X_train, y_train)

    y_pred_train= sentiment_fit.predict(X_train)
    y_pred_val = sentiment_fit.predict(X_val)
    y_pred_test = sentiment_fit.predict(X_test)
```

```
    y_pred_train = [1 if x>0.5 else 0 for x in y_pred_train]

    y_pred_val = [1 if x>0.5 else 0 for x in y_pred_val]

    y_pred_test = [1 if x>0.5 else 0 for x in y_pred_test]


    train_accuracy = np.round(accuracy_score(y_train, y_pred_train),4)*100
    train_precision    =    np.round(precision_score(y_train,    y_pred_train,
average='weighted'),4)
    train_recall       =      np.round(recall_score(y_train,     y_pred_train,
average='weighted'),4)
    train_F1        =           np.round(f1_score(y_train,       y_pred_train,
average='weighted'),4)
    train_kappa = np.round(cohen_kappa_score(y_train, y_pred_train),4)


    val_accuracy = np.round(accuracy_score(y_val, y_pred_val),4)*100
    val_precision      =      np.round(precision_score(y_val,      y_pred_val,
average='weighted'),4)
    val_recall         =       np.round(recall_score(y_val,        y_pred_val,
average='weighted'),4)
    val_F1 = np.round(f1_score(y_val, y_pred_val, average='weighted'),4)
    val_kappa = np.round(cohen_kappa_score(y_val, y_pred_val),4)


    test_accuracy = np.round(accuracy_score(y_test, y_pred_test),4)*100
    test_precision     =     np.round(precision_score(y_test,     y_pred_test,
average='weighted'),2)
    test_recall        =       np.round(recall_score(y_test,       y_pred_test,
average='weighted'),2)
    test_F1 = np.round(f1_score(y_test, y_pred_test, average='weighted'),2)
    test_kappa = np.round(cohen_kappa_score(y_test, y_pred_test),2)
    test_roc_auc       =        metrics.roc_auc_score(y_test,     y_pred_test
,multi_class='ovo', average='weighted')


    print()
    print('----------------------- Train Set Metrics------------------------
')
    print()
    print("Accuracy core : {}%".format(train_accuracy))
    confusion_mtx = confusion_matrix(y_train, y_pred_train)
    cm = plot_confusion_matrix(confusion_mtx, classes = list(labels.items()),
normalize=False)


    print('----------------------- Validation Set Metrics--------------------
```

```
-----')
    print()
    print("Accuracy score : {}%".format(val_accuracy))
    confusion_mtx = confusion_matrix(y_val, y_pred_val)
    cm = plot_confusion_matrix(confusion_mtx, classes = list(labels.items()),
normalize=False)


    print('----------------------- Test Set Metrics-----------------------
')
    print()
    print("Accuracy score : {}%".format(test_accuracy))
    print("F1_score : {}".format(test_F1))
    print("Kappa Score : {} ".format(test_kappa))
    print("Recall score: {}".format(test_recall))
    print("Precision score : {}".format(test_precision))
    print("ROC AUC score : {}".format(test_roc_auc))
    confusion_mtx = confusion_matrix(y_test, y_pred_test)
    cm = plot_confusion_matrix(confusion_mtx, classes = list(labels.items()),
normalize=False)


    print('\t\tClassification                              Report:\n',
metrics.classification_report(y_test, y_pred_test))
    print("-"*80)
    print()


def
classifier_comparator(X_train,y_train,X_val,y_val,X_test,y_test,classifier=zi
pped_clf):
    result = []
    for n,c in classifier:
        checker_pipeline = Pipeline([('Classifier', c)])
        print("----------------------------Fitting {} on input_data--------
----------------------- ".format(n))
        #print(c)
        classifier_summary(checker_pipeline,X_train,     y_train,     X_val,
y_val,X_test,y_test)
```

Deep Feature extraction with VGG-16 model.

```python
#create VGG-16 model
from keras.applications import VGG16
from sklearn import metrics

base_model = VGG16(input_shape=(224,224,3), include_top=False)
model=Sequential()
model.add(base_model)
model.add(Flatten())
model.add(BatchNormalization())
model.add(Dense(256,kernel_initializer='he_uniform'))
model.add(BatchNormalization())
model.add(Activation('relu'))
model.add(Dropout(0.5))
model.add(Dense(1,activation='sigmoid'))

for layer in base_model.layers:
    layer.trainable = False


model.compile(
    loss='binary_crossentropy',
    optimizer='adam',
    metrics=['accuracy' , 'AUC']
)

model.summary()



#extract features for training, testing & val data from VGG-16 model

train_features=model.predict(X_train_prep)

test_features=model.predict(X_test_prep)

val_features=model.predict(X_val_prep)


classifier_comparator(train_features,y_train,val_features,y_val,test_features
,y_test,classifier=zipped_clf)



# Define the LSTM model

lstm_model = Sequential()

lstm_model.add(LSTM(units=256,    input_shape=(train_features.shape[1],    1),
dropout=0.5, return_sequences=True))

lstm_model.add(Flatten())

lstm_model.add(Dense(1, activation='sigmoid'))
```

```
# Compile the LSTM model
lstm_model.compile(optimizer='adam',                    loss='binary_crossentropy',
metrics=['accuracy'])


# Train the LSTM model on the extracted features
lstm_model.fit(train_features, y_train, epochs=10, validation_split=0.2)


# Predict on train set and calculate train accuracy
y_pred_train = lstm_model.predict(train_features)
y_pred_train = (y_pred_train > 0.5).astype(int)
train_acc = np.round(metrics.accuracy_score(y_train, y_pred_train),4)*100
print('Train Accuracy:', train_acc)


# Predict on validation set and calculate validation accuracy
y_pred_val = lstm_model.predict(val_features)
y_pred_val = (y_pred_val > 0.5).astype(int)
val_acc = np.round(metrics.accuracy_score(y_val, y_pred_val),4)*100
print('Validation Accuracy:', val_acc)


# Predict on test set and calculate test accuracy
y_pred_test = lstm_model.predict(test_features)
y_pred_test = (y_pred_test > 0.5).astype(int)
test_acc = np.round(metrics.accuracy_score(y_test, y_pred_test),4)*100
print('Test Accuracy:', test_acc)


def print_performance_metrics(y_test,y_pred_test):
    print('ROC         Area:',        np.round(metrics.roc_auc_score(y_test,
y_pred_test),4))
    print('Precision:',             np.round(metrics.precision_score(y_test,
y_pred_test,average='weighted'),4))
    print('Recall:', np.round(metrics.recall_score(y_test, y_pred_test,
                                        average='weighted'),4))
    print('F1 Score:', np.round(metrics.f1_score(y_test, y_pred_test,
                                        average='weighted'),4))
    print('Cohen   Kappa   Score:',  np.round(metrics.cohen_kappa_score(y_test,
y_pred_test),4))
    print('\t\tClassification                              Report:\n',
metrics.classification_report(y_test, y_pred_test))
print_performance_metrics(y_test,y_pred_test)
```

```python
# Define the Bi-LSTM model
bilstm_model = Sequential()
bilstm_model.add(Bidirectional(LSTM(units=256,
input_shape=(train_features.shape[1],              1),                dropout=0.5,
return_sequences=True)))
bilstm_model.add(Bidirectional(LSTM(units=128, dropout=0.2)))
bilstm_model.add(Flatten())
bilstm_model.add(Dense(1, activation='sigmoid'))


# Compile the Bi-LSTM model
bilstm_model.compile(optimizer='adam',              loss='binary_crossentropy',
metrics=['accuracy'])


# Reshape the input data to be 3-dimensional
train_features        =         train_features.reshape(train_features.shape[0],
train_features.shape[1], 1)
val_features           =            val_features.reshape(val_features.shape[0],
val_features.shape[1], 1)
test_features          =          test_features.reshape(test_features.shape[0],
test_features.shape[1], 1)


# Train the Bi-LSTM model on the extracted features
bilstm_model.fit(train_features, y_train, epochs=10, validation_split=0.2)


# Predict on train set and calculate train accuracy
y_pred_train = bilstm_model.predict(train_features)
y_pred_train = (y_pred_train > 0.5).astype(int)
train_acc = np.round(metrics.accuracy_score(y_train, y_pred_train),4)*100
print('Train Accuracy:', train_acc)


# Predict on validation set and calculate validation accuracy
y_pred_val = bilstm_model.predict(val_features)
y_pred_val = (y_pred_val > 0.5).astype(int)
val_acc = np.round(metrics.accuracy_score(y_val, y_pred_val),4)*100
print('Validation Accuracy:', val_acc)


# Predict on test set and calculate test accuracy
y_pred_test = bilstm_model.predict(test_features)
```

```python
y_pred_test = (y_pred_test > 0.5).astype(int)
test_acc = np.round(metrics.accuracy_score(y_test, y_pred_test),4)*100
print('Test Accuracy:', test_acc)


print_performance_metrics(y_test,y_pred_test)
```

3.6 Methodology

This chapter utilized different deep learning models to evaluate their performance in detecting brain tumors. These models included various layers of CNNs and transfer learning techniques such as ResNet50, ResNet101, VGG16, VGG19, Inception v3, InceptionResNetV2, MobileNet, MobileNetV2, DenseNet169, and DenseNet121. They were chosen based on their popularity and success in medical image analysis. In addition to deep learning models, classical machine learning techniques including SVM, Random Forest, AdaBoost, KNN, XGBoost, Bagging, ANN, LSTM, and Bi-LSTM were also used with different deep feature extraction techniques. Performance was evaluated using accuracy, F1-score, recall, precision, receiver operating characteristic (ROC) area, and Kappa scores to comprehensively assess accuracy, sensitivity, specificity, and other relevant measures. The methodology employed in this study provides a comprehensive and robust framework for evaluating the performance of different deep learning models and techniques in medical image analysis.

4. Results and discussion

4.1 Experimental data

This study's results are based on analyzing a public dataset obtained from Kaggle available at https://www.kaggle.com/datasets/abhranta/brain-tumor-detection-mri, consisting of MRI images of brain tumors. The dataset comprised 3000 images, of which 1500 were classified with tumors, while 1500 were without tumors. Samples of the data are shown in Fig. 6.1.

The dataset was preprocessed to remove noise and artifacts and normalized to ensure image consistency. After preprocessing, the dataset was divided into training, validation, and test sets. The training set was used to train the AI models, while the validation set was used to tune the hyperparameters of the models and prevent overfitting. Finally, the test set was used to evaluate the performance of the models and report the results. Table 6.1 demonstrates the data split between the training, validation, and testing tests.

4.2 Performance evaluation measures

Performance evaluation measures are typically used to assess the effectiveness of the proposed models, as they are critical in determining the accuracy and reliability of the results. For example, models' performance in detecting brain tumors from MR images can be measured using sensitivity and specificity metrics (Mohamed Shakeel et al., 2019). Sensitivity is the model's efficiency in identifying positive results, while specificity is its ability to detect negative results (Subasi et al., 2022). Other

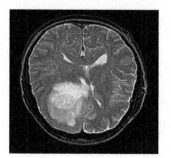

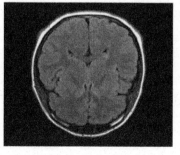

a) Tumor Sample   b) Non-Tumor Sample

FIGURE 6.1 Samples from the used dataset showing brain images (a) with tumors and (b) without tumors.

TABLE 6.1 Classification of the brain tumor data into training, validation, and testing sets.

	Tumor	No tumor	Total
Train	900	900	1800
Val	300	300	600
Test	300	300	600

commonly used measures include accuracy, F1-score, recall, precision, ROC area, and Kappa.

Accuracy is the amount of correctly classified instances in a dataset and is the most used metric. It is calculated from the ratio between the number of accurate predictions to the total number of predictions made by the model. While accuracy is valuable, it can be misleading when the dataset is imbalanced, when one class dominates others (Ilse et al., 2020; Kononenko & Kukar, 2007). Precision measures the proportion of true positives (TP) over the total predicted positives (TP + FP), while recall measures the proportion of true positives over the actual positives (TP + FN). A high precision score indicates that the model is less likely to produce false positives. In contrast, a high recall score suggests that the model is less likely to have false negatives. F1-score considers both precision and recall. It is calculated as the mean of precision and recall and provides a balanced performance measure for both classes (Kononenko & Kukar, 2007).

The ROC curve is a typical graphical depiction used to evaluate the performance of binary classification models. It illustrates the trade-off between true positive rate (sensitivity) and false positive rate (1-specificity) at various decision thresholds. The ROC curve illustrates how well a model can discriminate between positive and negative classes, with a higher area under the curve (AUC) suggesting stronger discriminating abilities. A perfect classifier has an AUC of 1, whereas a random classifier has an AUC of 0.5. The ROC curve provides vital insights into the overall performance of the model and aids in the selection of an acceptable decision threshold based on the desired balance of sensitivity and specificity.

Finally, Kappa accounts for the agreement between the predicted and actual labels of the model while considering the possibility of random agreement. It ranges from −1 to 1, where 1 indicates perfect agreement, 0 indicates no agreement beyond chance, and negative scores indicate agreement worse than chance (Mohamed Shakeel et al., 2019).

The choice of performance evaluation measures depends on the problem being solved and the dataset's characteristics. However, selecting the appropriate measure to avoid misleading results and ensure the model's accuracy and reliability is essential.

4.3 Experimental results

The results feature the performance of several models. The different layers of CNN results were first reported, where different layers from 2 to 8 layers were tested. Then, the results of varying transfer learning models were reported. The classifiers consisted of the base model and the dense layer of 256 units, followed by the dense output layer and the batch normalization and dropout layer between each layer. They were run for 50 epochs using "Adam" as an optimizer. Finally, the deep feature extraction was tested over several classifiers, and the results were reported.

4.3.1 Results of different CNN layers models

Table 6.2 presents the performance results of several layers of CNN models, ranging from 2 to 8 layers. These models demonstrated remarkable accuracy and efficiency across the entire CNN architecture. Specifically, the 7- and 8-layer models achieved the highest training, validation, and testing accuracy of 100%, 97.83%, and 98.67% and 100%, 98.17%, and 97.33%,

TABLE 6.2 Performance results of CNN different layers models.

Classifier	Training accuracy	Validation accuracy	Test accuracy	F1 measure	Kappa	ROC area
CNN 2 layer	100	93.67	93.50	0.93	0.87	0.98
CNN 3 layer	99.94	94.67	94.67	0.95	0.89	0.98
CNN 4 layer	93.22	85.33	82.33	0.82	0.65	0.97
CNN 5 layer	99.11	92	90.67	0.91	0.81	0.98
CNN 6 layer	99.39	92.83	92.17	0.92	0.84	0.99
CNN 7 layer	100	97.83	98.67	0.99	0.97	1
CNN 8 layer	100	98.17	97.33	0.97	0.95	1

respectively, followed by the 2- and 3-layer models. Conversely, the 4-layer model had the lowest training accuracy of 93.22%, validation accuracy of 85.33%, and test accuracy of 82.33%. Based on these results, the number of layers in the CNN models did not significantly impact their performance. However, the highest number of layers performed better in most metrics than the other models.

4.3.2 Results of different transfer learning models

Table 6.3 presents the performance results of several pretrained models, including ResNet50, VGG16, VGG19, InceptionV3, MobileNet, DenseNet169, DenseNet121, InceptionRes-NetV2, MobileNetV2, ResNet101. Most of the models showed remarkable closeness in their performance. However, the ResNet50 and ResNet101 models slightly outperformed the others, achieving high values for most performance measures evaluated with test accuracies of 98% and 97.50%, respectively. The DenseNet169 model, on the other hand, achieved the lowest accuracy of 88%.

4.3.3 Results of different deep feature extraction models

Table 6.4 contains the results of various models trained on deep features extracted from

TABLE 6.3 Performance results of different transfer learning models.

Classifier	Training accuracy	Validation accuracy	Test accuracy	F1 measure	Kappa	ROC area
ResNet50	99.83	97.33	98	0.98	0.96	1
VGG16	99	95.83	96	0.96	0.92	0.99
VGG19	99.44	96.17	97.50	0.97	0.95	0.99
InceptionV3	99.06	86.67	87.67	0.88	0.75	0.95
MobileNet	99.33	95.83	95.50	0.95	0.91	0.99
DenseNet169	94.94	89	88	0.88	0.76	0.95
DenseNet121	96.50	92.17	91.17	0.91	0.82	0.98
InceptionResNetV2	95.22	84.67	86.17	0.86	0.72	0.92
MobileNetV2	95.50	91.33	89.83	0.90	0.80	0.97
ResNet101	99.72	97.83	97.50	0.97	0.95	0.99
Xception	99.39	89.33	91.33	0.91	0.83	0.97

TABLE 6.4 Performance results of different deep feature-extracted models.

Feature-extracted: VGG16							
Model	Training accuracy	Validation accuracy	Test accuracy	F1 score	Kappa	Recall	Precision
SVM	99.22	95.50	96	0.96	0.92	0.96	0.96
Random Forest	100	95.17	95.83	0.96	0.92	0.96	0.96
AdaBoost	99.61	95.17	96	0.96	0.92	0.96	0.96
KNN	99.33	95.17	96	0.96	0.92	0.96	0.96
XGBoost	99.89	95.17	95.83	0.96	0.92	0.96	0.96
Bagging	99.89	95.17	95.50	0.95	0.91	0.96	0.96
ANN	99	95.67	96.17	0.96	0.92	0.96	0.96
LSTM	99	95.67	96.17	0.96	0.92	0.96	0.96
Bi-LSTM	99.06	95.50	96.17	0.96	0.92	0.96	0.96

Feature-extracted: VGG19							
Model	Training accuracy	Validation accuracy	Test accuracy	F1 score	Kappa	Recall	Precision
SVM	99.61	96.17	97.33	0.97	0.95	0.97	0.97
Random Forest	99.94	95.17	95.83	0.96	0.92	0.96	0.96
AdaBoost	99.94	95.17	95.83	0.96	0.92	0.96	0.96
KNN	99.67	95.83	97	0.97	0.94	0.97	0.97
XGBoost	99.89	95.33	95.83	0.96	0.92	0.96	0.96
Bagging	99.94	95.17	95.83	0.96	0.92	0.96	0.96
ANN	99.44	96	97.33	0.97	0.95	0.97	0.97
LSTM	99.44	96	97.33	0.97	0.95	0.97	0.97
Bi-LSTM	99.39	96.50	97.33	0.97	0.95	0.97	0.97

Feature-extracted: ResNet50							
Model	Training accuracy	Validation accuracy	Test accuracy	F1 score	Kappa	Recall	Precision
SVM	99.89	97.50	98.17	0.98	0.96	0.98	0.98
Random Forest	100	97.33	98.17	0.98	0.96	0.98	0.98
AdaBoost	100	97.33	98.17	0.98	0.96	0.98	0.98
KNN	99.89	97.50	98.17	0.98	0.96	0.98	0.98
XGBoost	99.94	97.50	98.17	0.98	0.96	0.98	0.98
Bagging	99.94	97.50	98.17	0.98	0.96	0.98	0.98
ANN	99.83	97.33	98	0.98	0.96	0.98	0.98
LSTM	99.83	97.33	98	0.98	0.96	0.98	0.98
Bi-LSTM	99.83	97.33	98	0.98	0.96	0.98	0.98

Feature-extracted: ResNet101

Model	Training accuracy	Validation accuracy	Test accuracy	F1 score	Kappa	Recall	Precision
SVM	99.72	96.83	97.67	0.98	0.95	0.98	0.98
Random Forest	100	95.67	96.33	0.96	0.93	0.96	0.96
AdaBoost	100	95.67	96.33	0.96	0.93	0.96	0.96
KNN	99.83	95.33	97.33	0.97	0.95	0.97	0.97
XGBoost	99.89	95	96.50	0.96	0.93	0.96	0.97
Bagging	100	95.67	96.33	0.96	0.93	0.96	0.96
ANN	99.72	97.67	97.50	0.97	0.95	0.98	0.98
LSTM	99.72	97.67	97.50	0.98	0.95	0.98	0.98
Bi-LSTM	99.72	97.67	97.50	0.98	0.95	0.98	0.98

Feature-extracted: MobileNetV2

Model	Training accuracy	Validation accuracy	Test accuracy	F1 score	Kappa	Recall	Precision
SVM	97.44	91.83	91.33	0.91	0.83	0.91	0.91
Random Forest	99.89	90.50	89.50	0.89	0.79	0.90	0.90
AdaBoost	97.94	91.33	90.50	0.90	0.81	0.90	0.91
KNN	97.78	91	90	0.90	0.80	0.90	0.90
XGBoost	98.83	91.17	89.33	0.89	0.79	0.89	0.89
Bagging	99.50	90.67	89.50	0.89	0.79	0.90	0.90
ANN	96.50	90.83	90.83	0.91	0.82	0.91	0.91
LSTM	97.44	91.67	92	0.92	0.84	0.92	0.92
Bi-LSTM	96.61	91	90.83	0.91	0.82	0.90	0.91

Feature-extracted: MobileNet

Model	Training accuracy	Validation accuracy	Test accuracy	F1 score	Kappa	Recall	Precision
SVM	99.67	95.83	95.67	0.96	0.91	0.96	0.96
Random Forest	100	96.50	96.17	0.96	0.92	0.96	0.96
AdaBoost	100	96.50	96.17	0.96	0.92	0.96	0.96
KNN	99.61	95.83	95.50	0.95	0.91	0.96	0.96
XGBoost	99.89	96.33	96	0.96	0.92	0.96	0.96
Bagging	99.89	96.50	95.83	0.96	0.92	0.96	0.96
ANN	99.28	95.83	95.33	0.95	0.91	0.95	0.95
LSTM	99.28	95.83	95.33	0.95	0.91	0.95	0.95
Bi-LSTM	99.33	95.83	95.50	0.96	0.91	0.96	0.96

Feature-extracted: InceptionV3							
Model	Training accuracy	Validation accuracy	Test accuracy	F1 score	Kappa	Recall	Precision
SVM	99.61	86.50	88.83	0.89	0.78	0.89	0.89
Random Forest	100	86.83	87.50	0.87	0.75	0.88	0.88
AdaBoost	100	86.83	87.50	0.87	0.75	0.88	0.88
KNN	99.67	86.50	88.67	0.89	0.77	0.89	0.89
XGBoost	99.89	86.67	87.50	0.87	0.75	0.88	0.88
Bagging	99.94	86.67	87.50	0.87	0.75	0.88	0.88
ANN	99.44	86.67	88.17	0.88	0.76	0.88	0.88
LSTM	99.39	86.33	88.17	0.88	0.76	0.88	0.88
Bi-LSTM	99.39	86.33	88.17	0.88	0.76	0.88	0.88

Feature-extracted: InceptionResNetV2							
Model	Training accuracy	Validation accuracy	Test accuracy	F1 score	Kappa	Recall	Precision
SVM	95.39	84.67	86.33	0.86	0.73	0.86	0.86
Random Forest	99.94	83.67	84.50	0.84	0.69	0.84	0.85
AdaBoost	95.89	84.67	86.83	0.87	0.74	0.87	0.87
KNN	95.67	84	85.33	0.85	0.71	0.85	0.85
XGBoost	97.44	83.67	84.67	0.85	0.69	0.85	0.85
Bagging	99.39	83.33	85.17	0.85	0.70	0.85	0.85
ANN	95.33	85.17	86.67	0.87	0.73	0.87	0.87
LSTM	91.56	82	82.67	0.82	0.65	0.83	0.84
Bi-LSTM	88	80.33	79.17	0.79	0.58	0.79	0.82

Feature-extracted: DenseNet169							
Model	Training accuracy	Validation accuracy	Test accuracy	F1 score	Kappa	Recall	Precision
SVM	96.44	91.17	87.5	0.87	0.75	0.88	0.88
Random Forest	99.83	89.67	87	0.87	0.74	0.87	0.87
AdaBoost	97	90.67	87	0.87	0.74	0.87	0.87
KNN	97.17	90.33	87.17	0.87	0.74	0.87	0.87
XGBoost	98.56	89.67	87.33	0.87	0.75	0.87	0.88
Bagging	99.22	89.33	86.83	0.87	0.74	0.87	0.87
ANN	96.06	90.67	88.17	0.88	0.76	0.88	0.88
LSTM	96.44	90.67	87.33	0.87	0.75	0.87	0.88
Bi-LSTM	96.44	90.83	87.83	0.88	0.76	0.88	0.88

Feature-extracted: DenseNet121

Model	Training accuracy	Validation accuracy	Test accuracy	F1 score	Kappa	Recall	Precision
SVM	98.83	93.83	92.50	0.92	0.85	0.92	0.93
Random Forest	100	93	91.50	0.91	0.83	0.92	0.92
AdaBoost	99.39	94	91.33	0.91	0.83	0.91	0.91
KNN	99.06	94.17	91.33	0.91	0.83	0.91	0.91
XGBoost	99.89	93	91.67	0.92	0.83	0.92	0.92
Bagging	99.83	93.17	91.50	0.91	0.83	0.92	0.92
ANN	98.06	93.50	92.50	0.92	0.85	0.92	0.93
LSTM	98.44	93.50	92.67	0.93	0.85	0.93	0.93
Bi-LSTM	97.67	92.83	91.83	0.92	0.84	0.92	0.92

Feature-extracted: Xception

Model	Training accuracy	Validation accuracy	Test accuracy	F1 score	Kappa	Recall	Precision
SVM	99.39	89.5	89.67	0.90	0.79	0.90	0.90
Random Forest	99.89	89	88.83	0.89	0.78	0.89	0.89
AdaBoost	99.50	89.33	89.33	0.89	0.79	0.89	0.90
KNN	99.39	89.33	89.5	0.89	0.79	0.90	0.90
XGBoost	99.78	89.33	89	0.89	0.78	0.89	0.89
Bagging	99.78	89	89.17	0.89	0.78	0.89	0.90
ANN	99.33	89.83	90.17	0.90	0.80	0.90	0.90
LSTM	99.39	89.83	90.33	0.90	0.81	0.90	0.90
Bi-LSTM	99.28	90	89.83	0.90	0.80	0.90	0.90

the pretrained models. The deep features were extracted from the last convolutional layer of these networks and were then used as inputs for training the models. Finally, the performance of the models was evaluated based on various metrics such as train accuracy, validation accuracy, test accuracy, F1 score, Kappa, recall, and precision.

Although the models achieved great performance across the different proposed deep feature extraction classifiers, the VGG16, VGG19, ResNet50, RestNet101, MobileNet, and DenseNet121 had accuracies of over 90% in the different classifiers. Nevertheless, the overall performance of the different classifiers was closely aligned within each architecture, despite the slight variations observed.

4.4 Discussion

This study evaluated several deep learning models' performance and assessed their efficiency in detecting brain tumors from MR images. The models included different CNNs architectures, transfer learning, and deep feature extraction models.

Regarding the CNN models, although the 7- and 8-layer models outperformed most of the others regarding training, validation, and testing accuracies, the number of layers did not significantly impact the overall performance. These findings align with previous studies that found that deeper networks do not necessarily lead to better performance (Goyal et al., 2021; He et al., 2015; Nichani et al., 2021). This is also confirmed by the closeness of the performance between these models. However, many factors affect the effectiveness of these models, such as the number of samples in each set as the optimal number of layers may vary depending on the available data and preprocessing methods. On the other hand, transfer learning models, ResNet50 and ResNet101, were the best-performing ones, achieving high values for most of the performance measures evaluated. These results are confirmed by other studies that have also reported their effectiveness in detecting brain tumors (e.g., Ghosal et al., 2019; Girshick, 2015; Krishnapriya & Karuna, 2023), while the lower performance of the DenseNet models aligns with the results of Huang et al. (2016). However, it is vital to mention that these models require significant computational resources and large amounts of data to train effectively and efficiently.

Finally, the deep feature extraction models evaluated in this study showed that Random Forest, AdaBoost, and Bagging consistently achieved high accuracy across the different architectures. These results complement previously reported deep feature extraction results and ensemble methods' effectiveness in tumor detection (Basir & Shantta, 2021; Kang et al., 2021; Rasool et al., 2022). Nevertheless, it is noted that the variations between the different classifiers were minimal within the models.

In addition to the performance results presented in the study, several other factors could affect the models' performance. For instance, the data size and quality used in training and testing the models could significantly impact their performance (Naseer et al., 2021). Another important consideration when interpreting the results is the evaluation metrics' limitations (Singh et al., 2021). In this study, the performance of the models was evaluated using accuracy, F1 score, Kappa, and ROC area. While these metrics are commonly used in machine learning, they have limitations in specific conditions. For example, accuracy can be misleading when the dataset is imbalanced, resulting in overfitting when the model is too complex. The F1 score is a better measure than accuracy when the dataset is imbalanced, but it can be sensitive to the threshold used to classify the samples. Kappa and ROC area are also commonly used metrics but have limitations in specific settings (Jeni et al., 2013). Therefore, it is essential to interpret the results with these limitations and consider using additional evaluation metrics in future studies.

Finally, the performance results reported in the study are based on a specific set of models and hyperparameters. Therefore, the performance of the models can be improved by optimizing the hyperparameters or using different architectures. Recent studies have reported that the performance of deep learning models can be enhanced by using attention mechanisms, which selectively attend to essential features in the input data (Niu et al., 2021).

5. Conclusion

Deep learning has evolved into an effective method for detecting brain tumors in MR images. It has demonstrated outstanding accuracy in identifying the existence, location, and size of brain tumors, which can assist clinicians plan treatment methods and improve patient outcomes. Furthermore, by exploiting the large quantity of imaging data available, deep learning algorithms can learn to spot tiny patterns and traits that the human eye may overlook or misinterpret, resulting in more accurate and reliable diagnoses.

However, several obstacles remain in deep learning for detecting brain cancers. One significant problem is the requirement for labeled data for training deep learning models. Furthermore, some brain tumors may have imaging properties comparable to normal brain tissue, making it challenging for even deep learning algorithms to differentiate between them. Finally, the interpretability of deep learning models remains a point of contention, and further study is required to ensure that the models are providing clinically meaningful predictions.

Overall, the current study's findings provide useful insight into the performance of various models in detecting brain tumors. However, depending on the task and dataset, the best model and architecture may differ. As a result, more study is needed to analyze the performance of these models in different scenarios and determine the best model and architecture for detecting brain tumors efficiently.

References

Abolanle, K., Amina, S., Muhammad, A., Hina, A., Omowumi T, K., Omowumi O, A., & Sunday O, O. (2020). Brain Tumor: An overview of the basic clinical manifestations and treatment. *Global Journal of Cancer Therapy*, 038–041. https://doi.org/10.17352/2581-5407.000034

Agus, M. E., Bagas, S. Y., Yuda, M., Hanung, N. A., & Ibrahim, Z. (2022). Convolutional neural network featuring VGG-16 model for glioma classification. *JOIV: International Journal on Informatics Visualization*, 6(3), 660. https://doi.org/10.30630/joiv.6.3.1230

Ahuja, S., Panigrahi, B. K., Gandhi, T., & Gautam, U. (2021). Deep learning-based computer-aided diagnosis tool for brain tumor classification. In *2021 11th international conference on cloud computing, data science and engineering (confluence)* (pp. 854–859). https://doi.org/10.1109/Confluence51648.2021.9377171

Al-jabery, K. K., Obafemi-Ajayi, T., Olbricht, G. R., & Wunsch Ii, D. C. (2020). Data preprocessing. In *Computational learning approaches to data analytics in biomedical applications* (pp. 7–27). Elsevier. https://doi.org/10.1016/B978-0-12-814482-4.00002-4

Alther, B., Mylius, V., Weller, M., & Gantenbein, A. (2020). From first symptoms to diagnosis: Initial clinical presentation of primary brain tumors. *Clinical and Translational Neuroscience*, 4(2). https://doi.org/10.1177/2514183X20968368, 2514183X2096836.

Alzubaidi, L., Zhang, J., Humaidi, A. J., Al-Dujaili, A., Duan, Y., Al-Shamma, O., Santamaría, J., Fadhel, M. A., Al-Amidie, M., & Farhan, L. (2021). Review of deep learning: Concepts, CNN architectures, challenges, applications, future directions. *Journal of Big Data*, 8(1), 53. https://doi.org/10.1186/s40537-021-00444-8

Arunkumar, N., Mohammed, M. A., Mostafa, S. A., Ibrahim, D. A., Rodrigues, J. J. P. C., & Albuquerque, V. H. C. (2020). Fully automatic model-based segmentation and classification approach for MRI brain tumor using artificial neural networks. *Concurrency and Computation: Practice and Experience*, 32(1). https://doi.org/10.1002/cpe.4962

Basir, O., & Shantta, K. (2021). Deep learning feature extraction for brain tumor characterization and detection. *IRA-International Journal of Applied Sciences*, 16(1), 1. https://doi.org/10.21013/jas.v16.n1.p1

Chaplot, S., Patnaik, L. M., & Jagannathan, N. R. (2006). Classification of magnetic resonance brain images using wavelets as input to support vector machine and neural network. *Biomedical Signal Processing and Control*, 1(1), 86–92. https://doi.org/10.1016/j.bspc.2006.05.002

Chen, J., Gui, W., Dai, J., Jiang, Z., Chen, N., & Li, X. (2021). A hybrid model combining mechanism with semi-supervised learning and its application for temperature prediction in roller hearth kiln. *Journal of Process Control*, 98, 18–29. https://doi.org/10.1016/j.jprocont.2020.11.012

Deshpande, A., Estrela, V. V., & Patavardhan, P. (2021). The DCT-CNN-ResNet50 architecture to classify brain tumors with super-resolution, convolutional neural network, and the ResNet50. *Neuroscience Informatics*, 1(4), 100013. https://doi.org/10.1016/j.neuri.2021.100013

Dhanwani, D. C., & Bartere, M. M. (2014). Survey on various techniques of brain tumor detection from MRI images. *International Journal of Computational Engineering Research*, 4(1), 24–26.

El-Dahshan, E.-S. A., Mohsen, H. M., Revett, K., & Salem, A.-B. M. (2014). Computer-aided diagnosis of human brain tumor through MRI: A survey and a new algorithm. *Expert Systems with Applications*, 41(11), 5526–5545. https://doi.org/10.1016/j.eswa.2014.01.021

Ghosal, P., Nandanwar, L., Kanchan, S., Bhadra, A., Chakraborty, J., & Nandi, D. (2019). Brain tumor classification using ResNet-101 based squeeze and excitation deep neural network. In *2019 second international conference on advanced computational and communication paradigms (ICACCP)*, 1–6. https://doi.org/10.1109/ICACCP.2019.8882973

Girshick, R. (2015). Fast R-CNN. In *2015 IEEE international conference on computer vision (ICCV)* (pp. 1440–1448). https://doi.org/10.1109/ICCV.2015.169

Gite, S., Patil, S., Dharrao, D., Yadav, M., Basak, S., Rajendran, A., & Kotecha, K. (2023). Textual feature extraction using ant colony optimization for hate speech classification. *Big Data and Cognitive Computing, 7*(1), 45. https://doi.org/10.3390/bdcc7010045

Goodfellow, I., Pouget-Abadie, J., Mirza, M., Xu, B., Warde-Farley, D., Ozair, S., Courville, A., & Bengio, Y. (2014). Generative adversarial nets. In Z. Ghahramani, M. Welling, C. Cortes, N. Lawrence, & K. Q. Weinberger (Eds.), *Advances in neural information processing systems* (Vol. 27). Curran Associates, Inc. https://proceedings.neurips.cc/paper_files/paper/2014/file/5ca3e9b122f61f8f06494c97b1afccf3-Paper.pdf.

Goyal, A., Bochkovskiy, A., Deng, J., & Koltun, V. (2021). *Non-deep networks.* https://doi.org/10.48550/ARXIV.21 10.07641

Greene, J., & Al-Dhahir, M. A. (2023). *Acoustic neuroma.* StatPearls Publishing.

Gupta, S., Tran, T., Luo, W., Phung, D., Kennedy, R. L., Broad, A., Campbell, D., Kipp, D., Singh, M., Khasraw, M., Matheson, L., Ashley, D. M., & Venkatesh, S. (2014). Machine-learning prediction of cancer survival: A retrospective study using electronic administrative records and a cancer registry. *BMJ Open, 4*(3), e004007. https://doi.org/10.1136/bmjopen-2013-004007

Hashemzehi, R., Mahdavi, S. J. S., Kheirabadi, M., & Kamel, S. R. (2020). Detection of brain tumors from MRI images base on deep learning using hybrid model CNN and NADE. *Biocybernetics and Biomedical Engineering, 40*(3), 1225–1232. https://doi.org/10.1016/j.bbe.2020.06.001

Havaei, M., Davy, A., Warde-Farley, D., Biard, A., Courville, A., Bengio, Y., Pal, C., Jodoin, P.-M., & Larochelle, H. (2017). Brain tumor segmentation with deep neural networks. *Medical Image Analysis, 35*, 18–31. https://doi.org/10.1016/j.media.2016.05.004

He, K., Zhang, X., Ren, S., & Sun, J. (2015). *Deep residual learning for image recognition.* https://doi.org/10.48550/ARXIV.1512.03385

Hicks, S. A., Strümke, I., Thambawita, V., Hammou, M., Riegler, M. A., Halvorsen, P., & Parasa, S. (2022). On evaluation metrics for medical applications of artificial intelligence. *Scientific Reports, 12*(1), 5979. https://doi.org/10.1038/s41598-022-09954-8

Huang, G., Liu, Z., van der Maaten, L., & Weinberger, K. Q. (2016). *Densely connected convolutional networks.* https://doi.org/10.48550/ARXIV.1608.06993

Ilse, M., Tomczak, J. M., & Welling, M. (2020). Deep multiple instance learning for digital histopathology. In *Handbook of medical image computing and computer assisted intervention* (pp. 521–546). Elsevier. https://doi.org/10.1016/B978-0-12-816176-0.00027-2

Jayanthi, P., & Iyyanki, M. (2020). Deep learning techniques for prediction, detection, and segmentation of brain tumors. In A. Suresh, R. Udendhran, & S. Vimal (Eds.), *Advances in bioinformatics and biomedical engineering* (pp. 118–154). IGI Global. https://doi.org/10.4018/978-1-7998-3591-2.ch009

Jeni, L. A., Cohn, J. F., & De La Torre, F. (2013). Facing imbalanced data—recommendations for the use of performance metrics. *Humaine Association Conference on Affective Computing and Intelligent Interaction*, 245–251. https://doi.org/10.1109/ACII.2013.47

Jia, Z., & Chen, D. (2020). Brain Tumor Identification and Classification of MRI images using deep learning techniques. *IEEE Access, 1–1.* https://doi.org/10.1109/ACCESS.2020.3016319

Jiang, F., Jiang, Y., Zhi, H., Dong, Y., Li, H., Ma, S., Wang, Y., Dong, Q., Shen, H., & Wang, Y. (2017). Artificial intelligence in healthcare: Past, present and future. *Stroke and Vascular Neurology, 2*(4), 230–243. https://doi.org/10.1136/svn-2017-000101

Kabitha, K., Moses, S. R., Karunakar, H., Sunil, K., & Ashok, S. (2013). A comprehensive review on brain tumor. *International Journal of Pharmaceutical, Chemical and Biological Sciences, 3*(4), 1165–1171.

Kadhim, Y. A., Khan, M. U., & Mishra, A. (2022). Deep learning-based computer-aided diagnosis (CAD): Applications for medical image datasets. *Sensors, 22*(22), 8999. https://doi.org/10.3390/s22228999

Kalan Farmanfarma, K. H., Mohammadian, M., Shahabinia, Z., Hassanipour, S., & Salehiniya, H. (2019). Brain cancer in the world: An epidemiological review. *World Cancer Research Journal, 6*(e1356).

Kang, J., Ullah, Z., & Gwak, J. (2021). MRI-based brain tumor classification using ensemble of deep features and machine learning classifiers. *Sensors, 21*(6), 2222. https://doi.org/10.3390/s21062222

Koh, D.-M., Papanikolaou, N., Bick, U., Illing, R., Kahn, C. E., Kalpathi-Cramer, J., Matos, C., Martí-Bonmatí, L., Miles, A., Mun, S. K., Napel, S., Rockall, A., Sala, E., Strickland, N., & Prior, F. (2022). Artificial intelligence and machine learning in cancer imaging. *Communications Medicine, 2*(1), 133. https://doi.org/10.1038/s43856-022-00199-0

Kononenko, I., & Kukar, M. (2007). Machine learning basics. In *Machine learning and data mining* (pp. 59–105). Elsevier. https://doi.org/10.1533/9780857099440.59

Krishnapriya, S., & Karuna, Y. (2023). Pre-trained deep learning models for brain MRI image classification. *Frontiers in Human Neuroscience, 17*, 1150120. https://doi.org/10.3389/fnhum.2023.1150120

Lapointe, S., Perry, A., & Butowski, N. A. (2018). Primary brain tumours in adults. *The Lancet, 392*(10145), 432–446. https://doi.org/10.1016/S0140-6736(18)30990-5

Li, L., Li, S., & Su, J. (2021). A multi-category brain tumor classification method bases on improved ResNet50. *Computers, Materials and Continua, 69*(2), 2355–2366. https://doi.org/10.32604/cmc.2021.019409

Li, Q., Yu, Z., Wang, Y., & Zheng, H. (2020). TumorGAN: A multi-modal data augmentation framework for brain tumor segmentation. *Sensors, 20*(15), 4203. https://doi.org/10.3390/s20154203

Linardatos, P., Papastefanopoulos, V., & Kotsiantis, S. (2020). Explainable AI: A review of machine learning interpretability methods. *Entropy, 23*(1), 18. https://doi.org/10.3390/e23010018

Liu, S., Liu, X., Xiao, Y., Chen, S., & Zhuang, W. (2019). Prognostic factors associated with survival in patients with anaplastic oligodendroglioma. *PLoS One, 14*(1), e0211513. https://doi.org/10.1371/journal.pone.0211513

Lu, S.-Y., Wang, S.-H., & Zhang, Y.-D. (2020). A classification method for brain MRI via MobileNet and feedforward network with random weights. *Pattern Recognition Letters, 140*, 252–260. https://doi.org/10.1016/j.patrec.2020.10.017

Mabray, M. C., Barajas, R. F., & Cha, S. (2015). Modern brain tumor imaging. *Brain Tumor Research and Treatment, 3*(1), 8. https://doi.org/10.14791/btrt.2015.3.1.8

Mahmud, M., Kaiser, M. S., Hussain, A., & Vassanelli, S. (2018). Applications of deep learning and reinforcement learning to biological data. *IEEE Transactions on Neural Networks and Learning Systems, 29*(6), 2063–2079. https://doi.org/10.1109/TNNLS.2018.2790388

Majib, M. S., Rahman, Md M., Sazzad, T. M. S., Khan, N. I., & Dey, S. K. (2021). VGG-SCNet: A VGG net-based deep learning framework for brain tumor detection on MRI images. *IEEE Access, 9*, 116942–116952. https://doi.org/10.1109/ACCESS.2021.3105874

Maqsood, S., Damaševičius, R., & Maskeliūnas, R. (2022). Multi-modal brain tumor detection using deep neural network and multiclass SVM. *Medicina, 58*(8), 1090. https://doi.org/10.3390/medicina58081090

Mehrotra, R., Ansari, M. A., Agrawal, R., & Anand, R. S. (2020). A Transfer Learning approach for AI-based classification of brain tumors. *Machine Learning with Applications, 2*, 100003. https://doi.org/10.1016/j.mlwa.2020.100003

Miller, K. D., Ostrom, Q. T., Kruchko, C., Patil, N., Tihan, T., Cioffi, G., Fuchs, H. E., Waite, K. A., Jemal, A., Siegel, R. L., & Barnholtz-Sloan, J. S. (2021). Brain and other central nervous system tumor statistics, 2021. *CA: A Cancer Journal for Clinicians, 71*(5), 381–406. https://doi.org/10.3322/caac.21693

Mohamed Shakeel, P., Tobely, T. E. El, Al-Feel, H., Manogaran, G., & Baskar, S. (2019). Neural network based brain tumor detection using wireless infrared imaging sensor. *IEEE Access, 7*, 5577–5588. https://doi.org/10.1109/ACCESS.2018.2883957

Nadeem, M. W., Ghamdi, M. A. A., Hussain, M., Khan, M. A., Khan, K. M., Almotiri, S. H., & Butt, S. A. (2020). Brain tumor analysis empowered with deep learning: A review, taxonomy, and future challenges. *Brain Sciences, 10*(2), 118. https://doi.org/10.3390/brainsci10020118

Naseer, A., Yasir, T., Azhar, A., Shakeel, T., & Zafar, K. (2021). Computer-aided brain tumor diagnosis: Performance evaluation of deep learner CNN using augmented brain MRI. *International Journal of Biomedical Imaging, 2021*, 1–11. https://doi.org/10.1155/2021/5513500

Nichani, E., Radhakrishnan, A., & Uhler, C. (2021). Do deeper convolutional networks perform better? In *International conference on learning representations*. https://openreview.net/forum?id=rYt0p0Um9r.

Niu, Z., Zhong, G., & Yu, H. (2021). A review on the attention mechanism of deep learning. *Neurocomputing, 452*, 48–62. https://doi.org/10.1016/j.neucom.2021.03.091

Ostrom, Q. T., Fahmideh, M. A., Cote, D. J., Muskens, I. S., Schraw, J. M., Scheurer, M. E., & Bondy, M. L. (2019). Risk factors for childhood and adult primary brain tumors. *Neuro-Oncology, 21*(11), 1357–1375. https://doi.org/10.1093/neuonc/noz123

Ostrom, Q. T., Francis, S. S., & Barnholtz-Sloan, J. S. (2021). Epidemiology of brain and other CNS tumors. *Current Neurology and Neuroscience Reports, 21*(12), 68. https://doi.org/10.1007/s11910-021-01152-9

Patil, Ms P., Pawar, Ms S., Patil, Ms S., & Nichal, Prof A. (2017). A review paper on brain tumor segmentation and detection. *IJIREEICE, 5*(1), 12–15. https://doi.org/10.17148/IJIREEICE.2017.5103

Rasool, M., Ismail, N. A., Al-Dhaqm, A., Yafooz, W. M. S., & Alsaeedi, A. (2022). A novel approach for classifying brain tumours combining a SqueezeNet model with SVM and fine-tuning. *Electronics, 12*(1), 149. https://doi.org/10.3390/electronics12010149

Sahaai, M. B., Jothilakshmi, G. R., Ravikumar, D., Prasath, R., & Singh, S. (2022). *ResNet-50 based deep neural network using transfer learning for brain tumor classification.* https://doi.org/10.1063/5.0082328

Sangeetha, M., Keerthika, P., Devendran, K., Sridhar, S., Raagav, S. S., & Vigneshwar, T. (2022). Brain tumor segmentation and prediction on MRI images using deep learning network. *International Journal of Health Sciences*, 13486–13503. https://doi.org/10.53730/ijhs.v6nS2.8542

Sarhan, A. M. (2020). Brain tumor classification in magnetic resonance images using deep learning and wavelet transform. *Journal of Biomedical Science and Engineering*,

13(06), 102–112. https://doi.org/10.4236/jbise.2020.136010

Sarker, I. H. (2021a). Deep learning: A comprehensive overview on techniques, taxonomy, applications and research directions. *SN Computer Science, 2*(6), 420. https://doi.org/10.1007/s42979-021-00815-1

Sarker, I. H. (2021b). Machine learning: Algorithms, real-world applications and research directions. *SN Computer Science, 2*(3), 160. https://doi.org/10.1007/s42979-021-00592-x

Schaff, L. R., & Mellinghoff, I. K. (2023). Glioblastoma and other primary brain malignancies in adults: A review. *JAMA, 329*(7), 574. https://doi.org/10.1001/jama.2023.0023

Sethuram Rao, G., & Vydeki, D. (2018). Brain tumor detection approaches: A review. In *2018 international conference on smart systems and inventive technology (ICSSIT)* (pp. 479–488). https://doi.org/10.1109/ICSSIT.2018.8748692

Simone, V., Rizzo, D., Cocciolo, A., Caroleo, A. M., Carai, A., Mastronuzzi, A., & Tornesello, A. (2021). Infantile brain tumors: A review of literature and future perspectives. *Diagnostics, 11*(4), 670. https://doi.org/10.3390/diagnostics11040670

Singh, V., Sharma, S., Goel, S., Lamba, S., & Garg, N. (2021). Brain tumor prediction by binary classification using VGG-16. In N. Gupta, P. Chatterjee, & T. Choudhury (Eds.), *Smart and sustainable intelligent systems* (1st ed., pp. 127–138). Wiley. https://doi.org/10.1002/9781119752134.ch9

Sravanthi Peddinti, A., Maloji, S., & Manepalli, K. (2021). Evolution in diagnosis and detection of brain tumor – review. *Journal of Physics: Conference Series, 2115*(1), 012039. https://doi.org/10.1088/1742-6596/2115/1/012039

Subasi, A., Kapadnis, M. N., & Kosal Bulbul, A. (2022). Alzheimer's disease detection using artificial intelligence. In *Augmenting neurological disorder prediction and rehabilitation using artificial intelligence* (pp. 53–74). Elsevier. https://doi.org/10.1016/B978-0-323-90037-9.00011-4

Taşcı, B. (2023). Attention deep feature extraction from brain MRIs in explainable mode: DGXAINet. *Diagnostics, 13*(5), 859. https://doi.org/10.3390/diagnostics13050859

Talabis, M. R. M., McPherson, R., Miyamoto, I., Martin, J. L., & Kaye, D. (2015). Analytics defined. In *Information security analytics* (pp. 1–12). Elsevier. https://doi.org/10.1016/B978-0-12-800207-0.00001-0

Taye, M. M. (2023a). Theoretical understanding of convolutional neural network: Concepts, architectures, applications, future directions. *Computation, 11*(3), 52. https://doi.org/10.3390/computation11030052

Taye, M. M. (2023b). Understanding of machine learning with deep learning: Architectures, workflow, applications and future directions. *Computers, 12*(5), 91. https://doi.org/10.3390/computers12050091

Tazin, T., Sarker, S., Gupta, P., Ayaz, F. I., Islam, S., Monirujjaman Khan, M., Bourouis, S., Idris, S. A., & Alshazly, H. (2021). A robust and novel approach for brain tumor classification using convolutional neural network. *Computational Intelligence and Neuroscience, 2021*, 1–11. https://doi.org/10.1155/2021/2392395

Yamashita, R., Nishio, M., Do, R. K. G., & Togashi, K. (2018). Convolutional neural networks: An overview and application in radiology. *Insights into Imaging, 9*(4), 611–629. https://doi.org/10.1007/s13244-018-0639-9

Zahid, U., Ashraf, I., Khan, M. A., Alhaisoni, M., Yahya, K. M., Hussein, H. S., & Alshazly, H. (2022). BrainNet: Optimal deep learning feature fusion for brain tumor classification. *Computational Intelligence and Neuroscience, 2022*, 1–13. https://doi.org/10.1155/2022/1465173

Zeinalkhani, L., Ali Jamaat, A., & Rostami, K. (2018). Diagnosis of brain tumor using combination of K-means clustering and genetic algorithm. *Iranian Journal of Medical Informatics, 7*, 6. https://doi.org/10.24200/ijmi.v7i0.159

Zhang, Y., Zhang, P., & Yan, Y. (2017). Attention-based LSTM with multi-task learning for distant speech recognition. *Interspeech, 2017*, 3857–3861. https://doi.org/10.21437/Interspeech.2017-805

Artificial intelligence—based fatty liver disease detection using ultrasound images

Safdar Wahid Inamdar[1], Abdulhamit Subasi[2,3]

[1]Indian Institute of Technology, Indore, Madhya Pradesh, India; [2]Institute of Biomedicine, Faculty of Medicine, University of Turku, Turku, Finland; [3]Department of Computer Science, College of Engineering, Effat University, Jeddah, Saudi Arabia

1. Introduction

Nonalcoholic fatty liver disease (NAFLD) is a fast-growing pathology worldwide, considered the most common chronic liver disease (Popa et al., 2021). It is diagnosed based on the presence of steatosis in more than 5% of hepatocytes without significant alcohol consumption (Byra et al., 2018). NAFLD has emerged as a major health concern worldwide, driven by the increasing prevalence of diabetes, obesity, and metabolic syndrome (Pouwels et al., 2022). NAFLD encompasses a spectrum of liver conditions ranging from simple steatosis to nonalcoholic steatohepatitis (NASH), fibrosis, cirrhosis, and hepatocellular carcinoma (Maurice & Manousou, 2018).

Early detection and accurate diagnosis of NAFLD are crucial for implementing timely interventions and preventing disease progression to advanced stages. While liver biopsy has long been considered the gold standard for NAFLD diagnosis, its invasiveness, expense, and inherent risks have stimulated the search for noninvasive alternatives (Sumida, 2014). Imaging modalities, particularly ultrasound, have gained considerable attention due to their accessibility, cost-effectiveness, and real-time capabilities. Moreover, recent advances in machine learning (ML) techniques have shown promise in automating the analysis of ultrasound images, facilitating efficient and accurate detection of NAFLD (Zhang et al., 2018).

Ultrasound imaging is widely used in clinical practice to evaluate liver morphology, identify hepatic steatosis, and assess liver fibrosis. This noninvasive technique utilizes sound waves to produce real-time images, providing valuable information about liver and kidney abnormalities (Liu et al., 2019). Ultrasonographic features associated with NAFLD include hepatomegaly, increased liver echogenicity, hepatorenal echo contrast, and hepatic vascular changes (Gerstenmaier & Gibson, 2014). However, the manual interpretation of ultrasound images is subjective and relies heavily on the operator's expertise. Therefore, the application of ML techniques to automate the analysis of ultrasound images

Applications of Artificial Intelligence in Healthcare and Biomedicine
https://doi.org/10.1016/B978-0-443-22308-2.00015-9

holds immense potential for improving the accuracy and efficiency of NAFLD diagnosis (Lupsor-Platon et al., 2021).

Machine learning algorithms, such as convolutional neural networks (CNNs), support vector machines (SVMs), and random forests (RFs), have demonstrated remarkable capabilities in image analysis and pattern recognition (Sarvamangala & Kulkarni, 2021). By training on large datasets of ultrasound images, ML models can learn complex relationships and features indicative of NAFLD, enabling automated detection and classification. CNNs have shown exceptional performance in various medical imaging tasks. These deep-learning models learn hierarchical representations of images and can extract subtle visual features that may be imperceptible to human observers (Sarvamangala & Kulkarni, 2021). Integration of ML algorithms with ultrasound imaging holds tremendous potential for improving NAFLD diagnosis accuracy, enhancing patient care, and reducing the burden on healthcare systems (Lupsor-Platon et al., 2021). However, several challenges must be addressed to ensure their successful integration into clinical practice. Large, diverse, and well-annotated datasets are essential for training robust ML models. Furthermore, standardization of imaging protocols and the development of reliable ground truth labels are necessary for generating high-quality training data (Willemink et al., 2020).

Unfortunately, it is challenging to generate large medical datasets of high quality to train such models. One of the ways to overcome this problem is to use transfer learning. Transfer learning is a technique to use a model that has been pretrained on some other dataset to extract essential features and train a classifier on the features extracted from new dataset. This is our intuition in carrying out this research. We use a dataset of ultrasound images of the liver and kidney and apply various pretrained models and classifiers on it, as shown in Fig. 7.1. Hence, our research aims to compare the diagnostic performance of various models using ultrasound images to detect NAFLD.

2. Literature review

Mihăilescu et al. (2013) evaluated the possibility of automated steatosis staging using ultrasound images analyzed by RFs and SVM

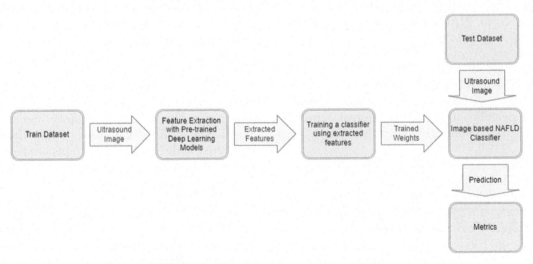

FIGURE 7.1 Proposed approach to detect NSFLD.

classifiers. The RFs approach demonstrated better accuracy than the SVM classifier, highlighting its potential for automated NAFLD staging. Nagy et al. (2015) evaluated the diagnosis values of hepatic ultrasound characteristics using an SVM classifier for steatosis in patients with chronic liver disorders. The SVM classifier utilized three ultrasound parameters. The study demonstrated good accuracy and an AUC of 0.923 for the division of liver steatosis into two stages. Acharya et al. (2016) developed an automated system based on the curvelet transform method to discriminate between normal liver, fatty liver disease, and cirrhosis using ultrasound images. The software utilized various features extracted from curvelet transform coefficients. The study reported high accuracy, sensitivity, and specificity for the automated system in diagnosing normal liver, fatty liver disease, and cirrhosis, demonstrating the availability of a precise diagnosis tool for chronic liver disorders. Bayes factor-based computer-aided diagnosis (CAD) systems have also been investigated for NAFLD detection. Ribeiro et al. (2014) developed a CAD system using objective liver ultrasound parameters. They achieved high accuracy, sensitivity, and specificity in steatosis classification, demonstrating the potential of automated systems for accurate NAFLD diagnosis. Subramanya et al. (2014) analyzed the efficiency of a CAD system for diagnosing and staging fatty liver disease on ultrasound images. The CAD system demonstrated superior performance with high accuracy, providing a convenient tool for diagnosing and staging fatty liver disease.

Biswas et al. (2018) developed the DL Symtosis System, a deep-learning model with a 22-layered neural network, for hepatic image analysis obtained with ultrasound. The system exhibited superior accuracy for NAFLD diagnosis and risk stratification compared to conventional ML protocols. The study concluded that the DL Symtosis System is an accurate tool for diagnosing and risk stratification of fatty liver disease. Cao et al. (2019) compared three automated diagnosis techniques for NAFLD using ultrasound images. They employed deep-learning algorithms and evaluated envelope signal, grayscale signal, and deep-learning index methods. The deep-learning index exhibited the best diagnostic capability, achieving high accuracy, sensitivity, and specificity for NAFLD staging. The study demonstrated the feasibility of a complete automated diagnosis of NAFLD using two-dimensional hepatic ultrasound images.

Graffy et al. (2019) developed an automated volume-based liver attenuation method using three-dimensional CNNs for CT-based NAFLD diagnosis. The automated measurements showed high agreement with manual evaluations, suggesting the feasibility of complete automation for NAFLD diagnosis using CT images. Jirapatnakul et al. (2020) utilized noncontrast low-dose chest CT scans for measuring liver attenuation and demonstrated the accuracy and reliability of automated measurements for NAFLD diagnosis. Huo et al. (2019) concluded that hepatic steatosis measured using an automatic CT-based automatic liver attenuation ROI-based measurement (ALARM) achieved a significant match with human radiologists' estimations for liver steatosis. The automated CT-based measurement provided a reliable tool for diagnosing NAFLD. De Rudder et al. (2020) developed an automated method for measuring steatosis in rodents based on quantification of macrovesicular steatosis area. The automated tool showed significant correlations with micro-CT liver density, hepatic fat content, steatosis scores, and gene expression. The study concluded that the automated method is a precise tool for monitoring fatty liver disease evolution in animal models.

Han et al. (2020) developed a DL system based on radiofrequency data obtained from MRI-derived proton density fat fraction. The DL algorithms exhibited high accuracy, sensitivity, specificity, positive predictive value (PPV), and negative predictive value (NPV) for NAFLD diagnosis using radiofrequency ultrasound data.

3. Dataset

The original dataset is taken from Byra et al. (2018). It consists of data from 55 severely obese patients admitted for bariatric surgery. Patients having more than 5% hepatocytes with steatosis were classified as suffering from NAFLD. From the dataset, 38 patients were positive for NAFLD. A sequence of B-mode ultrasound images corresponding to one heartbeat was taken for each patient. From each sequence, 10 consecutive images were used for further processing. Finally, the dataset contained 550 B-mode images of size 434 × 636 pixels. Of these, 380 images belonged to patients suffering from NAFLD, and 170 to patients not suffering from NAFLD. Images of the positive class were augmented to 1140 in number using the Keras Image data generator. The images were augmented using the hyperparameters rotation_range, horizontal_flip, brightness_range, and zoom_range. The images of the negative class were also augmented using the same method to 1020 in number. The images were converted to RGB, and their size was changed to 224 × 224 pixels. Images from the dataset are given in Fig. 7.2.

4. Proposed architecture

We experimented with 11 different CNNs pretrained on ImageNet as our base models. The top layer of all the pretrained models was removed. The head was replaced with a flattened layer instead. For transfer learning, the output of the flattened layer was used as features for our various classifiers. For the ML classifiers, all the extracted features were given as input, whereas for the artificial neural network (ANN) classifier, only the first 128 features were given as input and for LSTM and Bi-LSTM classifiers, the first 32 features were given as input. The default batch size of 32 was used.

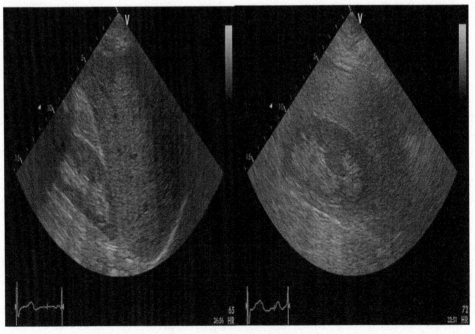

FIGURE 7.2   Ultrasound images. *Left*: healthy liver, *right*: sick liver.

The same training and validation datasets were selected for all networks to facilitate the performance comparison of networks. As mentioned before, the output of the flattened layer is used as features for the various classifiers for both training and testing.

4.1 ResNet

Residual blocks play a crucial role in mitigating the vanishing gradient problem without compromising the depth of the neural architecture. Residual networks, also known as ResNets, excel in retaining and augmenting the knowledge they accumulate by utilizing identity mapping. This mapping is fused with activations after a single step in the network. This approach ensures that the network continues to build upon its prior learning whenever it acquires new insights. In a residual block, the initial weights, denoted as "x," are duplicated to serve as a reference in the skip connection. Subsequently, an activation function is applied to these weights ("x") to yield "F(x)." Following the activation, the original weights ("x") are added to "F(x)" through the skip connection, completing the fundamental mechanism of the residual block. ResNets are constructed by assembling multiple such residual blocks, resulting in deep architectures that have demonstrated remarkable success in tasks like image recognition and various other domains (He et al., 2016).

4.2 DenseNet

In conventional deep-learning structures, every layer executes a sequence of actions. These actions encompass various operations, such as applying activations like ReLU or softmax and performing tasks like convolutions, pooling, and batch normalization. The equation for this would be

$$X_l = H_l(X_{l-1})$$

ResNets extended this behavior, including the skip connection, reformulating this equation into

$$X_l = H_l(X_{l-1}) + X_{l-1}$$

This is where DenseNets usually differ from ResNets. Instead of summing up the history of the weights learned, it is concatenated together. The equation now transforms into

$$X_l = H_l([X_o, X_1, ..., X_{l-1}])$$

DenseNets are constructed using dense blocks that capitalize on the concept of shared information. As feature maps are concatenated, the channel dimension progressively expands with each layer. If we make H_l to produce k feature maps every time, then for the lth layer:

$$k_l = k_o + k(l - 1)$$

This hyperparameter k, the growth rate, measures the amount of information being added to the network in the form of feature maps at each layer (Huang et al., 2017).

4.3 MobileNet

MobileNets rely on a simplified design that employs depthwise separable convolutions to construct compact and efficient deep neural networks. They are recognized for their low memory usage and lightweight nature, attributed to the limited trainable parameters. MobileNets effectively leverage dense blocks by generating numerous output feature maps through minimal convolution operations. Additionally, they maintain a modest growth rate "k," further enhancing computational efficiency and memory usage (Howard et al., 2017).

4.4 Xception

Like MobileNet, Xception adopts depthwise separable convolutions to create efficient deep-learning networks. This architecture was crafted by Google researchers. The essence of the inception module involves applying a depthwise convolution followed by a pointwise convolution. Consequently, the employed depthwise separable convolution resembles an inception module with an increased number of towers. The data's path involves initial traversal through the entry flow, subsequent progression through the middle flow (iterated eight times), and, ultimately, traversal through the exit flow (Chollet, 2017).

4.5 VGG

The VGG architecture is notably straightforward to comprehend. Initially, a 224×224 input image undergoes processing through five convolution blocks containing 333 kernels with a stride of one and ReLU activations. Following these convolutional stages, max-pooling is applied. The image then progresses through three fully connected layers, with the final layer's dimension being 1000 to match 1000 distinct image classes. VGG architectures have demonstrated substantial success in the ImageNet challenge as well (Simonyan & Zisserman, 2015).

4.6 Inception

In many conventional network architectures, it is often unclear when to apply max-pooling and convolutional operations. For example, in AlexNet, these operations are sequential, whereas, in VGGNet, there are clusters of three consecutive convolutional operations before a single max-pooling layer. The concept behind inception is to integrate all these operations simultaneously. It involves the parallel computation of diverse kernels with varying sizes on the same input map, and their outcomes are combined into a singular output. The underlying notion is that using convolution filters of different sizes enhances the network's capability to recognize objects at multiple scales. This approach is referred to as an inception module (Szegedy et al., 2015).

4.7 k-Nearest neighbors

The k-nearest neighbors (k-NN) algorithm is characterized by its simplicity. It involves identifying the k-nearest objects to the test object from the training set, determined using a specific distance metric. Subsequently, the class of the test object is determined through a majority vote count from these k neighbors. The choice of distance metric holds significance in this process. The metric is chosen such that smaller distances between two objects indicate a higher probability of them belonging to the same class. Consequently, when applying k-NN to classify documents, it might be more advantageous to employ cosine similarity as the distance metric rather than the Euclidean distance (Wu et al., 2007).

4.8 Support vector machine

The SVM stands out as a robust and highly accurate method among the well-established ML algorithms. SVM's primary objective is to identify a function that can effectively differentiate between two distinct classes in binary classification scenarios. In cases where the dataset is linearly separable, a linear classification function can be visualized as a hyperplane positioned between the two classes, effectively segregating them. Given the existence of multiple potential hyperplanes, SVM aids in selecting the hyperplane that optimally maximizes the separation between the classes. This separation, sometimes called the margin, can be defined as the distance between the closest instances of different classes as determined by the classifier. By adopting this margin definition, the classifier is compelled to identify the most optimal boundary from the available options. Consequently, this approach enhances the classifier's generalization ability for new, unseen data, as the margin is optimized to achieve the greatest possible extent (Wu et al., 2007).

4.9 Random forests

RFs is an ensemble ML technique that creates a "forest" by amalgamating multiple decision trees. The fundamental concept behind this algorithm is to ensure the independence of these trees, resulting in an overall reduction in error rate. Decision trees operate by segmenting the data iteratively and selecting the most advantageous divisions that lead to accurate categorization. Moreover, each tree employs a random subset of features, intensifying the level of unpredictability and generalization. This deliberate introduction of randomness works to lower collective errors. The amalgamation of predictions generated by multiple decision trees collectively elevates the accuracy of predictions across the dataset (Breiman, 2001; Han et al., 2012).

4.10 Bagging

Ensemble methods leverage the collective insights of multiple models to enhance predictions. An evident approach involves aggregating model decisions to generate a more accurate outcome. This involves taking a majority vote for classification tasks or calculating the average of model predictions for numeric outputs. Within ensemble methods, there exist distinctions between Bagging and boosting. Bagging assigns equal significance to all models, whereas boosting adjusts weights to amplify the influence of more successful models based on their historical performance. In the Bagging technique, multiple training datasets of equivalent size are randomly selected from the problem domain to construct a decision tree for each dataset. Despite their structural similarity, these trees tend to provide consistent predictions for new test instances. Nevertheless, this assumption does not hold universally, particularly when working with small training datasets. This can lead to inaccuracies in predictions for certain models.

To tackle this, Bagging addresses the instability of ML methods by creating new datasets of the same size through random sampling with replacement from the original dataset. This process inadvertently eliminates some instances while duplicating others. Consequently, each of these new datasets is employed in the learning algorithm, and the outcomes collectively contribute to voting for the predicted class (Witten et al., 2011).

4.11 AdaBoost

AdaBoost (Adaptive Boosting) is an elegantly simple ensemble learning method where data points receive weights based on the classifier's predictions. Initially, all data points have equal weights, and the classifier makes predictions. These predictions are then compared to the actual labels, and the weights of correctly predicted data points are decreased while those of incorrect predictions are increased. The next iteration constructs the classifier using this newly weighted dataset. Notably, the earlier misclassified data points carry higher weights, causing the classifier to focus more intently on them compared to data points with lower weights. Through successive iterations, certain challenging instances become even more intricate while others are correctly classified. This outcome varies per model, but the ensemble method's accuracy collectively improves. Due to its ability to adapt by assigning and modifying weights based on current predictions, this approach is named as Adaptive Boosting (Witten et al., 2011).

4.12 XGBoost

XGBoost, also referred to as gradient boosting, is a more contemporary ensemble boosting algorithm akin to AdaBoost. The distinguishing factor lies in its approach to enhancing predictions. In contrast to assigning weights based on predicted values, this algorithm advances through the outcomes of a loss function. This loss function gauges the disparity between current predictions and actual labels. The algorithm's objective is to minimize this loss by calculating gradients in relation to the loss and proceeding in their suggested direction. These steps are determined by a learning rate, a hyperparameter specific to the algorithm. The algorithm's precision improves through repetitive iterations, eventually converging the loss to its minimal value (Chen & Guestrin, 2016).

4.13 Artificial neural network

The concept of ANNs draws inspiration from biological neurons responsible for transmitting information between cells. ANNs mimic the behavior of biological neurons by relaying data across different layers. There are diverse types of ANNs, with the multilayer perceptron (MLP) being the most prevalent, also used in this study (Yilmaz et al., 2010). The MLP architecture has three kinds of layers: input, hidden, and output. Input layers receive input in vector form, hidden layers process this data to extract features, and the output layer presents the final output. The algorithm involves multiplying inputs by weights, adding a bias term, and passing them through a nonlinear function, continuing until reaching the output layer (Wang, 2003, pp. 81–100). Throughout this process, only one layer's neurons are active at a time, where the output of a hidden layer becomes the input for the subsequent one. Increasing the number of hidden layers escalates the algorithm's intricacy. The absence of a nonlinear function leads to an output layer expression that is linearly dependent on inputs, nullifying the purpose of hidden layers and reducing the system to a single-layer

network. Nonlinearity adds complexity to the model and facilitates learning intricate features. Incorporating a bias term enhances model generalization and adaptability. Weights undergo updates through the backpropagation algorithm (Skorpil et al., 2006). This process involves calculating differentials with respect to weights and multiplying them in successive steps. Two prominent challenges arise, the exploding gradient problem, where differentials become excessively large, leading to overflow and NaN values, and the vanishing gradient problem, wherein differentials turn extremely small, causing them to approach zero after successive multiplications, resulting in negligible weight updates.

A simple ANN classifier was used here. It had an input layer with 128 neurons with ReLU activation function. The hidden layer has two consecutive dense layers with 64 and 32 neurons, with ReLU activation functions. The output layer was a dense layer with a single neuron and sigmoid activation function. The model was trained with binary cross-entropy loss, Adam optimizer, and up to 100 epochs. While training, model checkpointing was used to save the model with the highest validation accuracy. Early stopping was also used, monitoring validation loss with a patience of 9 epochs.

4.14 Long short-term memory

LSTM, which stands for long short-term memory, is a specialized recurrent neural network (RNN). Unlike conventional neural networks, RNNs possess an internal memory for storing past states, where inputs are treated as independent entities. RNNs, by considering both input and previous state output, enable decision-making. This attribute makes RNNs well suited for tasks like speech recognition and music composition. However, a challenge arises when using RNNs, specifically in cases of long sequences during backpropagation in time. This can lead to vanishing and exploding gradient problems. In instances where derivatives become exceedingly large, subsequent derivatives may overflow. Conversely, tiny derivatives result in diminished information transfer across time steps, causing previous state effects to weaken. To address exploding gradient issues, they can be clipped within a designated range. To tackle vanishing gradients, the LSTM architecture was developed, enhancing the memory of preceding hidden states. LSTMs wield the capacity to selectively retain or discard information, facilitated by three gates: the input gate, which decides which input value updates the memory; the forget gate, which determines what information is retained or discarded; and the output gate, which controls output information flow by leveraging input and cell memory (Hochreiter & Schmidhuber, 1997).

A simple LSTM classifier was used here. The entire training dataset, as well as the test dataset, was passed in a single timestep. The model had an LSTM layer with 32 neurons, followed by a dense layer with 32 neurons and ReLU activation function. The output layer was a dense layer with a single neuron and sigmoid activation function. The model was trained with binary cross-entropy loss, Adam optimizer, and up to 200 epochs. While training, model checkpointing was used to save the model with the highest validation accuracy. Early stopping was also used, monitoring validation loss with a patience of 9 epochs.

4.15 Bidirectional long short-term memory

The bidirectional LSTM (Bi-LSTM) represents an enhanced iteration of conventional LSTMs. Unlike its predecessor, this design integrates information not just from prior states but also from forthcoming states. This empowers the current state to base decisions on both historical and immediate information. Bi-LSTM incorporates two LSTMs, directing inputs in both forward and backward directions. This augmentation enhances the model's resilience and decision-making efficacy by leveraging past and future hidden states. The other attributes of the Bi-LSTM remain analogous to those of the unidirectional LSTM (Graves et al., 2005).

A simple Bi-LSTM classifier was used here. The entire training dataset, as well as the test dataset, was passed in a single timestep. The model had a Bi-LSTM layer with 32 neurons, followed by a dense layer with 32 neurons and ReLU activation function. The output layer was a dense layer with a single neuron and sigmoid activation function. The model was trained with binary cross-entropy loss, Adam optimizer, and up to 200 epochs. While training, model checkpointing was used to save the model with the highest validation accuracy. Early stopping was also used, monitoring validation loss with a patience of 9 epochs.

5. NAFLD detection with artificial intelligence

We implemented nine different ML classifiers on the extracted features from our proposed architecture. For the k-NN classifier, the value of K was kept as 7. The hyperparameter "probability" was set to true for the support vector classifier. For all the other classifiers, default hyperparameters were used.

```python
import numpy as np
from tensorflow.keras.preprocessing.image import ImageDataGenerator
import cv2
from PIL import Image
import tensorflow as tf
from sklearn.model_selection import train_test_split
from sklearn.metrics import accuracy_score
from sklearn.metrics import f1_score
from sklearn.metrics import cohen_kappa_score
from sklearn.metrics import confusion_matrix
from sklearn import svm
from sklearn.ensemble import RandomForestClassifier
from sklearn.ensemble import AdaBoostClassifier
from sklearn.neighbors import KNeighborsClassifier
from xgboost import XGBClassifier
from sklearn.ensemble import BaggingClassifier
from tensorflow.keras.callbacks import ModelCheckpoint, EarlyStopping
import scipy.io
from matplotlib import pyplot as plt

fname='/content/drive/MyDrive/archive/dataset_liver_bmodes_steatosis_asses
sment_IJCARS.mat' #address of the dataset file
annots = scipy.io.loadmat(fname)
annots['data'][0][1]['images'][1,:,:].shape #initial image shape

plt.imshow(annots['data'][0][1]['images'][3,:,:], interpolation='nearest')
#printing an image
plt.show()
```

Storing images from .mat struct datatype to numpy array along with resizing to 224×224 and conversion to rgb.

```python
x_healthy_data=[]
y_healthy_data=[]
x_sick_data=[]
y_sick_data=[]
k=0
for i in range(55):
  for j in range(10):
    x=annots['data'][0][i]['images'][j,:,:]
    y=annots['data'][0][i]['class']
    x=cv2.resize(x,(224,224))
    x=x.reshape(224,224,1)
    x=tf.image.grayscale_to_rgb(tf.constant(x), name=None)
    x=np.array(x)
    if y==0:
      x_healthy_data.append(x)
      y_healthy_data.append(y)
    else:
      x_sick_data.append(x)
      y_sick_data.append(y)
    k=k+1
X_healthy=np.array(x_healthy_data)
Y_healthy=np.array(y_healthy_data)
X_sick=np.array(x_sick_data)
Y_sick=np.array(y_sick_data)
Y_sick=Y_sick.reshape((380,1))
Y_healthy=Y_healthy.reshape((170,1))
```

**Using data augmentation to increase images
to around 1000 in each class.**

```
IMG_SHAPE = 224
img_gen=          ImageDataGenerator(rescale=1./255,          rotation_range=40,
horizontal_flip=True, brightness_range = (0.5, 1.5), zoom_range = 0.1)


image_gen_healthy = img_gen.flow(X_healthy, Y_healthy, batch_size=170)
image_gen_sick = img_gen.flow(X_sick, Y_sick, batch_size=380)
batch=0

augmented_x_healthy=[]
augmented_y_healthy=[]
for x_batch, y_batch in image_gen_healthy:
  print(x_batch.shape)
  print(y_batch.shape)
  print()
  for a in x_batch:
    augmented_x_healthy.append(a)
    augmented_y_healthy.append(0)
  batch=batch+1
  if batch>5:
    break
batch=0

augmented_x_sick=[]
augmented_y_sick=[]
for x_batch, y_batch in image_gen_sick:
  print(x_batch.shape)
  print(y_batch.shape)
  print()
  for a in x_batch:
    augmented_x_sick.append(a)
    augmented_y_sick.append(1)
  batch=batch+1
  if batch>2:
    break
augmented_x=np.array(augmented_x_healthy+augmented_x_sick)
augmented_y=np.array(augmented_y_healthy+augmented_y_sick)
x_train,x_test,y_train,y_test=train_test_split(augmented_x,augmented_y,tes
t_size=0.1,shuffle=True)
```

Saving Augmented Dataset and loading saved version (Optional)

```
np.save('/content/drive/MyDrive/x_train.npy',x_train)
np.save('/content/drive/MyDrive/x_test.npy',x_test)
np.save('/content/drive/MyDrive/y_train.npy',y_train)
np.save('/content/drive/MyDrive/y_test.npy',y_test)


x_train1=np.load('/content/drive/MyDrive/x_train.npy')
y_train=np.load('/content/drive/MyDrive/y_train.npy')
x_test1=np.load('/content/drive/MyDrive/x_test.npy')
y_test=np.load('/content/drive/MyDrive/y_test.npy')
```

Loading all models used for feature extraction.

```
dl_models=[
    tf.keras.applications.VGG16(
    include_top=False,
    weights="imagenet",
    input_tensor=None,
    input_shape=(IMG_SHAPE, IMG_SHAPE, 3),
    pooling='avg'
),
    tf.keras.applications.VGG19(
    include_top=False,
    weights="imagenet",
    input_tensor=None,
    input_shape=(IMG_SHAPE, IMG_SHAPE, 3),
    pooling='avg'
),
    tf.keras.applications.ResNet50(
    include_top=False,
    weights="imagenet",
    input_tensor=None,
    input_shape=(IMG_SHAPE, IMG_SHAPE, 3),
    pooling='avg'
),
    tf.keras.applications.ResNet101(
    include_top=False,
    weights="imagenet",
    input_tensor=None,
    input_shape=(IMG_SHAPE, IMG_SHAPE, 3),
    pooling='avg'
),
    tf.keras.applications.MobileNetV2(
    include_top=False,
    weights="imagenet",
    input_tensor=None,
    input_shape=(IMG_SHAPE, IMG_SHAPE, 3),
    pooling='avg'
),
    tf.keras.applications.MobileNet(
    include_top=False,
    weights="imagenet",
    input_tensor=None,
```

```
    input_shape=(IMG_SHAPE, IMG_SHAPE, 3),
    pooling='avg'
),
    tf.keras.applications.InceptionV3(
    include_top=False,
    weights="imagenet",
    input_tensor=None,
    input_shape=(IMG_SHAPE, IMG_SHAPE, 3),
    pooling='avg'
),
    tf.keras.applications.InceptionResNetV2(
    include_top=False,
    weights="imagenet",
    input_tensor=None,
    input_shape=(IMG_SHAPE, IMG_SHAPE, 3),
    pooling='avg'
),
    tf.keras.applications.DenseNet169(
    include_top=False,
    weights="imagenet",
    input_tensor=None,
    input_shape=(IMG_SHAPE, IMG_SHAPE, 3),
    pooling='avg'
),
    tf.keras.applications.DenseNet121(
    include_top=False,
    weights="imagenet",
    input_tensor=None,
    input_shape=(IMG_SHAPE, IMG_SHAPE, 3),
    pooling='avg'
),
    tf.keras.applications.Xception(
    include_top=False,
    weights="imagenet",
    input_tensor=None,
    input_shape=(IMG_SHAPE, IMG_SHAPE, 3),
    pooling='avg'
)
]
```

Loading all ML classifiers.

```
models=[
    svm.SVC(probability=True),
    RandomForestClassifier(),
    AdaBoostClassifier(),
    KNeighborsClassifier(n_neighbors=7),
    XGBClassifier(),
    BaggingClassifier(),
]
```

Passing augmented data through all feature extractors and ML classifiers.

```python
count_dl=0
for dl_model in dl_models:
  dl_model.trainable=False

  extraction_model=tf.keras.Sequential([dl_model,
                          tf.keras.layers.Flatten()
                          ])
  x_train = extraction_model.predict(x_train1)
  x_test  = extraction_model.predict(x_test1)


  count_ml=0
  for model in models:
    classifier=model
    classifier.fit(x_train, y_train)
    y_pred = classifier.predict(x_test)
    y_probs=classifier.predict_proba(x_test)
    probability=[]
    for a, b in y_probs:
      probability.append(b)

    print(count_dl)
    print(count_ml)
    accuracy = accuracy_score(y_test, y_pred)
    print('Accuracy: %f' % accuracy)
    # precision tp / (tp + fp)
    precision = precision_score(y_test, y_pred)
    print('Precision: %f' % precision)
    # recall: tp / (tp + fn)
    recall = recall_score(y_test, y_pred)
    print('Recall: %f' % recall)
    # f1: 2 tp / (2 tp + fp + fn)
    f1 = f1_score(y_test, y_pred)
    print('F1 score: %f' % f1)


    # kappa
    kappa = cohen_kappa_score(y_test, y_pred)
    print('Cohens kappa: %f' % kappa)


    # confusion matrix
    matrix = confusion_matrix(y_test, y_pred)
    print(matrix)
    count_ml+=1
  count_dl+=1
```

Passing augmented data through all feature extractors and ANN classifier.

```python
count_dl=0
for dl_model in dl_models:
  dl_model.trainable=False
  extraction_model=tf.keras.Sequential([dl_model,
                          tf.keras.layers.Flatten()
                          ])
  x_train = extraction_model.predict(x_train1)
  x_test  = extraction_model.predict(x_test1)

  classifier=tf.keras.Sequential([tf.keras.layers.Dense(128,
activation='relu'),
                        tf.keras.layers.Dense(64, activation='relu'),
                        tf.keras.layers.Dense(32, activation='relu'),
                        tf.keras.layers.Dense(1, activation='sigmoid')
                        ])
  classifier.compile(loss='binary_crossentropy',
            optimizer='adam',
            metrics=['accuracy'])
  filepath="/content/drive/MyDrive/weights_best.hdf5"
  checkpoint    =    ModelCheckpoint(filepath,    monitor='val_accuracy',
verbose=1, save_best_only=True, mode='max')
  es = EarlyStopping(monitor='val_loss', patience=9)
  callbacks_list = [checkpoint, es]

  history=classifier.fit(x_train,                              y_train,
epochs=200,validation_split=0.1, callbacks=callbacks_list)

  classifier.load_weights("/content/drive/MyDrive/weights_best.hdf5")
  y_probs = classifier.predict(x_test)
  y_pred=(y_probs > 0.5).astype("int32")


  print(count_dl)
  accuracy = accuracy_score(y_test, y_pred)
  print('Accuracy: %f' % accuracy)

  # f1: 2 tp / (2 tp + fp + fn)
  f1 = f1_score(y_test, y_pred)
  print('F1 score: %f' % f1)
```

```
# kappa
kappa = cohen_kappa_score(y_test, y_pred)
print('Cohens kappa: %f' % kappa)

# confusion matrix
matrix = confusion_matrix(y_test, y_pred)
print(matrix)
count_dl+=1
```

Passing augmented data through all feature extractors and LSTM classifier.

```python
count_dl=0
for dl_model in dl_models:
  dl_model.trainable=False
  extraction_model=tf.keras.Sequential([dl_model,
                            tf.keras.layers.Flatten()
                            ])
  x_train = extraction_model.predict(x_train1)
  x_test  = extraction_model.predict(x_test1)

  num_timesteps = 1  # Set the number of timesteps
  x_train  =  np.reshape(x_train,  (x_train.shape[0],  num_timesteps,
x_train.shape[1]))
  x_test  =  np.reshape(x_test,  (x_test.shape[0],  num_timesteps,
x_test.shape[1]))

  classifier=tf.keras.Sequential([tf.keras.layers.LSTM(32),
                        tf.keras.layers.Dense(32, activation='relu'),
                        tf.keras.layers.Dense(1, activation='sigmoid')])
  classifier.compile(loss='binary_crossentropy',
              optimizer='adam',
              metrics=['accuracy'])
  filepath="/content/drive/MyDrive/weights_lstm_best.hdf5"
  checkpoint    =    ModelCheckpoint(filepath,    monitor='val_accuracy',
verbose=1, save_best_only=True, mode='max')
  es = EarlyStopping(monitor='val_loss', patience=9)
  callbacks_list = [checkpoint, es]

  history=classifier.fit(x_train,                              y_train,
epochs=200,validation_split=0.1, callbacks=callbacks_list)

  classifier.load_weights("/content/drive/MyDrive/weights_lstm_best.hdf5")
  y_probs = classifier.predict(x_test)
  y_pred=(y_probs > 0.5).astype("int32")


  print(count_dl)
  accuracy = accuracy_score(y_test, y_pred)
  print('Accuracy: %f' % accuracy)
```

```
# f1: 2 tp / (2 tp + fp + fn)
f1 = f1_score(y_test, y_pred)
print('F1 score: %f' % f1)

# kappa
kappa = cohen_kappa_score(y_test, y_pred)
print('Cohens kappa: %f' % kappa)

# confusion matrix
matrix = confusion_matrix(y_test, y_pred)
print(matrix)
count_dl+=1
```

Passing augmented data through all feature extractors and Bi-LSTM classifier.

```python
count_dl=0
for dl_model in dl_models:
  dl_model.trainable=False
  extraction_model=tf.keras.Sequential([dl_model,
                        tf.keras.layers.Flatten()
                        ])
  x_train = extraction_model.predict(x_train1)
  x_test  = extraction_model.predict(x_test1)


  num_timesteps = 1  # Set the number of timesteps
  x_train   =   np.reshape(x_train,   (x_train.shape[0],   num_timesteps,
x_train.shape[1]))
  x_test   =   np.reshape(x_test,   (x_test.shape[0],   num_timesteps,
x_test.shape[1]))


classifier=tf.keras.Sequential([tf.keras.layers.Bidirectional(tf.keras.lay
ers.LSTM(32)),
                        tf.keras.layers.Dense(32, activation='relu'),
                        tf.keras.layers.Dense(1, activation='sigmoid')])
  classifier.compile(loss='binary_crossentropy',
             optimizer='adam',
             metrics=['accuracy'])
  filepath="/content/drive/MyDrive/weights_bilstm_best.hdf5"
  checkpoint   =   ModelCheckpoint(filepath,   monitor='val_accuracy',
verbose=1, save_best_only=True, mode='max')
  es = EarlyStopping(monitor='val_loss', patience=9)
  callbacks_list = [checkpoint, es]


  history=classifier.fit(x_train,                        y_train,
epochs=200,validation_split=0.1, callbacks=callbacks_list)


classifier.load_weights("/content/drive/MyDrive/weights_bilstm_best.hdf5")
  y_probs = classifier.predict(x_test)
  y_pred=(y_probs > 0.5).astype("int32")


  print(count_dl)
  accuracy = accuracy_score(y_test, y_pred)
  print('Accuracy: %f' % accuracy)
```

```
# f1: 2 tp / (2 tp + fp + fn)
f1 = f1_score(y_test, y_pred)
print('F1 score: %f' % f1)


# kappa
kappa = cohen_kappa_score(y_test, y_pred)
print('Cohens kappa: %f' % kappa)


# confusion matrix
matrix = confusion_matrix(y_test, y_pred)
print(matrix)
count_dl+=1
```

6. Results and discussions

6.1 Performance evaluation measures

Evaluating a classifier's effectiveness using the training set does not reliably indicate its performance on an independent test set. The question of predicting performance based on limited data is an interesting yet controversial one. Classifier performance is quantified by the number of accurate predictions among the samples, contributing to its accuracy. A higher accuracy signifies more correctness. However, relying solely on training set accuracy is not a good enough measure to gauge classifier performance. Therefore, the need for a test set comes into the picture, which is also representative of the same dataset but which the classifier has never seen before. Formally, the training and test sets are distinct and separate. The classifier is trained on the training set to create a proficient classifier for the task and subsequently assessed by measuring its accuracy on the test set. The training set is kept large for effective training, while the test set remains small (Witten et al., 2011).

The true positives (TP) and true negatives (TN) refer to the number of samples for which the classification was correct. A false positive (FP) is when the outcome is incorrectly predicted

as yes (or positive) when it is actually no (negative). A false negative (FN) is when the outcome is incorrectly predicted as negative when it is actually positive (Witten et al., 2011). Information retrieval researchers define parameters called recall and precision:

$$\text{Recall} = \frac{TP}{TP + FN}$$

$$\text{Precision} = \frac{TP}{TP + FP}$$

Then F1 measure can be formulated as:

$$F1 = \frac{2 * \text{Precision} * \text{Recall}}{\text{Precision} + \text{Recall}}$$

Cohen's kappa statistic is a metric that factors in instances where the classifier's classifications were purely coincidental, utilizing the notion of expectation (Wu et al., 2007). This statistic (κ) inherently regards predictions as probabilistic and subject to chance. Probability calculations for correct chance-based predictions and confidence stem from the expectation of a random variable. The kappa statistic is the prevalent metric for assessing categorical data when independent means to evaluate the likelihood of chance agreement between multiple observers are unavailable. A kappa value of 0 signifies chance occurrence, while a value of 1 reflects perfect concurrence (Lantz & Nebenzahl, 1996; Viera & Garrett, 2005). Cohen (Cohen, 1960) defined the kappa statistic as an agreement index and defined it as the following:

$$K = \frac{P_o - P_e}{1 - P_e}$$

where P_0 is the observed agreement and P_e measures the agreement expected by chance (Yang & Zhou, 2015).

6.2 Experimental results

k-NN, SVM, Random Forest, AdaBoost, XGBoost, Bagging, ANN, LSTM, and Bi-LSTM classifiers were applied to features extracted from each pretrained model using the proposed architecture. For ANN, LSTM, and Bi-LSTM classifiers, the versions with the highest validation accuracy, as saved by model checkpoint, were used for calculating performance metrics. A total of 11 different pretrained models (VGG16, VGG19, MobileNet, MobileNetV2, InceptionV3, InceptionResNetV2, DenseNet121, DenseNet169, Xception, ResNet, and ResNet50) were for feature extraction.

In Table 7.1, VGG16 architecture was used for feature extraction. It can be seen that the "Bi-LSTM" classifier performed the best, with an F1 score of 0.942085. The "SVM" classifier gave the worst performance with an F1 score of 0.83004.

In Table 7.2, VGG19 architecture was used for feature extraction. Among the classifiers, the "Bi-LSTM" classifier performed the best with an F1 score of 0.924303, while the worst performance was given by the "SVM" classifier with an F1 score of 0.756554.

In Table 7.3, ResNet50 architecture was used for feature extraction. The classifiers that achieved the highest and the lowest F1 score were "XGBoost," with a score of 0.869919, and "SVM," with an F1 score of 0.743363, respectively.

In Table 7.4, ResNet101 architecture was used for feature extraction. It can be observed that the "XGBoost" classifier performed the best with an F1 score of 0.844961, and the worst performance

TABLE 7.1 Feature extraction with **VGG16**.

Model	Accuracy (%)	F1 score	Kappa
SVM	80.09%	0.83004	0.589825
Random forest	89.81%	0.912698	0.790476
AdaBoost	82.87%	0.845188	0.654743
KNN	82.41%	0.841667	0.700701
XGBoost	92.13%	0.930041	0.64486
Bagging	91.20%	0.921811	0.821596
ANN	92.59%	0.937008	0.847134
LSTM	92.59%	0.935484	0.84858
Bi-LSTM	93.06%	0.942085	0.855538

TABLE 7.2 Feature extraction with **VGG19**.

Model	Accuracy (%)	F1 score	Kappa
SVM	69.91%	0.756554	0.365854
Random forest	84.72%	0.847222	0.684211
AdaBoost	77.78%	0.806452	0.545741
KNN	81.94%	0.842105	0.631496
XGBoost	91.20%	0.92607	0.8176
Bagging	81.48%	0.84127	0.619048
ANN	88.43%	0.90566	0.756888
LSTM	90.28%	0.918919	0.797753
Bi-LSTM	91.20%	0.924303	0.819334

TABLE 7.3 Feature extraction with **ResNet50**.

Model	Accuracy (%)	F1 score	Kappa
SVM	59.72%	0.743363	0.038674
Random forest	81.48%	0.837398	0.622642
AdaBoost	70.83%	0.746988	0.402844
KNN	66.67%	0.692308	0.333333
XGBoost	85.19%	0.869919	0.698113
Bagging	76.85%	0.793388	0.53125
ANN	73.61%	0.755365	0.473035
LSTM	78.70%	0.825758	0.553398
Bi-LSTM	77.31%	0.803213	0.535545

was given by the "SVM" classifier with an F1 score of 0.701754.

In Table 7.5, MobileNetV2 architecture was used for feature extraction. It can be seen that the "Bi-LSTM" classifier performed the best, with an F1 score of 0.950413. The "AdaBoost" classifier gave the worst performance, with an F1 score of 0.837398.

In Table 7.6, MobileNet architecture was used for feature extraction. Among the classifiers, "ANN," "LSTM," and "Bi-LSTM" classifiers performed the best with an F1 score of 1, while the worst performance was given by the "Bagging" classifier with an F1 score of 0.90535.

In Table 7.7, InceptionV3 architecture was used for feature extraction. The classifiers that achieved the highest and the lowest F1 score were "Bi-LSTM," with a score of 0.955466, and "Bagging," with an F1 score of 0.859504, respectively.

In Table 7.8, InceptionResNetV2 architecture was used for feature extraction. It can be observed that the "LSTM" classifier performed the best with an F1 score of 0.946939, and the worst performance was given by the "Bagging" classifier with an F1 score of 0.806584.

In Table 7.9, DenseNet169 architecture was used for feature extraction. It can be seen that the "LSTM" classifier performed the best with an F1 score of 0.968. The "Bagging" classifier

TABLE 7.5 Feature extraction with **MobileNetV2**.

Model	Accuracy (%)	F1 score	Kappa
SVM	87.50%	0.890688	0.744882
Random forest	87.96%	0.896825	0.752381
AdaBoost	81.48%	0.837398	0.622642
KNN	87.96%	0.892562	0.75625
XGBoost	88.43%	0.896266	0.765991
Bagging	82.41%	0.841667	0.64486
ANN	93.06%	0.941634	0.856
LSTM	93.06%	0.941634	0.856
Bi-LSTM	94.44%	0.950413	0.8875

TABLE 7.6 Feature extraction with **MobileNet**.

Model	Accuracy (%)	F1 score	Kappa
SVM	96.76%	0.972112	0.933439
Random forest	96.76%	0.97166	0.933858
AdaBoost	87.96%	0.898438	0.750799
KNN	99.07%	0.992063	0.980952
XGBoost	96.30%	0.968254	0.92381
Bagging	89.35%	0.90535	0.784038
ANN	100.00%	1	1
LSTM	100.00%	1	1
Bi-LSTM	100.00%	1	1

TABLE 7.4 Feature extraction with **ResNet101**.

Model	Accuracy (%)	F1 score	Kappa
SVM	60.65%	0.701754	0.145729
Random forest	78.24%	0.817121	0.5488
AdaBoost	70.37%	0.746032	0.390476
KNN	68.06%	0.708861	0.35814
XGBoost	81.48%	0.844961	0.615385
Bagging	70.83%	0.731915	0.415765
ANN	75.93%	0.79845	0.5
LSTM	77.31%	0.798354	0.539906
Bi-LSTM	77.78%	0.801653	0.55

TABLE 7.7 Feature extraction with **InceptionV3**.

Model	Accuracy (%)	F1 score	Kappa
SVM	93.52%	0.943548	0.867508
Random forest	90.28%	0.915663	0.800948
AdaBoost	87.04%	0.890625	0.731629
KNN	90.28%	0.916996	0.799682
XGBoost	93.98%	0.947791	0.876777
Bagging	84.26%	0.859504	0.68125
ANN	93.98%	0.948207	0.876387
LSTM	93.98%	0.939815	0.877165
Bi-LSTM	94.91%	0.955466	0.896063

TABLE 7.8 Feature extraction with InceptionResNetV2.

Model	Accuracy (%)	F1 score	Kappa
SVM	84.72%	0.860759	0.693023
Random forest	85.19%	0.870968	0.870968
AdaBoost	79.17%	0.822134	0.570747
KNN	83.33%	0.85	0.663551
XGBoost	86.11%	0.877049	0.717868
Bagging	78.24%	0.806584	0.558685
ANN	93.52%	0.944	0.944
LSTM	93.98%	0.946939	0.877551
Bi-LSTM	93.06%	0.941634	0.856

TABLE 7.9 Feature extraction with DenseNet169.

Model	Accuracy (%)	F1 score	Kappa
SVM	91.20%	0.921811	0.821596
Random forest	91.67%	0.928	0.829114
AdaBoost	87.96%	0.898438	0.750799
KNN	92.13%	0.92827	0.84186
XGBoost	93.52%	0.944882	0.866242
Bagging	87.50%	0.893281	0.742448
ANN	94.91%	0.955823	0.895735
LSTM	96.30%	0.968	0.924051
Bi-LSTM	95.83%	0.964143	0.914422

TABLE 7.10 Feature extraction with DenseNet121.

Model	Accuracy (%)	F1 score	Kappa
SVM	87.96%	0.899225	0.75
Random forest	88.89%	0.90625	0.769968
AdaBoost	82.41%	0.846774	0.640379
KNN	86.57%	0.876596	0.731066
XGBoost	93.98%	0.94902	0.875598
Bagging	83.33%	0.85	0.663551
ANN	95.83%	0.965251	0.913323
LSTM	93.52%	0.946565	0.864516
Bi-LSTM	95.83%	0.964706	0.913876

TABLE 7.11 Feature extraction with Xception.

Model	Accuracy (%)	F1 score	Kappa
SVM	88.43%	0.897959	0.764521
Random forest	87.04%	0.888	0.734177
AdaBoost	76.85%	0.788136	0.535604
KNN	86.57%	0.878661	0.729393
XGBoost	88.43%	0.895397	0.766719
Bagging	78.24%	0.796537	0.56682
ANN	92.59%	0.937008	0.847134
LSTM	92.59%	0.936508	0.847619
Bi-LSTM	92.59%	0.936	0.848101

gave the worst performance, with an F1 score of 0.893281.

In Table 7.10, DenseNet121 architecture was used for feature extraction. Among the classifiers, the "ANN" classifier performed the best with an F1 score of 0.965251, while the worst performance was given by the "AdaBoost" classifier with an F1 score of 0.846774.

In Table 7.11, Xception architecture was used for feature extraction. It can be seen that the "ANN" classifier performed the best with an F1 score of 0.937008. The "AdaBoost" classifier gave the worst performance, with an F1 score of 0.788136.

7. Discussion

Because of its noninvasive nature, low cost, and widespread availability, the use of ultrasound images in the diagnosis of fatty liver disease has been a frequent practice for many years. However, medical practitioners' visual interpretation of ultrasound pictures can be subjective and time-consuming. Furthermore, the accuracy of the diagnosis may vary based on the radiologist's or physician's experience. Artificial intelligence (AI)-based techniques offer a novel approach to addressing these issues and improving the efficiency and accuracy of fatty liver disease identification.

Researchers can create automated systems that can successfully evaluate ultrasound images and deliver reliable diagnostic results by leveraging AI and ML methods. These AI models can be trained on massive datasets of tagged ultrasound images containing annotations indicating the presence or severity of fatty liver disease. Deep-learning techniques, such as CNNs, have demonstrated promising results in picture classification tasks and have been successfully employed in medical image analysis, including the diagnosis of liver disease.

One of the most significant advantages of AI-based fatty liver disease detection is its capacity to extract subtle details and patterns from ultrasound images that human observers may miss. Even modest changes in liver texture, echogenicity, and structural anomalies that are symptomatic of fatty liver disease can be detected by AI models. Furthermore, since AI models learn and improve over time with new data, their diagnostic accuracy can be enhanced, resulting in more trustworthy and consistent outcomes.

However, some obstacles and issues must be solved before AI-based fatty liver disease detection systems can be deployed. To begin, getting big and diverse datasets of tagged ultrasound images can be a challenge when training powerful AI models. Collaboration among medical organizations and data-sharing efforts can aid in addressing this problem. Furthermore, before being integrated into clinical practice, AI models must be rigorously examined and clinically tested to verify their safety and effectiveness.

From the obtained results, we observe that transfer learning is effective in the classification of NAFLD using ultrasound images. MobileNet performed especially well. ANN, LSTM, and Bi-LSTM classifiers achieved 100% test accuracy using features extracted by MobileNet. The classifiers also achieved a perfect F1 score and Kappa value. Other classifiers, such as Random Forest, XGBoost, and Bagging, also performed well. All classifiers trained on features extracted from VGG16, MobileNetV2, MobileNet, InceptionV3, InceptionResNetV2, DenseNet169, and DenseNet121 had an F1 score higher than 0.8.

Hence the use of AI and ML techniques in the detection of fatty liver disease using ultrasound images has significant promise for enhancing the accuracy, efficiency, and accessibility of diagnosis. As technology progresses and annotated data becomes more widely available, AI-based systems have the potential to become significant tools in the early detection and management of fatty liver disease, benefiting both patients and healthcare practitioners. Collaboration between the medical and AI research groups, on the other hand, is critical for overcoming difficulties and ensuring responsible and ethical AI application in clinical practice.

8. Conclusion

The use of AI in the identification of fatty liver disease using ultrasound images represents a significant achievement in the field of medical imaging and diagnostic technologies. Because fatty liver disease is a common and potentially fatal disorder, early and precise identification is critical for successful patient management and improved outcomes.

Using AI and ML systems to analyze ultrasound images has various advantages over traditional manual interpretation. AI models may be trained on enormous datasets, allowing them to distinguish complicated patterns and tiny changes in liver texture and echogenicity that human observers may find difficult to detect. As AI models learn from additional data over time, their diagnostic performance can improve, resulting in more trustworthy and consistent outcomes.

Furthermore, because ultrasonic imaging is noninvasive and inexpensive, it is an appealing choice for fatty liver disease screening and monitoring. By merging this imaging modality with AI-based algorithms, healthcare professionals can streamline the diagnosis process, reduce radiologists' burden, and perhaps enhance access to diagnostic services, particularly in areas where healthcare resources are limited.

As a result, incorporating AI in fatty liver disease diagnosis utilizing ultrasound images has enormous promise for improving diagnostic accuracy, speed, and accessibility. We can dramatically enhance patient outcomes and contribute to more effective management of fatty liver disease by combining the experience of healthcare professionals with the analytical capacity of AI. Continued research, collaboration, and responsible AI technology adoption are required to reap the full benefits of this disruptive approach in clinical practice. As AI technology advances, it is anticipated to play a larger role in the future of medical imaging and disease identification, ultimately benefiting people globally.

References

Acharya, U. R., Raghavendra, U., Fujita, H., Hagiwara, Y., Koh, J. E., Jen Hong, T., Sudarshan, V. K., Vijayananthan, A., Yeong, C. H., Gudigar, A., & Ng, K. H. (2016). Automated characterization of fatty liver disease and cirrhosis using curvelet transform and entropy features extracted from ultrasound images. *Computers in Biology and Medicine, 79*, 250—258. https://doi.org/10.1016/j.compbiomed.2016.10.022

Biswas, M., Kuppili, V., Edla, D. R., Suri, H. S., Saba, L., Marinhoe, R. T., Sanches, J. M., & Suri, J. S. (2018). Symtosis: A liver ultrasound tissue characterization and risk stratification in optimized deep learning paradigm. *Computer Methods and Programs in Biomedicine, 155*, 165—177. https://doi.org/10.1016/j.cmpb.2017.12.016

Breiman, L. (2001). Random forests. *Machine Learning, 45*(1), 5—32. https://doi.org/10.1023/a:1010933404324

Byra, M., Styczynski, G., Szmigielski, C., Kalinowski, P., Michałowski, Ł., Paluszkiewicz, R., Ziarkiewicz-Wróblewska, B., Zieniewicz, K., Sobieraj, P., & Nowicki, A. (2018). Transfer learning with deep convolutional neural network for liver steatosis assessment in ultrasound images. *International Journal of Computer Assisted Radiology and Surgery, 13*(12), 1895—1903. https://doi.org/10.1007/s11548-018-1843-2

Cao, W., An, X., Cong, L., Lyu, C., Zhou, Q., & Guo, R. (2019). Application of deep learning in quantitative analysis of 2-dimensional ultrasound imaging of nonalcoholic fatty liver disease. *Journal of Ultrasound in Medicine, 39*(1), 51—59. https://doi.org/10.1002/jum.15070

Chen, T., & Guestrin, C. (2016). XGBoost. In *Proceedings of the 22nd ACM SIGKDD international conference on knowledge discovery and data mining*. https://doi.org/10.1145/2939672.2939785

Chollet, F. (2017). Xception: Deep learning with depthwise separable convolutions. In *2017 IEEE conference on computer vision and pattern recognition (CVPR)*. https://doi.org/10.1109/cvpr.2017.195

Cohen, J. (1960). A coefficient of agreement for nominal scales. *Educational and Psychological Measurement, 20*(1), 37—46. https://doi.org/10.1177/001316446002000104

De Rudder, M., Bouzin, C., Nachit, M., Louvegny, H., Vande Velde, G., Julé, Y., & Leclercq, I. A. (2020). Automated computerized image analysis for the user-independent evaluation of disease severity in preclinical models of NAFLD/Nash. *Laboratory Investigation, 100*(1), 147—160. https://doi.org/10.1038/s41374-019-0315-9

Gerstenmaier, J. F., & Gibson, R. N. (2014). Ultrasound in chronic liver disease. *Insights into Imaging, 5*(4), 441—455. https://doi.org/10.1007/s13244-014-0336-2

Graffy, P. M., Sandfort, V., Summers, R. M., & Pickhardt, P. J. (2019). Automated liver fat quantification at nonenhanced abdominal CT for population-based steatosis assessment. *Radiology, 293*(2), 334—342. https://doi.org/10.1148/radiol.2019190512

Graves, A., Fernández, S., & Schmidhuber, J. (2005). Bidirectional LSTM networks for improved phoneme classification and recognition. *Lecture Notes in Computer Science*, 799—804. https://doi.org/10.1007/11550907_126

Han, A., Byra, M., Heba, E., Andre, M. P., Erdman, J. W., Loomba, R., Sirlin, C. B., & O'Brien, W. D. (2020). Noninvasive diagnosis of nonalcoholic fatty liver disease and quantification of liver fat with radiofrequency ultrasound data using one-dimensional convolutional neural networks. *Radiology, 295*(2), 342—350. https://doi.org/10.1148/radiol.2020191160

Han, J., Kamber, M., & Pei, J. (2012). *Data mining: Concepts and techniques*. Morgan Kaufmann: Elsevier.

He, K., Zhang, X., Ren, S., & Sun, J. (2016). Deep residual learning for image recognition. In *2016 IEEE conference* on computer vision and pattern recognition (CVPR). https://doi.org/10.1109/cvpr.2016.90

Hochreiter, S., & Schmidhuber, J. (1997). Long short-term memory. *Neural Computation, 9*(8), 1735–1780. https://doi.org/10.1162/neco.1997.9.8.1735

Howard, A. G., Zhu, M., Chen, B., Kalenichenko, D., Wang, W., Weyand, T., Andreetto, M., & Adam, H. (April 17, 2017). MobileNets: Efficient convolutional networks for mobile vision applications. *arXiv.org*. https://arxiv.org/abs/1704.04861.

Huang, G., Liu, Z., Van Der Maaten, L., & Weinberger, K. Q. (2017). Densely connected convolutional networks. In *2017 IEEE conference on computer vision and pattern recognition (CVPR)*. https://doi.org/10.1109/cvpr.2017.243

Huo, Y., Terry, J. G., Wang, J., Nair, S., Lasko, T. A., Freedman, B. I., Carr, J. J., & Landman, B. A. (2019). Fully automatic liver attenuation estimation combing CNN segmentation and morphological operations. *Medical Physics, 46*(8), 3508–3519. https://doi.org/10.1002/mp.13675

Jirapatnakul, A., Reeves, A. P., Lewis, S., Chen, X., Ma, T., Yip, R., Chin, X., Liu, S., Perumalswami, P. V., Yankelevitz, D. F., Crane, M., Branch, A. D., & Henschke, C. I. (2020). Automated measurement of liver attenuation to identify moderate-to-severe hepatic steatosis from chest CT scans. *European Journal of Radiology, 122*, 108723. https://doi.org/10.1016/j.ejrad.2019.108723

Lantz, C. A., & Nebenzahl, E. (1996). Behavior and interpretation of the κ statistic: Resolution of the two paradoxes. *Journal of Clinical Epidemiology, 49*(4), 431–434. https://doi.org/10.1016/0895-4356(95)00571-4

Liu, S., Wang, Y., Yang, X., Lei, B., Liu, L., Li, S. X., Ni, D., & Wang, T. (2019). Deep learning in medical ultrasound analysis: A review. *Engineering, 5*(2), 261–275. https://doi.org/10.1016/j.eng.2018.11.020

Lupsor-Platon, M., Serban, T., Silion, A. I., Tirpe, G. R., Tirpe, A., & Florea, M. (2021). Performance of ultrasound techniques and the potential of artificial intelligence in the evaluation of hepatocellular carcinoma and non-alcoholic fatty liver disease. *Cancers, 13*(4), 790. https://doi.org/10.3390/cancers13040790

Maurice, J., & Manousou, P. (2018). Non-alcoholic fatty liver disease. *Clinical Medicine, 18*(3), 245–250. https://doi.org/10.7861/clinmedicine.18-3-245

Mihăilescu, D. M., Sporea, I., Popescu, A., Toma, C. I., & Gui, V. (2013). Computer aided diagnosis method for steatosis rating in ultrasound images using random forests. *Medical Ultrasonography, 15*(3), 184–190. https://doi.org/10.11152/mu.2013.2066.153.dmm1vg2

Nagy, G., Munteanu, M., Gordan, M., Chira, R., Iancu, M., Crisan, D., & Mircea, P. A. (2015). Computerized ultrasound image analysis for noninvasive evaluation of hepatic steatosis. *Medical Ultrasonography, 17*(4). https://doi.org/10.11152/mu.2013.2066.174.cmu

Popa, S. L., Ismaiel, A., Cristina, P., Cristina, M., Chiarioni, G., David, L., & Dumitrascu, D. L. (2021). Non-alcoholic fatty liver disease: Implementing complete automated diagnosis and staging. A systematic review. *Diagnostics, 11*(6), 1078. https://doi.org/10.3390/diagnostics11061078

Pouwels, S., Sakran, N., Graham, Y., Leal, A., Pintar, T., Yang, W., Kassir, R., Singhal, R., Mahawar, K., & Ramnarain, D. (2022). Non-alcoholic fatty liver disease (NAFLD): A review of pathophysiology, clinical management and effects of weight loss. *BMC Endocrine Disorders, 22*(1). https://doi.org/10.1186/s12902-022-00980-1

Ribeiro, R. T., Tato Marinho, R., & Sanches, J. M. (2014). An ultrasound-based computer-aided diagnosis tool for steatosis detection. *IEEE Journal of Biomedical and Health Informatics, 18*(4), 1397–1403. https://doi.org/10.1109/jbhi.2013.2284785

Sarvamangala, D. R., & Kulkarni, R. V. (2021). Convolutional neural networks in medical image understanding: A survey. *Evolutionary Intelligence, 15*(1), 1–22. https://doi.org/10.1007/s12065-020-00540-3

Simonyan, K., & Zisserman, A. (April 10, 2015). Very deep convolutional networks for large-scale image recognition. *arXiv.org*. https://arxiv.org/abs/1409.1556.

Skorpil, V., & Stastny, J. (2006). Neural networks and back propagation algorithm. *Electron Bulg Sozopol, 20*(2).

Subramanya, M. B., Kumar, V., Mukherjee, S., & Saini, M. (2014). A CAD system for B-mode fatty liver ultrasound images using texture features. *Journal of Medical Engineering & Technology, 39*(2), 123–130. https://doi.org/10.3109/03091902.2014.990160

Sumida, Y. (2014). Limitations of liver biopsy and non-invasive diagnostic tests for the diagnosis of nonalcoholic fatty liver disease/nonalcoholic steatohepatitis. *World Journal of Gastroenterology, 20*(2), 475. https://doi.org/10.3748/wjg.v20.i2.475

Szegedy, C., Liu, W., Jia, Y., Sermanet, P., Reed, S., Anguelov, D., Erhan, D., Vanhoucke, V., & Rabinovich, A. (2015). Going deeper with convolutions. In *2015 IEEE conference on computer vision and pattern recognition (CVPR)*. https://doi.org/10.1109/cvpr.2015.7298594

Viera, A. J., & Garrett, J. M. (May 2005). Understanding interobserver agreement: The kappa statistic. *Family Medicine, 37*, 360. https://pubmed.ncbi.nlm.nih.gov/15883903/.

Wang, S.-C. (2003). *Artificial neural network. Interdisciplinary computing in Java programming*. https://doi.org/10.1007/978-1-4615-0377-4_5

Willemink, M. J., Koszek, W. A., Hardell, C., Wu, J., Fleischmann, D., Harvey, H., Folio, L. R., Summers, R. M., Rubin, D. L., & Lungren, M. P. (2020). Preparing medical imaging data for machine learning.

Radiology, 295(1), 4—15. https://doi.org/10.1148/radiol.2020192224

Witten, I. H., Frank, E., & Hall, M. A. (2011). *Data mining: Practical machine learning tools and techniques*. Elsevier.

Wu, X., Kumar, V., Ross Quinlan, J., Ghosh, J., Yang, Q., Motoda, H., McLachlan, G. J., Ng, A., Liu, B., Yu, P. S., Zhou, Z.-H., Steinbach, M., Hand, D. J., & Steinberg, D. (2007). Top 10 algorithms in data mining. *Knowledge and Information Systems, 14*(1), 1—37. https://doi.org/10.1007/s10115-007-0114-2

Yang, Z., & Zhou, M. (2015). Kappa statistic for clustered physician—patients polytomous data. *Computational Statistics & Data Analysis, 87*, 1—17. https://doi.org/10.1016/j.csda.2015.01.007

Yilmaz, L., Erik, N., & Kaynar, O. (2010). Different types of learning algorithms of artificial neural network (ANN) models for prediction of gross calorific value (GCV) of coals. *Scientific Research and Essays, 5*(2010), 2242—2249.

Zhang, Y. N., Fowler, K. J., Hamilton, G., Cui, J. Y., Sy, E. Z., Balanay, M., Hooker, J. C., Szeverenyi, N., & Sirlin, C. B. (2018). Liver fat imaging—A clinical overview of ultrasound, CT, and MR imaging. *British Journal of Radiology*, 20170959. https://doi.org/10.1259/bjr.20170959

Deep learning approaches for breast cancer detection using breast MRI

Tanisha Sahu[1], Abdulhamit Subasi[2,3]

[1]Department of Computer Science and Engineering, Indian Institute of Technology, Indore, Madhya Pradesh, India; [2]Institute of Biomedicine, Faculty of Medicine, University of Turku, Turku, Finland; [3]Department of Computer Science, College of Engineering, Effat University, Jeddah, Saudi Arabia

1. Introduction

Breast cancer is a significant health problem that affects millions of women globally. Detecting breast cancer early is crucial for improving treatment outcomes and increasing survival rates. While mammography and ultrasound are conventional methods of breast cancer detection, they have limitations in terms of accuracy and sensitivity. Therefore, researchers are increasingly interested in exploring alternative techniques, such as artificial intelligence (AI) and deep learning algorithms, for breast cancer detection from breast magnetic resonance imaging (MRI). MRI has shown promise in providing detailed images of breast tissue and detecting suspicious lesions. However, interpreting MRI scans requires expertise and can be time-consuming. This is where deep learning models, particularly convolutional neural networks (CNNs), come into play. CNNs have demonstrated remarkable capabilities in image recognition and analysis tasks. By training these models on a large dataset of annotated breast MRI images, they can learn to identify specific features and patterns associated with breast cancer. These features may include the shape, size, and texture of tumors or abnormal regions within the breast tissue.

Our research aims to investigate the diagnostic performance of various deep learning models in screening for breast cancer using breast MRI images. We developed a CNN-based algorithm that could accurately classify breast MRI scans as either malignant or benign. By analyzing the unique features captured by CNN, we aim to improve the accuracy of breast cancer detection and assist radiologists in making more informed decisions. The utilization of AI and deep learning in breast cancer detection holds great potential for improving the efficiency and reliability of diagnosis. If successful, this technology could aid in the early detection of breast cancer, leading to timely intervention and improved patient outcomes. Moreover, the widespread availability of MRI scans and the advancements in computing power make it feasible to deploy AI-based breast cancer detection

Applications of Artificial Intelligence in Healthcare and Biomedicine
https://doi.org/10.1016/B978-0-443-22308-2.00012-3

205

systems in clinical settings, benefiting a larger population of women at risk.

Our goal is to develop a robust and accurate tool for breast cancer detection by harnessing the power of deep learning and leveraging the information contained within breast MRI images. We hope that our research will contribute to the ongoing efforts in combating breast cancer and providing women with better healthcare options for early detection and treatment.

2. Literature review

Breast cancer is the most common cancer among women, and early detection plays a crucial role in improving survival rates. MRI is a highly sensitive imaging modality used for breast cancer detection, even in its early stages. However, the interpretation of MRI images can be challenging, necessitating the development of automated methods to assist radiologists in detecting breast cancer. In recent years, deep learning techniques have emerged as powerful tools for medical image analysis, including breast cancer detection from MRI images. This literature review explores recent advances in deep learning approaches for breast cancer detection. Several notable studies have made significant contributions to the development of automated methods for breast cancer detection using deep learning techniques.

Yang et al. (2022) proposed a novel multimodality relation attention network. Witowski et al. (2022) proposed a deep learning system to improve the accuracy of breast cancer diagnosis in DCE-MRI. Dar et al. (2022) provided a critical review of research on breast cancer detection and classification using various imaging modalities and AI techniques. Eskreis-Winkler et al. (2021) used a convolutional neural network to classify breast MRI slices into "cancer" and "no cancer" categories with high accuracy. Zhao et al.

(2023) presented the effectiveness of deep learning models in breast MRI for diagnosis, classification, and prediction. Rautela et al. (2022) explored breast cancer screening techniques. Altabella et al. (2022) proposed machine learning techniques to analyze multiparametric breast MRI data.

Alruwaili and Gouda (2022) used the ResNet50 and NASNet-Mobile network. Yin et al. (2022) used multiparametric MRI-based CNNs. Sun et al. (2023) propose a weakly supervised CNN based on self-transfer learning. Seemendra et al. (2020) developed a CNN with image augmentation. Ashfaq et al. (2022, pp. 2292–2297) proposed a transfer learning model based on AlexNet. Wang et al. (2019) used a deep learning (DL)-empowered diagnostic system that can be used to classify benign versus malignant tumors with high accuracy. Chiu et al. (2020) used a new processing method that combines PCA, MLP, and support vector machine (SVM) to predict breast cancer with high accuracy. Chugh et al. (2021) proposed machine learning and deep learning techniques to develop computer-aided diagnosis (CAD) systems for breast cancer diagnosis.

Aidossov et al. (2023) proposed a deep neural network with Bayesian Networks that can be used to detect breast cancer from thermograms. Khalil et al. (2023) fine-tuned 3D U-Net models with data augmentation. Cong et al. (2023) used deep learning methods to classify breast cancer lesions with high accuracy. Murtaza et al. (2020) used a deep learning-based breast cancer classification method that used medical imaging modalities. Dewangan et al. (2022) proposed a novel Back Propagation Boosting Recurrent Wienmed (BPBRW) model for detecting breast cancer in an earlier stage. Zhang et al. (2023) combined Mask R-CNN and ResNet50.

Abunasser et al. (2023) proposed a deep learning model to detect and classify breast cancers into eight classes. Rane et al. (2020) used six machine learning algorithms to compare the Wisconsin Diagnostic Breast Cancer (WDBC) dataset to determine the best algorithm for use in a breast cancer diagnosis website. Houssein et al. (2021) proposed machine learning and deep learning techniques to develop CAD systems for breast cancer diagnosis and classification. Fujioka et al. (2021) used CNNs to develop models that can discriminate between benign and malignant lesions on maximum intensity projections of dynamic contrast-enhanced breast MRI. Chung et al. (2022) proposed DL methods to perform medical imaging tasks. Altabella et al. (2022) utilized machine learning techniques to develop new methods for breast MRI segmentation.

Khan et al. (2019) developed a novel deep learning framework for the detection and classification of breast cancer in breast cytology images. Aljuaid et al. (2022) proposed a novel CAD method for breast cancer detection using a combination of deep neural networks (DNNs) and transfer learning. Singh et al. (2020) developed a framework based on transfer learning for breast cancer detection.

3. Subjects and data acquisition

The dataset for the experiments (Breast Cancer Patients MRI's, n.d.)[1] contained 1480 total breast MRI images, including 740 images of malignantly sick breasts and 740 images of healthy breasts. The images were normalized, and it was split into 80% training, 10% validation, and 10% test sets. The image used had a $224 \times 224 \times 3$ resolution. Fig. 8.1 displays images from the dataset, with the left image representing a breast MRI of a malignant breast and the right image representing a breast MRI of a healthy breast.

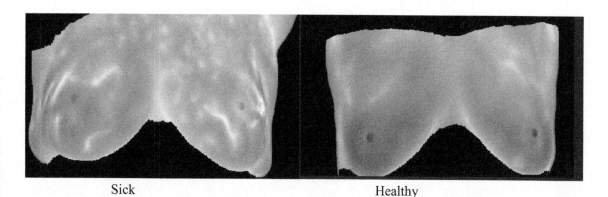

Sick   Healthy

FIGURE 8.1 Images from dataset.

[1] https://www.kaggle.com/datasets/uzairkhan45/breast-cancer-patients-mris.

4. Proposed architecture for custom convolutional neural networks

We tested seven distinct custom CNNs, with convolutional layer counts ranging from 2 to 8. The models were trained using Binary Cross entropy loss and Adam optimizer with a learning rate of 0.01 and batch size of 32 for 10 epochs. The architectures were trained from scratch on the dataset. Every epoch involved a shuffle of the dataset. To make network performance comparison easier, the same training and validation datasets were chosen for each network. The final output is obtained by passing the input image through a convolutional layer, followed by a sequence of pooling, batch normalization, and dense layers.

```
model1.compile(loss='binary_crossentropy',
               optimizer='adam',
               metrics=['accuracy','AUC','Precision','Recall'])

from tensorflow.keras.callbacks import ModelCheckpoint, EarlyStopping
es=EarlyStopping(patience=3,monitor='val_loss')
filepath='best_model.h51'
checkpoint = ModelCheckpoint(filepath, monitor='val_accuracy', verbose=1,
save_best_only=True, mode='max')

for j,(train_idx,val_idx) in enumerate(folds):
    print("Fold "+str(j+1))

    x_train=X_dataset[train_idx]
    y_train=Y_dataset[train_idx]
    x_val=X_dataset[val_idx]
    y_val=Y_dataset[val_idx]

history=model1.fit(train_datagen.flow(x_train,y_train),epochs=10,validatio
n_data=(x_val,y_val),callbacks=checkpoint)
    y_predict1_1=model1.predict(x_val)
    Y_predict1_1=np.float32(y_predict1_1>0.50)
    l=len(y_val)
    Y_predict1_1=np.reshape(Y_predict1_1,(l,1))
```

The custom model pipelines are as follows.

4.1 CNN 2 layers

```python
model1 = tf.keras.models.Sequential([
    tf.keras.layers.Conv2D(32,              (3,3),              activation='relu',
input_shape=(224,224,3)),
    tf.keras.layers.MaxPooling2D(pool_size=(2,2)),
    tf.keras.layers.Dropout(0.25),

    tf.keras.layers.Conv2D(32, (3,3), activation='relu'),
    tf.keras.layers.MaxPooling2D(pool_size=(2,2)),
    tf.keras.layers.Dropout(0.25),

    tf.keras.layers.Flatten(),
    tf.keras.layers.Dense(256, activation='relu'),
    tf.keras.layers.Dropout(0.5),

    tf.keras.layers.Dense(1, activation='sigmoid')
])
model1.summary()
```

4.2 CNN 3 layers

```python
model2 = tf.keras.models.Sequential([
    tf.keras.layers.Conv2D(32,            (3,3),            activation='relu',
input_shape=(224,224,3)),
    tf.keras.layers.MaxPooling2D(pool_size=(2,2)),
    tf.keras.layers.Dropout(0.25),

    tf.keras.layers.Conv2D(32, (3,3), activation='relu'),
    tf.keras.layers.MaxPooling2D(pool_size=(2,2)),
    tf.keras.layers.Dropout(0.25),

    tf.keras.layers.Conv2D(32, (3,3), activation='relu'),
    tf.keras.layers.MaxPooling2D(pool_size=(2,2)),
    tf.keras.layers.Dropout(0.25),

    tf.keras.layers.Flatten(),
    tf.keras.layers.Dense(256, activation='relu'),
    tf.keras.layers.Dropout(0.5),

    tf.keras.layers.Dense(1, activation='sigmoid')
])
model2.summary()
```

4.3 CNN 4 layers

```python
model3 = tf.keras.models.Sequential([
    tf.keras.layers.Conv2D(32,          (3,3),          activation='relu',
input_shape=(224,224,3)),
    tf.keras.layers.MaxPooling2D(pool_size=(2,2)),
    tf.keras.layers.Dropout(0.25),

    tf.keras.layers.Conv2D(32, (3,3), activation='relu'),
    tf.keras.layers.MaxPooling2D(pool_size=(2,2)),
    tf.keras.layers.Dropout(0.25),

    tf.keras.layers.Conv2D(32, (3,3), activation='relu'),
    tf.keras.layers.MaxPooling2D(pool_size=(2,2)),
    tf.keras.layers.Dropout(0.25),

    tf.keras.layers.Conv2D(32, (3,3), activation='relu'),
    tf.keras.layers.MaxPooling2D(pool_size=(2,2)),
    tf.keras.layers.Dropout(0.25),

    tf.keras.layers.Flatten(),
    tf.keras.layers.Dense(256, activation='relu'),
    tf.keras.layers.Dropout(0.5),

    tf.keras.layers.Dense(1, activation='sigmoid')
])
model3.summary()
```

4.4 CNN 5 layers

```
model4 = tf.keras.models.Sequential([
    tf.keras.layers.Conv2D(32,            (3,3),            activation='relu',
input_shape=(224,224,3)),
    tf.keras.layers.Dropout(0.25),

    tf.keras.layers.Conv2D(32, (3,3), activation='relu'),
    tf.keras.layers.MaxPooling2D(pool_size=(2,2)),
    tf.keras.layers.Dropout(0.25),

    tf.keras.layers.Conv2D(32, (3,3), activation='relu'),
    tf.keras.layers.Dropout(0.25),

    tf.keras.layers.Conv2D(32, (3,3), activation='relu'),
    tf.keras.layers.MaxPooling2D(pool_size=(2,2)),
    tf.keras.layers.Dropout(0.25),

    tf.keras.layers.Conv2D(32, (3,3), activation='relu'),
    tf.keras.layers.MaxPooling2D(pool_size=(2,2)),
    tf.keras.layers.Dropout(0.25),

    tf.keras.layers.Flatten(),
    tf.keras.layers.Dense(256, activation='relu'),

    tf.keras.layers.Dense(1, activation='sigmoid')
])
model4.summary()
```

4.5 CNN 6 layers

```
model5 = tf.keras.models.Sequential([
    tf.keras.layers.Conv2D(32,            (3,3),            activation='relu',
input_shape=(224,224,3)),
    tf.keras.layers.Conv2D(32, (3,3), activation='relu'),
    tf.keras.layers.MaxPooling2D(pool_size=(2,2)),

    tf.keras.layers.Conv2D(32, (3,3), activation='relu'),
    tf.keras.layers.Conv2D(32, (3,3), activation='relu'),
    tf.keras.layers.MaxPooling2D(pool_size=(2,2)),

    tf.keras.layers.Conv2D(32, (3,3), activation='relu'),
    tf.keras.layers.Conv2D(32, (3,3), activation='relu'),
    tf.keras.layers.MaxPooling2D(pool_size=(2,2)),

    tf.keras.layers.Flatten(),
    tf.keras.layers.Dense(256, activation='relu'),
    tf.keras.layers.Dropout(0.5),

    tf.keras.layers.Dense(1, activation='sigmoid')
])
model5.summary()
```

4.6 CNN 7 layers

```python
model6 = tf.keras.models.Sequential([
    tf.keras.layers.Conv2D(32,             (3,3),             activation='relu',
input_shape=(224,224,3)),
    tf.keras.layers.Conv2D(32, (3,3), activation='relu'),
    tf.keras.layers.MaxPooling2D(pool_size=(2,2)),

    tf.keras.layers.Conv2D(32, (3,3), activation='relu'),
    tf.keras.layers.Conv2D(32, (3,3), activation='relu'),
    tf.keras.layers.MaxPooling2D(pool_size=(2,2)),

    tf.keras.layers.Conv2D(32, (3,3), activation='relu'),
    tf.keras.layers.MaxPooling2D(pool_size=(2,2)),

    tf.keras.layers.Conv2D(32, (3,3), activation='relu'),
    tf.keras.layers.Conv2D(32, (3,3), activation='relu'),
    tf.keras.layers.MaxPooling2D(pool_size=(2,2)),

    tf.keras.layers.Flatten(),
    tf.keras.layers.Dense(256, activation='relu'),
    tf.keras.layers.Dropout(0.5),

    tf.keras.layers.Dense(1, activation='sigmoid')
])
model6.summary()
```

4.7 CNN 8 layers

```
model7 = tf.keras.models.Sequential([
    tf.keras.layers.Conv2D(32,           (3,3),           activation='relu',
input_shape=(224,224,3)),
    tf.keras.layers.Conv2D(32, (3,3), activation='relu'),
    tf.keras.layers.MaxPooling2D(pool_size=(2,2)),

    tf.keras.layers.Conv2D(32, (3,3), activation='relu'),
    tf.keras.layers.Conv2D(32, (3,3), activation='relu'),
    tf.keras.layers.MaxPooling2D(pool_size=(2,2)),

    tf.keras.layers.Conv2D(32, (3,3), activation='relu'),
    tf.keras.layers.Conv2D(32, (3,3), activation='relu'),
    tf.keras.layers.MaxPooling2D(pool_size=(2,2)),

    tf.keras.layers.Conv2D(32, (3,3), activation='relu'),
    tf.keras.layers.Conv2D(32, (3,3), activation='relu'),
    tf.keras.layers.MaxPooling2D(pool_size=(2,2)),

    tf.keras.layers.Flatten(),
    tf.keras.layers.Dense(256, activation='relu'),
    tf.keras.layers.Dropout(0.5),

    tf.keras.layers.Dense(1, activation='sigmoid')
])
model7.summary()
```

5. Proposed pipeline for transfer learning

Five different transfer learning models were used in our experiments, each of which had been pretrained using ImageNet as our base model. All of the pretrained models had their top layers removed. The final fully connected layer's "number of features" was maintained at 2 for transfer learning. The models were trained over 10 epochs with a learning rate of 0.01 using Binary Cross entropy loss and Adam optimizer. Every epoch involved a shuffle of the dataset. To make network performance comparison easier, the same training and validation datasets were

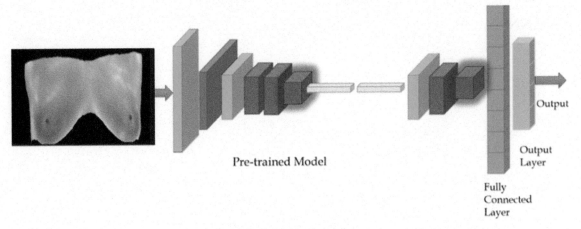

FIGURE 8.2   Proposed pipeline.

chosen for each network. Fig. 8.2 shows the architecture we suggest. A convolutional layer and a pretrained model are applied to the input image. To obtain the final output, the output is then subjected to a series of pooling, batch normalization, and dense layers.

```python
def select_pretrained_model(name):
# function to select one of the pre-trained ImageNet models
pretrainedbase = None
if name == "DenseNet121":
  pretrainedbase = densenet
if name == "VGG16":
  pretrainedbase = vgg16
if name == "VGG19":
  pretrainedbase = vgg19
if name == "ResNet50":
  pretrainedbase = resnet50
if name == "InceptionV3":
  pretrainedbase = inceptionv3
if name == "InceptionResNetV2":
  pretrainedbase = inceptionresnetv2
if name == "MobileNet":
  pretrainedbase = mobilenet
if name == "MobileNetV2":
  pretrainedbase = mobilenetv2
```

```
pretrainedbase.compile(loss='binary_crossentropy',
           optimizer='adam',
           metrics=['accuracy','AUC','Precision','Recall'])


from tensorflow.keras.callbacks import ModelCheckpoint, EarlyStopping
es=EarlyStopping(patience=3,monitor='val_loss')
filepath='best_model.h51'
checkpoint = ModelCheckpoint(filepath, monitor='val_accuracy', verbose=1,
save_best_only=True, mode='max')


for j,(train_idx,val_idx) in enumerate(folds):
    print("Fold "+str(j+1))

    x_train=X_dataset[train_idx]
    y_train=Y_dataset[train_idx]
    x_val=X_dataset[val_idx]
    y_val=Y_dataset[val_idx]

history=model1.fit(train_datagen.flow(x_train,y_train),epochs=10,validatio
n_data=(x_val,y_val),callbacks=checkpoint)
    y_predict1_1=pretrainedbase.predict(x_val)
    Y_predict1_1=np.float32(y_predict1_1>0.50)
    l=len(y_val)
    Y_predict1_1=np.reshape(Y_predict1_1,(l,1))
```

5.1 ResNet

Residual blocks assist in solving the vanishing gradient issue while maintaining the desired number of hidden layers in the architecture. Using an identity mapping that is added to the activations after one step, residual networks preserve what they have learned and continue to add to the preserved weights. As a result, the network continuously builds on its prior knowledge whenever it learns something new.

To be used in the skip connection, the residual block first creates a copy of the weights x and saves it. The weights x are then subjected to an activation to obtain F(x). The basic operation of a residual block is completed by adding the preserved weights x through the skip connection to F(x) following activation (He et al., 2015). ResNet is constructed from a number of these residual blocks that are very deep and have demonstrated excellent performance for tasks like image recognition.

5.1.1 ResNet101

```
resnet101= tf.keras.applications.resnet_rs.ResNetRS101(
    weights='imagenet',
    input_shape=(224, 224, 3),
    include_top=False)
resnet101.trainable = False
inputs = tf.keras.Input(shape=(224, 224, 3))
x = resnet101(inputs, training=False)
x = tf.keras.layers.GlobalAveragePooling2D()(x)
x = tf.keras.layers.Dropout(0.2)(x)
x = tf.keras.layers.Dense(1)(x)
outputs=tf.keras.layers.Activation(activation='sigmoid')(x)
model = tf.keras.Model(inputs, outputs)
model.summary()
```

5.1.2 ResNet50

```
resnet50= tf.keras.applications.resnet50.ResNet50(
weights='imagenet',
input_shape=(224, 224, 3),
include_top=False)
resnet50.trainable = False
inputs = tf.keras.Input(shape=(224, 224, 3))
x = resnet50(inputs, training=False)
x = tf.keras.layers.GlobalAveragePooling2D()(x)
x = tf.keras.layers.Dropout(0.2)(x)
x = tf.keras.layers.Dense(1)(x)
outputs=tf.keras.layers.Activation(activation='sigmoid')(x)
model = tf.keras.Model(inputs, outputs)
model.summary()
```

5.2 DenseNet

In a conventional deep learning architecture, each layer receives a different set of operations, such as activations like ReLU or softmax or operations like convolutions, pooling, or batch normalization. For this, the equation would be:

$$x_l = H_l(x_{l-1})$$

ResNets extended this behavior including the skip connection, reformulating this equation into:

$$x_l = H_l(x_{l-1}) + x_{l-1}$$

This step is where DenseNets and ResNets diverge the most. The history of the weights learned is concatenated together rather than being summarized. As a result, the equation changes once more to:

$$x_l = H_l([x_0, x_1, ..., x_{l-1}])$$

This concept of collective knowledge serves as the foundation for dense blocks called DenseNets. The channel dimension keeps growing at each layer as a result of the feature maps being concatenated. If we decide to regularly create feature maps, then the l-th layer would look like this:

$$k_l = k_0 + k(l - 1)$$

The growth rate, or hyperparameter k, measures how much information is being added to the network at each layer in the form of feature maps (Huang et al., 2017, pp. 4700–4708).

5.2.1 DenseNet121

```
densenet121= tf.keras.applications.densenet.DenseNet121(
    weights='imagenet',
    input_shape=(224, 224, 3),
    include_top=False)
densenet121.trainable = False
inputs = tf.keras.Input(shape=(224, 224, 3))
x = densenet121(inputs, training=False)
x = tf.keras.layers.GlobalAveragePooling2D()(x)
x = tf.keras.layers.Dropout(0.2)(x)
x = tf.keras.layers.Dense(1)(x)
outputs=tf.keras.layers.Activation(activation='sigmoid')(x)
model = tf.keras.Model(inputs, outputs)
model.summary()
```

5.2.2 DenseNet169

```
densenet169= tf.keras.applications.densenet.DenseNet169(
    weights='imagenet',
    input_shape=(224, 224, 3),
    include_top=False)
densenet169.trainable = False
inputs = tf.keras.Input(shape=(224, 224, 3))
x = densenet169(inputs, training=False)
x = tf.keras.layers.GlobalAveragePooling2D()(x)
x = tf.keras.layers.Dropout(0.2)(x)
x = tf.keras.layers.Dense(1)(x)
outputs=tf.keras.layers.Activation(activation='sigmoid')(x)
model = tf.keras.Model(inputs, outputs)
model.summary()
```

5.3 MobileNet

MobileNets are based on a simplified architecture that creates lightweight DNNs using depthwise separable convolutions. Due to the fact that there are fewer parameters to be trained on, they are renowned for being memory efficient and lightweight. By producing a large number of output feature maps with few convolution operations, they effectively utilize dense blocks. To further optimize the computation and increase memory efficiency, the growth rate k is also kept low (Howard et al., 2017).

5.3.1 MobileNet

```
from keras.applications import MobileNet
mobilenet= tf.keras.applications.mobilenet.MobileNet(
    weights='imagenet',
    input_shape=(224, 224, 3),
    include_top=False)
mobilenet.trainable = False
inputs = tf.keras.Input(shape=(224, 224, 3))
x5 = mobilenet(inputs, training=False)
x5 = tf.keras.layers.GlobalAveragePooling2D()(x5)
x5 = tf.keras.layers.Dropout(0.2)(x5)
x5 = tf.keras.layers.Dense(1)(x5)
outputs=tf.keras.layers.Activation(activation='sigmoid')(x5)
model5 = tf.keras.Model(inputs, outputs)
model.summary()
```

5.3.2 MobileNetV2

```
mobilenetv2= tf.keras.applications.mobilenet_v2.MobileNetV2(
    weights='imagenet',
    input_shape=(224, 224, 3),
    include_top=False)
mobilenetv2
inputs = tf.keras.Input(shape=(224, 224, 3))
x10 = mobilenetv2(inputs, training=False)
x10 = tf.keras.layers.GlobalAveragePooling2D()(x10)
x10 = tf.keras.layers.Dropout(0.2)(x10)
x10 = tf.keras.layers.Dense(1)(x10)
outputs=tf.keras.layers.Activation(activation='sigmoid')(x10)
model10 = tf.keras.Model(inputs, outputs)
model.summary()
```

5.3.3 MobileNet-11 layers

```
mobilenet= tf.keras.applications.mobilenet.MobileNet(
    weights='imagenet',
    input_shape=(224, 224, 3),
    include_top=False)
mobilenet.trainable = False
inputs = tf.keras.Input(shape=(224, 224, 3))
x12 = mobilenet(inputs, training=False)
x12 = tf.keras.layers.Flatten()(x12)
x12 = tf.keras.layers.BatchNormalization()(x12)
x12 = tf.keras.layers.Dense(256,kernel_initializer='he_uniform')(x12)
x12 = tf.keras.layers.BatchNormalization()(x12)
x12 = tf.keras.layers.Activation(activation='relu')(x12)
x12 = tf.keras.layers.Dense(128,kernel_initializer='he_uniform')(x12)
x12 = tf.keras.layers.BatchNormalization()(x12)
x12 = tf.keras.layers.Activation(activation='relu')(x12)
x12 = tf.keras.layers.Dropout(0.5)(x12)
x12 = tf.keras.layers.Dense(1)(x12)
outputs=tf.keras.layers.Activation(activation='sigmoid')(x12)
model12 = tf.keras.Model(inputs, outputs)
model.summary()
```

5.3.4 *MobileNet V2-11 layers*

```
mobilenetv2= tf.keras.applications.mobilenet_v2.MobileNetV2(
weights='imagenet',
input_shape=(224, 224, 3),
include_top=False)
mobilenetv2.trainable = False
inputs = tf.keras.Input(shape=(224, 224, 3))
x11 = mobilenetv2(inputs, training=False)
x11 = tf.keras.layers.Flatten()(x11)
x11 = tf.keras.layers.BatchNormalization()(x11)
x11 = tf.keras.layers.Dense(256,kernel_initializer='he_uniform')(x11)
x11 = tf.keras.layers.BatchNormalization()(x11)
x11 = tf.keras.layers.Activation(activation='relu')(x11)
x11 = tf.keras.layers.Dense(128,kernel_initializer='he_uniform')(x11)
x11 = tf.keras.layers.BatchNormalization()(x11)
x11 = tf.keras.layers.Activation(activation='relu')(x11)
x11 = tf.keras.layers.Dropout(0.5)(x11)
x11 = tf.keras.layers.Dense(1)(x11)
outputs=tf.keras.layers.Activation(activation='sigmoid')(x11)
model11 = tf.keras.Model(inputs, outputs)
model.summary()
```

5.4 VGG

Understanding the VGG architecture is very easy. The input image, which is 224 × 224 in size, is then passed through five convolution blocks made up of 3 × 3 kernels activated with stride 1 and ReLU. Max pooling comes after the convolution blocks. The final layer, which has a dimension of 1000 and corresponds to 1000 different image classes, is the last of three fully connected layers that it passes through (Simonyan & Zisserman, 2014). In the ImageNet test, VGG architectures have also shown to be very effective.

5.4.1 VGG16

```
from keras.applications.vgg16 import VGG16
vgg16= VGG16(
    weights='imagenet',
    input_shape=(224, 224, 3),
    include_top=False)
vgg16.trainable = False
inputs = tf.keras.Input(shape=(224, 224, 3))
x2 = vgg16(inputs, training=False)
x2 = tf.keras.layers.GlobalAveragePooling2D()(x2)
x2 = tf.keras.layers.Dropout(0.2)(x2)
x2 = tf.keras.layers.Dense(1)(x2)
outputs=tf.keras.layers.Activation(activation='sigmoid')(x2)
model2 = tf.keras.Model(inputs, outputs)
model2.summary()
```

5.4.2 VGG19

```
from keras.applications.vgg19 import VGG19
vgg19= VGG19(
    weights='imagenet',
    input_shape=(224, 224, 3),
    include_top=False)
vgg19.trainable = False
inputs = tf.keras.Input(shape=(224, 224, 3))
x3 = vgg19(inputs, training=False)
x3 = tf.keras.layers.GlobalAveragePooling2D()(x3)
x3 = tf.keras.layers.Dropout(0.2)(x3)
x3 = tf.keras.layers.Dense(1)(x3)
outputs=tf.keras.layers.Activation(activation='sigmoid')(x3)
model3 = tf.keras.Model(inputs, outputs)
model.summary()
```

5.5 Inception/GoogLeNet

In the majority of common network architectures, it is not intuitively obvious when to use the convolutional operation and when to perform the max-pooling operation. As an illustration, in AlexNet, the convolution and max-pooling operations come after one another, whereas in VGGNet, there are three consecutive convolutional operations before the first max-pooling layer. Thus, using all operations simultaneously is the idea behind GoogleNet. It performs parallel computations on various kernels of varying sizes over the same input map, concatenating the results into a single output. It makes sense that convolution filters of various sizes would handle objects at various scales more effectively. It is referred to as an Inception module (Szegedy et al., 2015).

5.5.1 InceptionV3

```
inceptionv3= tf.keras.applications.inception_v3.InceptionV3(
    weights='imagenet',
    input_shape=(224, 224, 3),
    include_top=False)
inceptionv3.trainable = False
inputs = tf.keras.Input(shape=(224, 224, 3))
x1 = inceptionv3(inputs, training=False)
x1 = tf.keras.layers.GlobalAveragePooling2D()(x1)
x1 = tf.keras.layers.Dropout(0.2)(x1)
x1 = tf.keras.layers.Dense(1)(x1)
outputs=tf.keras.layers.Activation(activation='sigmoid')(x1)
model1 = tf.keras.Model(inputs, outputs)
model.summary()
```

5.5.2 Inception ResNetV2

```
inceptionresnetv2= tf.keras.applications.InceptionResNetV2(
    weights='imagenet',
    input_shape=(224, 224, 3),
    include_top=False)
inceptionresnetv2.trainable = False
inputs = tf.keras.Input(shape=(224, 224, 3))
x4 = inceptionresnetv2(inputs, training=False)
x4 = tf.keras.layers.GlobalAveragePooling2D()(x4)
x4 = tf.keras.layers.Dropout(0.2)(x4)
x4 = tf.keras.layers.Dense(1)(x4)
outputs=tf.keras.layers.Activation(activation='sigmoid')(x4)
model4 = tf.keras.Model(inputs, outputs)
model4.summary()
```

6. Breast cancer detection with deep feature extraction

Additionally, we experimented with six different machine learning classifiers on the features that were extracted from our suggested architecture. To extract features, the images are first preprocessed and sent through a convolutional neural network that has already been trained on the ImageNet Dataset. A classifier head is then attached and used to classify the breast MRI images (Fig. 8.3).

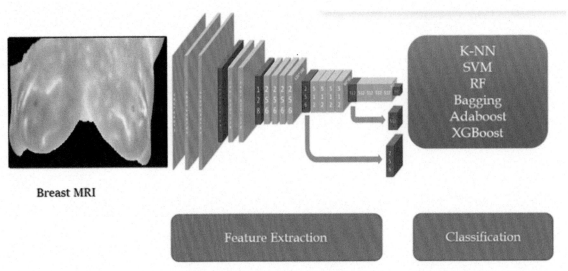

FIGURE 8.3   Breast cancer detection using deep feature extraction and conventional machine learning.

```python
from keras.preprocessing.image import ImageDataGenerator

train_datagen = ImageDataGenerator(  rotation_range=8,
                                     width_shift_range=0.03,
                                     height_shift_range=0.03,
                                     shear_range=0.25,
                                     zoom_range=0.45,
                                     horizontal_flip=True)
val_datagen = ImageDataGenerator(zoom_range=0.45)

#Defining the Classifiers
from sklearn.pipeline import make_pipeline
from sklearn.pipeline import Pipeline
from sklearn.neural_network import MLPClassifier
from sklearn import metrics
names = [
        "K Nearest Neighbour Classifier",
        'SVM',
        "Random Forest Classifier",
        "AdaBoost Classifier",
        "XGB Classifier",
        "ANN Classifier",
        "BaggingClassifier"
        ]
classifiers = [
    KNeighborsClassifier(),
    SVC(probability = True),
    RandomForestClassifier(),
    AdaBoostClassifier(),
    XGBClassifier(),
    MLPClassifier(),
    BaggingClassifier()
        ]
zipped_clf = zip(names,classifiers)
```

6.1 k-nearest neighbors

The k-nearest neighbors (k-NN) algorithm follows a simple methodology. Based on a predetermined distance metric, it involves locating the k training set objects that are closest to a given test object. The test object's class is then determined by a majority vote count among its k neighbors. The effectiveness of the algorithm heavily depends on the selection of an appropriate distance metric. Smaller distances between objects indicate a higher likelihood that they belong to the same class because the metric is chosen in a way that makes this assumption. For instance, using cosine similarity as the distance metric instead of Euclidean distance when applying k-NN for document classification may produce better results (Wu et al., 2008).

```
#Import k-NN Model
from sklearn.neighbors import KNeighborsClassifier
#Create the Model
clf = KNeighborsClassifier(n_neighbors=3)
#Train the model with Training Dataset
clf.fit(X_train,Y_train)
#Test the model with Testset
Y_pred = clf.predict(X_test)
#Print performance
print(clf.score(X_train,Y_train))
print(clf.score(X_val,Y_val))
print(clf.score(X_test,Y_test))
print_confusion_matrix(Y_test,Y_pred)
print_scores(Y_test,Y_pred)
```

6.2 Support vector machine

SVM stands out as one of the most reliable and accurate machine learning algorithms. Its main goal is to solve a binary classification problem by finding a function that can effectively distinguish between two different classes. A linear classification function can be visualized as a separating hyperplane positioned between the two classes, effectively dividing them, when working with a linearly separable dataset. SVM assists in choosing the hyperplane that maximizes the separation, also known as the margin, between the classes given that multiple hyperplanes can complete this task. The margin can be characterized as the separation between the closest objects belonging to various classes as determined by the SVM. This margin definition forces the classifier to select the best boundary out of all potential candidates, resulting in a classifier that is more general and performs well on untested data. The goal is to increase the classifier's robustness by setting the margin to its maximum value (Wu et al., 2008).

```
#Import SVM Model
from sklearn import svm
#Create the Model
clf = make_pipeline(StandardScaler(), SVC())
#Train the model with Training set
clf.fit(X_train, Y_train)
#Test the model with Test set
Y_pred = clf.predict(X_test)
#Print performance
print(clf.score(X_train,Y_train))
print(clf.score(X_val,Y_val))
print(clf.score(X_test,Y_test))
print_confusion_matrix(Y_test,Y_pred)
print_scores(Y_test,Y_pred)
```

6.3 Random forests

The ensemble machine learning algorithm known as Random Forests (RFs), as its name implies, combines various decision trees to create a "forest." The underlying idea of this algorithm is that the trees in a forest are autonomous from one another and that their combination lowers the overall error rate. To accurately classify the data, decision trees divide it at each stage and choose the best splits. Additionally, RFs introduce additional randomness and encourage generalization by training each tree on a random subset of features. Errors are decreased overall by this process. When used on a dataset, RFs increase the model's overall accuracy by combining predictions from various decision trees (Breiman, 2001; Han et al., 2012).

```python
#Import Random Forest Ensemble Model
from sklearn.ensemble import RandomForestClassifier
#Create the Model
clf = RandomForestClassifier(n_estimators=100)
#Train the model with Training set
clf.fit(X_train,Y_train)
#Test the model with Test set
Y_pred = clf.predict(X_test)
#Print performance
print(clf.score(X_train,Y_train))
print(clf.score(X_val,Y_val))
print(clf.score(X_test,Y_test))
print_confusion_matrix(Y_test,Y_pred)
print_scores(Y_test,Y_pred)
```

6.4 Bagging

By combining the judgments made by various models into a single prediction, ensemble methods seek to enhance results. Different methods can be used to achieve this. While the average of all model predictions can be used to calculate numeric outputs, the majority vote method is frequently used for classification tasks. Different methods of combining model decisions are used in ensemble methods like bagging and boosting. Each model is given an equal amount of weight when bagging. To create a decision tree for each training dataset, several datasets of the same size are randomly chosen from the problem domain. These trees are made to predict the same outcomes for fresh test cases. This presumption is not always accurate, particularly when training datasets are modest in size. As a result, for some test instances, some models may produce incorrect predictions. By minimizing the instability of machine learning techniques, bagging solves this problem. This is done by creating new datasets of the same size by randomly selecting instances from the original dataset and replacing them. Unavoidably, some examples are lost during this sampling process, and others are duplicated. The outputs produced from each of these resulting datasets are then fed into the learning algorithm, which selects the predicted class as the winner (Witten & Frank, 2002).

```python
#Import Bagging Ensemble Model
from sklearn.ensemble import BaggingClassifier
from sklearn import tree
#Create a Bagging Ensemble Model
clf = BaggingClassifier(n_estimators=1000)
#Train the model using the training set
clf.fit(X_train,Y_train)
#Predict the response for test set
Y_pred = clf.predict(X_test)
#Print performance
print(clf.score(X_train,Y_train))
print(clf.score(X_val,Y_val))
print(clf.score(X_test,Y_test))
print_confusion_matrix(Y_test,Y_pred)
print_scores(Y_test,Y_pred)
```

6.5 AdaBoost

The ensemble learning method known as AdaBoost, or Adaptive Boosting, gives weights to data points based on the predictions made by the classifier. The classifier first assigns equal weights to each data point before making predictions. The weights for correctly predicted instances are then reduced, while those for incorrectly predicted instances are increased. These predictions are then compared to the actual labels. The classifier is trained on this newly weighted dataset in the following iteration. The primary distinction is that the classifier now gives higher weights to instances that were previously misclassified and pays more attention to those instances. Through a series of iterations, some instances that were initially difficult become even more difficult while other instances are correctly classified, increasing the accuracy of the ensemble method as a whole. The algorithm's capacity to adapt to the dataset by allocating and modifying weights based on current predictions gives rise to the name "Adaptive Boosting" (Witten & Frank, 2002).

```python
#Import Adaboost ensemble model
from sklearn.ensemble import AdaBoostClassifier
#Create an Adaboost Ensemble Model
clf = AdaBoostClassifier(n_estimators=1000)
#Train the model using the training set
clf.fit(X_train,Y_train)
#Predict the response for test set
Y_pred = clf.predict(X_test)
#Print performance
print(clf.score(X_train,Y_train))
print(clf.score(X_val,Y_val))
print(clf.score(X_test,Y_test))
print_confusion_matrix(Y_test,Y_pred)
print_scores(Y_test,Y_pred)
```

6.6 XGBoost

A more recent ensemble boosting algorithm comparable to AdaBoost is XGBoost, also referred to as gradient boosting. The approach used to improve predictions is where the main difference lies. The progress of XGBoost is based on the results of a loss function rather than the weights being assigned based on expected values. The difference between the present predictions and the labeled ground truth is measured using this loss function. By calculating gradients with respect to it and moving in the suggested direction, the algorithm seeks to minimize this loss function. A learning rate, an algorithm hyperparameter, governs the size of these steps. The algorithm's accuracy rises with each iteration as the loss decreases to its minimum value (Chen & Guestrin, 2016, pp. 785–794).

```python
#Import XGBoost ensemble model
from xgboost import XGBClassifier
# Create XGB model
clf = GradientBoostingClassifier(n_estimators=1000,learning_rate=1)
#Train the model using the training set
clf.fit(X_train,Y_train)
# make predictions for test data
Y_pred = clf.predict(X_test)
#Print performance
print(clf.score(X_train,Y_train))
print(clf.score(X_val,Y_val))
print(clf.score(X_test,Y_test))
print_confusion_matrix(Y_test,Y_pred)
print_scores(Y_test,Y_pred)
```

7. Results and discussions

7.1 Performance evaluation measures

XGBoost, also known as gradient boosting, is a more recent ensemble boosting algorithm that is comparable to AdaBoost. The main distinction is in the method used to enhance predictions. Instead of assigning weights based on expected values, XGBoost bases its progress on the outcomes of a loss function. This loss function is used to calculate the discrepancy between the current predictions and the labeled ground truth. The algorithm aims to minimize this loss function by computing gradients with respect to it and moving in the suggested direction. The size of these steps is controlled by an algorithm hyperparameter called learning rate. With each iteration, the algorithm's accuracy increases and the loss gets closer and closer to zero.

The classifier is trained on one set while being evaluated on another thanks to the formal separation of the training and test sets. The dataset may occasionally be split into three separate sets known as train, test, and validation. While the train and test sets have the same function as described earlier, the validation set is used to fine-tune the hyperparameters used in training the classifier. To facilitate effective training, the training set is typically larger, while the test set is kept smaller (Witten & Frank, 2002).

The number of samples for which the classification was accurate is referred to as the true positives (TP) and true negatives (TN). When an outcome is incorrectly predicted as yes (or positive) when it is actually no (negative), that situation is known as a false positive (FP). When an outcome is incorrectly predicted as negative when it is actually positive, it is called a false negative (FN) (Witten & Frank, 2002). Researchers in information retrieval define variables called recall and precision:

$$Recall = \frac{TP}{TP + FN}$$

$$Precision = \frac{TP}{TP + FP}$$

Then F1 measure can be formulated as:

$$F1 = \frac{2 * Precision * Recall}{Precision + Recall}$$

Another important measure is the confusion matrix which can be formulated as:

$$Confusion\ Matrix = \begin{bmatrix} TP & FP \\ FN & TN \end{bmatrix}$$

ROC curves represent a classifier's performance without accounting for the actual error rate or cost. Plotting the "true positive rate" on the y-axis and the "true negative rate" on the x-axis yields the curve. Formally:

$$TP\ Rate = 100\ x\ TP/(TP\ x\ FN)$$

$$FP\ Rate = 100\ x\ FP/(FP\ x\ TN)$$

The probability that the classifier ranks a positively predicted instance ahead of a negatively predicted instance is measured by the area under the ROC curve (AUC). As a result, in a variety of classification tasks, this area is sometimes used as a performance metric. Several techniques are frequently used to calculate the AUC (Witten & Frank, 2002).

The Cohen kappa statistic is a measure that accounts for samples that the classifier classified solely by chance using the ideas of expectation (Witten & Frank, 2002). The kappa statistic () always assumes that the predictions are the result of chance and uncertainty. The concept of expectation of a random variable is used to calculate the probabilities of receiving a correct prediction by chance and confidence. When there is no other independent way to determine the likelihood of chance agreement between two or more observers, the kappa statistic is the statistic that is most frequently used to evaluate categorical data. An agreement that is equal to chance is designated by a kappa value of 0, while perfect agreement is designated by a kappa value of 1 (Lantz & Nebenzahl, 1996; Viera & Garrett, 2005). Cohen (1960) defined the kappa statistic as an agreement index and defined as the following:

$$K = \frac{P_0 - P_e}{1 - P_e}$$

where P_0 is observed agreement and P_e measures the agreement expected by chance (Viera & Garrett, 2005).

7.2 Experimental results

7.2.1 Transfer learning

7.2.1.1 For pretrained models

Table 8.1 displays the outcomes of experiments on various pretrained CNNs using our suggested architecture for transfer learning. With MobileNet, we were able to achieve a high-test accuracy of 97.30%. Additionally, it was discovered that MobileNet had the highest F1 score, AUC, and Cohen kappa with values of 0.937, 0.973, and 0.94 respectively.

TABLE 8.1 Experiment results for pretrained CNN models.

Model	Accuracy (%)	F1 score	Kappa	AUC
ResNet50	60.81	0.5397	0.2104	0.6045
ResNet101	63.51	0.6197	0.2775	0.6408
VGG16	66.22	0.7126	0.3029	0.6508
VGG19	70.27	0.7556	0.3782	0.6857
DenseNet121	79.73	0.8101	0.5964	0.7997
InceptionResNetV2	82.43	0.8506	0.6392	0.8149
MobileNetV2	82.43	0.8415	0.6451	0.8208
DenseNet169	85.14	0.8333	0.7001	0.8563
MobileNet	89.19	0.8750	0.7799	0.8935
MobileNetV2 (11 layers)	93.92	0.9379	0.8783	0.9391
MobileNet (11 layers)	97.30	0.9730	0.9460	0.9732

7.2.1.2 Results for convolutional neural networks

The experiment results for transfer learning conducted on various custom CNNs with

```
from sklearn.metrics import auc,roc_curve
from sklearn.metrics import f1_score
from sklearn.metrics import cohen_kappa_score
from sklearn.metrics import accuracy_score
f1 = f1_score(y_test, Y_predict)
test_accuracy = accuracy_score(y_test, Y_predict)
kappa = cohen_kappa_score(y_test, Y_predict)
auc = roc_auc_score(y_test, Y_predict)

print(auc1)
print(f11)
print(kappa1)
print(test_accuracy1)
```

TABLE 8.2 Experimental results for custom CNN.

Model	Accuracy (%)	F1 score	Kappa	AUC
CNN 2 layer	67.57	0.6757	0.3518	0.6762
CNN 4 layer	58.78	0.7081	0.1575	0.5771
CNN 5 layer	62.84	0.7264	0.2045	0.5966
CNN 3 layer	64.19	0.7006	0.2767	0.6371
CNN 8 layer	65.54	0.6222	0.3080	0.6535
CNN 7 layer	66.22	0.6154	0.3350	0.6707
CNN 6 layer	72.30	0.7092	0.4504	0.7277

various numbers of convolutional layers are displayed in Table 8.2. With the convolutional network having six layers, we were able to achieve a high-test accuracy of 72.30%. Convolutional neural networks with six layers were found to have the highest F1 score, AUC, and Cohen kappa values, each with a value of 0.709, 0.727, and 0.450.

7.2.2 Feature extraction with pretrained models

In the second phase of our experiments, features extracted from each pretrained model were applied to KNN, SVM, Random Forest, AdaBoost, XGBoost, and Bagging classifiers using the proposed architecture. After careful adjustment, the extracted features' dimension was maintained at 256.

7.2.2.1 k-nearest neighbors

One of the simplest classification algorithms is the k-NN algorithm. Even with its simplicity, it produced results that were fiercely competitive. With DenseNet169 and VGG16, the k-NN classifier's highest test accuracy was 97.97%. DenseNet169 and VGG16 had F1 scores of 0.98, 0.98, 0.98, and 0.96, respectively, which is also fairly comparable to the other architectures. All models were successful in exceeding the 95% test accuracy threshold. F1 score greater than 0.95, a precision score greater than 0.95, a recall

value greater than 0.95, and a kappa value greater than 0.90 are all exceptions to this rule. MobileNetV2 performed admirably and on par with our top candidates, VGG16 and DenseNet169 (Table 8.3).

7.2.2.2 Support vector machine

As the accuracy dropped from 97.97% to 95.95%, we saw that the SVM classifier's performance for DenseNet169 was marginally worse than the k-NN classifier. DenseNet169 performed best with SVM classifiers, which had the highest test accuracy of 95.95% of any other method. Additionally, MobileNet and DensetNet121 provided an accuracy of 95.27%. In the case of SVM, DenseNet169 obtained the highest test accuracy of 95.95%. DenseNet169 also had the highest F1 score (0.96), recall and precision (0.96), and kappa (0.92). Compared to its performance in k-NN, ResNet50 underperformed (Table 8.4).

7.2.2.3 Random Forest

With regard to Random Forest, VGG16 outperformed DenseNet121 with an accuracy of 95.95%, edging it out by a relatively small margin (difference of about 1.36%). The F1 score, AUC, and kappa values were also marginally higher in the case of VGG16 compared to DenseNet121. All models performed with accuracy higher than 87%, F1 score higher than 0.88, precision and recall higher than 0.88, and kappa score higher than 0.76 (Table 8.5).

7.2.2.4 AdaBoost and XGBoost

The two well-known and frequently used boosting algorithms are XGBoost and AdaBoost. After each prediction round, AdaBoost penalizes incorrectly predicted samples by giving them a heavier weight. Decision trees are sequentially growing as weak learners in this method. In this manner, the algorithm is improving upon prior errors. Contrarily, in XGBoost, also referred to as gradient boosting, the boosting is carried out concurrently with the gradient

TABLE 8.3 Performance of k-NN classifier for deep feature extraction models.

Model	Accuracy (%)	F1 score	Kappa	Recall	Precision
InceptionV3	93.92	0.94	0.88	0.94	0.94
ResNet50	95.27	0.95	0.90	0.95	0.95
InceptionResNetV2	95.27	0.95	0.90	0.95	0.95
Xception	95.27	0.95	0.90	0.95	0.95
MobileNet	95.95	0.96	0.92	0.96	0.96
ResNet101	96.62	0.97	0.93	0.97	0.97
DenseNet121	96.62	0.97	0.93	0.97	0.97
MobileNetV2	97.30	0.97	0.95	0.97	0.97
VGG19	97.97	0.98	0.96	0.98	0.98
DenseNet169	97.97	0.98	0.96	0.98	0.98

TABLE 8.4 Performance of SVM classifier for deep feature extraction models.

Model	Accuracy (%)	F1 score	Kappa	Recall	Precision
ResNet50	65.54	0.65	0.3	0.66	0.65
ResNet101	70.95	0.71	0.41	0.71	0.71
VGG19	81.08	0.81	0.62	0.81	0.81
VGG16	86.49	0.87	0.73	0.86	0.87
InceptionResNetV2	89.19	0.89	0.78	0.89	0.89
InceptionV3	89.86	0.9	0.79	0.9	0.9
Xception	90.54	0.91	0.81	0.91	0.91
MobileNet	95.27	0.95	0.9	0.95	0.95
DenseNet121	95.27	0.95	0.9	0.95	0.95
DenseNet169	95.95	0.96	0.92	0.96	0.96

adjustments. Tables 8.6 and 8.7 display the results of the comparison between AdaBoost and XGBoost. The results showed that XGBoost performed significantly better than AdaBoost with a margin of 10%–15% for each, demonstrating that XGBoost is more effective at classifying the CT scan features obtained from our suggested model. With an accuracy of 83.78%, an F1 score of 0.84, a recall and precision of 0.84, and a kappa score of 0.67, Xception provided the best performance for AdaBoost. On the other hand, VGG19 displayed the best performance in XGBoost, achieving accuracy of 95.95, F1 score of 0.96, precision and recall of 0.96, and kappa of 0.92.

TABLE 8.5 Performance of Random Forest classifier for deep feature extraction models.

Model	Accuracy (%)	F1 score	Kappa	Recall	Precision
InceptionV3	87.84	0.88	0.76	0.88	0.88
ResNet101	88.51	0.89	0.77	0.89	0.89
MobileNet	91.22	0.91	0.82	0.91	0.91
DenseNet169	91.89	0.92	0.84	0.92	0.92
Xception	91.89	0.92	0.83	0.92	0.92
ResNet50	94.59	0.95	0.89	0.95	0.95
InceptionResNetV2	94.59	0.95	0.89	0.95	0.95
DenseNet121	94.59	0.95	0.89	0.95	0.95
VGG16	95.95	0.96	0.92	0.96	0.96
MobileNetV2	95.95	0.96	0.92	0.96	0.96

TABLE 8.6 Performance of AdaBoost classifier for deep feature extraction models.

Model	Accuracy (%)	F1 score	Kappa	Recall	Precision
InceptionV3	73.65	0.74	0.47	0.74	0.74
DenseNet169	73.65	0.74	0.47	0.74	0.74
ResNet101	75.68	0.76	0.51	0.76	0.76
MobileNetV2	77.70	0.78	0.56	0.78	0.78
MobileNet	79.73	0.8	0.59	0.8	0.8
VGG16	80.41	0.8	0.61	0.8	0.8
ResNet50	81.08	0.81	0.62	0.81	0.81
DenseNet121	81.76	0.82	0.63	0.82	0.82
InceptionResNetV2	82.43	0.82	0.65	0.82	0.83
Xception	83.78	0.84	0.67	0.84	0.84

7.2.2.5 Bagging

The results of the Bagging classifier, which performed marginally better than the AdaBoost algorithm, are shown in Table 8.8. In comparison to AdaBoost, which achieved a test accuracy of 83.78%, VGG19 achieved the highest test accuracy of 90.54%, with an F1 score of 0.91, precision and recall of 0.91, and a kappa score of 0.81. The ROC AUC, kappa, and F1 score are all higher than those of AdaBoost. However, when compared to XGBoost, the performance of the Bagging classifier is subpar.

TABLE 8.7 Performance of XGBoost classifier for deep feature extraction models.

Model	Accuracy (%)	F1 score	Kappa	Recall	Precision
InceptionV3	89.86	0.9	0.8	0.9	0.9
ResNet50	91.22	0.91	0.82	0.91	0.91
Xception	91.22	0.91	0.82	0.91	0.91
ResNet101	91.89	0.92	0.84	0.92	0.92
MobileNet	91.89	0.92	0.84	0.92	0.92
InceptionResNetV2	93.24	0.93	0.86	0.93	0.93
VGG16	93.92	0.94	0.88	0.94	0.94
DenseNet121	94.59	0.95	0.89	0.95	0.95
MobileNetV2	95.27	0.95	0.9	0.95	0.95
VGG19	95.95	0.96	0.92	0.96	0.96

TABLE 8.8 Performance of Bagging classifier for deep feature extraction models.

Model	Accuracy (%)	F1 score	Kappa	Recall	Precision
InceptionV3	81.76	0.82	0.64	0.82	0.83
Xception	83.11	0.83	0.66	0.83	0.83
ResNet101	87.16	0.87	0.74	0.87	0.88
ResNet50	87.84	0.88	0.76	0.88	0.88
InceptionResNetV2	88.51	0.89	0.77	0.89	0.89
VGG16	89.19	0.89	0.78	0.89	0.9
DenseNet121	89.19	0.89	0.78	0.89	0.89
MobileNetV2	89.86	0.9	0.8	0.9	0.91
DenseNet169	89.86	0.9	0.79	0.9	0.9
VGG19	90.54	0.91	0.81	0.91	0.91

7.2.2.6 ANN

Table 8.9 displays the performance of the ANN classifier, which outperforms all other classification algorithms with the exception of the k-NN classifier, which achieved the best test accuracy of 96.62%, an F1 score of 0.97, precision and recall of 0.97, as well as a kappa score of 0.93, as determined by MobileNetV2.

DenseNet169, MobileNetV2, and VGG19 achieved the highest test accuracy among all the models for almost all machine learning classifiers. With the k-NN classifier, 97.97% was the highest value attained. In this instance, the F1 score, AUC, and Cohen kappa were also the highest. k-NN appeared to perform the best among the machine learning classifiers, followed

TABLE 8.9 Performance of ANN classifier for deep feature extraction models.

Model	Accuracy (%)	F1 score	Kappa	Recall	Precision
ResNet50	85.81	0.86	0.71	0.86	0.86
ResNet101	88.51	0.89	0.77	0.89	0.89
InceptionV3	91.22	0.91	0.82	0.91	0.91
DenseNet169	91.22	0.91	0.82	0.91	0.91
MobileNet	94.59	0.95	0.89	0.95	0.95
DenseNet121	94.59	0.95	0.89	0.95	0.95
VGG16	95.27	0.95	0.91	0.95	0.96
InceptionResNetV2	95.95	0.96	0.92	0.96	0.96
VGG19	96.62	0.97	0.93	0.97	0.97
MobileNetV2	96.62	0.97	0.93	0.97	0.97

by ANN, XGBoost, Random Forest, SVM, Bagging, and AdaBoost, in that order. Additionally, it should be mentioned that DenseNet169 and VGG19 produced excellent results using the k-NN classifier and a maximum value of 97.97%. The worst performers were InceptionV3 and Resnet50, while MobileNet and Dense-Net121's performance was comparable to that of VGG16 and VGG19.

7.3 Discussion

Our suggested approaches and architecture outperformed a number of other deep learning models. In comparison to Fujioka et al. (2021) accuracy of 96%, our transfer learning strategy produced an accuracy of 97.30% and an AUC of 0.9732. With an AUC value of 0.9732 as opposed to 0.895 by Fujioka et al. (2021), our model also outperformed it. Similar to the deep learning approaches, our feature extraction method produced an accuracy of 97.97%, which is significantly higher. The precision and recall obtained by applying k-NN to our feature extractor are 0.98, demonstrating the effectiveness of our suggested architecture for feature extraction. These findings lend support to other distinct feature extraction-based approaches and cutting-edge methods in this field.

8. Conclusion

In this study, we proposed a deep learning model for breast cancer detection using breast MRI. The model was trained on a dataset of breast MRI images and achieved an accuracy of 97.3% and an AUC of 0.9732. This is comparable to the state-of-the-art methods and suggests that our model is a promising approach for breast cancer detection and diagnosis. The model was based on transfer learning, which allowed us to use a pretrained model that had already been trained on a large dataset of natural images. This helped to improve the performance of our model on breast cancer images. We also used a machine learning-based approach to extract features from the images. This helped to identify the features that were most important for distinguishing between benign and malignant tumors.

The model was evaluated on a held-out dataset of breast cancer images. The results showed that the model was able to accurately distinguish between benign and malignant tumors. This suggests that the model could be used to improve the accuracy and efficiency of breast cancer screening and diagnosis. The model has several limitations, including the need for large datasets of breast cancer images to train the model and the need for accurate and consistent annotation of breast cancer images. However, we believe that the model is a promising approach for breast cancer detection and diagnosis, and we are currently working on improving the performance of the model and making it more accessible to researchers and clinicians.

The results of our study are consistent with the results of other studies that have used deep learning for breast cancer detection. These studies have shown that deep learning models can achieve high accuracy in distinguishing between benign and malignant tumors. We believe that our model is a promising addition to the body of work on deep learning for breast cancer detection. We are currently working on improving the performance of the model and making it more accessible to researchers and clinicians.

Funding

This work was supported by Effat University, Jeddah, Saudi Arabia.

References

Abunasser, B. S., Al-Hiealy, M. R. J., Zaqout, I. S., & Abu-Naser, S. S. (2023). Convolution neural network for breast cancer detection and classification using deep learning. *Asian Pacific Journal of Cancer Prevention, 24*(2), 531.

Aidossov, N., Zarikas, V., Mashekova, A., Zhao, Y., Ng, E. Y. K., Midlenko, A., & Mukhmetov, O. (2023). Evaluation of integrated CNN, transfer learning, and BN with thermography for breast cancer detection. *Applied Sciences, 13*(1), 600.

Aljuaid, H., Alturki, N., Alsubaie, N., Cavallaro, L., & Liotta, A. (2022). Computer-aided diagnosis for breast cancer classification using deep neural networks and transfer learning. *Computer Methods and Programs in Biomedicine, 223*, 106951.

Alruwaili, M., & Gouda, W. (2022). Automated breast cancer detection models based on transfer learning. *Sensors, 22*(3), 876.

Altabella, L., Benetti, G., Camera, L., Cardano, G., Montemezzi, S., & Cavedon, C. (2022). Machine learning for multi-parametric breast MRI: Radiomics-based approaches for lesion classification. *Physics in Medicine and Biology, 67*(15), Article 15TR01. https://doi.org/10.1088/1361-6560/ac7d8f

Ashfaq, A., Wenhui, Y., Jinhai, S., & Nasir, M. U. (2022). *Breast cancer diagnosing empowered with transfer learning.*

Breiman, L. (2001). Random forests. *Machine Learning, 45*, 5–32.

Chen, T., & Guestrin, C. (2016). *Xgboost: A scalable tree boosting system.*

Chiu, H.-J., Li, T.-H. S., & Kuo, P.-H. (2020). Breast cancer–detection system using PCA, multilayer perceptron, transfer learning, and support vector machine. *IEEE Access, 8*, 204309–204324.

Chugh, G., Kumar, S., & Singh, N. (2021). Survey on machine learning and deep learning applications in breast cancer diagnosis. *Cognitive Computation*, 1–20.

Chung, M., Calabrese, E., Mongan, J., Ray, K. M., Hayward, J. H., Kelil, T., Sieberg, R., Hylton, N., Joe, B. N., & Lee, A. Y. (2022). Deep learning to simulate contrast-enhanced breast MRI of invasive breast cancer. *Radiology, 306*(3), e213199.

Cohen, J. (1960). A coefficient of agreement for nominal scales. *Educational and Psychological Measurement, 20*(1), 37–46.

Cong, C., Li, X., Zhang, C., Zhang, J., Sun, K., Liu, L., Ambale-Venkatesh, B., Chen, X., & Wang, Y. (2023). MRI-Based breast cancer classification and localization by multiparametric feature extraction and combination using deep learning. *Journal of Magnetic Resonance Imaging*. https://doi.org/10.1002/jmri.28713. In press.

Dar, R. A., Rasool, M., & Assad, A. (2022). Breast cancer detection using deep learning: Datasets, methods, and challenges ahead. *Computers in Biology and Medicine*, 106073.

Dewangan, K. K., Dewangan, D. K., Sahu, S. P., & Janghel, R. (2022). Breast cancer diagnosis in an early stage using novel deep learning with hybrid optimization technique. *Multimedia Tools and Applications, 81*(10), 13935–13960.

Eskreis-Winkler, S., Onishi, N., Pinker, K., Reiner, J. S., Kaplan, J., Morris, E. A., & Sutton, E. J. (2021). Using deep learning to improve nonsystematic viewing of breast cancer on MRI. *Journal of Breast Imaging, 3*(2), 201–207. https://doi.org/10.1093/jbi/wbaa102

Fujioka, T., Yashima, Y., Oyama, J., Mori, M., Kubota, K., Katsuta, L., Kimura, K., Yamaga, E., Oda, G., & Nakagawa, T. (2021). Deep-learning approach with convolutional neural network for classification of maximum intensity projections of dynamic contrast-enhanced breast magnetic resonance imaging. *Magnetic Resonance Imaging, 75*, 1–8.

Han, J., Kamber, M., & Pei, J. (2012). *Data mining concepts and techniques* (3rd ed.). University of Illinois at Urbana-Champaign Micheline Kamber Jian Pei Simon Fraser University.

He, K., Zhang, X., Ren, S., & Sun, J. (2015). *Deep residual learning for image recognition (arXiv: 1512.03385)*. ArXiv.

Houssein, E. H., Emam, M. M., Ali, A. A., & Suganthan, P. N. (2021). Deep and machine learning techniques for medical imaging-based breast cancer: A comprehensive review. *Expert Systems with Applications, 167*, 114161.

Howard, A. G., Zhu, M., Chen, B., Kalenichenko, D., Wang, W., Weyand, T., Andreetto, M., & Adam, H. (2017). *Mobilenets: Efficient convolutional neural networks for mobile vision applications*. ArXiv Preprint ArXiv:1704.04861.

Huang, G., Liu, Z., Van Der Maaten, L., & Weinberger, K. Q. (2017). *Densely connected convolutional networks*.

Khalil, S., Nawaz, U., Zubariah, Mushtaq, Z., Arif, S., ur Rehman, M. Z., Qureshi, M. F., Malik, A., Aleid, A., & Alhussaini, K. (2023). Enhancing ductal carcinoma classification using transfer learning with 3D U-Net models in breast cancer imaging. *Applied Sciences, 13*(7), 4255.

Khan, S., Islam, N., Jan, Z., Din, I. U., & Rodrigues, J. J. C. (2019). A novel deep learning based framework for the detection and classification of breast cancer using transfer learning. *Pattern Recognition Letters, 125*, 1–6.

Lantz, C. A., & Nebenzahl, E. (1996). Behavior and interpretation of the κ statistic: Resolution of the two paradoxes. *Journal of Clinical Epidemiology, 49*(4), 431–434.

Murtaza, G., Shuib, L., Abdul Wahab, A. W., Mujtaba, G., Mujtaba, G., Nweke, H. F., Al-garadi, M. A., Zulfiqar, F., Raza, G., & Azmi, N. A. (2020). Deep learning-based breast cancer classification through medical imaging modalities: State of the art and research challenges. *Artificial Intelligence Review, 53*, 1655–1720.

Rane, N., Sunny, J., Kanade, R., & Devi, S. (2020). Breast cancer classification and prediction using machine learning. *International Journal of Engineering Research and Technology, 9*(2), 576–580.

Rautela, K., Kumar, D., & Kumar, V. (2022). A systematic review on breast cancer detection using deep learning techniques. *Archives of Computational Methods in Engineering, 29*(7), 4599–4629.

Seemendra, A., Singh, R., & Singh, S. (2020). Breast cancer classification using transfer learning. In *Evolving technologies for computing, communication and smart world: Proceedings of ETCCS 2020* (pp. 425–436). Springer.

Simonyan, K., & Zisserman, A. (2014). *Very deep convolutional networks for large-scale image recognition*. ArXiv Preprint ArXiv:1409.1556.

Singh, R., Ahmed, T., Kumar, A., Singh, A. K., Pandey, A. K., & Singh, S. K. (2020). Imbalanced breast cancer classification using transfer learning. *IEEE/ACM Transactions on Computational Biology and Bioinformatics, 18*(1), 83–93.

Sun, R., Zhang, X., Xie, Y., & Nie, S. (2023). Weakly supervised breast lesion detection in DCE-MRI using self-transfer learning. *Medical Physics*. https://doi.org/10.1002/mp.16296. In press.

Szegedy, C., Wei, L., & Yangqing, J. (2015). *Going deeper with convolutions in 2015 IEEE Conf. Comput. Vis. Pattern Recognition*.

Viera, A. J., & Garrett, J. M. (2005). Understanding interobserver agreement: The kappa statistic. *Family Medicine, 37*(5), 360–363.

Wang, K., Patel, B. K., Wang, L., Wu, T., Zheng, B., & Li, J. (2019). A dual-mode deep transfer learning (D2TL) system for breast cancer detection using contrast enhanced digital mammograms. *IISE Transactions on Healthcare Systems Engineering, 9*(4), 357–370.

Witowski, J., Heacock, L., Reig, B., Kang, S. K., Lewin, A., Pysarenko, K., Patel, S., Samreen, N., Rudnicki, W., & Łuczyńska, E. (2022). Improving breast cancer diagnostics with deep learning for MRI. *Science Translational Medicine, 14*(664), eabo4802.

Witten, I. H., & Frank, E. (2002). Data mining: Practical machine learning tools and techniques with Java implementations. *Acm Sigmod Record, 31*(1), 76–77.

Wu, X., Kumar, V., Ross Quinlan, J., Ghosh, J., Yang, Q., Motoda, H., McLachlan, G. J., Ng, A., Liu, B., & Yu, P. S. (2008). Top 10 algorithms in data mining. *Knowledge and Information Systems, 14*, 1–37.

Yang, X., Xi, X., Yang, L., Xu, C., Song, Z., Nie, X., Qiao, L., Li, C., Shi, Q., & Yin, Y. (2022). Multi-modality relation attention network for breast tumor classification. *Computers in Biology and Medicine, 150*, 106210.

Yin, H., Bai, L., Jia, H., & Lin, G. (2022). Noninvasive assessment of breast cancer molecular subtypes on multiparametric MRI using convolutional neural network with transfer learning. *Thoracic Cancer, 13*(22), 3183–3191.

Zhang, Y., Liu, Y.-L., Nie, K., Zhou, J., Chen, Z., Chen, J.-H., Wang, X., Kim, B., Parajuli, R., & Mehta, R. S. (2023). Deep learning-based automatic diagnosis of breast cancer on MRI using mask R-CNN for detection followed by ResNet50 for classification. *Academic Radiology, 30*(2), S161–S171. https://doi.org/10.1016/j.acra.2022.12.038

Zhao, X., Bai, J.-W., Guo, Q., Ren, K., & Zhang, G.-J. (2023). Clinical applications of deep learning in breast MRI. *Biochimica et Biophysica Acta (BBA) - Reviews on Cancer, 1878*(2), 188864. https://doi.org/10.1016/j.bbcan.2023.188864

Automated detection of colon cancer from histopathological images using deep neural networks

Mirka Suominen[1], Muhammed Enes Subasi[2], Abdulhamit Subasi[1,3]

[1]Institute of Biomedicine, Faculty of Medicine, University of Turku, Turku, Finland; [2]Faculty of Medicine, Izmir Katip Celebi University, Izmir, Turkey; [3]College of Engineering, Effat University, Jeddah, Saudi Arabia

1. Introduction

Colon cancer, also called colorectal cancer, is a cancerous growth that starts in the colon or rectum when abnormal cells in the colon's lining begin to multiply uncontrollably, resulting in tumors. It is one of the most widespread types of cancer globally and can affect both men and women, usually diagnosed in those over 50 years old. However, nowadays it is not uncommon for it to occur at any age. In males and females, the mortality rate ranges from 15 to 20 out of 100,000, and 9 to 14 out of 100,000 in the European Union. The 5-year survival rates exhibit a range from 28.5% to 57% in men and 30.9% to 60% in women. The pooling estimation yielded 46.8% for men and 48.4% for women in 23 countries (Argilés et al., 2020).

Detecting colon cancer early on and treating it appropriately can significantly increase survival chances. Regular screening tests, typically colonoscopies, are advised for early detection because the survival rate for stage 0 is over 93% while the rates for later stages I, II, and III are 87%, 74%, and 18% respectively. Symptoms and risk factors associated with colon cancer should therefore be taken seriously. Diagnostic procedures can detect cancerous growth at an early stage. Unfortunately, the early symptoms are usually vague or nonexistent, and cancer is detected when treatment options are already limited (Masud et al., 2021).

Several risk factors are associated with an individual's susceptibility to colon cancer, including lifestyle, family history, inheritance of colon cancer or polyps, age of over 50 years, a sedentary lifestyle, smoking, and obesity. Moreover, the modern diet consisting of red meat and processed foods with reduced fiber and excessive fat content increases the risk of colon cancer (Katsaounou et al., 2022).

Applications of Artificial Intelligence in Healthcare and Biomedicine
https://doi.org/10.1016/B978-0-443-22308-2.00014-7

243

Several diagnostic methods are used to diagnose colon cancer, including imaging tests, magnetic resonance imaging (MRI), computed tomography (CT), ultrasound, positron emission tomography (PET) (Nasseri & Langenfeld, 2017), biopsies, and blood tests (Burt, 2000). Despite being the most effective approach for preventing colon cancer, colonoscopy solely identifies morphological changes in the intestinal lining and lacks the ability to identify early precursors of tumorigenesis and dysplasia. Numerous studies have documented significant rates of polyp misses, with reported figures ranging widely from 14% to 30% (Häfner, 2007). In a meta-analysis conducted by van Rijn et al. (2006), the overall miss rate was determined to be 21%.

Deep learning is a subset of machine learning that uses neural networks to learn and make predictions or decisions based on input data. Deep learning models are constructed using artificial neural networks, which are composed of multiple layers of interconnected nodes called neurons. These networks are trained on a large number of data to recognize patterns and make predictions or classifications on new, unseen data. The ability of deep learning algorithms to directory extract high-level features from raw images makes them potential tools for cancer detection, diagnosis, and tumor segmentation (Pacal et al., 2020). For instance, deep learning algorithms can be used to analyze medical imaging data from colonoscopies to identify abnormalities and tumors. Additionally, it can be used to analyze patient data, such as genetic information and medical histories to identify if an individual has a higher risk of developing colon cancer. This can help healthcare providers develop new personalized screening and prevention for high-risk patients (Gupta et al., 2022). Physicians can benefit from the use of deep learning methods as they provide additional insights and can highlight specific areas in images. Studies have shown that a single deep learning model can outperform other

medical methods in diagnostics accuracy (Sakr et al., 2022). Finally, deep learning can be used to analyze large amounts of research data to identify new biomarkers or genetic signatures associated with colon cancer, which can aid in developing new treatments and improving patient outcomes (Ben Hamida et al., 2021).

2. Background/literature review

Convolutional neural network (CNN) is a type of deep learning technology that has been introduced by Azer (2019) as a potential solution to enhance the detection rate of colonic polyps and early cancerous lesions from histopathological images. CNN offers several advantages, including polyp classification, detection, segmentation, polyp tracking, and increased accuracy.

These state-of-the-art CNN models and digital image processing (DIP) techniques have been used in a study by Masud et al. (2021) to differentiate five types of lung and colon cancers with a maximum accuracy of 96.33%. LC200 dataset was used for training and validation. Four sets of features were extracted using domain transformation and combined for classification. State-of-the-art CNN models were also used in the study conducted by Ben Hamida et al. (2021). AlexNet, VGG, ResNet, DenseNet, and InceptionNet were reviewed and compared to classify colon cancer images. Transfer learning was used to overcome the shortage of whole slide image (WSI) datasets, with IMAGENET used for training. Testing with the AiCOLO dataset showed up to 96.98% accuracy with ResNet. Pixelwise segmentation using UNET and SEGNET models was presented, with a multistep training strategy for parse annotation. Accuracy rates of up to 76.18% and 81.22% are achieved, respectively.

Comparison of spatially convolution neural networks to classical machine learning algorithms showing outperformance of CNN over

Random Forest and k-nearest neighbor algorithms in terms of accuracy for colon segmentation (87% for CNN vs. 85% for RF and 83% for KNN) and polyp detection (88% for CNN vs. 85% for RF and 80% for KNN) of CT colonography images was studied by Godkhindi and Gowda (2017).

Likewise, CNN networks are analyzed in studies by Gessert et al. (2019) and Mohalder et al. (2022). The first one used various transfer learning strategies to analyze a small dataset of confocal laser microscopy images to classify them into different tissue types. Overall, 97.1% for peritoneal metastases and 73.1% for colon primaries AUC values were achieved. The latter study used histopathological images with the same dimensions as colon cancer tissues. The deep neural network consisted of an input layer, four hidden layers, and an output layer. The rectified linear unit (ReLU) activation function was used in hidden layers, and the SoftMax function was used in the output layer resulting in an accuracy of 99.70%.

Furthermore, a study by Tasnim et al. (2021) utilized CNN-based networks for deep learning in predicting colon cancer. The research employed CNN to analyze image data of colon cells. For classification, max pooling and average pooling layers, as well as the MobileNetV2 model, were utilized. The findings revealed that the accuracy of max pooling and average pooling layers were 97.49% and 95.48%, respectively. Notably, MobileNetV2 outperformed the other two methods, achieving an accuracy of 99.67%.

Sakr et al. (2022) normalized histopathological images before inputting them into the CNN model for cancer detection. A study utilized the LC5000 lung and colon histopathological image dataset, which consists of 10,000 images of colon tissue classified into two categories: benign colon (5000 images) and colon adenocarcinomas (5000 images). The proposed system's efficiency was evaluated using publicly available databases and compared to existing methods for colon cancer detection. The analysis of results showed that the proposed deep learning model achieved an accuracy of 99.50%.

Deep convolutional neural networks (CNNs) can be used to detect and classify adenocarcinomas. Hasan et al. (2022) introduced a model that achieved the highest maximum accuracy of 99.80%. It outperforms other models by replacing the sigmoid function in the output activation layer for binary classification. A training and evaluation technique was proposed to maintain the high resolution of textured images without transforming them into low-resolution images.

Accurate and efficient classification of histopathological cell nuclei holds significant implications in the domain of medical image analysis. Shabbeer Basha et al. (2018) used RCCNet to classify histological nuclei of routine colon cancer images. The proposed method is evaluated and compared to five other state-of-the-art CNN models in terms of accuracy, weighted average F1 score, and training time, resulting in an accuracy of 80.61%. Sirinukunwattana et al. (2016) introduced a novel approach called spatially constrained convolutional neural network (SC−CNN) for nucleus detection. SC-CNN uses regression to estimate the likelihood of pixel being in the center of a nucleus, with high probability values constrained to the vicinity of nucleus centers. For nucleus classification, neighboring ensemble predictor (NEP) combined CNN accurately detected nuclei.

Biomarkers in deep learning are studied by Sarker et al. (2021). They proved that biomarkers can be prognostic predictors for stage III colon cancer. Routinely stained hematoxylin and eosin slides were used in a convolutional neural network machine classifier with the selected model of Gradient Boosting. HR values obtained from multivariate Cox proportional hazard analysis were statistically significant, indicating that Gradient Boosting is an independent prognostic predictor. The previously mentioned study applies deep learning to identify cells from immunohistochemistry-stained slides for quantification of nuclear biomarkers, with a specific focus on immune checkpoint biomarker inducible T-cell costimulator (ICOS). The workflow consisted of two parts: simplified robust annotation and application of cell identification models. Results were compared to ground-truth data provided by pathologists and therefore proved that the detection method can be used for survival analysis for stage II and III cancer patients.

Biomarkers were used to train Bayesian binary classifiers to predict 5-year survival. Hsu and Lin (2021) provided more accurate predictions compared to non-Bayesian models, with the Bayesian bimodal neural network (B-Bimodal) classifier outperforming others. Another study by Sarwinda et al. (2021) used ResNet architecture to identify colon gland images. Results show that ResNet-50 consistently outperformed ResNet-18 in terms of accuracy, sensitivity, and specificity across three datasets. The best results were achieved with 20% and 25% datasets, with classification accuracy above 80%, sensitivity above 87%, and specificity above 83%.

Tumor-stroma ratio (TSR) is a prognostic parameter in colon cancer. Smit et al. (2023) analyzed 75 colon cancer images visually and the results were compared to semiautomated and fully automated deep learning algorithms. The result was that the Spearman correlation was above 0.70 and visual examination had the highest accuracy. Hyperspectral images acquired through micro-FTIR absorbance spectroscopy were modeled into hyperspectral data in a voxel format by Muniz et al. (2023). A fully connected deep neural network was used to detect patterns of each voxel. Experiments using K-fold cross-validation achieved an overall accuracy of 99% with a deep neural network, and 96% with a linear support vector machine.

Shapcott et al. (2019) utilized a deep learning algorithm trained with images from TCGA (Cancer Genome Atlas) by tiling them and identifying cells in each patch defined by the tiling. Each cell

was assigned a label indicating its location and classification. Systematic random spatial sampling was employed, selecting 100 tiles from the whole slide to process them, with only a 4% decrease in performance.

Gupta et al. (2021) utilized a method to localize abnormal regions in whole slide images (WSI) of colon cancer patients with localization models. The proposed models were pretrained Inception-v3 achieving an F-score of 0.97 and customized Inception-ResNet-v2 Type 5 model achieving 0.99. The pretrained model performed better compared to other models, while the customized achieved superior performance when trained from scratch. YOLOv3 Multiscale Framework (YOLOv3-MSF) was used to detect various stages of colon cancer based on tumor length by Murugesan et al. (2023). Transfer learning was adopted using YOLOv3, and the K-Medoids algorithm was used to choose anchor boxes during object detection in the training stage.

In a study by Schiele et al. (2021), a deep learning classifier Blq-CoMet was developed based on InceptionResNetV2 for classifying cancer patients into risk groups. Binary images were used to detect metastasis. Blq-CoMet achieved a positive predictive value of 80%.

Hage Chehade et al. (2022) compared six machine learning models (XGBoost, SVM, RF, LDA, MLP, and LightGMP) to overcome interpretation difficulties with deep learning. The dataset images were preprocessed and then three feature sets were extracted. The resulting features were concatenated to create a combined feature set for the machine learning algorithm. The XGBoost showed the best performance with an accuracy of 99% and an F1-score of 98.8%.

Toğaçar (2021) employed the DarkNet-19 deep learning model in their research. They utilized Equilibrium and Manta Ray Foraging optimization algorithms for feature extraction. The resulting selected features were combined and classified using the support vector machine method. The overall accuracy rate of the process was 99.69%. These study findings highlight the significant improvement in classification performance through the integration of optimization algorithms. Similarly, Talukder et al. (2022) integrated a hybrid feature extraction model by integrating deep feature extraction, ensemble learning, and high-performance filtering techniques. The model incorporated preprocessing, k-fold cross-validation, and feature extraction. MobileNet was used as a transfer learning model resulting in 99.05% accuracy of colon cancer detection.

A framework based on multiple lightweight DL models is proposed to reduce computational complexity by Attallah et al. (2022). The framework used principal component analysis, fast Walsh–Hadamard transform, and discrete wavelet transform to reduce the dimensionality of features extracted from SuffleNet, MobileNet, and SqueezeNet models achieving an accuracy of 99.6%. Domain-agnostic handcrafted features were investigated in a study by Tripathi and Singh (2020). The performance of different DL models was analyzed. The study focuses on nuclei classification and demonstrates that combining DL methods in different ways may be more effective than combining domain-agnostic handcrafted features as the background, and artifacts can create challenges in feature extraction.

Tsirikoglou et al. (2020) researched mitigation strategies for limited data access scenarios in cancer detection. The study came to two conclusions: (1) Adding primary tumor data in a small lymph node set can improve performance and (2) transferring histopathology information from one cancer type to another is possible, opening possibilities for cross-domain cancer detection. Deep learning methods show promise in diagnosing colon cancer, but there is a need for explainable methods (Kavitha et al., 2022). Hybrid, end-to-end, and transfer learning techniques achieve high accuracy, whereas attention-based networks improve decision-making.

3. Machine learning techniques

3.1 Deep learning

Deep learning is a subfield of machine learning that involves training artificial neural networks to recognize patterns of data. Neural networks are composed of layers of interconnected nodes or "neurons" that process and transform the input data in a hierarchical manner. Deep learning is a type of network that has multiple hidden layers between the input and output layers, and they progressively integrate features from layer to layer. By adjusting the strengths of the connections between neurons during training, the network can learn to recognize complex patterns in the data and make predictions. This is achieved by using a method called backpropagation, which instructs the machine to adjust its internal parameters to calculate the presentation of each layer from the previous layers. This process allows the machine to learn and identify intricate structures within the data. The network's performance is assessed by the weighted input/outputs. Deep Learning models can learn from vast amounts of data and automatically identify complex patterns and features that can be impossible for humans to identify. This can lead to significantly higher accuracy compared to the traditional machine learning models. Deep learning models can automatically identify and extract relevant features from data, which can save a lot of data and effort. There is no need for manual feature engineering. Deep learning models can be easily scaled up to handle large datasets and complex problems, making them well suited for big data applications. Deep learning has shown a great success in a wide range of applications, including medical imaging (Suzuki, 2017).

However, deep learning has its downsides. Deep learning models are typically trained using large datasets and require a lot of computational power (Suzuki, 2017). Training a deep learning model can take hours, days, or even weeks on high-end hardware. Moreover, it can be difficult to interpret how a deep learning model makes decisions, which can be problematic in some applications where transparency is important. Additionally, deep learning requires large amounts of data (Fernandez-Quilez, 2023).

3.2 Convolutional neural networks

Convolutional neural networks (CNNs) are neural networks used for analyzing images and classification tasks. The fundamental concept behind CNNs is the inclusion of convolutional filters, which aim to extract features from images. These extracted features are further used for image classification. Filters perform mathematical operations at each location of the image, systematically traversing the entire image. As a result, a feature map is generated, containing the most relevant information of the image. Future maps are often passed to the following layers for further processing (CS231n Convolutional Neural Networks for Visual Recognition, n.d.).

The basic building block of CNN is a convolutional layer, which consists of a set of filters. These filters are often called kernels. Each filter consists of a matrix of numbers, which is learned during training, and represents a feature such as a horizontal edge. The output of each convolutional layer is typically passed through a nonlinear activation function such as the rectified linear unit (ReLU), which helps to introduce nonlinearity into the network and improve its ability to capture complex patterns of the data. CNNs also often include pooling layers, which downsample the feature maps by taking the maximum or average value in each local neighborhood. This helps to reduce the size of the feature maps and makes the network more computationally efficient. CNNs typically end with one or more fully connected layers, which take the output from the previous layers and perform the final classification or regression task. The final layer may use the activation function such as the Softmax function to produce a probability distribution over a possible output class (CS231n Convolutional Neural Networks for Visual Recognition, n.d.).

The depth and complexity of CNN can vary. Moreover, 2-layer CNN is a simple architecture that may be suitable for simpler patterns and smaller datasets. In this chapter, CNNs from 2 to 8 layers were trained. The more layers the CNN contains, the more complex the model gets; 8-layer CNN has a larger number of parameters and a deeper network improving the performance in feature extraction and classification tasks. Overall, CNNs have shown remarkable performance in a wide range of computer vision tasks and are widely used in industry and academia.

```
#CNN 8 Layer


import numpy as np
import pandas as pd
from sklearn.metrics import f1_score, roc_auc_score, cohen_kappa_score, precision_score,
recall_score, accuracy_score, confusion_matrix
from keras.layers import Dense, Flatten, Conv2D, MaxPooling2D, Dropout, Input
from keras.models import Sequential
import tensorflow as tf
import matplotlib.pyplot as plt
import cv2
from keras.layers.normalization import BatchNormalization
from keras.models import Model, Sequential
from keras.applications.xception import Xception
from keras.applications import *
import matplotlib.pyplot as plt
from sklearn.pipeline import make_pipeline
from sklearn.pipeline import Pipeline
from PIL import Image
import random
import os
import cv2
from sklearn.model_selection import train_test_split
from sklearn.neighbors import KNeighborsClassifier
from sklearn.svm import SVC
from sklearn.ensemble import RandomForestClassifier, AdaBoostClassifier
from xgboost import XGBClassifier
from keras.callbacks import EarlyStopping
import seaborn as sns
from sklearn.preprocessing import StandardScaler
from tqdm import tqdm
from sklearn.decomposition import PCA
from tensorflow.keras.utils import to_categorical
from keras.layers import BatchNormalization


data_dir = "C:/Users/mirka/Downloads/archive/lung_colon_image_set/colon_image_sets"
SIZE_X = SIZE_Y = 128
datagen = tf.keras.preprocessing.image.ImageDataGenerator(validation_split = 0.3)
train_it = datagen.flow_from_directory(data_dir,
                                        class_mode = "categorical",
                                        target_size = (SIZE_X,SIZE_Y),
                                        color_mode="rgb",
                                        batch_size = 12,
                                        shuffle = False,
                                        subset='training',
                                        seed = 42)
```

```python
validate_it = datagen.flow_from_directory(data_dir,
                                          class_mode = "categorical",
                                          target_size = (SIZE_X, SIZE_Y),
                                          color_mode="rgb",
                                          batch_size = 12,
                                          shuffle = False,
                                          subset='validation',
                                          seed = 42)


def get_accuracy_metrics(model, train_it, validate_it):
  y_val = validate_it.classes
  val_pred_proba = model.predict(validate_it)
  val_pred_proba, predicted_proba, y_val, y_test = train_test_split(val_pred_proba, y_val,
test_size = 0.5, shuffle = True)
  val_pred = np.argmax(val_pred_proba, axis = 1)
  predicted = np.argmax(predicted_proba, axis = 1)
  print("Train accuracy Score------------>")
  print ("{0:.3f}".format(accuracy_score(train_it.classes, np.argmax(model.predict(train_it),
axis = 1))*100), "%")
  print("Val accuracy Score--------->")
  print("{0:.3f}".format(accuracy_score(y_val, val_pred)*100), "%")
  print("Test accuracy Score--------->")
  print("{0:.3f}".format(accuracy_score(y_test, predicted)*100), "%")
  print("F1 Score---------------->")
  print("{0:.3f}".format(f1_score(y_test, predicted, average = 'weighted')*100), "%")
  print("Cohen Kappa Score------------->")
  print("{0:.3f}".format(cohen_kappa_score(y_test, predicted)*100), "%")
  print("ROC AUC Score------------->")
  print("{0:.3f}".format(roc_auc_score(y_test, predicted_proba[:, 1])*100), "%")
  print("Recall-------------->")
  print("{0:.3f}".format(recall_score(y_test, predicted, average = 'weighted')*100), "%")
  print("Precision-------------->")
  print("{0:.3f}".format(precision_score(y_test, predicted, average = 'weighted')*100), "%")


inp = Input(shape = (128, 128, 3))
model = BatchNormalization()(inp)
model = Conv2D(filters = 64, kernel_size = (3, 3), padding = 'same', activation='relu')(model)
model = BatchNormalization()(model)
model = Conv2D(filters = 128, kernel_size = (3, 3), padding = 'same',
activation='relu')(model)
model = MaxPooling2D()(model)
model = Dropout(0.2)(model)
model = Conv2D(filters = 128, kernel_size = (3, 3), padding = 'same',
activation='relu')(model)
model = MaxPooling2D()(model)
model = Dropout(0.2)(model)
model = Conv2D(filters = 64, kernel_size = (3, 3), padding = 'same', activation='relu')(model)
```

```
model = MaxPooling2D()(model)
model = Dropout(0.2)(model)
model = Conv2D(filters = 32, kernel_size = (3, 3), padding = 'same', activation='relu')(model)
model = Conv2D(filters = 32, kernel_size = (3, 3), padding = 'same', activation='relu')(model)
model = MaxPooling2D()(model)
model = Conv2D(filters = 16, kernel_size = (3, 3), padding = 'same', activation='relu')(model)
model = Conv2D(filters = 16, kernel_size = (3, 3), padding = 'same', activation='relu')(model)
model = Flatten()(model)
output = Dense(units = len(train_it.class_indices), activation = 'softmax')(model)

model = Model(inputs=inp, outputs=output)
model.summary()

model.compile(loss='categorical_crossentropy', optimizer ='adam', metrics=['accuracy'])
es = EarlyStopping(monitor='val_loss', mode='min', verbose=1, patience=5)
history = model.fit(train_it, validation_data = validate_it, verbose = 1, epochs = 10,
callbacks = [es])

get_accuracy_metrics(model, train_it, validate_it)
```

3.3 Transfer learning techniques

Transfer learning is a deep learning technique that leverages knowledge gained from one model to enhance the performance of another model. In other words, the goal of transfer learning is to transfer knowledge learned from a source domain to a target domain, where the target domain may have less data or different characteristics. Pretrained models have been trained with a large dataset for a specific task, such as image classification, object detection, or natural language processing. Instead of training model from scratch, we can use a pretrained model as a starting point and fine-tune it on a smaller dataset (Weiss et al., 2016).

In feature extraction, a pretrained model is taken and its internal representations are used as features for a new model. This is particularly useful when only a small dataset is available, and it is not large enough to train deep neural networks from scratch. By using a pretrained model as a feature extractor, we can take advantage of the high-level features learned from a

larger dataset (Weiss et al., 2016). Transfer learning has become an important technique in deep learning, enabling the development of models that can achieve state-of-the-art performance with less data and computational resources. It has been successfully applied to a variety of domains, including computer vision, speech recognition, and natural language processing (Weiss et al., 2016).

3.3.1 VGG16 and VGG19

VGG16 architecture consists of 16 layers: 13 convolutional and 3 fully connected layers. It follows a sequential arrangement of convolutional layers with smaller 3×3 filters, followed by max pooling layers. VGG16 maintains a uniform structure throughout the network by using the same filter size as the same stride for convolutional layers. This design choice helps to maintain simplicity and enables the network to learn hierarchical representations. With multiple small 3×3 filters instead of a single larger filter, VGG16 aims to capture finer details more effectively.

The usage of smaller filters also reduces the number of parameters and helps to prevent overfitting. VGG16 incorporates max-pooling layers after every two or three convolutional layers: max pooling helps downsampling the spatial dimensions of the feature maps, reducing the computational cost and providing some degree of translation invariance. The last three VGG16 layers are fully connected layers. They combine high-level features extracted by the convolutional layers and map them to the appropriate class labels (Boesch, 2021).

VGG19 has similar design principles to VGG16 and it utilizes 3×3 convolutional filters, with max-pooling layers for downsampling. The convolutional layers are stacked together, and fully connected are appended at the end to perform classification. The difference is that VGG19 has a deeper architecture consisting of 19 layers, additional layers being extra convolutional layers allowing more detailed and expressive feature representation. Due to additional layers, VGG19 has a higher number of trainable parameters than VGG16. The greater number of parameters enhances the model's capacity to learn complex relationships; however, it also means that the computational costs are higher. In general, VGG19 tends to have higher accuracy than VGG16 on image classification tasks. The additional layers capture more fine-grained details and extract richer features from the input images resulting in improved discrimination power (Boesch, 2021; Sec, 2021).

```
#VGG19
import numpy as np
import pandas as pd
from sklearn.metrics import f1_score, roc_auc_score, cohen_kappa_score, precision_score,
recall_score, accuracy_score, confusion_matrix
from keras.layers import Dense, Flatten, Conv2D, MaxPooling2D, Dropout
from keras.models import Sequential
import tensorflow as tf
import matplotlib.pyplot as plt
import cv2
from keras.layers.normalization import BatchNormalization
from keras.models import Model, Sequential
from keras.applications.xception import Xception
from keras.applications import *
import matplotlib.pyplot as plt
from sklearn.pipeline import make_pipeline
from sklearn.pipeline import Pipeline
from PIL import Image
import random
import os
import cv2
from sklearn.model_selection import train_test_split
from sklearn.neighbors import KNeighborsClassifier
from sklearn.svm import SVC
from sklearn.ensemble import RandomForestClassifier, AdaBoostClassifier
from xgboost import XGBClassifier
from keras.callbacks import EarlyStopping
import seaborn as sns
from sklearn.preprocessing import StandardScaler
from tqdm import tqdm
from sklearn.decomposition import PCA
from keras.layers import BatchNormalization


data_dir = "C:/Users/mirka/Downloads/archive/lung_colon_image_set/colon_image_sets"
SIZE_X = SIZE_Y = 128
datagen = tf.keras.preprocessing.image.ImageDataGenerator(validation_split = 0.3)
train_it = datagen.flow_from_directory(data_dir,
                                       class_mode = "categorical",
                                       target_size = (SIZE_X,SIZE_Y),
                                       color_mode="rgb",
                                       batch_size = 12,
                                       shuffle = False,
                                       subset='training',
                                       seed = 42)
```

```
validate_it = datagen.flow_from_directory(data_dir,
                                          class_mode = "categorical",
                                          target_size = (SIZE_X, SIZE_Y),
                                          color_mode="rgb",
                                          batch_size = 12,
                                          shuffle = False,
                                          subset='validation',
                                          seed = 42)
def fit_model(model, train_it, validate_it, epochs = 10):
  es = EarlyStopping(monitor='val_loss', mode='min', verbose=1, patience=5)
    for layer in model.layers:
      layer.trainable = False
  flat1 = Flatten()(model.layers[-1].output)
    output = Dense(len(train_it.class_indices), activation='softmax')(flat1)
    model = Model(inputs=model.inputs, outputs=output)
    print(model.summary())
    model.compile(loss='categorical_crossentropy', optimizer ='adam', metrics=['accuracy'])
    history = model.fit(train_it, validation_data=validate_it, epochs=epochs, verbose=1,
callbacks=[es])
    return model



def get_accuracy_metrics(model, train_it, validate_it):
    y_val = validate_it.classes
    val_pred_proba = model.predict(validate_it)
    val_pred_proba, predicted_proba, y_val, y_test = train_test_split(val_pred_proba, y_val,
test_size = 0.5, shuffle = True)
    val_pred = np.argmax(val_pred_proba, axis = 1)
    predicted = np.argmax(predicted_proba, axis = 1)
    print("Train accuracy Score------------>")
    print ("{0:.3f}".format(accuracy_score(train_it.classes,
np.argmax(model.predict(train_it), axis = 1))*100), "%")
    print("Val accuracy Score--------->")
    print("{0:.3f}".format(accuracy_score(y_val, val_pred)*100), "%")
    print("Test accuracy Score--------->")
    print("{0:.3f}".format(accuracy_score(y_test, predicted)*100), "%")
    print("F1 Score--------------->")
    print("{0:.3f}".format(f1_score(y_test, predicted, average = 'weighted')*100), "%")
    print("Cohen Kappa Score------------->")
    print("{0:.3f}".format(cohen_kappa_score(y_test, predicted)*100), "%")
    print("ROC AUC Score------------->")
    print("{0:.3f}".format(roc_auc_score(y_test, predicted_proba[:, 1])*100), "%")
    print("Recall-------------->")
    print("{0:.3f}".format(recall_score(y_test, predicted, average = 'weighted')*100), "%")
    print("Precision-------------->")
    print("{0:.3f}".format(precision_score(y_test, predicted, average = 'weighted')*100), "%")
```

```
model = VGG19(include_top=False, input_shape=(SIZE_X, SIZE_Y, 3), weights='imagenet')
model = fit_model(model, train_it, validate_it)

get_accuracy_metrics(model, train_it, validate_it)
```

3.3.2 InceptionV3

InceptionV3 is a deep convolutional neural networks, consisting of multiple stacked Inception modules that are connected in a sequential manner. It was originally designed and developed by Google researchers to improve the efficiency and performance of CNNs for image classification and other computer vision tasks. It is based on its predecessor InceptionV1 (Shabbeer Basha et al., 2018).

The key concept behind InceptionV3 is the use of inception modules. It differs from VGG16 and VGG19 in a way that each inception module incorporates different filter sizes in the same layer. The output of one Inception module serves as an input to the next model. Utilization of multiple filter sizes allows the network to capture both local and global features at different scales enabling more complex and diverse representations (Shabbeer Basha et al., 2018).

The inceptionV3 model has been trained on the large-scale ImageNet dataset, which consists of millions of labeled images across thousands of categories. As a result, it has learned to recognize a wide variety of objects and features, making it useful for tasks such as image classification, object detection, and image segmentation (Raghu et al., 2019).

InceptionV3 includes reduction modules at certain stages to reduce the spatial dimension of feature maps. These reduction modules use a combination of 3×3 max pooling and 1×1 convolutions to down sample the feature maps and reduce their dimensionality. It also incorporates auxiliary classifiers at intermediate layers during training. Auxiliary classifiers are additional branches attached to the network to learn more discriminative features to improve the overall training process (Dong et al., 2020; Wahed & Nivrito, 2016). InceptionV3 has been a popular choice for deep learning applications in the field of computer vision since it has achieved state-of-the-art performance on various computer vision benchmarks (Morid et al., 2021).

```
#Inception V3
import numpy as np
import pandas as pd
from sklearn.metrics import f1_score, roc_auc_score, cohen_kappa_score, precision_score,
recall_score, accuracy_score, confusion_matrix
from keras.layers import Dense, Flatten, Conv2D, MaxPooling2D, Dropout
from keras.models import Sequential
import tensorflow as tf
import matplotlib.pyplot as plt
import cv2
from keras.layers.normalization import BatchNormalization
from keras.models import Model, Sequential
from keras.applications.xception import Xception
from keras.applications import *
import matplotlib.pyplot as plt
from sklearn.pipeline import make_pipeline
from sklearn.pipeline import Pipeline
from PIL import Image
import random
import os
import cv2
from sklearn.model_selection import train_test_split
from sklearn.neighbors import KNeighborsClassifier
from sklearn.svm import SVC
from sklearn.ensemble import RandomForestClassifier, AdaBoostClassifier
from xgboost import XGBClassifier
from keras.callbacks import EarlyStopping
import seaborn as sns
from sklearn.preprocessing import StandardScaler
from tqdm import tqdm
from sklearn.decomposition import PCA
from keras.layers import BatchNormalization


data_dir = "C:/Users/mirka/Downloads/archive/lung_colon_image_set/colon_image_sets"
SIZE_X = SIZE_Y = 128
datagen = tf.keras.preprocessing.image.ImageDataGenerator(validation_split = 0.3)
train_it = datagen.flow_from_directory(data_dir,
                                       class_mode = "categorical",
                                       target_size = (SIZE_X,SIZE_Y),
                                       color_mode="rgb",
                                       batch_size = 12,
                                       shuffle = False,
                                       subset='training',
                                       seed = 42)
```

```
validate_it = datagen.flow_from_directory(data_dir,
                                          class_mode = "categorical",
                                          target_size = (SIZE_X, SIZE_Y),
                                          color_mode="rgb",
                                          batch_size = 12,
                                          shuffle = False,
                                          subset='validation',
                                          seed = 42)
def fit_model(model, train_it, validate_it, epochs = 10):
  es = EarlyStopping(monitor='val_loss', mode='min', verbose=1, patience=5)
    for layer in model.layers:
      layer.trainable = False
  flat1 = Flatten()(model.layers[-1].output)
    output = Dense(len(train_it.class_indices), activation='softmax')(flat1)
    model = Model(inputs=model.inputs, outputs=output)
    print(model.summary())
    model.compile(loss='categorical_crossentropy', optimizer ='adam', metrics=['accuracy'])
    history = model.fit(train_it, validation_data=validate_it, epochs=epochs, verbose=1,
callbacks=[es])
    return model
def get_accuracy_metrics(model, train_it, validate_it):
    y_val = validate_it.classes
    val_pred_proba = model.predict(validate_it)
    val_pred_proba, predicted_proba, y_val, y_test = train_test_split(val_pred_proba, y_val,
test_size = 0.5, shuffle = True)
    val_pred = np.argmax(val_pred_proba, axis = 1)
    predicted = np.argmax(predicted_proba, axis = 1)
    print("Train accuracy Score------------>")
    print ("{0:.3f}".format(accuracy_score(train_it.classes,
np.argmax(model.predict(train_it), axis = 1))*100), "%")
    print("Val accuracy Score--------->")
    print("{0:.3f}".format(accuracy_score(y_val, val_pred)*100), "%")
    print("Test accuracy Score--------->")
    print("{0:.3f}".format(accuracy_score(y_test, predicted)*100), "%")
    print("F1 Score--------------->")
    print("{0:.3f}".format(f1_score(y_test, predicted, average = 'weighted')*100), "%")
    print("Cohen Kappa Score------------->")
    print("{0:.3f}".format(cohen_kappa_score(y_test, predicted)*100), "%")
    print("ROC AUC Score------------->")
    print("{0:.3f}".format(roc_auc_score(y_test, predicted_proba[:, 1])*100), "%")
    print("Recall-------------->")
    print("{0:.3f}".format(recall_score(y_test, predicted, average = 'weighted')*100), "%")
    print("Precision-------------->")
    print("{0:.3f}".format(precision_score(y_test, predicted, average = 'weighted')*100), "%")
model = InceptionV3(include_top=False, input_shape=(SIZE_X, SIZE_Y, 3), weights='imagenet')
model = fit_model(model, train_it, validate_it)
get_accuracy_metrics(model, train_it, validate_it)
```

3.3.3 MobileNet

MobileNet is a deep learning architecture specifically designed for mobile and embedded devices. It is a lightweight and efficient convolutional neural network model that aims to achieve a good balance between model size and accuracy. MobileNetV1 was the original version introduced by Google. It consists of 28 layers and is trained on the ImageNet dataset. MobileNeVt2 is its improved version, and it utilizes a concept called "inverted residual blocks" that includes shortcut connections and linear bottlenecks to improve the flow of information and capture richer representations. It offers better accuracy and efficiency compared to MobileNetV2. MobileNetV3 is the latest version of the MobileNet series. It further improves performance and efficiency (Abd Elaziz et al., 2021).

The main goal of MobileNet is to enable deep learning applications on resource-constrained devices with limited computational power, memory, and energy requirements. It addresses these challenges by utilizing depthwise separable convolutions, which significantly reduce the number of parameters and computational cost compared to traditional convolutional layers. MobileNet splits the convolution into two separate operations: a depthwise convolution and a pointwise convolution. This reduces computational complexity and model size. MobileNet uses a width multiplier and a resolution multiplier to reduce the channels of each layer and input image. These parameters enable trade-offs between model size, inference speed, and accuracy (Howard et al., 2017).

```python
#MobileNetV2
import numpy as np
import pandas as pd
from sklearn.metrics import f1_score, roc_auc_score, cohen_kappa_score, precision_score,
recall_score, accuracy_score, confusion_matrix
from keras.layers import Dense, Flatten, Conv2D, MaxPooling2D, Dropout
from keras.models import Sequential
import tensorflow as tf
import matplotlib.pyplot as plt
import cv2
from keras.layers.normalization import BatchNormalization
from keras.models import Model, Sequential
from keras.applications.xception import Xception
from keras.applications import *
import matplotlib.pyplot as plt
from sklearn.pipeline import make_pipeline
from sklearn.pipeline import Pipeline
from PIL import Image
import random
import os
import cv2
from sklearn.model_selection import train_test_split
from sklearn.neighbors import KNeighborsClassifier
from sklearn.svm import SVC
from sklearn.ensemble import RandomForestClassifier, AdaBoostClassifier
from xgboost import XGBClassifier
from keras.callbacks import EarlyStopping
import seaborn as sns
from sklearn.preprocessing import StandardScaler
from tqdm import tqdm
from sklearn.decomposition import PCA
from keras.layers import BatchNormalization


data_dir = "C:/Users/mirka/Downloads/archive/lung_colon_image_set/colon_image_sets"
SIZE_X = SIZE_Y = 128
datagen = tf.keras.preprocessing.image.ImageDataGenerator(validation_split = 0.3)
train_it = datagen.flow_from_directory(data_dir,
                                       class_mode = "categorical",
                                       target_size = (SIZE_X,SIZE_Y),
                                       color_mode="rgb",
                                       batch_size = 12,
                                       shuffle = False,
                                       subset='training',
                                       seed = 42)
```

```python
validate_it = datagen.flow_from_directory(data_dir,
                                class_mode = "categorical",
                                target_size = (SIZE_X, SIZE_Y),
                                color_mode="rgb",
                                batch_size = 12,
                                shuffle = False,
                                subset='validation',
                                seed = 42)
def fit_model(model, train_it, validate_it, epochs = 10):
  es = EarlyStopping(monitor='val_loss', mode='min', verbose=1, patience=5)
    for layer in model.layers:
      layer.trainable = False
  flat1 = Flatten()(model.layers[-1].output)
    output = Dense(len(train_it.class_indices), activation='softmax')(flat1)
    model = Model(inputs=model.inputs, outputs=output)
    print(model.summary())
    model.compile(loss='categorical_crossentropy', optimizer ='adam', metrics=['accuracy'])
    history = model.fit(train_it, validation_data=validate_it, epochs=epochs, verbose=1,
callbacks=[es])
    return model
def get_accuracy_metrics(model, train_it, validate_it):
    y_val = validate_it.classes
    val_pred_proba = model.predict(validate_it)
    val_pred_proba, predicted_proba, y_val, y_test = train_test_split(val_pred_proba, y_val,
test_size = 0.5, shuffle = True)
    val_pred = np.argmax(val_pred_proba, axis = 1)
    predicted = np.argmax(predicted_proba, axis = 1)
    print("Train accuracy Score------------>")
    print ("{0:.3f}".format(accuracy_score(train_it.classes,
np.argmax(model.predict(train_it), axis = 1))*100), "%")
    print("Val accuracy Score--------->")
    print("{0:.3f}".format(accuracy_score(y_val, val_pred)*100), "%")
    print("Test accuracy Score--------->")
    print("{0:.3f}".format(accuracy_score(y_test, predicted)*100), "%")
    print("F1 Score-------------->")
    print("{0:.3f}".format(f1_score(y_test, predicted, average = 'weighted')*100), "%")
    print("Cohen Kappa Score------------->")
    print("{0:.3f}".format(cohen_kappa_score(y_test, predicted)*100), "%")
    print("ROC AUC Score------------->")
    print("{0:.3f}".format(roc_auc_score(y_test, predicted_proba[:, 1])*100), "%")
    print("Recall-------------->")
    print("{0:.3f}".format(recall_score(y_test, predicted, average = 'weighted')*100), "%")
    print("Precision-------------->")
    print("{0:.3f}".format(precision_score(y_test, predicted, average = 'weighted')*100), "%")

model = MobileNetV2(include_top=False, input_shape=(SIZE_X, SIZE_Y, 3), weights='imagenet')
model = fit_model(model, train_it, validate_it)
get_accuracy_metrics(model, train_it, validate_it)
```

3.3.4 DenseNet

DenseNet169 and DenseNet121 are different variations of the DenseNet architecture. As the name suggests, the DenseNet architecture consists of densely connected layers that are connected in a feed-forward manner. The deeply connected layer-type architecture allows direct information flow between the layers, enabling feature reuse and gradient flow throughout the network. This solves the vanishing gradient problem and enables feature propagation, leading to improved model performance. The DenseNet architecture consists of several dense blocks where dense layers are located. In each block, the output feature maps from all preceding layers are concatenated and serve as inputs to the current layer, allowing the network to capture fine-grained features at different scales. DenseNet169 consists of six dense blocks and 169 dense layers, hence the name. In contrast, DenseNet121 has 121 dense layers distributed among four dense blocks. Due to the higher number of layers, DenseNet169 has a deeper architecture and can accommodate a larger number of parameters. However, this comes with a higher computational power requirement (Huang et al., 2018).

Transitional layers are located between dense blocks to downsample the spatial dimensions and reduce the number of different feature maps. These transitional layers typically consist of batch normalization layers, a 1×1 convolutional layer for dimensionality reduction, and a downsampling operation such as average pooling. This helps to control the complexity of the model and improve computational efficiency. The growth rate in DenseNet is a hyperparameter that determinates the number of feature maps added to each dense layer. Both DenseNet169 and DenseNet121 have a growth rate of 32, meaning that each layer in dense blocks adds 32 feature maps to the network (Huang et al., 2018).

```
#DenseNet169
import numpy as np
import pandas as pd
from sklearn.metrics import f1_score, roc_auc_score, cohen_kappa_score, precision_score,
recall_score, accuracy_score, confusion_matrix
from keras.layers import Dense, Flatten, Conv2D, MaxPooling2D, Dropout
from keras.models import Sequential
import tensorflow as tf
import matplotlib.pyplot as plt
import cv2
from keras.layers.normalization import BatchNormalization
from keras.models import Model, Sequential
from keras.applications.xception import Xception
from keras.applications import *
import matplotlib.pyplot as plt
from sklearn.pipeline import make_pipeline
from sklearn.pipeline import Pipeline
from PIL import Image
import random
import os
import cv2
from sklearn.model_selection import train_test_split
from sklearn.neighbors import KNeighborsClassifier
from sklearn.svm import SVC
from sklearn.ensemble import RandomForestClassifier, AdaBoostClassifier
from xgboost import XGBClassifier
from keras.callbacks import EarlyStopping
import seaborn as sns
from sklearn.preprocessing import StandardScaler
from tqdm import tqdm
from sklearn.decomposition import PCA
from keras.layers import BatchNormalization

data_dir = "C:/Users/mirka/Downloads/archive/lung_colon_image_set/colon_image_sets"
SIZE_X = SIZE_Y = 128
datagen = tf.keras.preprocessing.image.ImageDataGenerator(validation_split = 0.3)
train_it = datagen.flow_from_directory(data_dir,
                                        class_mode = "categorical",
                                        target_size = (SIZE_X,SIZE_Y),
                                        color_mode="rgb",
                                        batch_size = 12,
                                        shuffle = False,
                                        subset='training',
                                        seed = 42)
```

```
validate_it = datagen.flow_from_directory(data_dir,
                                class_mode = "categorical",
                                target_size = (SIZE_X, SIZE_Y),
                                color_mode="rgb",
                                batch_size = 12,
                                shuffle = False,
                                subset='validation',
                                seed = 42)
def fit_model(model, train_it, validate_it, epochs = 10):
  es = EarlyStopping(monitor='val_loss', mode='min', verbose=1, patience=5)
    for layer in model.layers:
      layer.trainable = False
  flat1 = Flatten()(model.layers[-1].output)
    output = Dense(len(train_it.class_indices), activation='softmax')(flat1)
    model = Model(inputs=model.inputs, outputs=output)
    print(model.summary())
    model.compile(loss='categorical_crossentropy', optimizer ='adam', metrics=['accuracy'])
    history = model.fit(train_it, validation_data=validate_it, epochs=epochs, verbose=1,
callbacks=[es])
    return model
def get_accuracy_metrics(model, train_it, validate_it):
    y_val = validate_it.classes
    val_pred_proba = model.predict(validate_it)
    val_pred_proba, predicted_proba, y_val, y_test = train_test_split(val_pred_proba, y_val,
test_size = 0.5, shuffle = True)
    val_pred = np.argmax(val_pred_proba, axis = 1)
    predicted = np.argmax(predicted_proba, axis = 1)
    print("Train accuracy Score------------>")
    print ("{0:.3f}".format(accuracy_score(train_it.classes,
np.argmax(model.predict(train_it), axis = 1))*100), "%")
    print("Val accuracy Score--------->")
    print("{0:.3f}".format(accuracy_score(y_val, val_pred)*100), "%")
    print("Test accuracy Score--------->")
    print("{0:.3f}".format(accuracy_score(y_test, predicted)*100), "%")
    print("F1 Score--------------->")
    print("{0:.3f}".format(f1_score(y_test, predicted, average = 'weighted')*100), "%")
    print("Cohen Kappa Score------------>")
    print("{0:.3f}".format(cohen_kappa_score(y_test, predicted)*100), "%")
    print("ROC AUC Score------------>")
    print("{0:.3f}".format(roc_auc_score(y_test, predicted_proba[:, 1])*100), "%")
    print("Recall-------------->")
    print("{0:.3f}".format(recall_score(y_test, predicted, average = 'weighted')*100), "%")
    print("Precision-------------->")
    print("{0:.3f}".format(precision_score(y_test, predicted, average = 'weighted')*100), "%")

model = DenseNet169(include_top=False, input_shape=(SIZE_X, SIZE_Y, 3), weights='imagenet')
model = fit_model(model, train_it, validate_it)
get_accuracy_metrics(model, train_it, validate_it)
```

3.3.5 ResNet

ResNet101 and ResNet50 are deep neural network architectures that belong to the ResNet family. The ResNet architecture introduced the concept of residual learning, which helps address the degradation problem in very deep neural networks. ResNet101 consists of residual blocks containing 101 layers, while ResNet50, as the name suggests, has 50 layers. They both include convolutional layers, pooling layers, fully connected layers, and shortcut connections. The shortcut connections skip one or more layers and add original input to the output, allowing the network to learn residual mappings. Skip connections allow the gradient to flow directly through the network without passing through multiple layers. These connections provide a shortcut path for information to flow, facilitating the training of deep networks and improving their performance (Huang et al., 2018).

Similar to MobileNet, ResNet incorporates a bottleneck architecture in its residual blocks. This reduces the need for computational power by using a 1×1 convolutional layer architecture. These layers reduce the dimensionality of the input feature maps before applying 3×3 convolutions. This approach improves the efficiency and effectiveness of the network. ResNet architecture has shown superior classification in image classification benchmarks, achieving state-of-the-art accuracy on various datasets. Its depth and skip connections make it particularly effective for complex visual recognition tasks (He et al., 2021).

```
#ResNet101
import numpy as np
import pandas as pd
from sklearn.metrics import f1_score, roc_auc_score, cohen_kappa_score, precision_score,
recall_score, accuracy_score, confusion_matrix
from keras.layers import Dense, Flatten, Conv2D, MaxPooling2D, Dropout
from keras.models import Sequential
import tensorflow as tf
import matplotlib.pyplot as plt
import cv2
from keras.layers.normalization import BatchNormalization
from keras.models import Model, Sequential
from keras.applications.xception import Xception
from keras.applications import *
import matplotlib.pyplot as plt
from sklearn.pipeline import make_pipeline
from sklearn.pipeline import Pipeline
from PIL import Image
import random
import os
import cv2
from sklearn.model_selection import train_test_split
from sklearn.neighbors import KNeighborsClassifier
from sklearn.svm import SVC
from sklearn.ensemble import RandomForestClassifier, AdaBoostClassifier
from xgboost import XGBClassifier
from keras.callbacks import EarlyStopping
import seaborn as sns
from sklearn.preprocessing import StandardScaler
from tqdm import tqdm
from sklearn.decomposition import PCA
from keras.layers import BatchNormalization


data_dir = "C:/Users/mirka/Downloads/archive/lung_colon_image_set/colon_image_sets"
SIZE_X = SIZE_Y = 128
datagen = tf.keras.preprocessing.image.ImageDataGenerator(validation_split = 0.3)
train_it = datagen.flow_from_directory(data_dir,
                                       class_mode = "categorical",
                                       target_size = (SIZE_X,SIZE_Y),
                                       color_mode="rgb",
                                       batch_size = 12,
                                       shuffle = False,
                                       subset='training',
                                       seed = 42)
```

```python
validate_it = datagen.flow_from_directory(data_dir,
                                    class_mode = "categorical",
                                    target_size = (SIZE_X, SIZE_Y),
                                    color_mode="rgb",
                                    batch_size = 12,
                                    shuffle = False,
                                    subset='validation',
                                    seed = 42)
def fit_model(model, train_it, validate_it, epochs = 10):
  es = EarlyStopping(monitor='val_loss', mode='min', verbose=1, patience=5)
    for layer in model.layers:
      layer.trainable = False
  flat1 = Flatten()(model.layers[-1].output)
    output = Dense(len(train_it.class_indices), activation='softmax')(flat1)
    model = Model(inputs=model.inputs, outputs=output)
    print(model.summary())
    model.compile(loss='categorical_crossentropy', optimizer ='adam', metrics=['accuracy'])
    history = model.fit(train_it, validation_data=validate_it, epochs=epochs, verbose=1,
callbacks=[es])
    return model
def get_accuracy_metrics(model, train_it, validate_it):
    y_val = validate_it.classes
    val_pred_proba = model.predict(validate_it)
    val_pred_proba, predicted_proba, y_val, y_test = train_test_split(val_pred_proba, y_val,
test_size = 0.5, shuffle = True)
    val_pred = np.argmax(val_pred_proba, axis = 1)
    predicted = np.argmax(predicted_proba, axis = 1)
    print("Train accuracy Score------------>")
    print ("{0:.3f}".format(accuracy_score(train_it.classes,
np.argmax(model.predict(train_it), axis = 1))*100), "%")
    print("Val accuracy Score--------->")
    print("{0:.3f}".format(accuracy_score(y_val, val_pred)*100), "%")
    print("Test accuracy Score--------->")
    print("{0:.3f}".format(accuracy_score(y_test, predicted)*100), "%")
    print("F1 Score--------------->")
    print("{0:.3f}".format(f1_score(y_test, predicted, average = 'weighted')*100), "%")
    print("Cohen Kappa Score------------->")
    print("{0:.3f}".format(cohen_kappa_score(y_test, predicted)*100), "%")
    print("ROC AUC Score------------>")
    print("{0:.3f}".format(roc_auc_score(y_test, predicted_proba[:, 1])*100), "%")
    print("Recall------------->")
    print("{0:.3f}".format(recall_score(y_test, predicted, average = 'weighted')*100), "%")
    print("Precision-------------->")
    print("{0:.3f}".format(precision_score(y_test, predicted, average = 'weighted')*100), "%")

model = ResNet101(include_top=False, input_shape=(SIZE_X, SIZE_Y, 3), weights='imagenet')
model = fit_model(model, train_it, validate_it)
get_accuracy_metrics(model, train_it, validate_it)
```

3.3.6 *InceptionResNet*

InceptionResNet is a fusion of the Inception and ResNet models, developed by Google researchers to enhance the performance and accuracy of image recognition tasks. It builds upon the Inception architecture by incorporating the advantages of multi-scale feature extraction and ResNet's ability to train deep networks. InceptionResNet also incorporates residual connections from the ResNet architecture to address to solve the vanishing gradient problem with auxiliary classifiers. The network connections enable effective gradient propagation, facilitating the training of deeper architectures with improved optimization (Muhammad et al., 2023).

Similar to the original Inception models, InceptionreResNetV2 utilized Inception modules that combine different filter sizes to extract features at various scales. Additionally, it includes residual connections, also known as skip connections, that allow the model to bypass one or more layers. These connections enable the network to learn residual mappings and aid in optimizing the training process. Inception-ResNetV2 includes a stem block at the network's beginning, performing initial convolutions and pooling operations to extract basic features from input images. It also incorporates reduction blocks, which reduce spatial dimensions while increasing the number of channels. This compression of information facilitates efficient flow throughout the network (Schmarje et al., 2019).

Overall, InceptionResNet consists of approximately 572 layers, making it a powerful model for various computer vision tasks (Schmarje et al., 2019). It has been pretrained with large-scale image datasets like ImageNet, enabling it to excel in image classification and object recognition tasks. Its depth and complex architecture make it suitable for highly accurate and detailed feature representations. However, it should be noted that InceptionResNet requires substantial computational resources even if it adds a bottleneck layer of a 1×1 convolutional filter (Muhammad et al., 2023).

```python
#InceptionResNetV2
import numpy as np
import pandas as pd
from sklearn.metrics import f1_score, roc_auc_score, cohen_kappa_score, precision_score,
recall_score, accuracy_score, confusion_matrix
from keras.layers import Dense, Flatten, Conv2D, MaxPooling2D, Dropout
from keras.models import Sequential
import tensorflow as tf
import matplotlib.pyplot as plt
import cv2
from keras.layers.normalization import BatchNormalization
from keras.models import Model, Sequential
from keras.applications.xception import Xception
from keras.applications import *
import matplotlib.pyplot as plt
from sklearn.pipeline import make_pipeline
from sklearn.pipeline import Pipeline
from PIL import Image
import random
import os
import cv2
from sklearn.model_selection import train_test_split
from sklearn.neighbors import KNeighborsClassifier
from sklearn.svm import SVC
from sklearn.ensemble import RandomForestClassifier, AdaBoostClassifier
from xgboost import XGBClassifier
from keras.callbacks import EarlyStopping
import seaborn as sns
from sklearn.preprocessing import StandardScaler
from tqdm import tqdm
from sklearn.decomposition import PCA
from keras.layers import BatchNormalization


data_dir = "C:/Users/mirka/Downloads/archive/lung_colon_image_set/colon_image_sets"
SIZE_X = SIZE_Y = 128
datagen = tf.keras.preprocessing.image.ImageDataGenerator(validation_split = 0.3)
train_it = datagen.flow_from_directory(data_dir,
                                       class_mode = "categorical",
                                       target_size = (SIZE_X,SIZE_Y),
                                       color_mode="rgb",
                                       batch_size = 12,
                                       shuffle = False,
                                       subset='training',
                                       seed = 42)
```

```python
validate_it = datagen.flow_from_directory(data_dir,
                                    class_mode = "categorical",
                                    target_size = (SIZE_X, SIZE_Y),
                                    color_mode="rgb",
                                    batch_size = 12,
                                    shuffle = False,
                                    subset='validation',
                                    seed = 42)
def fit_model(model, train_it, validate_it, epochs = 10):
  es = EarlyStopping(monitor='val_loss', mode='min', verbose=1, patience=5)
    for layer in model.layers:
      layer.trainable = False
  flat1 = Flatten()(model.layers[-1].output)
    output = Dense(len(train_it.class_indices), activation='softmax')(flat1)
    model = Model(inputs=model.inputs, outputs=output)
    print(model.summary())
    model.compile(loss='categorical_crossentropy', optimizer ='adam', metrics=['accuracy'])
    history = model.fit(train_it, validation_data=validate_it, epochs=epochs, verbose=1,
callbacks=[es])
    return model
def get_accuracy_metrics(model, train_it, validate_it):
    y_val = validate_it.classes
    val_pred_proba = model.predict(validate_it)
    val_pred_proba, predicted_proba, y_val, y_test = train_test_split(val_pred_proba, y_val,
test_size = 0.5, shuffle = True)
    val_pred = np.argmax(val_pred_proba, axis = 1)
    predicted = np.argmax(predicted_proba, axis = 1)
    print("Train accuracy Score----------->")
    print ("{0:.3f}".format(accuracy_score(train_it.classes,
np.argmax(model.predict(train_it), axis = 1))*100), "%")
    print("Val accuracy Score--------->")
    print("{0:.3f}".format(accuracy_score(y_val, val_pred)*100), "%")
    print("Test accuracy Score--------->")
    print("{0:.3f}".format(accuracy_score(y_test, predicted)*100), "%")
    print("F1 Score--------------->")
    print("{0:.3f}".format(f1_score(y_test, predicted, average = 'weighted')*100), "%")
    print("Cohen Kappa Score------------->")
    print("{0:.3f}".format(cohen_kappa_score(y_test, predicted)*100), "%")
    print("ROC AUC Score------------->")
    print("{0:.3f}".format(roc_auc_score(y_test, predicted_proba[:, 1])*100), "%")
    print("Recall-------------->")
    print("{0:.3f}".format(recall_score(y_test, predicted, average = 'weighted')*100), "%")
    print("Precision-------------->")
    print("{0:.3f}".format(precision_score(y_test, predicted, average = 'weighted')*100), "%")

model = InceptionResNetV2(include_top=False, input_shape=(SIZE_X, SIZE_Y, 3),
weights='imagenet')
model = fit_model(model, train_it, validate_it)
get_accuracy_metrics(model, train_it, validate_it)
```

3.3.7 *Xception*

Xception is a model derived from the Inception architecture. Much like MobileNet, Xception employs depthwise separable convolutions, but it goes a step further by replacing conventional Inception modules with these depthwise separable convolutions. These depthwise separable convolutions are intricately stacked together, resulting in a highly interconnected network. What sets them apart is their ability to disentangle spatial and cross-channel filtering, which leads to a reduction in the number of parameters and computational requirements when compared to standard convolutions. The Xception model has demonstrated exceptional performance across a range of computer vision tasks, including image classification, object detection, and image segmentation. By amalgamating the advantages of depthwise separable convolutions and a more intricate network design, Xception proves to be an efficient and effective choice for feature extraction (Chollet, 2017).

3.4 Deep feature extraction

Deep feature extraction is a technique used in machine learning and computer vision to extract meaningful features from raw data such as images and audio signals. The term "deep" refers to the fact that these features are extracted from multiple layers of a deep neural network, typically a convolutional neural network (CNN) (Dong et al., 2020). In image recognition tasks, the initial layers of CNN are responsible for detecting elementary features like edges and corners, while the subsequent layers capture more intricate features such as shapes and textures. By extracting these high-level features, the network can represent the input image in a concise and informative manner, which proves beneficial for tasks like classification and object recognition. Deep feature extraction can also be applied to other types of data such as texts and audio. In natural language processing, for example, a pretrained model such as BERT can be used to extract features from text, which then can be used for tasks such as sentiment analysis and named entity recognition. Overall, deep feature extraction has become an important technique in machine learning and computer vision, enabling more accurate and efficient analysis of complex data (Subakti et al., 2022).

```
#Feature extraction with MobileNet
import numpy as np
import pandas as pd
from sklearn.metrics import f1_score, roc_auc_score, cohen_kappa_score, precision_score,
recall_score, accuracy_score, confusion_matrix
from keras.layers import Dense, Flatten, Conv2D, MaxPooling2D, Dropout
from keras.models import Sequential
import tensorflow as tf
import matplotlib.pyplot as plt
import cv2
from keras.layers.normalization import BatchNormalization
from keras.models import Model, Sequential
from keras.applications.xception import Xception
from keras.applications import *
import matplotlib.pyplot as plt
from sklearn.pipeline import make_pipeline
from sklearn.pipeline import Pipeline
from PIL import Image
import random
import os
import cv2
from sklearn.model_selection import train_test_split
from sklearn.neighbors import KNeighborsClassifier
from sklearn.svm import SVC
from sklearn.ensemble import RandomForestClassifier, AdaBoostClassifier
from xgboost import XGBClassifier
from keras.callbacks import EarlyStopping
import seaborn as sns
from sklearn.preprocessing import StandardScaler
from tqdm import tqdm
from sklearn.decomposition import PCA
from keras.layers import BatchNormalization


data_dir = "C:/Users/mirka/Downloads/archive/lung_colon_image_set/colon_image_sets"
SIZE_X = SIZE_Y = 128
datagen = tf.keras.preprocessing.image.ImageDataGenerator(validation_split = 0.3)
train_it = datagen.flow_from_directory(data_dir,
                                       class_mode = "categorical",
                                       target_size = (SIZE_X,SIZE_Y),
                                       color_mode="rgb",
                                       batch_size = 12,
                                       shuffle = False,
                                       subset='training',
                                       seed = 42)
```

```
validate_it = datagen.flow_from_directory(data_dir,
                                  class_mode = "categorical",
                                  target_size = (SIZE_X, SIZE_Y),
                                  color_mode="rgb",
                                  batch_size = 12,
                                  shuffle = False,
                                  subset='validation',
                                  seed = 42)


def get_features(base_model, train, validate):
    X_train = base_model.predict(train)
    y_train = train.classes
    X_val = base_model.predict(validate)
    y_val = validate.classes
    X_val, X_test, y_val, y_test = train_test_split(X_val, y_val, test_size = 0.5, shuffle =
True)
    print('Shape of X_train----->', str(X_train.shape))
    print('Shape of X_val----->', str(X_val.shape))
    print('Shape of X_test----->', str(X_test.shape))
    return (X_train, X_val, X_test, y_train, y_val, y_test)


def get_models():
  ANN = Sequential()
  ANN.add(Dense(128, input_dim = X_train.shape[1], activation = 'relu'))
  ANN.add(BatchNormalization())
  ANN.add(Dropout(0.2))
  ANN.add(Dense(64, activation='relu'))
  ANN.add(Dense(32, activation='relu'))
  ANN.add(Dense(16, activation='relu'))
  ANN.add(Dense(8, activation='relu'))
  ANN.add(Dense(len(train_it.class_indices), activation='softmax'))
  ANN.compile(loss='sparse_categorical_crossentropy', optimizer='adam', metrics=['accuracy'])
  KNN = KNeighborsClassifier()
  SVM = SVC(kernel = 'linear')
  RF = RandomForestClassifier(n_estimators = 50)
  ADB = AdaBoostClassifier()
  XGB = XGBClassifier(n_estimators = 50, use_label_encoder=False)
  print("Defined------->")
  print("ANN -------->", "(128x64x32x16x8)")
  print("KNeighborsClassifier()")
  print("SVC(kernel = 'linear')")
  print("RandomForestClassifier(n_estimators = 50)")
  print("AdaBoostClassifier()")
  print("XGBClassifier(n_estimators = 50)")
  return (ANN, KNN, SVM, RF, ADB, XGB)
def reshape_data(X_train, X_val, X_test):
```

```
        X_train = X_train.reshape(X_train.shape[0], -1)
        X_val = X_val.reshape(X_val.shape[0], -1)
        X_test = X_test.reshape(X_test.shape[0], -1)
        print("Shape after reshaping------->")
        print("X train------->", str(X_train.shape))
        print("X val-------->", str(X_val.shape))
        print("X test-------->", str(X_test.shape))
        return (X_train, X_val, X_test)


def fit_ANN(model, X_train, y_train, X_val, y_test):
        es = EarlyStopping(monitor='val_loss', mode='min', verbose=1, patience=5)
        history = model.fit(X_train, y_train, validation_data=(X_val, y_test), epochs=10,
verbose=1, callbacks=[es])
        return model


def fit_model(model, X_train, y_train):
    model.fit(X_train, y_train)
    return model


def get_accuracy_metrics_for_ANN(model, X_train, y_train, X_val, y_val, X_test, y_test):
        print("Train accuracy Score------------>")
        print ("{0:.3f}".format(accuracy_score(y_train, np.argmax(model.predict(X_train), axis =
1))*100), "%")
        print("Val accuracy Score--------->")
        val_pred = np.argmax(model.predict(X_val), axis = 1)
        print("{0:.3f}".format(accuracy_score(y_val, val_pred)*100), "%")
        predicted =  np.argmax(model.predict(X_test), axis = 1)
        print("Test accuracy Score--------->")
        print("{0:.3f}".format(accuracy_score(y_test, predicted)*100), "%")
        print("F1 Score--------------->")
        print("{0:.3f}".format(f1_score(y_test, predicted, average = 'weighted')*100), "%")
        print("Cohen Kappa Score------------->")
        print("{0:.3f}".format(cohen_kappa_score(y_test, predicted)*100), "%")
        print("Recall-------------->")
        print("{0:.3f}".format(recall_score(y_test, predicted, average = 'weighted')*100), "%")
        print("Precision-------------->")
        print("{0:.3f}".format(precision_score(y_test, predicted, average = 'weighted')*100), "%")


def fit_KNN_metrics(model, X_train, y_train, X_val, y_val, X_test, y_test):
        pca = PCA(n_components=7000)
        randlist = random.sample(range(0, X_train.shape[0]), 1000)
        if(X_train.shape[1] > 10000):
            X_train = pca.fit_transform(X_train)
            X_val = pca.transform(X_val)
            X_test = pca.transform(X_test)
        model.fit(X_train, y_train)
        get_accuracy_metrics(model, X_train[randlist, :], y_train[randlist], X_val, y_val, X_test,
y_test)
```

```python
def get_accuracy_metrics(model, X_train, y_train, X_val, y_val, X_test, y_test):
    print("Train accuracy Score------------>")
    print ("{0:.3f}".format(accuracy_score(y_train, model.predict(X_train))*100), "%")
    print("Val accuracy Score--------->")
    val_pred = model.predict(X_val)
    print("{0:.3f}".format(accuracy_score(y_val, val_pred)*100), "%")
    predicted =  model.predict(X_test)
    print("Test accuracy Score--------->")
    print("{0:.3f}".format(accuracy_score(y_test, predicted)*100), "%")
    print("F1 Score--------------->")
    print("{0:.3f}".format(f1_score(y_test, predicted, average = 'weighted')*100), "%")
    print("Cohen Kappa Score------------->")
    print("{0:.3f}".format(cohen_kappa_score(y_test, predicted)*100), "%")
    print("Recall-------------->")
    print("{0:.3f}".format(recall_score(y_test, predicted, average = 'weighted')*100), "%")
    print("Precision-------------->")
    print("{0:.3f}".format(precision_score(y_test, predicted, average = 'weighted')*100), "%")

base_model = MobileNet(include_top=False, input_shape=(SIZE_X, SIZE_Y, 3), weights='imagenet')


for layer in base_model.layers:
    layer.trainable = False


model = Model(inputs=base_model.input, outputs=base_model.layers[-1].output)
model.summary()


X_train, X_val, X_test, y_train, y_val, y_test = get_features(model, train_it, validate_it)
X_train, X_val, X_test = reshape_data(X_train, X_val, X_test)
ANN, KNN, SVM, RF, ADB, XGB = get_models()


ANN = fit_ANN(ANN, X_train, y_train, X_val, y_val)


get_accuracy_metrics_for_ANN(ANN, X_train, y_train, X_val, y_val, X_test, y_test)


scaler = StandardScaler()
X_train_scaled = scaler.fit_transform(X_train)
X_val_scaled = scaler.transform(X_val)
X_test_scaled = scaler.transform(X_test)
fit_KNN_metrics(KNN, X_train_scaled, y_train, X_val_scaled, y_val, X_test_scaled, y_test)


SVM = fit_model(SVM, X_train_scaled, y_train)
get_accuracy_metrics(SVM, X_train_scaled, y_train, X_val_scaled, y_val, X_test_scaled, y_test)
```

```
RF = fit_model(RF, X_train, y_train)
get_accuracy_metrics(RF,  X_train, y_train, X_val, y_val, X_test, y_test)


ADB = fit_model(ADB, X_train, y_train)
get_accuracy_metrics(ADB,  X_train, y_train, X_val, y_val, X_test, y_test)


XGB = fit_model(XGB, X_train, y_train)
get_accuracy_metrics(XGB,  X_train, y_train, X_val, y_val, X_test, y_test)
```

3.5 Colon cancer detection using deep learning

Deep learning techniques are increasingly being used in colon cancer detection to improve the accuracy and efficiency of diagnosis. Various ways to apply deep learning to this field exist, including image analysis, classification, predictive modeling, and treatment planning. For instance, deep learning models can analyze colonoscopy videos, and CT scans to detect potentially cancerous lesions by recognizing specific patterns and shapes related to polyps (Taha et al., 2022). Deep learning can also be used to classify colon cancer based on analyzing patient's health data, which could facilitate early intervention and prevention. Finally, deep learning models can assist treatment planning by predicting the efficacy of various treatment options and identifying patients who are most likely to benefit from certain therapies (Kennion et al., 2022). The usage of deep learning can improve diagnosis accuracy, reduce the cost and time required for interpretation, and help identify those patients who need further testing and treatment. Overall, deep learning has shown great potential for improving colon cancer detection and treatment, and its applications in this field are expected to grow further in the future (Kennion et al., 2022).

The deep learning algorithms can be trained on large datasets of colonoscopy images and can analyze images at various levels, such as identifying polyps or lesions, characterizing their size, shape, and color, and predicting whether the lesions are malignant or benign. Several studies have demonstrated that deep learning algorithms have high accuracy in detecting colon cancer from medical images (Usher-Smith et al., 2016).

Classification can help with the treatment of colon cancer by providing doctors with important information about the stage and the type of cancer. This information can help doctors develop a personalized treatment plan that is tailored to the individual patient's needs. For example, if the cancer is localized, the primary treatment option might be surgery, whereas chemotherapy and radiation therapy may be recommended if the cancer has spread (Usher-Smith et al., 2016).

Furthermore, classification can help identify the specific type of colon cancer for a more effective treatment plan. It is also helpful for monitoring the progress of the disease and making informed decisions about ongoing treatment. By using classification systems, such as imaging studies and tumor marker tests, doctors can track the effectiveness of treatment and determine if additional treatment or follow-up is necessary. Overall, deep learning has shown great potential for improving colon cancer detection and treatment, and its applications in this field are expected to grow further in the future (Kennion et al., 2022).

4. Results and discussion

4.1 Experimental data

In this study, the Lung and Colon Cancer Histopathological Images LC25000 dataset was used to train to validate the methods. A colon cancer dataset was extracted from this dataset, and it is publicly available in Kaggle (Lung and Colon Cancer Histopathological Images, n.d.). The dataset included 5000 images of benign colon cancer and 5000 of adenocarcinomas. Moreover, 30% of the original dataset was used as test data, and the rest as training data. Test and training data were automatically split; 30% of the training data was automatically detached from the dataset for the test set Therefore, the training data included 7000 images, the test set 3000 images, and the validation data 1400 images, respectively. The dataset was balanced.

4.2 Performance evaluation measures

The performance of deep learning models was evaluated based on training accuracy, validation accuracy, test accuracy, F1 measure, Cohen's kappa score, ROC area score, recall, and precision.

Training accuracy is a measure of how well a machine learning model performs on the training dataset. It is calculated by comparing the predictions made by the model to the actual labels or targets of the training data. Validation accuracy is similar to training accuracy, but it assesses the model's performance on data that it has not seen before. High validation accuracy indicates that the model is performing well on the validation data and is likely to perform well on new unseen data. Training accuracy evaluates the models' capability to successfully capture the patterns and relationships present in the training dataset (Brownlee, 2017).

F1 score is a metric that combines precision and recall into a single value. It provides a balanced measure of a model's performance by considering both the ability to correctly identify positive instances and the ability to capture all positive instances. Cohen's kappa is a statistic that measures agreement between predicted and true class labels. It is used when evaluating the performance of classification models especially when the classes are imbalanced. The area under the receiving operating characteristic (ROC) curve is a metric that quantifies the overall performance of the model across all possible thresholds. It measures the model's ability to distinguish between positive and negative classes. Recall is also known as sensitivity or true positive rate, and it measures the proportion of true positive predictions out of all positive samples in the dataset whereas precision measures the proportion of true positive predictions made by the model (Jin, 2020).

VGG16 and VGG19 models outperformed other models in this experiment. VGG16 and VGG19 both have a deep architecture with multiple layers, including convolutional layers, pooling layers, and fully connected layers, which allow to capture and learn features and enabling better representation of complex patterns and structures. They are also sensitive to capture small details because they use smaller representative fields for convolutional layers (Boesch, 2021).

4.3 Experimental results

In this study, a comparison between 16 different classifiers was made (Tables 9.1 and 9.2). Kaggle data (Lung and Colon Cancer Histopathological Images, n.d.) and platform were used as a base, with small adjustments, to perform the experiment. TensorFlow and Keras libraries were used in this study. Also, a comparison between different machine learning models, which were trained on various feature-extracted datasets, was made (Tables 9.3—9.6).

4.3.1 CNN classifier results

The images were resized to a resolution of 128×128 pixels. A batch size of 12 was chosen for training the model, which means that each iteration of the training process used 12 images.

To prepare the images for input into the CNN model, they were transformed into a one-

TABLE 9.1 Performance of CNN models.

Classifier	Training accuracy	Validation accuracy	Test accuracy	F1 measure	Kappa	ROC area
CNN 2 layer	95.614	93.867	93.333	93.321	86.617	98.426
CNN 3 layer	48.562	48.9	49.5	43.916	−1.769	48.742
CNN 4 layer	62.613	60.9	62.2	55.586	94.711	62.2
CNN 5 layer	50	49.4	50.6	50	50	50.6
CNN 6 layer	86.171	86.067	85.932	85.932	71.865	92.742
CNN 7 layer	50	49	50	34.002	0	49.934
CNN 8 layer	50	50.4	49.6	32.89	0	55.362

TABLE 9.2 Performance of transfer learning models.

Classifier	Training accuracy	Validation accuracy	Test accuracy	F1 measure	Kappa	ROC area
ResNet50	99.957	99.6	99.677	99.677	99.333	99.963
VGG16	99.843	99	99.533	99.533	99.067	99.794
VGG19	99.942	99.4	99.733	99.733	99.467	99.781
Inception_v3	67.986	68	66.133	62.08	32.03	71.619
MobileNet	99.3	93.8	94.533	94.528	89.064	99.146
DenseNet169	91.171	90	90.133	90.042	80.287	91.707
DenseNet121	98.586	96.867	95.865	97.591	91.732	95.93
InceptionResNetV2	58.714	59.267	59.267	54.472	18.617	63.719
MobileNetV2	98.143	93.267	94.067	94.066	88.134	98.672
ResNet101	99.133	99.2	99.2	99.2	99.862	99.209

TABLE 9.3 Feature extraction with VGG16.

Model	Training accuracy	Validation accuracy	Test accuracy	F1 measure	Kappa	Recall	Precision
SVM	100.000	99.467	99.733	99.733	99.467	99.733	99.733
Random Forest	100.000	98.800	98.600	98.600	97.200	98.600	98.604
AdaBoost	99.286	97.993	98.400	96.400	96.800	98.400	98.400
KNN	90.400	84.800	84.800	84.469	69.517	84.800	87.796
XGBoost	100.000	99.333	98.733	98.733	97.466	98.733	98.737
ANN	99.943	99.733	99.400	99.400	98.800	99.400	99.400

TABLE 9.4 Feature extraction with VGG19.

Model	Training accuracy	Validation accuracy	Test accuracy	F1 measure	Kappa	Recall	Precision
SVM	100.000	99.467	99.667	99.667	99.333	99.667	99.667
Random Forest	100.000	98.333	97.133	97.132	94.265	97.133	97.170
AdaBoost	98.857	97.867	97.733	97.733	95.466	97.733	97.733
KNN	91.300	89.400	88.267	88.105	76.482	88.267	90.271
XGBoost	100.000	99.000	97.933	97.933	95.865	97.933	97.958
ANN	99.957	99.200	99.067	99.067	98.133	99.067	99.075

TABLE 9.5 Feature extraction with ResNet50.

Model	Training accuracy	Validation accuracy	Test accuracy	F1 measure	Kappa	Recall	Precision
SVM	100.000	99.800	99.867	99.867	99.733	99.867	99.867
Random Forest	100.000	98.200	97.600	97.600	95.199	97.600	97.605
AdaBoost	99.686	98.267	97.867	95.733	97.867	95.733	97.867
KNN	98.100	96.867	95.933	95.933	91.867	95.933	95.952
XGBoost	100.000	99.267	98.533	98.533	97.066	98.533	98.536
ANN	99.943	99.800	99.800	99.800	99.600	99.800	99.801

TABLE 9.6 Feature extraction with ResNet101.

Model	Training accuracy	Validation accuracy	Test accuracy	F1 measure	Kappa	Recall	Precision
SVM	100.000	99.267	99.400	99.400	98.800	99.400	99.400
Random Forest	100.000	97.467	97.200	97.200	94.399	97.200	97.212
AdaBoost	99.057	96.867	96.800	96.800	93.600	96.800	96.801
KNN	96.800	95.533	95.067	95.066	90.135	95.067	95.107
XGBoost	100.000	98.400	98.733	98.733	97.466	98.733	98.733
ANN	99.986	99.267	99.400	99.400	98.800	99.400	99.401

dimensional array using the flatten function. This operation reshapes the image data into a linear format, allowing easier processing by subsequent layers.

The CNN 2-layer model, used in this report, consists of 2 convolutional layers. The ReLU activation function was applied to each convolutional layer. Before the convolutional layers, a Batch normalization layer was included to enhance the model's performance by normalizing the input data.

A max-pooling layer was employed after the convolutional layers. This layer selects the maximum value within each defined are of the feature map, reducing the spatial dimensions of the output. This pooling operation helps to reduce the complexity and dimensionality of the image representations. To prevent overfitting, early stopping was implemented during the training process. This technique monitors the model's performance on a validation set and stops the training as soon as the model's performance starts to deteriorate, thus preventing further training that could lead to overfitting. Additionally, a dropout rate of 0.2 was applied, which randomly drops out a fraction of the neural network units during each training step, further reducing the overfitting risk.

The dense layer of the model utilizes the SoftMax function. This function calculates the probabilities of each class being the correct classification by normalizing the output values. It is commonly used for multiclass classification problems. Also, the Adam optimizer was used. For the convolutional layers, a kernel size 3×3 was set. It defines the dimensions of the filters used to extract features from the input data. A larger kernel size would capture more complex patterns, while a smaller kernel size focuses on capturing small details.

This chapter also explored different architectures with varying convolutional layers. This allowed for a comparison of the performance of the models with different complexity and depth.

From CNN models, the worst performer was the 3-layer CNN classifier with a training accuracy of 48.562, validation accuracy of 48.900, test accuracy of 49.500, FI-measure of 43.916, kappa score of −1.769, and ROC area of 48.742. CNN 5 layer, CNN 7 layer, and CNN 8 layer all gave training accuracy of 50.000, and these models were trained again to evaluate the results, but they stayed the same.

The best performance was acquired with the CNN-2 layer classifier with a training accuracy of 95.614, validation accuracy of 93.867, test accuracy of 93.333, F1 measure of 93.321, kappa score of 86.617, and ROC area of 98.426. CNN 6-layer got the second best results with a training

accuracy of 86.171, validation accuracy of 86.067, test accuracy of 85.932, F1 measure of 85.932, kappa score of 71.865, and ROC area of 92.742. Other CNN models performed poorly.

4.3.2 Transfer learning models

All other classifiers used the same functions and parameters as CNN models.

The ResNet50 classifier outperformed others with a training accuracy of 99.957, validation accuracy of 99.600, kappa score of 99.333, and an ROC area of 99.963. VGG19 had the best test accuracy and F1-measure, both being 99.733. Also, VGG16, MobileNetV2, DenseNet121, ResNet101, and MobileNet all performed notably well with every measure.

The worst performer was InceptionResNetV2 with a training accuracy of 58.714, validation accuracy of 59.267, test accuracy of 59.267, and F1 score of 54.472. Kappa score and ROC area score were also low with values 18.617 and 63.719, respectively. InceptionV3 gave also relatively poor accuracy results with a training accuracy of 67.986, validation accuracy of 68.000, test accuracy of 66.133, F1 score of 62.080, kappa score of 32.030, and ROC area score of 71.619.

4.3.3 Feature extraction with pretrained models

Also, a comparison between different machine learning models, which were trained on various feature-extracted datasets, was made. Please refer to Attachment 1 for the detailed results. Each model was evaluated based on various performance metrics. The code for feature-extracted models was available on Kaggle, which was used with small modifications. The code for Bagging feature extraction was constructed by the writer.

The functions and parameters used are the same as in other classifiers. The validation data were automatically split from the dataset with the ImageDataGeneration function. However, dense layers of 128, 64, 32, 16, and 8 units with ReLU activation function were added with the addition of batch normalization layer to ANN. In VGG16, SVM, Random Forest, and XGBoost all achieved the highest training accuracy of 100.000. ANN had the best validation accuracy of 99.733, whereas the best test accuracy (99.733), F1 score (99.467), kappa coefficient (99.733), recall (99.733), and precision (99.733) were gained with SVM. Random Forest, AdaBoost, XGBoost, Bagging, and ANN also overall performed well. KNN showed lower performance compared to other models with a training accuracy of 90.400, validation and test accuracy of 84.800, F1 score of 84.469, kappa score of 69.517, recall of 84.800, and precision of 87.796.

Similarly, VGG19, SVM, Random Forest, and XCBoost all achieved the highest training accuracy of 100.000. SVM in addition had the best validation accuracy (99.467), test accuracy (99.667), F1 score (99.667), kappa coefficient (99.333), recall (99.733), and precision (99.733). As in VGG16, Random Forest, AdaBoost, and XGB boost all performed well with high accuracies. Likewise, the worst performer was KNN with a training accuracy of 91.300, validation accuracy of 89.400, test accuracy of 88.267, F1 score of 88.105, kappa score of 76.482, recall of 88.267, and precision of 90.271.

In both ResNet50 and ResNet101 SVM, Random Forest, and XGBoost achieved 100.00 training accuracy. In ResNet50, the best validation accuracy of 99.80 was gained with both SVM and ANN. Moreover, SVM got the best measures with test accuracy (99.867), kappa score (99.733, recall (99.867), and precision (99.867). In ResNet101, SVM and ANN gave the same results: validation accuracy 99.267, test accuracy, F1 score, recall, and precision all 99.400, and kappa score 98.800. Moreover, Random Forest, AdaBoost, and XGBoost all showed good performance with both ResNet models. KNN also performed well but compared to algorithms, got the worst results of both models. The following results were acquired with ResNet50: train accuracy 98.100, validation accuracy 96.867, test accuracy 95.933, F1 score

95.933 kappa coefficient 91.867, recall 95.933, and precision 95.952. The ResNet101 results for KNN have train accuracy 96.80, validation accuracy 95.533, test accuracy 95.067, F1 score 95.066, kappa coefficient 90.135, recall 95.067, and precision 95.107.

In both, MobileNet and MobileNetV2, SVM and XGBoost got training accuracy of 100.000. In MobileNet, Random Forest also got a training accuracy of 100%. Overall, the best performer of MobileNetV2 was SVM with validation accuracy, test accuracy, F1 score, recall, and precision all being 92.267 and kappa score 84.532. The worst performer was this time AdaBoost with a training accuracy of 89.329, validation accuracy of 84.533, test accuracy of 86.867, F1 score of 86.863, kappa coefficient of 73.723, recall of 86.867, and precision of 96.887. With MobileNet, ANN gave the highest validation accuracy of 95.133, whereas SVM gave the best test accuracy, F1 score, recall, and precision all of 95.667 and the best kappa score of 91.224. AdaBoost gave the worst training accuracy of 91.300. KNN had worse performance with other parameters: validation accuracy 86.000, test accuracy 85.800, F1 score 85.668, kappa score 71.651, recall 85.800, and precision 87.334.

SVM was the best performer also with InceptionV3, DenseNet169, and DenseNet121 with validation accuracy of 91.067, 98.933, and 98.867, test accuracy of 88.867, 99.133, and 98.867, F1 score of 88.867, 99.133, and 98.867, kappa score of 77.733, 98.267, and 97.733, recall of 88.867, 99.133, and 98.867, and precision of 88.867, 80.445, and 98.869, respectively.

Also Xception model behaved poorly with KNN, with a validation accuracy of 76.133, test accuracy of 76.733, F2 score of 76.671, kappa score of 53.421, recall of 76.733, and precision of 76.955. AdaBoost gave the worst training accuracy of 83.786. ANN was the best-performing algorithm of Xception with a validation accuracy of 89.200, test accuracy of 88.133, F1 score of 88.085, kappa score of 76.292, recall of 8.133, and precision of 88.728. However,

ANN showed the worst performance with InceptionV3, InceptionResnetV2, and Dense-Net169 with the results being 79.414, 62.586, and 84.514 for training accuracy, 78.267, 61.333, and 83.400 for validation accuracy, 77.400, 62.800, and 84.800 for test accuracy, and 76.470, 25.501, and 83.986 for F1 score, respectively. Moreover, the DenseNet and InceptionResnet got the worst recall (62.800 and 84.400) and precision (62.985 and 88.084) with ANN. The worst recall of inceptionV3 was acquired with Random Forest with the result of 73.207, and the lowest precision with Bagging with the result of 79.444.

InceptionV3, DenseNet169, DenseNet121, and Xception all got training accuracy of 100.000 with SVM, Random Forest, and XGBoost Algorithms. For InceptionResNetV2, the best training accuracy was acquired with Random Forest (99.986). Otherwise, Bagging

was the best performer for InceptionResNetV2 with a validation accuracy of 79.643, test accuracy of 80.300, F1 score of 80.270, kappa score of 60.557, recall of 80.300, and precision of 80.445. As an opposite, Bagging was the worst performer with DenseNet121 with a validation accuracy of 91.429, test accuracy, F1 score, and recall of 92.333, kappa score of 84.665, and precision of 87.334. However, AdaBoost had the worst training accuracy of 94.871 (Tables 9.7–9.13).

In summary, SVM consistently performs well across most feature-extracted models, achieving the best accuracy, F1 score, kappa, recall, and precision. Random Forest, AdaBoost, and XGBoost also demonstrated good performance in several models. KNN and ANN generally have lower performance compared to other models.

TABLE 9.7 Feature extraction with MobileNetV2.

Model	Training accuracy	Validation accuracy	Test accuracy	F1 measure	Kappa	Recall	Precision
SVM	100.000	92.267	92.267	92.267	84.532	92.267	92.267
Random Forest	99.986	87.267	87.133	87.131	74.274	87.133	87.194
AdaBoost	89.329	84.533	86.867	86.863	73.725	86.867	86.887
KNN	91.900	87.533	89.267	89.250	78.517	89.267	89.457
XGBoost	100.000	90.733	90.800	90.800	81.600	90.800	90.803
ANN	92.929	88.533	89.867	89.782	79.768	89.867	91.406

TABLE 9.8 Feature extraction with MobileNet.

Model	Training accuracy	Validation accuracy	Test accuracy	F1 measure	Kappa	Recall	Precision
SVM	100.000	94.800	95.667	95.667	91.224	95.667	95.667
Random Forest	100.000	87.733	87.667	87.654	75.348	87.667	87.879
AdaBoost	91.300	86.600	88.600	88.600	77.202	88.600	88.618
KNN	92.400	86.000	85.800	85.668	71.651	85.800	87.334
XGBoost	100.000	92.067	92.400	92.399	84.804	92.400	92.453
ANN	99.157	95.133	95.333	95.330	90.662	95.333	95.414

TABLE 9.9 Feature extraction with InceptionV3.

Model	Training accuracy	Validation accuracy	Test accuracy	F1 measure	Kappa	Recall	Precision
SVM	100.000	91.067	88.867	88.867	77.733	88.867	88.869
Random Forest	100.000	86.600	86.600	86.597	73.207	73.207	86.660
AdaBoost	84.714	83.333	82.267	82.267	64.533	82.267	82.270
KNN	92.100	85.200	86.533	86.515	73.048	86.533	86.676
XGBoost	100.000	89.867	88.400	88.395	76.809	88.400	88.504
ANN	79.414	78.267	77.400	76.470	54.632	77.400	82.201

TABLE 9.10 Feature extraction with DenseNet169.

Model	Training accuracy	Validation accuracy	Test accuracy	F1 measure	Kappa	Recall	Precision
SVM	100.000	98.933	99.133	99.133	98.267	99.133	99.138
Random Forest	100.000	96.933	95.400	95.400	90.802	95.400	95.444
AdaBoost	97.314	95.133	93.667	93.667	87.332	93.667	93.667
KNN	97.500	94.733	96.133	96.133	92.266	96.133	96.134
XGBoost	100.000	98.067	97.400	94.800	94.800	94.400	97.415
ANN	84.514	83.400	84.400	83.986	68.706	84.400	88.084

TABLE 9.11 Feature extraction with DenseNet121.

Model	Training accuracy	Validation accuracy	Test accuracy	F1 measure	Kappa	Recall	Precision
SVM	100.000	98.867	98.867	98.867	97.733	98.867	98.869
Random Forest	100.000	94.733	93.667	93.667	87.334	93.667	93.681
AdaBoost	94.871	92.933	92.467	92.467	84.933	92.467	92.469
KNN	96.500	93.667	93.733	93.731	87.471	93.733	93.849
XGBoost	100.000	97.400	97.067	97.066	94.134	97.067	97.108
ANN	99.686	98.267	98.400	98.400	96.799	98.400	98.405

4.4 Discussion and conclusion

The initial hypothesis of the writer before starting the project evaluated that from all CNN models, 8-layer CNN would yield the best performance, due to its deeper architecture containing the largest number of parameters. Similarly, ResNet101 would perform better than ResNet50, VGG19 better than VGG16, InceptionResNetV2 would overperform InceptionV3, and DenseNet169 outperform DenseNet121. However, only the CNN 2-layer and CNN 6-layer models showed satisfactory performance, while the remaining CNN models exhibited poor performance. InceptionResNetV2

TABLE 9.12 Feature extraction with InceptionResNetV2.

Model	Training accuracy	Validation accuracy	Test accuracy	F1 measure	Kappa	Recall	Precision
SVM	82.086	76.800	76.000	76.000	51.995	76.000	76.000
Random Forest	99.986	73.467	74.533	74.530	49.079	74.533	74.571
AdaBoost	74.500	71.733	70.933	70.934	41.869	70.933	70.941
KNN	78.400	67.400	69.667	69.563	39.267	69.667	69.868
XGBoost	99.243	74.467	73.933	73.930	47.878	73.933	73.968
ANN	62.586	61.333	62.800	62.610	25.501	62.800	62.985

TABLE 9.13 Feature extraction with Xception.

Model	Training accuracy	Validation accuracy	Test accuracy	F1 measure	Kappa	Recall	Precision
SVM	100.000	89.000	87.067	87.066	74.138	87.067	87.096
Random Forest	100.000	82.467	81.600	81.600	63.202	81.600	81.611
AdaBoost	83.786	79.867	77.333	77.333	54.671	77.333	77.347
KNN	86.100	76.133	76.733	76.671	53.421	76.733	76.955
XGBoost	100.000	86.800	85.067	85.058	70.147	85.067	85.196
ANN	92.700	89.200	88.133	88.095	76.292	88.133	88.728

and InceptionV3 yielded disappointing results, suggesting that the dataset employed may not be suitable for these models. This can be attributed to either the dataset being too small or the models being overly complex for the specific task, as evidenced by the successful performance of other models.

Notably, the ResNet50 model with fewer layers exhibited slightly better performance than ResNet101. A similar pattern emerged by comparing the results of DenseNet121 and DenseNet169, with a simpler model performing better. Conversely, the enhanced MobileNet121 model achieved slightly better performance compared to the basic MobileNet model, although both models are characterized by their lightweight nature. VGG16 and VGG19 displayed nearly identical performance measures. The dataset used in this study consisted of histopathological, cell-level microscopy images stained with hematoxylin-eosin, resulting in a pinkish hue to the images. The resolution of the data was not particularly high, and some images were photobleached or out of focus, thereby complicating the training process. These factors may account for challenges encountered by certain CNN models during training, as the models were expected to discern abnormal growth, irregular shapes, and disorganized cellular structures commonly observed in adenocarcinoma cells.

The results indicate that this type of dataset can be effectively trained and yield accurate outcomes using various deep learning models. Feature extraction also proved successful, except for Xception, InceptionV3, and InceptionResNet models, which produced mediocre results. Particularly, SVM's ability to work with outliers and extreme intensities makes it an appropriate algorithm for this type of data. Additionally, the Random Forest classifier showed successful results by effectively handling missing data and exhibition robustness against outliers and photobleached pixels.

While adenocarcinoma is generally easily visible to the human eye in microscopy images, deep learning algorithms could be beneficial in borderline cases where pathologists may have uncertainties regarding the right diagnosis. Furthermore, deep learning models could be employed in cancer screenings involving a large number of patients, significantly alleviating the workload burden on pathologists. Such screenings can aid in the early detection of potential cancer cases in patients who do not yet exhibit any symptoms. Consequently, cancer treatments could be initiated before the onset of symptoms when the likelihood of mortality is lower. It is important, however, not to rely solely on Deep Learning as the diagnostic tool for colon cancer, as it is a highly lethal form of cancer where errors in classifications should be avoided, or at the very least, accuracy should match that of human classification by visual inspection. This also raises an ethical question: when the technique is good enough, that it could be used as a diagnosis tool for high-mortality cancer? What happens if it leads to an error?

In conclusion, this chapter aligns with the results of previous research related to deep learning and colon cancer detection suggesting that deep learning could be potentially used as a tool in cancer diagnosis. However, it is not yet widely used in clinical settings outside of research.

References

Abd Elaziz, M., Dahou, A., Alsaleh, N. A., Elsheikh, A. H., Saba, A. I., & Ahmadein, M. (2021). Boosting COVID-19 image classification using MobileNetV3 and aquila optimizer algorithm. *Entropy, 23*(11). https://doi.org/10.3390/e23111383. Article 11.

Argilés, G., Tabernero, J., Labianca, R., Hochhauser, D., Salazar, R., Iveson, T., Laurent-Puig, P., Quirke, P., Yoshino, T., & Taieb, J. (2020). Localised colon cancer: ESMO clinical practice guidelines for diagnosis, treatment and follow-up. *Annals of Oncology, 31*(10), 1291−1305.

Attallah, O., Aslan, M. F., & Sabanci, K. (2022). A framework for lung and colon cancer diagnosis via lightweight deep learning models and transformation methods. *Diagnostics, 12*(12). https://doi.org/10.3390/diagnostics12122926. Article 12.

Azer, S. A. (2019). Challenges facing the detection of colonic polyps: What can deep learning do? *Medicina, 55*(8). https://doi.org/10.3390/medicina55080473. Article 8.

Ben Hamida, A., Devanne, M., Weber, J., Truntzer, C., Derangère, V., Ghiringhelli, F., Forestier, G., & Wemmert, C. (2021). Deep learning for colon cancer histopathological images analysis. *Computers in Biology and Medicine, 136,* 104730. https://doi.org/10.1016/j.compbiomed.2021.104730

Boesch, G. (October 6, 2021). *VGG Very Deep Convolutional Networks (VGGNet)—What you need to know.* Viso.Ai. https://viso.ai/deep-learning/vgg-very-deep-convolutional-networks/.

Brownlee, J. (July 13, 2017). *What is the difference between test and validation datasets?* MachineLearningMastery.Com. https://machinelearningmastery.com/difference-test-validation-datasets/.

Burt, R. W. (2000). Colon cancer screening. *Gastroenterology, 119*(3), 837−853. https://doi.org/10.1053/gast.2000.16508

Chollet, F. (2017). Xception: Deep learning with depthwise separable convolutions. In *2017 IEEE conference on computer vision and pattern recognition (CVPR)* (pp. 1800−1807). https://doi.org/10.1109/CVPR.2017.195

CS231n Convolutional Neural Networks for Visual Recognition. (n.d.). Retrieved June 2, 2023, from https://cs231n.github.io/convolutional-networks/.

Dong, N., Zhao, L., Wu, C. H., & Chang, J. F. (2020). Inception v3 based cervical cell classification combined with artificially extracted features. *Applied Soft Computing, 93,* 106311. https://doi.org/10.1016/j.asoc.2020.106311

Fernandez-Quilez, A. (2023). Deep learning in radiology: Ethics of data and on the value of algorithm transparency, interpretability and explainability. *AI and Ethics, 3*(1), 257−265. https://doi.org/10.1007/s43681-022-00161-9

Gessert, N., Bengs, M., Wittig, L., Drömann, D., Keck, T., Schlaefer, A., & Ellebrecht, D. B. (2019). Deep transfer learning methods for colon cancer classification in confocal laser microscopy images. *International Journal of Computer Assisted Radiology and Surgery, 14*(11), 1837−1845. https://doi.org/10.1007/s11548-019-02004-1

Godkhindi, A. M., & Gowda, R. M. (2017). Automated detection of polyps in CT colonography images using deep learning algorithms in colon cancer diagnosis. In *2017 international conference on energy, communication, data analytics and Soft computing (ICECDS)* (pp. 1722−1728). https://doi.org/10.1109/ICECDS.2017.8389744

Gupta, P., Huang, Y., Sahoo, P. K., You, J.-F., Chiang, S.-F., Onthoni, D. D., Chern, Y.-J., Chao, K.-Y., Chiang, J.-M., Yeh, C.-Y., & Tsai, W.-S. (2021). Colon tissues classification and localization in whole slide images using deep learning. *Diagnostics, 11*(8). https://doi.org/10.3390/diagnostics11081398. Article 8.

Gupta, S., Kalaivani, S., Rajasundaram, A., Ameta, G. K., Oleiwi, A. K., & Dugbakie, B. N. (2022). Prediction performance of deep learning for colon cancer survival prediction on SEER data. *BioMed Research International, 2022*, e1467070. https://doi.org/10.1155/2022/1467070

Hage Chehade, A., Abdallah, N., Marion, J.-M., Oueidat, M., & Chauvet, P. (2022). Lung and colon cancer classification using medical imaging: A feature engineering approach. *Physical and Engineering Sciences in Medicine, 45*(3), 729–746. https://doi.org/10.1007/s13246-022-01139-x

Hasan, M. I., Ali, M. S., Rahman, M. H., & Islam, M. K. (2022). Automated detection and characterization of colon cancer with deep convolutional neural networks. *Journal of Healthcare Engineering, 2022*, e5269913. https://doi.org/10.1155/2022/5269913

He, H., Yan, S., Lyu, D., Xu, M., Ye, R., Zheng, P., Lu, X., Wang, L., & Ren, B. (2021). Deep learning for biospectroscopy and biospectral imaging: State-of-the-art and perspectives. *Analytical Chemistry, 93*(8), 3653–3665. https://doi.org/10.1021/acs.analchem.0c04671

Häfner, M. (2007). Conventional colonoscopy: Technique, indications, limits. *European Journal of Radiology, 61*(3), 409–414. https://doi.org/10.1016/j.ejrad.2006.07.034

Howard, A. G., Zhu, M., Chen, B., Kalenichenko, D., Wang, W., Weyand, T., Andreetto, M., & Adam, H. (2017). MobileNets: Efficient convolutional neural networks for mobile vision applications (arXiv:1704.04861). *arXiv.* http://arxiv.org/abs/1704.04861.

Hsu, T.-C., & Lin, C. (2021). Training with small medical data: Robust Bayesian neural networks for colon cancer overall survival prediction. In *2021 43rd annual international conference of the IEEE Engineering in Medicine & Biology Society (EMBC)* (pp. 2030–2033). https://doi.org/10.1109/EMBC46164.2021.9630698

Huang, G., Liu, Z., van der Maaten, L., & Weinberger, K. Q. (2018). Densely connected convolutional networks (arXiv:1608.06993). *arXiv.* http://arxiv.org/abs/1608.06993.

Jin. (June 8, 2020). *Everything you need about evaluating classification models.* Analytics Vidhya. https://medium.com/analytics-vidhya/everything-you-need-about-evaluating-classification-models-dfb89c60e643.

Katsaounou, K., Nicolaou, E., Vogazianos, P., Brown, C., Stavrou, M., Teloni, S., Hatzis, P., Agapiou, A., Fragkou, E., Tsiaoussis, G., Potamitis, G., Zaravinos, A., Andreou, C., Antoniades, A., Shiammas, C., & Apidianakis, Y. (2022). Colon cancer: From epidemiology to prevention. *Metabolites, 12*(6). https://doi.org/10.3390/metabo12060499. Article 6.

Kavitha, M. S., Gangadaran, P., Jackson, A., Venmathi Maran, B. A., Kurita, T., & Ahn, B.-C. (2022). Deep neural network models for colon cancer screening. *Cancers, 14*(15). https://doi.org/10.3390/cancers14153707. Article 15.

Kennion, O., Maitland, S., & Brady, R. (2022). Machine learning as a new horizon for colorectal cancer risk prediction? A systematic review. *Health Sciences Review, 4*, 100041. https://doi.org/10.1016/j.hsr.2022.100041

Lung and Colon Cancer Histopathological Images. (n.d.). Retrieved June 6, 2023, from https://www.kaggle.com/datasets/andrewmvd/lung-and-colon-cancer-histopathological-images.

Masud, M., Sikder, N., Nahid, A.-A., Bairagi, A. K., & AlZain, M. A. (2021). A machine learning approach to diagnosing lung and colon cancer using a deep learning-based classification framework. *Sensors, 21*(3). https://doi.org/10.3390/s21030748. Article 3.

Mohalder, R. D., Bin Ali, F., Paul, L., & Hasan Talukder, K. (2022). Deep learning-based colon cancer tumor prediction using histopathological images. In *2022 25th international conference on computer and information technology (ICCIT)* (pp. 629–634). https://doi.org/10.1109/ICCIT57492.2022.10054766

Morid, M. A., Borjali, A., & Del Fiol, G. (2021). A scoping review of transfer learning research on medical image analysis using ImageNet. *Computers in Biology and Medicine, 128*, 104115. https://doi.org/10.1016/j.compbiomed.2020.104115

Muhammad, W., Bhutto, Z., Masroor, S., Shaikh, M., Shah, J., & Hussain, A. (2023). IRMIRS: Inception-ResNet-based network for MRI image super-resolution. *Computer Modeling in Engineering and Sciences, 136*(2), 1121–1142. https://doi.org/10.32604/cmes.2023.021438

Muniz, F. B., de Baffa, M. F. O., Garcia, S. B., Bachmann, L., & Felipe, J. C. (2023). Histopathological diagnosis of colon cancer using micro-FTIR hyperspectral imaging and deep learning. *Computer Methods and Programs in Biomedicine, 231*, 107388. https://doi.org/10.1016/j.cmpb.2023.107388

Murugesan, M., Madonna Arieth, R., Balraj, S., & Nirmala, R. (2023). Colon cancer stage detection in colonoscopy images using YOLOv3 MSF deep learning architecture. *Biomedical Signal Processing and Control, 80*, 104283. https://doi.org/10.1016/j.bspc.2022.104283

Nasseri, Y., & Langenfeld, S. J. (2017). Imaging for colorectal cancer. *Surgical Clinics of North America, 97*(3), 503–513. https://doi.org/10.1016/j.suc.2017.01.002

Pacal, I., Karaboga, D., Basturk, A., Akay, B., & Nalbantoglu, U. (2020). A comprehensive review of deep learning in colon cancer. *Computers in Biology and*

Medicine, 126, 104003. https://doi.org/10.1016/j.comp biomed.2020.104003

Raghu, M., Zhang, C., Kleinberg, J., & Bengio, S. (2019). Transfusion: Understanding transfer learning for medical imaging. *Advances in Neural Information Processing Systems, 32.* https://proceedings.neurips.cc/paper_files/paper/2019/hash/eb1e78328c46506b46a4ac4a1e378b91-Abstract.html.

Sakr, A. S., Soliman, N. F., Al-Gaashani, M. S., Pławiak, P., Ateya, A. A., & Hammad, M. (2022). An efficient deep learning approach for colon cancer detection. *Applied Sciences, 12*(17). https://doi.org/10.3390/app12178450. Article 17.

Sarker, M. M. K., Makhlouf, Y., Craig, S. G., Humphries, M. P., Loughrey, M., James, J. A., Salto-Tellez, M., O'Reilly, P., & Maxwell, P. (2021). A means of assessing deep learning-based detection of ICOS protein expression in colon cancer. *Cancers, 13*(15). https://doi.org/10.3390/cancers13153825. Article 15.

Sarwinda, D., Paradisa, R. H., Bustamam, A., & Anggia, P. (2021). Deep learning in image classification using residual network (ResNet) variants for detection of colorectal cancer. *Procedia Computer Science, 179,* 423–431. https://doi.org/10.1016/j.procs.2021.01.025

Schiele, S., Arndt, T. T., Martin, B., Miller, S., Bauer, S., Banner, B. M., Brendel, E.-M., Schenkirsch, G., Anthuber, M., Huss, R., Märkl, B., & Müller, G. (2021). Deep learning prediction of metastasis in locally advanced colon cancer using binary histologic tumor images. *Cancers, 13*(9). https://doi.org/10.3390/cancers13092074. Article 9.

Schmarje, L., Zelenka, C., Geisen, U., Glüer, C.-C., & Koch, R. (2019). 2D and 3D segmentation of uncertain local collagen fiber orientations in SHG microscopy. In G. A. Fink, S. Frintrop, & X. Jiang (Eds.), *Pattern recognition* (pp. 374–386). Springer International Publishing. https://doi.org/10.1007/978-3-030-33676-9_26

Sec, D. I. (March 6, 2021). *VGG-19 convolutional neural network. All about machine learning.* https://blog.techcraft.org/vgg-19-convolutional-neural-network/.

Shabbeer Basha, S. H., Ghosh, S., Kishan Babu, K., Ram Dubey, S., Pulabaigari, V., & Mukherjee, S. (2018). RCCNet: An efficient convolutional neural network for histological routine colon cancer nuclei classification. In *2018 15th international conference on control, automation, robotics and vision (ICARCV)* (pp. 1222–1227). https://doi.org/10.1109/ICARCV.2018.8581147

Shapcott, M., Hewitt, K. J., & Rajpoot, N. (2019). Deep learning with sampling in colon cancer histology. *Frontiers in Bioengineering and Biotechnology, 7.* https://www.frontiersin.org/articles/10.3389/fbioe.2019.00052.

Sirinukunwattana, K., Raza, S. E. A., Tsang, Y.-W., Snead, D. R. J., Cree, I. A., & Rajpoot, N. M. (2016). Locality sensitive deep learning for detection and classification of nuclei in routine colon cancer histology images. *IEEE Transactions on Medical Imaging, 35*(5), 1196–1206. https://doi.org/10.1109/TMI.2016.2525803

Smit, M. A., Ciompi, F., Bokhorst, J.-M., van Pelt, G. W., Geessink, O. G. F., Putter, H., Tollenaar, R. A. E. M., van Krieken, J. H. J. M., Mesker, W. E., & van der Laak, J. A. W. M. (2023). Deep learning based tumor–stroma ratio scoring in colon cancer correlates with microscopic assessment. *Journal of Pathology Informatics, 14,* 100191. https://doi.org/10.1016/j.jpi.2023.100191

Subakti, A., Murfi, H., & Hariadi, N. (2022). The performance of BERT as data representation of text clustering. *Journal of Big Data, 9*(1), 15. https://doi.org/10.1186/s40537-022-00564-9

Suzuki, K. (2017). Overview of deep learning in medical imaging. *Radiological Physics and Technology, 10*(3), 257–273. https://doi.org/10.1007/s12194-017-0406-5

Taha, D., Alzu'bi, A., Abuarqoub, A., Hammoudeh, M., & Elhoseny, M. (2022). Automated colorectal polyp classification using deep neural networks with colonoscopy images. *International Journal of Fuzzy Systems, 24*(5), 2525–2537. https://doi.org/10.1007/s40815-021-01182-y

Talukder, Md. A., Islam, Md. M., Uddin, M. A., Akhter, A., Hasan, K. F., & Moni, M. A. (2022). Machine learning-based lung and colon cancer detection using deep feature extraction and ensemble learning. *Expert Systems with Applications, 205,* 117695. https://doi.org/10.1016/j.eswa.2022.117695

Tasnim, Z., Chakraborty, S., Shamrat, F. M. J. M., Chowdhury, A. N., Nuha, H. A., Karim, A., Zahir, S. B., & Billah, Md. M. (2021). Deep learning predictive model for colon cancer patient using CNN-based classification. *International Journal of Advanced Computer Science and Applications, 12*(8). https://doi.org/10.14569/IJACSA.2021.0120880

Toğaçar, M. (2021). Disease type detection in lung and colon cancer images using the complement approach of inefficient sets. *Computers in Biology and Medicine, 137,* 104827. https://doi.org/10.1016/j.compbiomed.2021.104827

Tripathi, S., & Singh, S. K. (2020). Ensembling handcrafted features with deep features: An analytical study for classification of routine colon cancer histopathological nuclei images. *Multimedia Tools and Applications, 79*(47), 34931–34954. https://doi.org/10.1007/s11042-020-08891-w

Tsirikoglou, A., Stacke, K., Eilertsen, G., Lindvall, M., & Unger, J. (2020). A study of deep learning colon cancer detection in limited data access scenarios (arXiv: 2005.10326; version 2). *arXiv*. http://arxiv.org/abs/2005.10326.

Usher-Smith, J. A., Walter, F. M., Emery, J. D., Win, A. K., & Griffin, S. J. (2016). Risk prediction models for colorectal cancer: A systematic review. *Cancer Prevention Research, 9*(1), 13–26. https://doi.org/10.1158/1940-6207.CAPR-15-0274

van Rijn, J. C., Reitsma, J. B., Stoker, J., Bossuyt, P. M., van Deventer, S. J., & Dekker, E. (2006). Polyp miss rate determined by tandem colonoscopy: A systematic review. *Official Journal of the American College of Gastroenterology | ACG, 101*(2), 343. https://journals.lww.com/ajg/Abstract/2006/02000/Polyp_Miss_Rate_Determined_by_Tandem_Colonoscopy_.25.aspx.

Wahed, R. B., & Nivrito, A. (2016). *Comparative analysis between inception-v3 and other learning systems using facial expressions detection* (p. 35). Department of Computer Science & Engineering, BRAC University.

Weiss, K., Khoshgoftaar, T. M., & Wang, D. (2016). A survey of transfer learning. *Journal of Big Data, 3*(1), 9. https://doi.org/10.1186/s40537-016-0043-6

Optical coherence tomography image classification for retinal disease detection using artificial intelligence

Muhammed Enes Subasi[1], Sohan Patnaik[2], Abdulhamit Subasi[3,4]

[1]Faculty of Medicine, Izmir Katip Celebi University, Izmir, Turkey; [2]Department of Mechanical Engineering, Indian Institute of Technology Kharagpur, Kharagpur, West Bengal, India; [3]Faculty of Medicine, Institute of Biomedicine, University of Turku, Turku, Finland; [4]Department of Computer Science, College of Engineering, Effat University, Jeddah, Saudi Arabia

1. Introduction

Retinal disorders are major causes of vision impairment and blindness around the world. The early detection and proper diagnosis of these disorders are critical for timely intervention and good therapy to prevent irreversible vision loss. Optical coherence tomography (OCT) has evolved as a strong imaging technology that allows for high-resolution observation of the retinal layers, allowing abnormality diagnosis and disease progression tracking. With the rapid advancement of artificial intelligence (AI) methods, there is growing interest in utilizing AI algorithms to evaluate OCT images and aid in the classification of retinal diseases. Early diagnosis of retinal disorders is critical for effective treatment and improved visual results. Many retinal disorders are asymptomatic in the early stages, and permanent damage may have occurred by the time symptoms appear. Early intervention enables disease progression to be slowed, complications to be avoided, and patient outcomes to be improved. AI-based retinal disease classification based on OCT images has the potential to help ophthalmologists make more accurate and quicker diagnosis, resulting in better patient care (Patnaik & Subasi, 2023).

OCT imaging is a noninvasive, high-resolution imaging method that has completely changed the field of ophthalmology. It generates cross-sectional images of the retina, allowing for the detection of microstructural changes in the retinal layers. OCT scans provide extensive information about retinal thickness, shape, and the presence of fluid or lesions, all of which are important for diagnosing and monitoring retinal diseases. Because of the extensive use of OCT technology, enormous databases of OCT images have been developed, providing a great resource for building AI algorithms for retinal disease classification (Shelke & Subasi, 2023).

AI approaches, particularly machine learning and deep learning algorithms, have proven to be extremely effective in medical image processing tasks. AI systems can learn complicated patterns and features from big datasets to automatically detect and categorize distinct retinal pathologies in the context of retinal disease classification utilizing OCT images. Deep learning methods, such as convolutional neural networks (CNNs), have demonstrated greater performance by learning hierarchical representations directly from raw OCT images (Rajan & Kumar, 2023).

AI approaches combined with OCT imaging have enormous potential for improving the diagnosis and management of retinal disorders. Ophthalmologists can use AI algorithms to help them analyze OCT images, allowing for more accurate and rapid illness classification. By boosting diagnostic capabilities, optimizing treatment plans, and improving patient outcomes, the combination of AI and OCT imaging has the potential to change the profession of ophthalmology. However, further research is needed to address issues like data availability, model interpretability, and clinical validation, which will eventually lead to the creation of strong and clinically feasible AI systems for retinal disease classification.

2. Related work

Age-related macular degeneration (AMD) and diabetic retinopathy (DR) are the primary causes of vision impairment and blindness globally. OCT has evolved as a powerful imaging tool for the diagnosis and monitoring of retinal disorders. Recent advances in AI have shown considerable promise in the classification of retinal diseases using OCT images. A critical stage in the classification of retinal illnesses is the accurate segmentation of retinal layers and abnormalities. Several AI-based segmentation techniques have been developed to detect and outline retinal structures and abnormalities in OCT images automatically. These techniques include traditional machine learning algorithms, such as support vector machines (SVMs) and random forests, as well as deep learning models, such as CNN and U-net architectures. The application of these segmentation approaches has increased the accuracy and efficiency of retinal disease classification dramatically.

After segmentation, informative characteristics from OCT images must be retrieved for disease categorization. Statistical metrics, texture descriptors, and morphological characteristics of the retinal layers are examples of features. Wavelet transforms and Gabor filters, for example, have long been employed to extract features. Deep learning—based techniques, specifically CNNs, have shown greater performance in automatically learning discriminative features directly from OCT images. Transfer learning, which involves fine-tuning pretrained CNN models for retinal disease classification, has also gained popularity due to its ability to leverage knowledge from large-scale datasets.

To categorize retinal disorders using OCT images, various classification models have been used. Decision trees, SVMs, ensemble approaches, and deep neural networks are examples of these models. To analyze the success of classification models, performance evaluation metrics such as accuracy, sensitivity, specificity, and area under the receiver operating characteristic curve (AUC) are often utilized. High accuracy and AUC values have been reported in studies, showing the potential of AI-based techniques for accurate and reliable retinal disease classification.

The computer-aided diagnosis of retinopathy is a prominent area of research within medical image classification. Within this domain, diabetic macular edema (DME) and AMD are prevalent ocular diseases that can lead to partial or complete vision loss. OCT imaging is widely utilized for diagnosing such ocular conditions, including DME and AMD. In this study, we propose an automated deep learning–based method for detecting lesions associated with both DME and AMD. To address the challenge of limited data availability and to enhance adaptability to variations in different datasets, we employed two publicly accessible OCT datasets of retinal images. We developed a network model that effectively reused features extracted from the data. We conducted a comparative analysis of various network models, all equipped with the capability to reuse features effectively, and we explored transfer learning using pretrained models. Our results demonstrate that CliqueNet outperforms other network models, achieving an accuracy of over 0.98 and an area under the curve (AUC) value of 0.99 in classification (Wang & Wang, 2019).

OCT is a widely adopted technology for the identification and categorization of retinal diseases. Nevertheless, manual detection of OCT images by ophthalmologists is susceptible to errors and subjectivity. Consequently, various automated approaches have been suggested, but there remains a need for improved detection precision. Particularly, the development of deep learning–based automated methods utilizing OCT images is in progress to enable the early detection of a range of retinal disorders. Khan et al. (2023) introduced an automatic approach based on deep learning for the detection and classification of retinal diseases using OCT images. These diseases encompass AMD, branch retinal vein occlusion, central retinal vein occlusion, central serous chorioretinopathy, and DME. The proposed method involves four primary stages: first, three pretrained models, namely DenseNet-201, InceptionV3, and ResNet-50, are adapted to suit the dataset's characteristics, and feature extraction is carried out through transfer learning. These extracted features are subsequently refined, and the best ones are selected using ant colony optimization. Finally, the selected features are fed into the k-nearest neighbors and support vector machine algorithms for the ultimate classification. The proposed method, evaluated using

OCT retinal images obtained from Soonchunhyang University Bucheon Hospital, achieves an accuracy of 99.1% when incorporating ACO. Without ACO, the achieved accuracy stands at 97.4%. Moreover, our method demonstrates state-of-the-art performance and surpasses existing techniques in terms of accuracy.

Rong et al. (2019) presented a surrogate-assisted classification approach for the automatic categorization of retinal OCT images using CNNs. They begin by reducing image noise through denoising techniques. Subsequently, they employed thresholding and morphological dilation to extract masks from the images. These denoised images and masks are then utilized to generate numerous surrogate images, which serve as training data for the CNN model. Finally, when classifying a test image, the prediction is based on the average of the outputs from the trained CNN model applied to the surrogate images. The proposed method has undergone assessment using various databases, and the results indicate its high promise as an automated tool for classifying retinal OCT images, achieving an AUC of 0.9783 in the local database and an AUC of 0.9856 in the Duke database.

Wang et al. (2020) employed a CNN-based instance-level classifier, which is continually improved through an uncertainty-driven deep multiple instance learning approach that we propose. Next, they employed a recurrent neural network (RNN) that takes instance features from the same group as input. It generates the ultimate group-level prediction by considering both individually localized instance information and globally aggregated group-level representation. For thorough validation, they established two extensive datasets for DME OCT, comprising 30,151 B-scans across 1396 volumes from 274 patients (Heidelberg-DME dataset) and 38,976 B-scans across 3248 volumes from 490 patients (Triton-DME dataset), each with volume-level labels. The proposed method achieved volume-level accuracy, F1-score, and area under the receiver operating characteristic curve (AUC) of 95.1%, 0.939, and 0.990 on Heidelberg-DME and 95.1%, 0.935, and 0.986 on Triton-DME, respectively.

Lu et al. (2018) trained 101 one-layer convolutional neural networks (ResNet) for image categorization. To refine and optimize our algorithms, they employed a 10-fold cross-validation technique. The intelligent system achieved an AUC of 0.984 and an accuracy of 0.959 in detecting conditions such as macular hole, cystoid macular edema, epiretinal membrane, and serous macular detachment. Specifically, it displayed accuracies of 0.973 for normal images, 0.848 for cystoid macular edema, 0.947 for serous macular detachment, 0.957 for epiretinal membrane, and 0.978 for macular hole. The system exhibited a kappa value of 0.929, surpassing the individual kappa values of the two physicians, which were 0.882 and 0.889, respectively.

Fang et al. (2019) designed a lesion detection network that generates a soft attention map covering the entire OCT image. This attention map is then integrated into a classification network to weigh the significance of local convolutional features. By following the guidance of the lesion attention map, the classification network can effectively utilize information from local regions associated with lesions. This approach accelerates the network training process and enhances the accuracy of OCT image classification. Our experimental results, based on two sets of clinically acquired OCT data, demonstrate the efficacy and efficiency of the proposed LACNN method for classifying retinal OCT images.

Rajan and Kumar (2023) introduced a novel deep learning architecture called OCT Deep Net2, designed for classifying OCT images. The study focuses on the classification of four disease categories. OCT Deep Net2 represents an extension of the earlier OCT Deep Net1 and is composed of 50 layers, characterized as a dense architecture featuring three recurrent modules. The performance evaluation demonstrates the efficacy of OCT Deep Net2, showcasing strong results with an accuracy of 98% when trained for 100 epochs with a batch size of 32.

Wang et al. (2023) presented a novel deep semisupervised multiple instance learning framework. This framework leverages a small amount of coarsely labeled data and a substantial amount of unlabeled data. They introduced several modules to enhance performance based on the availability and granularity of labels. To initiate training, they propagated bag labels to corresponding instances and introduced a self-correction strategy to handle label noise in positive bags. This strategy utilizes confidence-based pseudolabeling with consistency regularization. The model generates pseudolabels for weakly augmented inputs only when it is highly confident in its predictions. These pseudolabels are then used to supervise strongly augmented versions of the same input, and this learning scheme is also applicable to unlabeled data. The results demonstrated that the proposed method enhances DME classification by incorporating unlabeled data and significantly outperforms competing MIL methods.

Zhang et al. (2021) introduced a novel approach called twin self-supervision—based semisupervised learning (TS-SSL). This approach incorporates two types of self-supervised strategies, namely generative self-supervised learning and discriminative self-supervised learning, into a semisupervised framework. TS-SSL enables simultaneous learning from a limited number of labeled images and a vast pool of unlabeled images. It is an end-to-end classification model, allowing semisupervision and self-supervision to be jointly trained. They applied TS-SSL to the task of classifying retinal anomalies using spectral-domain optical coherence tomography (SD-OCT) images. Experimental results demonstrated that TS-SSL achieves strong classification performance even with only 10% labeled data on a public SD-OCT dataset and two private SD-OCT datasets.

Lo et al. (2021) involved a retrospective analysis of clinical OCT and OCTA scans from individuals with and without diabetes. OCTA images for microvasculature segmentation were manually delineated and reviewed by retina experts. DR severity was categorized into non-RDR and RDR by retinal specialists. The federated learning setup was simulated using four clients for microvasculature segmentation and compared with other collaborative training methods. Federated learning was then applied across multiple institutions for RDR classification, and its performance was compared with models trained and tested on data from the same institution (internal models) and different institutions (external models). For both applications, federated learning demonstrated performance similar to that of internal models. In microvasculature segmentation, the federated learning model achieved comparable performance (mean Dice similarity coefficient, 0.793) to models trained on a fully centralized dataset (mean Dice similarity coefficient, 0.807). For RDR classification, federated learning achieved mean AUROC values of 0.954 and 0.960, while internal models achieved mean AUROC values of 0.956 and 0.973. Similar trends were observed in other evaluation metrics.

Jacoba et al. (2023) developed an AutoML model for the classification of DR images. The dataset consisted of 17,829 deidentified retinal images from 3566 eyes of individuals with diabetes, captured using handheld retinal cameras during a community-based DR screening program. The AutoML models were trained on 5-field retinal images (centered on macula, centered on disc, superior, inferior, and temporal macula). The prevalence of referable DR was 17.3%, 39.1%, and 48.0% in the training, internal validation, and external validation sets, respectively. The model achieved an area under the precision-recall curve (AUPRC) of 0.995 with a precision and recall of 97% at a threshold of 0.5. Internal validation results showed sensitivity (SN), specificity (SP), positive predictive value (PPV), negative predictive value (NPV), accuracy, and F1 scores of 0.96, 0.98, 0.96, 0.98, 0.97, and 0.96, respectively. External validation results indicated SN, SP, PPV, NPV, accuracy, and F1 scores of 0.94, 0.97, 0.96, 0.95, 0.97, and 0.96, respectively.

Altan (2022) proposed a DeepOCT model, which focuses on achieving high classification performance while maintaining simplicity compared to popular pretrained architectures. DeepOCT incorporates the block-matching and 3D filtering (BM3D) algorithm to flatten retinal layers, ensuring consistent analysis regardless of macula position, and excludes nonretinal layers through cropping. The model aims to identify OCT images with ME and achieves accuracy rates of 99.20%, sensitivity of 100%, and specificity of 98.40%.

Qi et al. (2023) developed an FSL model based on a student—teacher learning framework to categorize the images. The dataset included a total of 2317 images from 189 participants, comprising 1126 images with IRDs, 533 normal samples, and 658 control samples. The FSL model demonstrated impressive performance, achieving a total accuracy ranging from 0.974 to 0.983, total sensitivity between 0.934 and 0.957, total specificity ranging from 0.984 and 0.990, and total F1 score between 0.935 and 0.957. These results surpassed the performance of the baseline model, which had a total accuracy between 0.943 and 0.954, total sensitivity ranging from 0.866 to 0.886, total specificity between 0.962 and 0.971, and total F1 score from 0.859 to 0.885. The FSL model also exhibited superior performance in most subclassifications, with the higher AUC in the receiver operating characteristic (ROC) curves for the majority of subclassifications.

Mathews and Anzar (2023) evaluated the model's performance using two datasets: the publicly available Mendeley OCT dataset and the Duke dataset. The evaluation includes a comprehensive analysis of pretrained models such as LeNet, AlexNet, VGG-16, ResNet50, and SE-ResNet. The proposed model is designed to have a significantly lower number of trainable parameters while outperforming existing methods. The proposed model achieved impressive results, with a classification accuracy of 99.5% on the Mendeley test dataset and 94.9% on the Duke dataset. These outcomes surpass the performance of other pretrained models.

3. Dataset

The dataset used in this study was created in 2018 by Kermany, Goldbaum et al. (2018); Kermany, Zhang, and Goldbau (2018). The dataset is divided into three folders (train, test, and val) with subfolders for each image category (NORMAL, CNV, DME, and DRUSEN). There are 84,495 JPEG X-ray images and four categories (NORMAL, CNV, DME, and DRUSEN). Images are labeled with (disease)-(randomized patient ID)-(image number assigned by this patient) and organized into four directories: CNV, DME, DRUSEN, and NORMAL.

Before Kermany, Goldbaum et al. (2018) published the data, each image was subjected to a tiered grading framework comprising multiple tiers of trained graders of increasing skill for label check and verification. Each image added to the collection began with a label that corresponded to the patient's most recent diagnosis. Undergraduate and medical students who had taken and passed an OCT interpretation course review comprised the first tier of graders. The first tier of graders performed preliminary quality control, excluding OCT images with severe artifacts or considerable image resolution reductions. The second tier of graders comprised four ophthalmologists who graded each image that had cleared the first tier individually. On the OCT scan, the presence or absence of choroidal neovascularization (active or in the form of subretinal fibrosis), macular edema, drusen, and other diseases apparent were noted. Finally, the genuine labels for each image were validated by a third level of two qualified experts and

retinal specialists, each with over 20 years of medical retina experience. A validation subgroup of 993 scans was scored separately by two ophthalmologist graders to account for human error in grading, with clinical label disagreements arbitrated by a senior retinal specialist. A few images are included here.

4. Implementation details

In this section, we provide detailed explanations of the three methods we employed to achieve the task of categorizing images as either containing one of the three mentioned diseases or being normal. We implemented the codebase in Python and utilized the TensorFlow library to create the neural network architecture. For machine learning—based classification, we relied on the sklearn package. All neural network models underwent training for 10 epochs, with training times ranging from 45 min to 1 h and 20 min, depending on the model's size. To ensure consistency during training, we set the random seed to 0.

4.1 CNN-based classification

For medical image classification applications, CNNs have emerged as a powerful and frequently used deep learning architecture. In our study, we used CNNs to categorize medical images based on the presence of specific diseases or normal conditions. CNNs are ideal for this task because they can automatically learn hierarchical features from images, collecting both low-level details and high-level patterns. Convolutional layers, pooling layers, fully linked layers, and output layers are all part of our CNN design. Local patterns and characteristics in the input images are detected by the convolutional layers. We used max-pooling layers to reduce computational complexity by downsampling the spatial dimensions of feature maps. The fully interconnected layers combine the previously learned information to create final predictions. To provide class probabilities, the output layer employs an appropriate activation function, such as softmax (Fig. 10.1).

We used labeled medical image datasets to train our CNN models. To implement and train the models, we used TensorFlow, a prominent deep learning framework. We fine-tuned the network's parameters during training using backpropagation and optimization approaches such as the Adam optimizer. To update the model's weights and biases, we minimized a suitable loss function, categorical cross-entropy. To reduce overfitting and improve generalization performance, we used approaches like dropout. In our work, CNNs provided an efficient solution for identifying medical images, demonstrating their adaptability and robustness in image analysis applications.

CNNs have transformed image recognition, and their use in OCT recognition is no exception. CNNs are a type of deep learning model that learns patterns and features from visual data automatically and adaptively. CNNs have proven to be quite useful for classifying and diagnosing various retinal disorders in OCT recognition. This section presents an overview of CNN models used in OCT recognition and their importance in the field's advancement. While CNNs are well known for their exceptional performance, understanding their decisions in medical applications such as OCT identification is critical. Researchers are working on approaches to improve the interpretability of CNNs by displaying the features they learn and creating heatmaps indicating regions of interest in OCT scans. This interpretability helps clinicians comprehend and trust the model's predictions.

In this study, we used networks of 2−8-layered CNNs and several max-pooling layers placed between the convolutional layers. The 7-layered CNN, whose design is shown in Fig. 10.1, produced the best accuracy and F1-score on the test set. The input image was mapped to $150 \times 150 \times 3$, where 3 represents the number of channels (RGB) and 150×150 represents the image's spatial arrangement. To obtain a vector representation of the image for the classification stage, the convolutional network's output feature map was flattened and then sent to a fully connected layer. Finally, the probabilities of the four classes (3 illnesses and normal) were calculated using a softmax layer with four neurons.

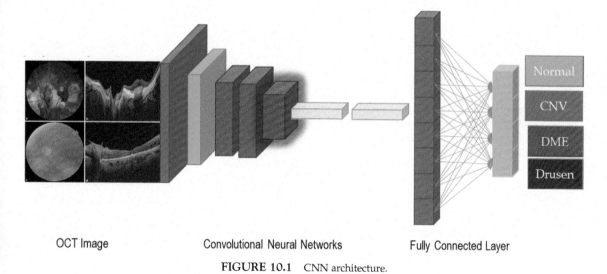

OCT Image   Convolutional Neural Networks   Fully Connected Layer

FIGURE 10.1   CNN architecture.

```
# Import the required libraries

import os
import random

import numpy as np
import matplotlib.pyplot as plt
import seaborn as sns

import tensorflow as tf
import tensorflow_addons as tfa
from tensorflow.keras.preprocessing.image import ImageDataGenerator,
load_img
# Store the base directory path

base_dir = os.path.join("/kaggle/input/kermany2018/oct2017/OCT2017 /")
print('Base directory --> ', os.listdir(base_dir))
# Store the train, validation and test directory paths

train_dir = os.path.join(base_dir + "train/")
print("Train Directory --> ", os.listdir(train_dir))

validation_dir = os.path.join(base_dir + "val/")
print("Validation Directory --> ", os.listdir(validation_dir))

test_dir = os.path.join(base_dir + "test/")
print("Test Directory --> ", os.listdir(test_dir))
# Plot each type of image in the dataset

fig, ax = plt.subplots(1, 4, figsize=(15, 10))

drusen = random.choice(os.listdir(train_dir + "DRUSEN"))
drusen_image = load_img(train_dir + "DRUSEN/" + drusen)
ax[0].imshow(drusen_image)
ax[0].set_title("DRUSEN")
ax[0].axis("Off")

dme = random.choice(os.listdir(train_dir + "DME"))
dme_image = load_img(train_dir + "DME/" + dme)
ax[1].imshow(dme_image)
ax[1].set_title("DME")
ax[1].axis("Off")
```

```python
cnv = random.choice(os.listdir(train_dir + "CNV"))
cnv_image = load_img(train_dir + "CNV/" + cnv)
ax[2].imshow(cnv_image)
ax[2].set_title("CNV")
ax[2].axis("Off")

normal = random.choice(os.listdir(train_dir + "NORMAL"))
normal_image = load_img(train_dir + "NORMAL/" + normal)
ax[3].imshow(normal_image)
ax[3].set_title("NORMAL")
ax[3].axis("Off")

plt.show()
INPUT_SHAPE = (150, 150, 3)
model = tf.keras.models.Sequential([

    tf.keras.layers.Conv2D(16, (3, 3), activation = 'relu', input_shape =
INPUT_SHAPE),
    tf.keras.layers.MaxPooling2D(2, 2),

    tf.keras.layers.Conv2D(16, (3, 3), activation = 'relu'),
    tf.keras.layers.Conv2D(32, (3, 3), activation = 'relu'),
    tf.keras.layers.MaxPooling2D(2, 2),

    tf.keras.layers.Conv2D(62, (3, 3), activation = 'relu'),
    tf.keras.layers.MaxPooling2D(2, 2),

    tf.keras.layers.Conv2D(32, (3, 3), activation = 'relu'),
    tf.keras.layers.MaxPooling2D(2, 2),

    tf.keras.layers.Conv2D(16, (3, 3), activation = 'relu'),
    tf.keras.layers.MaxPooling2D(2, 2),

    tf.keras.layers.Flatten(),

    tf.keras.layers.Dense(4, activation = 'softmax')
])
metrics_list = ['accuracy',
                tf.keras.metrics.AUC(),
                tfa.metrics.CohenKappa(num_classes = 4),
                tfa.metrics.F1Score(num_classes = 4)]
model.compile(loss = 'categorical_crossentropy', optimizer = 'adam',
metrics = metrics_list)
```

```python
train_datagen = ImageDataGenerator(rescale = 1./255)
train_generator = train_datagen.flow_from_directory(train_dir, target_size
= (150, 150), class_mode = 'categorical', batch_size = 100)

validation_datagen = ImageDataGenerator(rescale = 1./255)
validation_generator =
validation_datagen.flow_from_directory(validation_dir, target_size = (150,
150), class_mode = 'categorical', batch_size = 16)

test_datagen = ImageDataGenerator(rescale = 1./255)
test_generator = test_datagen.flow_from_directory(test_dir, target_size =
(150, 150), class_mode = 'categorical', batch_size = 44)

history = model.fit_generator(
    train_generator,
    steps_per_epoch = (83484/100),
    epochs = 10,
    validation_data = validation_generator,
    validation_steps = (32/16),
    verbose = 1)
acc = history.history['accuracy']
val_acc = history.history['val_accuracy']
loss = history.history['loss']
val_loss = history.history['val_loss']


epochs = range(len(acc))


plt.figure(figsize=(7,7))


plt.plot(epochs, acc, 'r', label = 'Training accuracy')
plt.plot(epochs, val_acc, 'b', label = 'Validation accuracy')
plt.title('Training and validation accuracy')
plt.legend()


plt.figure(figsize = (7,7))


plt.plot(epochs, loss, 'r', label = 'Training Loss')
plt.plot(epochs, val_loss, 'b', label = 'Validation Loss')
plt.title('Training and validation loss')
plt.legend()


plt.show()
model.predict(test_generator, steps = int(968/44))
model.evaluate(test_generator)
```

```
from sklearn.metrics import confusion_matrix, classification_report
import pandas as pd

Y_pred = model.predict(test_generator, int(968/44))
y_pred = np.argmax(Y_pred, axis = 1)

cm = confusion_matrix(test_generator.classes, y_pred)
df_cm = pd.DataFrame(cm, list(test_generator.class_indices.keys()),
list(test_generator.class_indices.keys()))

fig, ax = plt.subplots(figsize = (10,8))
sns.set(font_scale = 1.4) # for label size
sns.heatmap(df_cm, annot = True, annot_kws = {"size": 16}, cmap =
plt.cm.Blues)
plt.title('Confusion Matrix\n')
plt.savefig('confusion_matrix.png', transparent = False, bbox_inches =
'tight', dpi = 400)
plt.show()

print('Classification Report\n')
target_names = list(test_generator.class_indices.keys())
print(classification_report(test_generator.classes, y_pred, target_names =
target_names))
```

4.2 Transfer learning–based classification

Transfer learning has proven to be a highly effective strategy for exploiting pretrained deep neural networks to perform challenging classification tasks in the field of OCT image recognition. Transfer learning enables us to use the knowledge gained by a model on one task and apply it to a similar but distinct task, frequently with limited labeled data. Transfer learning is especially useful in the situation of OCT recognition, where gathering a large annotated dataset can be difficult and time-consuming (Fig. 10.2).

We used pretrained CNNs that were originally trained on large-scale image datasets like ImageNet to harness transfer learning for OCT recognition. These pretrained models have learned to extract hierarchical features such as basic edges, textures, and high-level object representations from images. We used these models as feature extractors, leveraging the useful information embodied in their learned weights. We used a procedure called as fine-tuning to adapt the pretrained models to our OCT recognition objective. Fine-tuning entails retraining the pretrained model's top layers (completely linked layers) while freezing the lower layers. This enables the model to specialize in extracting features pertinent to our particular OCT image identification problem. Because it begins with characteristics that are already effective for image recognition tasks, fine-tuning decreases the possibility of overfitting.

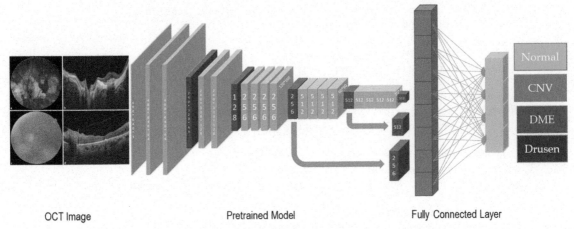

OCT Image   Pretrained Model   Fully Connected Layer

FIGURE 10.2   Transfer learning model.

Transfer learning, as previously described, has emerged as a prominent strategy in OCT recognition. Researchers frequently begin with pretrained CNN models learned on big image datasets such as ImageNet and fine-tune them for OCT tasks. This method enables the models to harness knowledge collected from millions of images and customize it to the specific features and patterns found in OCT scans. Transfer learning eliminates the need for substantial data annotation and dramatically speeds up model training. For OCT recognition, many transfer learning architectures have been investigated, including basic models such as LeNet, VGGNet, and ResNet, as well as more advanced forms such as DenseNet and Inception. The depth, complexity, and design principles of these structures differ. CNN architectures are chosen by researchers based on the unique OCT dataset, task complexity, and computer resources available. Deeper networks frequently outperform shallow structures on difficult OCT recognition tasks, although they may necessitate bigger datasets and longer training cycles.

The capacity of transfer learning in OCT recognition to attain high classification accuracy with relatively limited labeled datasets is its key advantage. Transfer learning helps models generalize better across this diversity in terms of quality, orientation, and disease in OCT images. Transfer learning also accelerates the training process and minimizes the computer resources required.

Transfer learning methods have emerged as an important tool in OCT recognition. We can overcome the obstacles provided by limited labeled data, enhance classification accuracy, and accelerate the development of OCT image recognition systems by using pretrained models and fine-tuning them for specific OCT datasets.

We employed 10 pretrained convolutional neural network-based models trained on the ImageNet dataset in our experiment. ResNet50, VGG16, VGG19, Inception v3, MobileNet, DenseNet169, DenseNet121, InceptionResNetV2, MobileNetV2, ResNet101 were the models utilized. Before the softmax layer, the feature maps generated by these pretrained models were sent through various fully linked layers. The following is the implementation of pretrained models.

VGG16 Model for Retinal Disease Classification.

```python
import sys
from keras.utils import to_categorical
from keras.models import Sequential ###
from keras.layers import Conv2D      ####
from keras.layers import MaxPooling2D ###
from keras.layers import Dense       ###
from keras.layers import Flatten ###
from keras.optimizers import SGD
from keras.optimizers import Adam ####
import tensorflow as tf
INPUT_SHAPE = (128, 128, 1)

vgg16 = tf.keras.applications.VGG16(
    include_top = False,
    weights = 'imagenet',
    input_tensor = None,
    input_shape = INPUT_SHAPE,
    pooling = None,
    classes = 1000
)
vgg16.trainable = False
model = tf.keras.models.Sequential([

    vgg16,
    tf.keras.layers.Conv2D(64, (3, 3), activation = 'relu'),
    tf.keras.layers.Flatten(),
    tf.keras.layers.Dense(100, activation = 'relu'),
    tf.keras.layers.Dense(4, activation = 'softmax')
])
metrics_list = ['accuracy',
                tf.keras.metrics.AUC(),
                tfa.metrics.CohenKappa(num_classes = 4),
                tfa.metrics.F1Score(num_classes = 4)]
model.compile(loss = 'categorical_crossentropy', optimizer = 'adam',
metrics = metrics_list)
train_datagen = ImageDataGenerator(rescale = 1./255)
train_generator = train_datagen.flow_from_directory(train_dir, target_size
= (150, 150), class_mode = 'categorical', batch_size = 100)

validation_datagen = ImageDataGenerator(rescale = 1./255)
validation_generator =
validation_datagen.flow_from_directory(validation_dir, target_size = (150,
150), class_mode = 'categorical', batch_size = 16)

test_datagen = ImageDataGenerator(rescale = 1./255)
test_generator = test_datagen.flow_from_directory(test_dir, target_size =
(150, 150), class_mode = 'categorical', batch_size = 44)
```

```python
history = model.fit_generator(
    train_generator,
    steps_per_epoch = (83484/100),
    epochs = 10,
    validation_data = validation_generator,
    validation_steps = (32/16),
    verbose = 1)
acc = history.history['accuracy']
val_acc = history.history['val_accuracy']
loss = history.history['loss']
val_loss = history.history['val_loss']

epochs = range(len(acc))

plt.figure(figsize=(7,7))

plt.plot(epochs, acc, 'r', label = 'Training accuracy')
plt.plot(epochs, val_acc, 'b', label = 'Validation accuracy')
plt.title('Training and validation accuracy')
plt.legend()

plt.figure(figsize = (7,7))

plt.plot(epochs, loss, 'r', label = 'Training Loss')
plt.plot(epochs, val_loss, 'b', label = 'Validation Loss')
plt.title('Training and validation loss')
plt.legend()

plt.show()
model.predict(test_generator, steps = int(968/44))
model.evaluate(test_generator)
from sklearn.metrics import confusion_matrix, classification_report
import pandas as pd

Y_pred = model.predict(test_generator, int(968/44))
y_pred = np.argmax(Y_pred, axis = 1)

cm = confusion_matrix(test_generator.classes, y_pred)
df_cm = pd.DataFrame(cm, list(test_generator.class_indices.keys()),
list(test_generator.class_indices.keys()))

fig, ax = plt.subplots(figsize = (10,8))
sns.set(font_scale = 1.4) # for label size
sns.heatmap(df_cm, annot = True, annot_kws = {"size": 16}, cmap =
plt.cm.Blues)
plt.title('Confusion Matrix\n')
plt.savefig('confusion_matrix.png', transparent = False, bbox_inches =
'tight', dpi = 400)
plt.show()

print('Classification Report\n')
target_names = list(test_generator.class_indices.keys())
print(classification_report(test_generator.classes, y_pred, target_names =
target_names))
```

VGG19 Model for Retinal Disease Classification.

```python
import sys
from keras.utils import to_categorical
from keras.models import Sequential ###
from keras.layers import Conv2D      ####
from keras.layers import MaxPooling2D ###
from keras.layers import Dense      ###
from keras.layers import Flatten ###
from keras.optimizers import SGD
from keras.optimizers import Adam ####
import tensorflow as tf
INPUT_SHAPE = (128, 128, 1)

vgg19 = tf.keras.applications.VGG19(
    include_top = False,
    weights = 'imagenet',
    input_tensor = None,
    input_shape = INPUT_SHAPE,
    pooling = None,
    classes = 1000
)
vgg19.trainable = False
model = tf.keras.models.Sequential([

    vgg19,
    tf.keras.layers.Conv2D(64, (3, 3), activation = 'relu'),
    tf.keras.layers.Flatten(),
    tf.keras.layers.Dense(100, activation = 'relu'),
    tf.keras.layers.Dense(4, activation = 'softmax')
])
metrics_list = ['accuracy',
                tf.keras.metrics.AUC(),
                tfa.metrics.CohenKappa(num_classes = 4),
                tfa.metrics.F1Score(num_classes = 4)]
model.compile(loss = 'categorical_crossentropy', optimizer = 'adam',
metrics = metrics_list)
train_datagen = ImageDataGenerator(rescale = 1./255)
train_generator = train_datagen.flow_from_directory(train_dir, target_size
= (150, 150), class_mode = 'categorical', batch_size = 100)
validation_datagen = ImageDataGenerator(rescale = 1./255)
validation_generator =
validation_datagen.flow_from_directory(validation_dir, target_size = (150,
150), class_mode = 'categorical', batch_size = 16)
test_datagen = ImageDataGenerator(rescale = 1./255)
test_generator = test_datagen.flow_from_directory(test_dir, target_size =
(150, 150), class_mode = 'categorical', batch_size = 44)
history = model.fit_generator(
```

```
    train_generator,
    steps_per_epoch = (83484/100),
    epochs = 10,
    validation_data = validation_generator,
    validation_steps = (32/16),
    verbose = 1)
acc = history.history['accuracy']
val_acc = history.history['val_accuracy']
loss = history.history['loss']
val_loss = history.history['val_loss']

epochs = range(len(acc))

plt.figure(figsize=(7,7))

plt.plot(epochs, acc, 'r', label = 'Training accuracy')
plt.plot(epochs, val_acc, 'b', label = 'Validation accuracy')
plt.title('Training and validation accuracy')
plt.legend()

plt.figure(figsize = (7,7))

plt.plot(epochs, loss, 'r', label = 'Training Loss')
plt.plot(epochs, val_loss, 'b', label = 'Validation Loss')
plt.title('Training and validation loss')
plt.legend()

plt.show()
model.predict(test_generator, steps = int(968/44))
model.evaluate(test_generator)
from sklearn.metrics import confusion_matrix, classification_report
import pandas as pd

Y_pred = model.predict(test_generator, int(968/44))
y_pred = np.argmax(Y_pred, axis = 1)

cm = confusion_matrix(test_generator.classes, y_pred)
df_cm = pd.DataFrame(cm, list(test_generator.class_indices.keys()),
list(test_generator.class_indices.keys()))

fig, ax = plt.subplots(figsize = (10,8))
sns.set(font_scale = 1.4) # for label size
sns.heatmap(df_cm, annot = True, annot_kws = {"size": 16}, cmap =
plt.cm.Blues)
plt.title('Confusion Matrix\n')
plt.savefig('confusion_matrix.png', transparent = False, bbox_inches =
'tight', dpi = 400)
plt.show()

print('Classification Report\n')
target_names = list(test_generator.class_indices.keys())
print(classification_report(test_generator.classes, y_pred, target_names =
target_names))
```

DenseNet121 Model for Retinal Disease Classification.

```python
import sys
from keras.utils import to_categorical
from keras.models import Sequential ###
from keras.layers import Conv2D      ####
from keras.layers import MaxPooling2D  ###
from keras.layers import Dense      ###
from keras.layers import Flatten ###
from keras.optimizers import SGD
from keras.optimizers import Adam ####
import tensorflow as tf
INPUT_SHAPE = (128, 128, 1)

dense_net_121 = tf.keras.applications.DenseNet121(
    include_top = False,
    weights = 'imagenet',
    input_tensor = None,
    input_shape = INPUT_SHAPE,
    pooling = None,
    classes = 1000
)
dense_net_121.trainable = False
model = tf.keras.models.Sequential([

    dense_net_121,
    tf.keras.layers.Conv2D(64, (3, 3), activation = 'relu'),
    tf.keras.layers.Flatten(),
    tf.keras.layers.Dense(100, activation = 'relu'),
    tf.keras.layers.Dense(4, activation = 'softmax')
])
metrics_list = ['accuracy',
                tf.keras.metrics.AUC(),
                tfa.metrics.CohenKappa(num_classes = 4),
                tfa.metrics.F1Score(num_classes = 4)]
model.compile(loss = 'categorical_crossentropy', optimizer = 'adam',
metrics = metrics_list)
train_datagen = ImageDataGenerator(rescale = 1./255)
train_generator = train_datagen.flow_from_directory(train_dir, target_size
= (150, 150), class_mode = 'categorical', batch_size = 100)
validation_datagen = ImageDataGenerator(rescale = 1./255)
validation_generator =
validation_datagen.flow_from_directory(validation_dir, target_size = (150,
150), class_mode = 'categorical', batch_size = 16)
test_datagen = ImageDataGenerator(rescale = 1./255)
test_generator = test_datagen.flow_from_directory(test_dir, target_size =
(150, 150), class_mode = 'categorical', batch_size = 44)
history = model.fit_generator(
    train_generator,
    steps_per_epoch = (83484/100),
    epochs = 10,
```

```python
    validation_data = validation_generator,
    validation_steps = (32/16),
    verbose = 1)
acc = history.history['accuracy']
val_acc = history.history['val_accuracy']
loss = history.history['loss']
val_loss = history.history['val_loss']

epochs = range(len(acc))

plt.figure(figsize=(7,7))

plt.plot(epochs, acc, 'r', label = 'Training accuracy')
plt.plot(epochs, val_acc, 'b', label = 'Validation accuracy')
plt.title('Training and validation accuracy')
plt.legend()

plt.figure(figsize = (7,7))

plt.plot(epochs, loss, 'r', label = 'Training Loss')
plt.plot(epochs, val_loss, 'b', label = 'Validation Loss')
plt.title('Training and validation loss')
plt.legend()
plt.show()
model.predict(test_generator, steps = int(968/44))
model.evaluate(test_generator)
from sklearn.metrics import confusion_matrix, classification_report
import pandas as pd

Y_pred = model.predict(test_generator, int(968/44))
y_pred = np.argmax(Y_pred, axis = 1)

cm = confusion_matrix(test_generator.classes, y_pred)
df_cm = pd.DataFrame(cm, list(test_generator.class_indices.keys()),
list(test_generator.class_indices.keys()))

fig, ax = plt.subplots(figsize = (10,8))
sns.set(font_scale = 1.4) # for label size
sns.heatmap(df_cm, annot = True, annot_kws = {"size": 16}, cmap =
plt.cm.Blues)
plt.title('Confusion Matrix\n')
plt.savefig('confusion_matrix.png', transparent = False, bbox_inches =
'tight', dpi = 400)
plt.show()

print('Classification Report\n')
target_names = list(test_generator.class_indices.keys())
print(classification_report(test_generator.classes, y_pred, target_names =
target_names))
```

DenseNet169 Model for Retinal Disease Classification.

```python
import sys
from keras.utils import to_categorical
from keras.models import Sequential ###
from keras.layers import Conv2D      ####
from keras.layers import MaxPooling2D ###
from keras.layers import Dense    ###
from keras.layers import Flatten ###
from keras.optimizers import SGD
from keras.optimizers import Adam ####
import tensorflow as tf
INPUT_SHAPE = (128, 128, 1)
dense_net_169 = tf.keras.applications.DenseNet169(
    include_top = False,
    weights = 'imagenet',
    input_tensor = None,
    input_shape = INPUT_SHAPE,
    pooling = None,
    classes = 1000
)
dense_net_169.trainable = False

model = tf.keras.models.Sequential([

    dense_net_169,
    tf.keras.layers.Conv2D(64, (3, 3), activation = 'relu'),
    tf.keras.layers.Flatten(),
    tf.keras.layers.Dense(100, activation = 'relu'),
    tf.keras.layers.Dense(4, activation = 'softmax')
])
metrics_list = ['accuracy',
                tf.keras.metrics.AUC(),
                tfa.metrics.CohenKappa(num_classes = 4),
                tfa.metrics.F1Score(num_classes = 4)]
model.compile(loss = 'categorical_crossentropy', optimizer = 'adam',
metrics = metrics_list)
train_datagen = ImageDataGenerator(rescale = 1./255)
train_generator = train_datagen.flow_from_directory(train_dir, target_size
= (150, 150), class_mode = 'categorical', batch_size = 100)
validation_datagen = ImageDataGenerator(rescale = 1./255)
validation_generator =
validation_datagen.flow_from_directory(validation_dir, target_size = (150,
150), class_mode = 'categorical', batch_size = 16)
test_datagen = ImageDataGenerator(rescale = 1./255)
test_generator = test_datagen.flow_from_directory(test_dir, target_size =
(150, 150), class_mode = 'categorical', batch_size = 44)
history = model.fit_generator(
    train_generator,
```

```
        steps_per_epoch = (83484/100),
        epochs = 10,
        validation_data = validation_generator,
        validation_steps = (32/16),
        verbose = 1)
acc = history.history['accuracy']
val_acc = history.history['val_accuracy']
loss = history.history['loss']
val_loss = history.history['val_loss']

epochs = range(len(acc))

plt.figure(figsize=(7,7))

plt.plot(epochs, acc, 'r', label = 'Training accuracy')
plt.plot(epochs, val_acc, 'b', label = 'Validation accuracy')
plt.title('Training and validation accuracy')
plt.legend()

plt.figure(figsize = (7,7))

plt.plot(epochs, loss, 'r', label = 'Training Loss')
plt.plot(epochs, val_loss, 'b', label = 'Validation Loss')
plt.title('Training and validation loss')
plt.legend()

plt.show()
model.predict(test_generator, steps = int(968/44))
model.evaluate(test_generator)
from sklearn.metrics import confusion_matrix, classification_report
import pandas as pd

Y_pred = model.predict(test_generator, int(968/44))
y_pred = np.argmax(Y_pred, axis = 1)

cm = confusion_matrix(test_generator.classes, y_pred)
df_cm = pd.DataFrame(cm, list(test_generator.class_indices.keys()),
list(test_generator.class_indices.keys()))

fig, ax = plt.subplots(figsize = (10,8))
sns.set(font_scale = 1.4) # for label size
sns.heatmap(df_cm, annot = True, annot_kws = {"size": 16}, cmap =
plt.cm.Blues)
plt.title('Confusion Matrix\n')
plt.savefig('confusion_matrix.png', transparent = False, bbox_inches =
'tight', dpi = 400)
plt.show()

print('Classification Report\n')
target_names = list(test_generator.class_indices.keys())
print(classification_report(test_generator.classes, y_pred, target_names =
target_names))
```

InceptionNet Model for Retinal Disease Classification.

```
import sys
from keras.utils import to_categorical
from keras.models import Sequential ###
from keras.layers import Conv2D      ####
from keras.layers import MaxPooling2D ###
from keras.layers import Dense    ###
from keras.layers import Flatten ###
from keras.optimizers import SGD
from keras.optimizers import Adam ####
import tensorflow as tf
INPUT_SHAPE = (128, 128, 1)

inception = tf.keras.applications.InceptionV3(
    include_top = False,
    weights = 'imagenet',
    input_tensor = None,
    input_shape = INPUT_SHAPE,
    pooling = None,
    classes = 1000
)
inception.trainable = False
model = tf.keras.models.Sequential([

    inception,
    tf.keras.layers.Conv2D(64, (3, 3), activation = 'relu'),
    tf.keras.layers.Flatten(),
    tf.keras.layers.Dense(100, activation = 'relu'),
    tf.keras.layers.Dense(4, activation = 'softmax')
])

metrics_list = ['accuracy',
                tf.keras.metrics.AUC(),
                tfa.metrics.CohenKappa(num_classes = 4),
                tfa.metrics.F1Score(num_classes = 4)]

model.compile(loss = 'categorical_crossentropy', optimizer = 'adam',
metrics = metrics_list)
train_datagen = ImageDataGenerator(rescale = 1./255)
train_generator = train_datagen.flow_from_directory(train_dir, target_size
= (150, 150), class_mode = 'categorical', batch_size = 100)
validation_datagen = ImageDataGenerator(rescale = 1./255)
validation_generator =
validation_datagen.flow_from_directory(validation_dir, target_size = (150,
150), class_mode = 'categorical', batch_size = 16)
test_datagen = ImageDataGenerator(rescale = 1./255)
test_generator = test_datagen.flow_from_directory(test_dir, target_size =
(150, 150), class_mode = 'categorical', batch_size = 44)
history = model.fit_generator(
```

```
    train_generator,
    steps_per_epoch = (83484/100),
    epochs = 10,
    validation_data = validation_generator,
    validation_steps = (32/16),
    verbose = 1)
acc = history.history['accuracy']
val_acc = history.history['val_accuracy']
loss = history.history['loss']
val_loss = history.history['val_loss']


epochs = range(len(acc))


plt.figure(figsize=(7,7))


plt.plot(epochs, acc, 'r', label = 'Training accuracy')
plt.plot(epochs, val_acc, 'b', label = 'Validation accuracy')
plt.title('Training and validation accuracy')
plt.legend()


plt.figure(figsize = (7,7))


plt.plot(epochs, loss, 'r', label = 'Training Loss')
plt.plot(epochs, val_loss, 'b', label = 'Validation Loss')
plt.title('Training and validation loss')
plt.legend()


plt.show()
model.predict(test_generator, steps = int(968/44))
model.evaluate(test_generator)
from sklearn.metrics import confusion_matrix, classification_report
import pandas as pd


Y_pred = model.predict(test_generator, int(968/44))
y_pred = np.argmax(Y_pred, axis = 1)


cm = confusion_matrix(test_generator.classes, y_pred)
df_cm = pd.DataFrame(cm, list(test_generator.class_indices.keys()),
list(test_generator.class_indices.keys()))


fig, ax = plt.subplots(figsize = (10,8))
sns.set(font_scale = 1.4) # for label size
sns.heatmap(df_cm, annot = True, annot_kws = {"size": 16}, cmap =
plt.cm.Blues)
plt.title('Confusion Matrix\n')
plt.savefig('confusion_matrix.png', transparent = False, bbox_inches =
'tight', dpi = 400)
plt.show()


print('Classification Report\n')
target_names = list(test_generator.class_indices.keys())
print(classification_report(test_generator.classes, y_pred, target_names =
target_names))
```

InceptionResnet Model for Retinal Disease Classification.

```python
import sys
from keras.utils import to_categorical
from keras.models import Sequential ###
from keras.layers import Conv2D    ####
from keras.layers import MaxPooling2D  ###
from keras.layers import Dense    ###
from keras.layers import Flatten ###
from keras.optimizers import SGD
from keras.optimizers import Adam ####
import tensorflow as tf
INPUT_SHAPE = (128, 128, 1)

inception_resnet_v2 = tf.keras.applications.InceptionResNetV2(
    include_top = False,
    weights = 'imagenet',
    input_tensor = None,
    input_shape = INPUT_SHAPE,
    pooling = None,
    classes = 1000
)

inception_resnet_v2.trainable = False
model = tf.keras.models.Sequential([

    inception_resnet_v2,
    tf.keras.layers.Conv2D(64, (3, 3), activation = 'relu'),
    tf.keras.layers.Flatten(),
    tf.keras.layers.Dense(100, activation = 'relu'),
    tf.keras.layers.Dense(4, activation = 'softmax')
])
metrics_list = ['accuracy',
                tf.keras.metrics.AUC(),
                tfa.metrics.CohenKappa(num_classes = 4),
                tfa.metrics.F1Score(num_classes = 4)]
model.compile(loss = 'categorical_crossentropy', optimizer = 'adam',
metrics = metrics_list)
train_datagen = ImageDataGenerator(rescale = 1./255)
train_generator = train_datagen.flow_from_directory(train_dir, target_size
= (150, 150), class_mode = 'categorical', batch_size = 100)
validation_datagen = ImageDataGenerator(rescale = 1./255)
validation_generator =
validation_datagen.flow_from_directory(validation_dir, target_size = (150,
150), class_mode = 'categorical', batch_size = 16)
test_datagen = ImageDataGenerator(rescale = 1./255)
test_generator = test_datagen.flow_from_directory(test_dir, target_size =
(150, 150), class_mode = 'categorical', batch_size = 44)
history = model.fit_generator(
    train_generator,
```

```
    steps_per_epoch = (83484/100),
    epochs = 10,
    validation_data = validation_generator,
    validation_steps = (32/16),
    verbose = 1)
acc = history.history['accuracy']
val_acc = history.history['val_accuracy']
loss = history.history['loss']
val_loss = history.history['val_loss']

epochs = range(len(acc))

plt.figure(figsize=(7,7))

plt.plot(epochs, acc, 'r', label = 'Training accuracy')
plt.plot(epochs, val_acc, 'b', label = 'Validation accuracy')
plt.title('Training and validation accuracy')
plt.legend()

plt.figure(figsize = (7,7))

plt.plot(epochs, loss, 'r', label = 'Training Loss')
plt.plot(epochs, val_loss, 'b', label = 'Validation Loss')
plt.title('Training and validation loss')
plt.legend()

plt.show()
model.predict(test_generator, steps = int(968/44))
model.evaluate(test_generator)
from sklearn.metrics import confusion_matrix, classification_report
import pandas as pd

Y_pred = model.predict(test_generator, int(968/44))
y_pred = np.argmax(Y_pred, axis = 1)

cm = confusion_matrix(test_generator.classes, y_pred)
df_cm = pd.DataFrame(cm, list(test_generator.class_indices.keys()),
list(test_generator.class_indices.keys()))

fig, ax = plt.subplots(figsize = (10,8))
sns.set(font_scale = 1.4) # for label size
sns.heatmap(df_cm, annot = True, annot_kws = {"size": 16}, cmap =
plt.cm.Blues)
plt.title('Confusion Matrix\n')
plt.savefig('confusion_matrix.png', transparent = False, bbox_inches =
'tight', dpi = 400)
plt.show()

print('Classification Report\n')
target_names = list(test_generator.class_indices.keys())
print(classification_report(test_generator.classes, y_pred, target_names =
target_names))
```

MobileNet Model for Retinal Disease Classification.

```python
import sys
from keras.utils import to_categorical
from keras.models import Sequential ###
from keras.layers import Conv2D     ####
from keras.layers import MaxPooling2D  ###
from keras.layers import Dense    ###
from keras.layers import Flatten ###
from keras.optimizers import SGD
from keras.optimizers import Adam ####
import tensorflow as tf
INPUT_SHAPE = (128, 128, 1)

mobile_net = tf.keras.applications.MobileNet(
    include_top = False,
    weights = 'imagenet',
    input_tensor = None,
    input_shape = INPUT_SHAPE,
    pooling = None,
    classes = 1000
)
mobile_net.trainable = False
model = tf.keras.models.Sequential([

    mobile_net,
    tf.keras.layers.Conv2D(64, (3, 3), activation = 'relu'),
    tf.keras.layers.Flatten(),
    tf.keras.layers.Dense(100, activation = 'relu'),
    tf.keras.layers.Dense(4, activation = 'softmax')
])
metrics_list = ['accuracy',
                tf.keras.metrics.AUC(),
                tfa.metrics.CohenKappa(num_classes = 4),
                tfa.metrics.F1Score(num_classes = 4)]
model.compile(loss = 'categorical_crossentropy', optimizer = 'adam',
metrics = metrics_list)
train_datagen = ImageDataGenerator(rescale = 1./255)
train_generator = train_datagen.flow_from_directory(train_dir, target_size
= (150, 150), class_mode = 'categorical', batch_size = 100)
validation_datagen = ImageDataGenerator(rescale = 1./255)
validation_generator =
validation_datagen.flow_from_directory(validation_dir, target_size = (150,
150), class_mode = 'categorical', batch_size = 16)
test_datagen = ImageDataGenerator(rescale = 1./255)
test_generator = test_datagen.flow_from_directory(test_dir, target_size =
(150, 150), class_mode = 'categorical', batch_size = 44)
history = model.fit_generator(
    train_generator,
    steps_per_epoch = (83484/100),
```

```
      epochs = 10,
      validation_data = validation_generator,
      validation_steps = (32/16),
      verbose = 1)
acc = history.history['accuracy']
val_acc = history.history['val_accuracy']
loss = history.history['loss']
val_loss = history.history['val_loss']

epochs = range(len(acc))

plt.figure(figsize=(7,7))

plt.plot(epochs, acc, 'r', label = 'Training accuracy')
plt.plot(epochs, val_acc, 'b', label = 'Validation accuracy')
plt.title('Training and validation accuracy')
plt.legend()

plt.figure(figsize = (7,7))

plt.plot(epochs, loss, 'r', label = 'Training Loss')
plt.plot(epochs, val_loss, 'b', label = 'Validation Loss')
plt.title('Training and validation loss')
plt.legend()

plt.show()
model.predict(test_generator, steps = int(968/44))
model.evaluate(test_generator)
from sklearn.metrics import confusion_matrix, classification_report
import pandas as pd

Y_pred = model.predict(test_generator, int(968/44))
y_pred = np.argmax(Y_pred, axis = 1)

cm = confusion_matrix(test_generator.classes, y_pred)
df_cm = pd.DataFrame(cm, list(test_generator.class_indices.keys()),
list(test_generator.class_indices.keys()))

fig, ax = plt.subplots(figsize = (10,8))
sns.set(font_scale = 1.4) # for label size
sns.heatmap(df_cm, annot = True, annot_kws = {"size": 16}, cmap =
plt.cm.Blues)
plt.title('Confusion Matrix\n')
plt.savefig('confusion_matrix.png', transparent = False, bbox_inches =
'tight', dpi = 400)
plt.show()

print('Classification Report\n')
target_names = list(test_generator.class_indices.keys())
print(classification_report(test_generator.classes, y_pred, target_names =
target_names))
```

MobileNetV2 Model for Retinal Disease Classification.

```python
import sys
from keras.utils import to_categorical
from keras.models import Sequential ###
from keras.layers import Conv2D     ####
from keras.layers import MaxPooling2D  ###
from keras.layers import Dense      ###
from keras.layers import Flatten ###
from keras.optimizers import SGD
from keras.optimizers import Adam ####
import tensorflow as tf
INPUT_SHAPE = (128, 128, 1)

mobile_net_v2 = tf.keras.applications.MobileNetV2(
    include_top = False,
    weights = 'imagenet',
    input_tensor = None,
    input_shape = INPUT_SHAPE,
    pooling = None,
    classes = 1000
)
mobile_net_v2.trainable = False
model = tf.keras.models.Sequential([

    mobile_net_v2,
    tf.keras.layers.Conv2D(64, (3, 3), activation = 'relu'),
    tf.keras.layers.Flatten(),
    tf.keras.layers.Dense(100, activation = 'relu'),
    tf.keras.layers.Dense(4, activation = 'softmax')
])
metrics_list = ['accuracy',
                tf.keras.metrics.AUC(),
                tfa.metrics.CohenKappa(num_classes = 4),
                tfa.metrics.F1Score(num_classes = 4)]
model.compile(loss = 'categorical_crossentropy', optimizer = 'adam',
metrics = metrics_list)
train_datagen = ImageDataGenerator(rescale = 1./255)
train_generator = train_datagen.flow_from_directory(train_dir, target_size
= (150, 150), class_mode = 'categorical', batch_size = 100)
validation_datagen = ImageDataGenerator(rescale = 1./255)
validation_generator =
validation_datagen.flow_from_directory(validation_dir, target_size = (150,
150), class_mode = 'categorical', batch_size = 16)
test_datagen = ImageDataGenerator(rescale = 1./255)
test_generator = test_datagen.flow_from_directory(test_dir, target_size =
(150, 150), class_mode = 'categorical', batch_size = 44)
history = model.fit_generator(
    train_generator,
```

```python
    steps_per_epoch = (83484/100),
    epochs = 10,
    validation_data = validation_generator,
    validation_steps = (32/16),
    verbose = 1)
acc = history.history['accuracy']
val_acc = history.history['val_accuracy']
loss = history.history['loss']
val_loss = history.history['val_loss']

epochs = range(len(acc))

plt.figure(figsize=(7,7))

plt.plot(epochs, acc, 'r', label = 'Training accuracy')
plt.plot(epochs, val_acc, 'b', label = 'Validation accuracy')
plt.title('Training and validation accuracy')
plt.legend()

plt.figure(figsize = (7,7))

plt.plot(epochs, loss, 'r', label = 'Training Loss')
plt.plot(epochs, val_loss, 'b', label = 'Validation Loss')
plt.title('Training and validation loss')
plt.legend()

plt.show()
model.predict(test_generator, steps = int(968/44))
model.evaluate(test_generator)
from sklearn.metrics import confusion_matrix, classification_report
import pandas as pd

Y_pred = model.predict(test_generator, int(968/44))
y_pred = np.argmax(Y_pred, axis = 1)

cm = confusion_matrix(test_generator.classes, y_pred)
df_cm = pd.DataFrame(cm, list(test_generator.class_indices.keys()),
list(test_generator.class_indices.keys()))

fig, ax = plt.subplots(figsize = (10,8))
sns.set(font_scale = 1.4) # for label size
sns.heatmap(df_cm, annot = True, annot_kws = {"size": 16}, cmap =
plt.cm.Blues)
plt.title('Confusion Matrix\n')
plt.savefig('confusion_matrix.png', transparent = False, bbox_inches =
'tight', dpi = 400)
plt.show()

print('Classification Report\n')
target_names = list(test_generator.class_indices.keys())
print(classification_report(test_generator.classes, y_pred, target_names =
target_names))
```

ResNet Model for Retinal Disease Classification.

```python
import sys
from keras.utils import to_categorical
from keras.models import Sequential ###
from keras.layers import Conv2D      ####
from keras.layers import MaxPooling2D  ###
from keras.layers import Dense     ###
from keras.layers import Flatten ###
from keras.optimizers import SGD
from keras.optimizers import Adam ####
import tensorflow as tf
INPUT_SHAPE = (128, 128, 1)
resnet = tf.keras.applications.ResNet50(
    include_top = False,
    weights='imagenet',
    input_tensor = None,
    input_shape = INPUT_SHAPE,
    pooling = None,
    classes=1000
)
resnet.trainable = False
model = tf.keras.models.Sequential([resnet,
    tf.keras.layers.Conv2D(64, (3, 3), activation = 'relu'),
    tf.keras.layers.Flatten(),
    tf.keras.layers.Dense(100, activation = 'relu'),
    tf.keras.layers.Dense(4, activation = 'softmax')
])
metrics_list = ['accuracy',
                tf.keras.metrics.AUC(),
                tfa.metrics.CohenKappa(num_classes = 4),
                tfa.metrics.F1Score(num_classes = 4)]
model.compile(loss = 'categorical_crossentropy', optimizer = 'adam',
metrics = metrics_list)
train_datagen = ImageDataGenerator(rescale = 1./255)
train_generator = train_datagen.flow_from_directory(train_dir, target_size
= (150, 150), class_mode = 'categorical', batch_size = 100)
validation_datagen = ImageDataGenerator(rescale = 1./255)
validation_generator =
validation_datagen.flow_from_directory(validation_dir, target_size = (150,
150), class_mode = 'categorical', batch_size = 16)
test_datagen = ImageDataGenerator(rescale = 1./255)
test_generator = test_datagen.flow_from_directory(test_dir, target_size =
(150, 150), class_mode = 'categorical', batch_size = 44)
history = model.fit_generator(
    train_generator,
    steps_per_epoch = (83484/100),
    epochs = 10,
    validation_data = validation_generator,
    validation_steps = (32/16),
```

```
    verbose = 1)
acc = history.history['accuracy']
val_acc = history.history['val_accuracy']
loss = history.history['loss']
val_loss = history.history['val_loss']

epochs = range(len(acc))

plt.figure(figsize=(7,7))

plt.plot(epochs, acc, 'r', label = 'Training accuracy')
plt.plot(epochs, val_acc, 'b', label = 'Validation accuracy')
plt.title('Training and validation accuracy')
plt.legend()

plt.figure(figsize = (7,7))

plt.plot(epochs, loss, 'r', label = 'Training Loss')
plt.plot(epochs, val_loss, 'b', label = 'Validation Loss')
plt.title('Training and validation loss')
plt.legend()

plt.show()
model.predict(test_generator, steps = int(968/44))
from sklearn.metrics import confusion_matrix, classification_report
import pandas as pd

Y_pred = model.predict(test_generator, int(968/44))
y_pred = np.argmax(Y_pred, axis = 1)

cm = confusion_matrix(test_generator.classes, y_pred)
df_cm = pd.DataFrame(cm, list(test_generator.class_indices.keys()),
list(test_generator.class_indices.keys()))

fig, ax = plt.subplots(figsize = (10,8))
sns.set(font_scale = 1.4) # for label size
sns.heatmap(df_cm, annot = True, annot_kws = {"size": 16}, cmap =
plt.cm.Blues)
plt.title('Confusion Matrix\n')
plt.savefig('confusion_matrix.png', transparent = False, bbox_inches =
'tight', dpi = 400)
plt.show()

model.predict(test_generator, steps = int(968/44))
model.evaluate(test_generator)
from sklearn.metrics import confusion_matrix, classification_report
import pandas as pd
```

```python
Y_pred = model.predict(test_generator, int(968/44))
y_pred = np.argmax(Y_pred, axis = 1)

cm = confusion_matrix(test_generator.classes, y_pred)
df_cm = pd.DataFrame(cm, list(test_generator.class_indices.keys()),
list(test_generator.class_indices.keys()))

fig, ax = plt.subplots(figsize = (10,8))
sns.set(font_scale = 1.4) # for label size
sns.heatmap(df_cm, annot = True, annot_kws = {"size": 16}, cmap =
plt.cm.Blues)
plt.title('Confusion Matrix\n')
plt.savefig('confusion_matrix.png', transparent = False, bbox_inches =
'tight', dpi = 400)
plt.show()

print('Classification Report\n')
target_names = list(test_generator.class_indices.keys())
print(classification_report(test_generator.classes, y_pred, target_names =
target_names))
```

ResNet101 Model for Retinal Disease Classification.

```python
import sys
from keras.utils import to_categorical
from keras.models import Sequential ###
from keras.layers import Conv2D      ####
from keras.layers import MaxPooling2D ###
from keras.layers import Dense    ###
from keras.layers import Flatten ###
from keras.optimizers import SGD
from keras.optimizers import Adam ####
import tensorflow as tf
INPUT_SHAPE = (128, 128, 1)
resnet_101 = tf.keras.applications.ResNet101(
    include_top = False,
    weights = 'imagenet',
    input_tensor = None,
    input_shape = INPUT_SHAPE,
    pooling = None,
    classes = 1000
)
resnet_101.trainable = False
model = tf.keras.models.Sequential([
```

```
    resnet_101,
    tf.keras.layers.Conv2D(64, (3, 3), activation = 'relu'),
    tf.keras.layers.Flatten(),
    tf.keras.layers.Dense(100, activation = 'relu'),
    tf.keras.layers.Dense(4, activation = 'softmax')
])
metrics_list = ['accuracy',
                tf.keras.metrics.AUC(),
                tfa.metrics.CohenKappa(num_classes = 4),
                tfa.metrics.F1Score(num_classes = 4)]
model.compile(loss = 'categorical_crossentropy', optimizer = 'adam',
metrics = metrics_list)
train_datagen = ImageDataGenerator(rescale = 1./255)
train_generator = train_datagen.flow_from_directory(train_dir, target_size
= (150, 150), class_mode = 'categorical', batch_size = 100)
validation_datagen = ImageDataGenerator(rescale = 1./255)
validation_generator =
validation_datagen.flow_from_directory(validation_dir, target_size = (150,
150), class_mode = 'categorical', batch_size = 16)
test_datagen = ImageDataGenerator(rescale = 1./255)
test_generator = test_datagen.flow_from_directory(test_dir, target_size =
(150, 150), class_mode = 'categorical', batch_size = 44)
history = model.fit_generator(
    train_generator,
    steps_per_epoch = (83484/100),
    epochs = 10,
    validation_data = validation_generator,
    validation_steps = (32/16),
    verbose = 1)
acc = history.history['accuracy']
val_acc = history.history['val_accuracy']
loss = history.history['loss']
val_loss = history.history['val_loss']


epochs = range(len(acc))


plt.figure(figsize=(7,7))


plt.plot(epochs, acc, 'r', label = 'Training accuracy')
plt.plot(epochs, val_acc, 'b', label = 'Validation accuracy')
plt.title('Training and validation accuracy')
plt.legend()


plt.figure(figsize = (7,7))


plt.plot(epochs, loss, 'r', label = 'Training Loss')
plt.plot(epochs, val_loss, 'b', label = 'Validation Loss')
plt.title('Training and validation loss')
plt.legend()


plt.show()
```

```
model.predict(test_generator, steps = int(968/44))
model.evaluate(test_generator)
from sklearn.metrics import confusion_matrix, classification_report
import pandas as pd

Y_pred = model.predict(test_generator, int(968/44))
y_pred = np.argmax(Y_pred, axis = 1)

cm = confusion_matrix(test_generator.classes, y_pred)
df_cm = pd.DataFrame(cm, list(test_generator.class_indices.keys()),
list(test_generator.class_indices.keys()))

fig, ax = plt.subplots(figsize = (10,8))
sns.set(font_scale = 1.4) # for label size
sns.heatmap(df_cm, annot = True, annot_kws = {"size": 16}, cmap =
plt.cm.Blues)
plt.title('Confusion Matrix\n')
plt.savefig('confusion_matrix.png', transparent = False, bbox_inches =
'tight', dpi = 400)
plt.show()

print('Classification Report\n')
target_names = list(test_generator.class_indices.keys())
print(classification_report(test_generator.classes, y_pred, target_names =
target_names))
```

5. Results and discussions

The outcomes for the OCT dataset in the classification of retinal diseases using a straightforward convolutional neural network and pretrained model architectures are presented in Table 10.1. It is evident from the table that both accuracy and the F1-score are exceptionally high, standing at 0.9804 for the 6-layer convolutional neural network. This remarkable accuracy can be attributed to the effectiveness of CNNs in capturing intricate image features with great precision. Notably, the VGG-19 model achieved the highest accuracy and F1-score, both measuring at 0.9628 among pretrained models. These favorable results can be attributed to the advantages of pretraining, which enables the models to grasp and encode essential low-level features like edges, color gradients, and shapes, thus enhancing their performance.

6. Discussion

Using OCT pictures, AI approaches have become effective tools for classifying retinal diseases, providing great opportunities to increase diagnostic precision and patient outcomes. Enhancing early detection and intervention is one of the main effects of AI-based retinal disease classification. AI models can evaluate enormous amounts of OCT data with great efficiency and accuracy by utilizing machine learning and deep learning techniques, making it possible to identify subtle symptoms of the disease that may be missed by human observers. Early detection makes it easier to administer timely interventions and treatments, which can greatly enhance patient outcomes, especially in situations of degenerative retinal illnesses.

Additionally, AI methods have the potential to solve the problem of subjectivity and

TABLE 10.1 Results of deep learning models.

Classifier	Training accuracy	Validation accuracy	Test accuracy	F1-score	Kappa	ROC area
6-Layer CNN	0.9393	1	0.9804	0.9804	0.9738	0.9995
ResNet50	0.7551	0.6875	0.75	0.7299	0.6667	0.9338
VGG16	0.9346	0.875	0.939	0.9388	0.9187	0.9944
VGG19	0.9197	0.9688	0.9628	0.9628	0.9504	0.9962
InceptionV3	0.915	0.9375	0.8988	0.8999	0.865	0.9854
MobileNet	0.971	0.875	0.9473	0.9478	0.9298	0.9931
DenseNet169	0.9386	0.9375	0.9421	0.9423	0.9229	0.9935
DenseNet121	0.929	0.9062	0.9525	0.9527	0.9366	0.9942
InceptionResNetV2	0.9159	0.9375	0.9329	0.932	0.9105	0.9937
MobileNetV2	0.9471	0.9375	0.9576	0.9574	0.9435	0.9956
ResNet101	0.7514	0.75	0.6849	0.6541	0.5799	0.9133

interobserver variability in the diagnosis of retinal diseases. AI models can deliver consistent and trustworthy classification findings by offering automated and objective assessments, minimizing the reliance on individual expertise, and enhancing the reproducibility of diagnoses. This is particularly useful in areas where access to ophthalmology specialists is limited, as AI-based solutions can help medical personnel make accurate and early diagnoses, improving patient management and care.

7. Conclusion

The use of AI approaches in retinal disease classification utilizing OCT images offers considerable promise for improving retinal disease diagnosis and therapy. AI models powered by machine learning and deep learning algorithms have excelled in accurately classifying various eye disorders. AI-based solutions provide the potential for early identification, objective assessments, and improved treatment results through the automated analysis of massive amounts of OCT data. Ophthalmologists can overcome problems related to interobserver variability and subjective interpretation by employing AI algorithms, resulting in improved diagnostic accuracy and effective patient care. Furthermore, AI models can help eliminate the shortage of ophthalmic doctors in impoverished areas, allowing for more fast and accurate diagnoses.

Future research and development on the classification of retinal diseases based on AI and OCT images have enormous promise. The accuracy, effectiveness, and clinical usefulness of AI systems will continue to increase as deep learning techniques advance. As a result, the combination of AI methods and OCT imaging offers a revolutionary method for classifying retinal diseases in the field of ophthalmology. AI-based technologies have the potential to revolutionize the field, helping patients worldwide and easing the strain on healthcare systems through early detection, precise diagnoses, and better patient management.

References

Altan, G. (2022). DeepOCT: An explainable deep learning architecture to analyze macular edema on OCT images. *Engineering Science and Technology, an International Journal, 34*, 101091. https://doi.org/10.1016/j.jestch.2021.101091

Fang, L., Wang, C., Li, S., Rabbani, H., Chen, X., & Liu, Z. (2019). Attention to lesion: Lesion-aware convolutional neural network for retinal optical coherence tomography image classification. *IEEE Transactions on Medical Imaging, 38*(8), 1959–1970. https://doi.org/10.1109/TMI.2019.2898414

Jacoba, C. M. P., Doan, D., Salongcay, R. P., Aquino, L. A. C., Silva, J. P. Y., Salva, C. M. G., Zhang, D., Alog, G. P., Zhang, K., Locaylocay, K. L. R. B., Saunar, A. V., Ashraf, M., Sun, J. K., Peto, T., Aiello, L. P., & Silva, P. S. (2023). Performance of automated machine learning for diabetic retinopathy image classification from multi-field handheld retinal images. *Ophthalmology Retina, 7*(8), 703–712. https://doi.org/10.1016/j.oret.2023.03.003

Kermany, D. S., Goldbaum, M., Cai, W., Valentim, C. C. S., Liang, H., Baxter, S. L., McKeown, A., Yang, G., Wu, X., Yan, F., Dong, J., Prasadha, M. K., Pei, J., Ting, M. Y. L., Zhu, J., Li, C., Hewett, S., Dong, J., Ziyar, I., … Zhang, K. (2018). Identifying medical diagnoses and treatable diseases by image-based deep learning. *Cell, 172*(5). https://doi.org/10.1016/j.cell.2018.02.010. Article 5.

Kermany, D., Zhang, K., & Goldbaum, M. (2018). Labeled optical coherence tomography (oct) and chest x-ray images for classification. *Mendeley Data, 2*(2), 651.

Khan, A., Pin, K., Aziz, A., Han, J. W., & Nam, Y. (2023). Optical coherence tomography image classification using hybrid deep learning and ant colony optimization. *Sensors, 23*(15). https://doi.org/10.3390/s23156706

Lo, J., Yu, T. T., Ma, D., Zang, P., Owen, J. P., Zhang, Q., Wang, R. K., Beg, M. F., Lee, A. Y., Jia, Y., & Sarunic, M. V. (2021). Federated learning for microvasculature segmentation and diabetic retinopathy classification of OCT data. *Ophthalmology Science, 1*(4), 100069. https://doi.org/10.1016/j.xops.2021.100069

Lu, W., Tong, Y., Yu, Y., Xing, Y., Chen, C., & Shen, Y. (2018). Deep learning-based automated classification of multi-categorical abnormalities from optical coherence tomography images. *Translational Vision Science and Technology, 7*(6), 41.

Mathews, M. R., & Anzar, S. T. M. (2023). A lightweight deep learning model for retinal optical coherence tomography image classification. *International Journal of Imaging Systems and Technology, 33*(1), 204–216.

Patnaik, S., & Subasi, A. (2023). Artificial intelligence-based retinal disease classification using optical coherence tomography images. In *Applications of artificial intelligence in medical imaging* (pp. 305–319). Elsevier.

Qi, Z., Si Wei, M., Qian, L., Guan Chong, H., Ming Chen, G., Wen Li, Y., Ge, W., Ya, M., Lei, L., & Xiao Yan, P. (2023). Automated classification of inherited retinal diseases in optical coherence tomography images using few-shot learning. *Biomedical and Environmental Sciences, 36*(5), 431–440. https://doi.org/10.3967/bes2023.052

Rajan, R., & Kumar, S. N. (2023). IoT based optical coherence tomography retinal images classification using OCT Deep Net2. *Measurement: Sensors, 25*, 100652. https://doi.org/10.1016/j.measen.2022.100652

Rong, Y., Xiang, D., Zhu, W., Yu, K., Shi, F., Fan, Z., & Chen, X. (2019). Surrogate-assisted retinal OCT image classification based on convolutional neural networks. *IEEE Journal of Biomedical and Health Informatics, 23*(1), 253–263. https://doi.org/10.1109/JBHI.2018.2795545

Shelke, S., & Subasi, A. (2023). Detection and classification of Diabetic Retinopathy Lesions using deep learning. In *Applications of artificial intelligence in medical imaging* (pp. 241–264). Elsevier.

Wang, X., Tang, F., Chen, H., Cheung, C. Y., & Heng, P.-A. (2023). Deep semi-supervised multiple instance learning with self-correction for DME classification from OCT images. *Medical Image Analysis, 83*, 102673. https://doi.org/10.1016/j.media.2022.102673

Wang, X., Tang, F., Chen, H., Luo, L., Tang, Z., Ran, A.-R., Cheung, C. Y., & Heng, P.-A. (2020). UD-MIL: Uncertainty-Driven deep multiple instance learning for OCT image classification. *IEEE Journal of Biomedical and Health Informatics, 24*(12), 3431–3442. https://doi.org/10.1109/JBHI.2020.2983730

Wang, D., & Wang, L. (2019). On OCT image classification via deep learning. *IEEE Photonics Journal, 11*(5), 1–14. https://doi.org/10.1109/JPHOT.2019.2934484

Zhang, Y., Li, M., Ji, Z., Fan, W., Yuan, S., Liu, Q., & Chen, Q. (2021). Twin self-supervision based semi-supervised learning (TS-SSL): Retinal anomaly classification in SD-OCT images. *Neurocomputing, 462*, 491–505. https://doi.org/10.1016/j.neucom.2021.08.051

Heart muscles inflammation (myocarditis) detection using artificial intelligence

Rupal Shah[1], Abdulhamit Subasi[2,3]

[1]Department of Electrical Engineering, Indian Institute of Technology, Indore, Madhya Pradesh, India;
[2]Institute of Biomedicine, Faculty of Medicine, University of Turku, Turku, Finland; [3]Department of
Computer Science, College of Engineering, Effat University, Jeddah, Saudi Arabia

1. Introduction

Myocarditis, an inflammatory condition of the heart muscles, poses a significant health concern worldwide due to its potential to lead to severe cardiac dysfunction and life-threatening complications. The muscle myocardium contracts and relaxes to pump blood to the entire body. Its inflammation can reduce its ability to pump blood and can cause blood clots leading to heart attack or stroke, damage to the heart, or even death. Such inflammations are primarily caused due to an infection. Sometimes the pathogens or the microbes damage the heart muscle directly otherwise it is caused as a result of the immune response of the body to the infection. The major symptoms of myocarditis include chest pain, shortness of breath, fatigue, and weakness. The symptoms do not show up and it becomes the biggest challenge to cure the disease. Myocarditis is very difficult to diagnose. Early and accurate detection of myocarditis is crucial for timely intervention and appropriate management, as delayed diagnosis can result in irreversible damage to the heart and increased mortality rates. In recent years, the emergence of machine learning techniques has presented a promising avenue for enhancing myocarditis detection and classification, revolutionizing the landscape of cardiovascular diagnostics. Traditional diagnostic methods for myocarditis, such as electrocardiography (ECG), cardiac biomarkers, and imaging modalities like echocardiography and cardiac MRI, while valuable, can sometimes yield inconclusive or challenging-to-interpret results. Moreover, the reliance on these methods often depends on the expertise of the interpreting physician, leading to variations in diagnostic accuracy. This has spurred the exploration of alternative approaches that can complement and augment conventional diagnostic practices (Woodruff, 1980).

Machine learning techniques have shown remarkable capabilities in analyzing large amounts of medical data, extracting complex patterns from the images, and giving accurate predictions. At the initial layers of a convolutional neural network (CNN) (LeCun & Bengio, 1995), the filters or kernels primarily focus on identifying simple features, such as edges and corners. As the information flows through subsequent layers, the network combines these simple features to recognize more complex patterns, such as textures or specific shapes. This hierarchical feature learning process is critical for the network's ability to comprehend intricate structures and nuances present in the input data. Moreover, CNNs can adapt their learned features based on the specific task at hand. Through a process called backpropagation, the network fine-tunes its internal parameters during training to minimize the difference between its predicted outputs and the ground truth labels. As a result, the network becomes highly specialized in extracting relevant features specific to the target classification task.

The hierarchical nature of CNNs not only enables them to achieve exceptional performance on image-related tasks but also makes them transferable to other domains. Pretrained CNN models, obtained from vast image datasets, can be fine-tuned or utilized as feature extractors for tasks in various fields, such as medical image analysis, natural language processing, and robotics. This transfer learning approach has proven to be especially valuable in situations where labeled data is scarce or challenging to obtain. In the context of medical image analysis, CNNs have revolutionized the field by significantly improving the accuracy and efficiency of disease detection, including myocarditis. By automatically learning discriminative features from medical imaging data, CNNs can assist doctors in identifying subtle indicators of diseases like myocarditis at an early stage, enhancing the prospects for timely intervention and better patient outcomes.

Furthermore, in this chapter, we will explore the preprocessing steps required to prepare medical imaging data for deep learning models, addressing issues like data augmentation and normalization. Subsequently, we will discuss the creation and curation of datasets for training and validation, highlighting the significance of large, diverse, and well-labeled datasets to ensure the generalizability and reliability of the machine learning models. To evaluate the efficacy of machine learning models for myocarditis detection, we will present a comparative analysis of different architectures and techniques, emphasizing their strengths and limitations. Furthermore, we will explore the concept of transfer learning, wherein pretrained models on similar tasks are adapted for myocarditis detection. This approach can mitigate the data limitations often encountered in medical settings and accelerate the development of accurate diagnostic models.

2. Literature review

Myocarditis is an uncommon cardiovascular condition caused mostly by infectious agents, particularly viruses. Its symptoms are frequently chest pain or heart failure. Cardiovascular magnetic resonance (CMR) imaging is routinely used for diagnosis. However, difficulties develop as a result of elements such as low contrast, noise, and the intricacy of CMR slices. To address this, Shoeibi et al. (2022, pp. 145–155) proposed a deep learning (DL)-based AI-driven strategy for myocarditis diagnosis. The procedure begins with preprocessing CMR images, followed by image enhancement using a cycle generative adversarial network (GAN) model. For classifying the input data, various pretrained DL models such as EfficientNetV2, HrNet, and ResNet50 are used. The results reveal that EfficientNetV2 achieves an amazing accuracy of 99.33% among the pretrained models. This illustrates the effectiveness of the suggested AI-based method in detecting myocarditis from CMR pictures, making it a promising tool for accurate and automated diagnosis.

In recent years, the field of artificial intelligence (AI) and machine learning (ML) has shown promising advancements in various medical disciplines, including the study of myocarditis (Caforio et al., 2012). AI and ML techniques are being applied to aid in the early detection, accurate diagnosis, and personalized treatment of myocarditis. One area of active research involves the development of AI-powered algorithms that can analyze large datasets of cardiac imaging, biomarkers, and patient history to identify subtle patterns indicative of myocardial inflammation. These algorithms can assist clinicians in making more precise and timely diagnoses, potentially leading to better patient outcomes.

Sharifrazi et al. (2020) proposed automatic myocarditis diagnosis using a CNN combined with k-means clustering. Fenoglio Jr et al. (1983) use endomyocardial biopsy to classify the disease. Moreover, AI-driven predictive models are being explored to assess the risk of disease progression and identify individuals at higher risk of developing severe complications. As the availability of electronic health records and real-time data continues to expand, AI and ML applications hold significant promise in revolutionizing the field of myocarditis by facilitating more effective and patient-centric approaches to diagnosis and treatment. However, further research and validation studies are essential to ensure the reliability and generalizability of these AI-driven solutions in the clinical setting (Sagar et al., 2012). Moravvej et al. (2022) examined the increased incidence of myocarditis, an inflammation of the heart muscle, in relation to COVID-19 in particular. It underlines the need of employing noninvasive CMR imaging to diagnose myocarditis, despite the fact that image interpretation takes time and requires specialist clinicians. To solve this, they proposed Reinforcement Learning-Based Myocarditis Diagnosis paired with Population-Based Algorithm (RLMD-PA). To categorize myocarditis using CMR images, the proposed model leverages deep reinforcement learning, a technique that involves sequential decision making. The model is designed to manage the unbalanced nature of CMR datasets and is based on a CNN architecture. The artificial bee colony (ABC) algorithm is used to generate initial weights for the model. The agent in the RLMD-PA model processes each CMR image and classifies it, getting incentives from the environment for each categorization. Notably, the payment mechanism is meant to reward more properly classifying the minority class, addressing the dataset's underlying class imbalance.

Ghareeb et al. (2022) used CMR to identify and estimate the risk of myocardial inflammation in patients with acute viral myocarditis. The researchers used unsupervised ML to examine patterns of CMR inflammation in a heterogeneous population of acute myocarditis. They looked at 169 persons with diagnosed acute myocarditis of diverse races and ethnicities. The primary goal was a composite clinical endpoint that included cardiac mortality, arrhythmia, and dilated cardiomyopathy. For exploratory analysis, ML was used to find inflammatory patterns in CMR images. Patients who achieved the combined endpoint had higher late gadolinium enhancement (LGE) in the anterior and septal areas of the heart. The researchers detected two distinct patterns of CMR inflammation using unsupervised ML and factor analysis. One pattern showed increased LGE, whereas the other showed increased myocardial T1/T2. These findings shed information on the peculiarities of myocarditis in patients from various countries, as well as the possibility of ML for detecting unique inflammatory patterns in CMR images.

3. Subjects and data acquisition

The dataset (Z Alizadeh Sani Myocarditis Dataset) used for the experiments consisted of 7135 total images of heart with 4686 images of heart which are abnormal and 2449 images of normal heart. The images from the dataset are divided

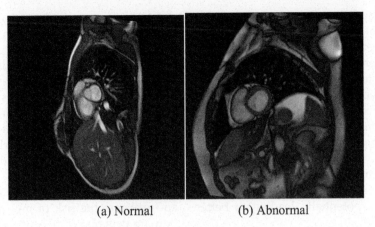

(a) Normal   (b) Abnormal

FIGURE 11.1   Images from dataset.

across two classes namely "Normal" Fig. 11.1a and "Abnormal" Fig. 11.1b.

For further experiments, the dataset is split into a training set, validation set, and test set containing 80% of the entire dataset, two-thirds of the remaining 20%, and one-third of the remaining 20% dataset, respectively. The size of the target images that were fed to the network is 300 × 300. The code used for the split is as follows.

```
import os
import random
print(len(os.listdir('./Abnormal')))
random.seed(2022)
abn_list = random.sample(os.listdir('./Abnormal'),
                         len(os.listdir('./Abnormal'))*4//5)
n_list = random.sample(os.listdir('./Normal'),
                       len(os.listdir('./Normal'))*4//5)
for i in range(len(abn_list)):
    os.rename(f'./Abnormal/{abn_list[i]}',f'./Train/Abnormal/{i}.jpg')


for i in range(len(n_list)):
    os.rename(f'./Normal/{n_list[i]}', f'./Train/Normal/{i}.jpg')
abn_list = random.sample(os.listdir('./Abnormal'),
                         len(os.listdir('./Abnormal'))*2//3)
n_list = random.sample(os.listdir('./Normal'),
                       len(os.listdir('./Normal'))*2//3)
for i in range(len(abn_list)):
    os.rename(f'./Abnormal/{abn_list[i]}', f'./Validation/Abnormal/{i}.jpg')


for i in range(len(n_list)):
    os.rename(f'./Normal/{n_list[i]}',f'./Validation/Normal/{i}.jpg')
```

Now, we will rename the images.

```python
random.seed(2022)
abn_list = random.sample(os.listdir('./Abnormal'),
                         len(os.listdir('./Abnormal')))
n_list = random.sample(os.listdir('./Normal'),
                       len(os.listdir('./Normal')))
for i in range(len(abn_list)):
    os.rename(f'./Abnormal/{abn_list[i]}', f'./Test/Abnormal/{i}.jpg')

for i in range(len(n_list)):
    os.rename(f'./Normal/{n_list[i]}', f'./Test/Normal/{i}.jpg')
```

As a result, the dataset will be ready with the training, validation, and test split.

Import the data in the following way to perform various experiments on it further:

```
from tensorflow.keras.preprocessing.image import ImageDataGenerator
// data imports
train_datagen = ImageDataGenerator()
val_datagen = ImageDataGenerator()
train_dir= './myocarditis/Train'
val_dir= './myocarditis/Validation'
test_dir= './myocarditis/Test'
train_generator = train_datagen.flow_from_directory(
    directory=train_dir,
    target_size=(300, 300),
    color_mode="rgb",
    batch_size=32,
    class_mode="binary",
    shuffle=True,
    seed=2023
)
valid_generator = val_datagen.flow_from_directory(
    directory=val_dir,
    target_size=(300, 300),
    color_mode="rgb",
    batch_size=32,
    class_mode="binary",
    shuffle=True,
    seed=2023
)
test_generator = val_datagen.flow_from_directory(
    directory=test_dir,
    target_size=(300, 300),
    color_mode="rgb",
    batch_size=1,
    class_mode="binary",
    shuffle=True,
    seed=2023
)
```

4. Proposed architecture for custom convolutional neural networks

This section discusses the experiments performed on the dataset using CNNs, the number of layers ranging from 2 to 8. On experimentation, it was found that a filter size of 64 and subsequent layers with a filter size of 32 gave the best results with a kernel size of 3. The stack of convolution layers followed by the flatten layer uses the RELU activation function. The last layer comprises a dense layer with one neuron and activation function. The optimizer used is Adam (Kingma & Ba, 2014) with a learning rate of 0.0001 and binary cross-entropy loss. The result is reported on the following metrics:

(1) Accuracy
(2) Cohen kappa (Kvålseth, 1989)
(3) AUC
(4) Precision
(5) Recall
(6) F1 score

Proposed Pipeline-

```
import tensorflow as tf
from model import CNN2, CNN3, CNN3, CNN4, CNN5, CNN6, CNN7, CNN8
from tensorflow.keras.metrics import AUC, Accuracy,Recall, Precision
from tensorflow_addons.metrics import CohenKappa

// Using CNN2 as the model here. You can use CNN2-8 for experimentation.
model = CNN2()
model.compile(optimizer = tf.keras.optimizers.Adam(lr=0.0001), loss =
'binary_crossentropy', metrics =
['acc',CohenKappa(num_classes=2),AUC(),Precision(),Recall()])
cnn = model.fit(train_generator, validation_data = valid_generator,
steps_per_epoch = 100, epochs = 10)
score = model.evaluate(test_generator, verbose=0)
d = dict(zip(model.metrics_names, score))
precision =d['precision']
recall = d['recall']
f1_score = 2*precision*recall/(precision+recall)

// Note: create a model.py to create custom CNN models and import it in the
pipeline as mentioned above
```

A new layer of filter size 32 is introduced as a subsequent layer to know how the results change with the addition of a CNN layer.

The custom model classes used in the above pipeline are as follows:

4.1 CNN 2 layers

```
from keras.models import Sequential
from keras.layers import Dense, Conv2D, Flatten
class CNN2():
    def __init__():
        model = Sequential()
        model.add(Conv2D(64, kernel_size=3, activation='relu',
input_shape=(300, 300, 3)))
        model.add(Conv2D(32, kernel_size=3, activation='relu'))
        model.add(Flatten())
        model.add(Dense(1, activation='sigmoid'))
        return model
```

4.2 CNN 3 layers

```
from keras.models import Sequential
from keras.layers import Dense, Conv2D, Flatten
class CNN3():
    def __init__():
        model = Sequential()
        model.add(Conv2D(64, kernel_size=3, activation='relu',
input_shape=(300, 300, 3)))
        model.add(Conv2D(32, kernel_size=3, activation='relu'))
        model.add(Conv2D(32, kernel_size=3, activation='relu'))
        model.add(Flatten())
        model.add(Dense(1, activation='sigmoid'))
        return model
```

4.3 CNN 4 layers

```python
from keras.models import Sequential
from keras.layers import Dense, Conv2D, Flatten
class CNN4():
    def __init__():
        model = Sequential()
        model.add(Conv2D(64, kernel_size=3, activation='relu',
input_shape=(300, 300, 3)))
        model.add(Conv2D(32, kernel_size=3, activation='relu'))
        model.add(Conv2D(32, kernel_size=3, activation='relu'))
        model.add(Conv2D(32, kernel_size=3, activation='relu'))
        model.add(Flatten())
        model.add(Dense(1, activation='sigmoid'))
        return model
```

4.4 CNN 5 layers

```python
from keras.models import Sequential
from keras.layers import Dense, Conv2D, Flatten
class CNN5():
    def __init__():
        model = Sequential()
        model.add(Conv2D(64, kernel_size=3, activation='relu',
input_shape=(300, 300, 3)))
        model.add(Conv2D(32, kernel_size=3, activation='relu'))
        model.add(Conv2D(32, kernel_size=3, activation='relu'))
        model.add(Conv2D(32, kernel_size=3, activation='relu'))
        model.add(Conv2D(32, kernel_size=3, activation='relu'))
        model.add(Flatten())
        model.add(Dense(1, activation='sigmoid'))
        return model
```

4.5 CNN 6 layers

```
from keras.models import Sequential
from keras.layers import Dense, Conv2D, Flatten
class CNN6():
    def __init__():
        model = Sequential()
        model.add(Conv2D(64, kernel_size=3, activation='relu',
input_shape=(300, 300, 3)))
        model.add(Conv2D(32, kernel_size=3, activation='relu'))
        model.add(Conv2D(32, kernel_size=3, activation='relu'))
        model.add(Conv2D(32, kernel_size=3, activation='relu'))
        model.add(Conv2D(32, kernel_size=3, activation='relu'))
        model.add(Conv2D(32, kernel_size=3, activation='relu'))
        model.add(Flatten())
        model.add(Dense(1, activation='sigmoid'))
        return model
```

4.6 CNN 7 layers

```
from keras.models import Sequential
from keras.layers import Dense, Conv2D, Flatten
class CNN7():
    def __init__():
        model = Sequential()
        model.add(Conv2D(64, kernel_size=3, activation='relu',
input_shape=(300, 300, 3)))
        model.add(Conv2D(32, kernel_size=3, activation='relu'))
        model.add(Conv2D(32, kernel_size=3, activation='relu'))
        model.add(Conv2D(32, kernel_size=3, activation='relu'))
        model.add(Conv2D(32, kernel_size=3, activation='relu'))
        model.add(Conv2D(32, kernel_size=3, activation='relu'))
        model.add(Conv2D(32, kernel_size=3, activation='relu'))
        model.add(Flatten())
        model.add(Dense(1, activation='sigmoid'))
        return model
```

4.7 CNN 8 layers

```python
from keras.models import Sequential
from keras.layers import Dense, Conv2D, Flatten
class CNN8():
    def __init__():
        model = Sequential()
        model.add(Conv2D(64, kernel_size=3, activation='relu',
input_shape=(300, 300, 3)))
        model.add(Conv2D(32, kernel_size=3, activation='relu'))
        model.add(Conv2D(32, kernel_size=3, activation='relu'))
        model.add(Conv2D(32, kernel_size=3, activation='relu'))
        model.add(Conv2D(32, kernel_size=3, activation='relu'))
        model.add(Conv2D(32, kernel_size=3, activation='relu'))
        model.add(Conv2D(32, kernel_size=3, activation='relu'))
        model.add(Conv2D(32, kernel_size=3, activation='relu'))
        model.add(Flatten())
        model.add(Dense(1, activation='sigmoid'))
        return model
```

5. Proposed pipeline for transfer learning

This section discusses the transfer learning technique, which involves the use of pretrained models to evaluate the results of the classification. Transfer learning is a ML technique that involves leveraging knowledge gained from one task or domain and applying it to another related or similar task. In traditional ML approaches, models are trained from scratch on a specific dataset for a particular task. However, transfer learning takes advantage of the fact that the features learned by a model from one task can be useful for solving a different but related task, even if the target domain has different data distributions or characteristics. Fig. 11.2 shows the pipeline pictorially, the pretrained models take in the input in the form of the

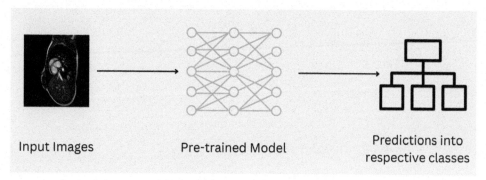

Input Images   Pre-trained Model   Predictions into
respective classes

FIGURE 11.2   Proposed pipeline.

images and the feature extraction takes place in these layers to give us the final output at a fully connected layer with one neuron. In the experiments, the hyperparameters like epochs, learning rate, and batch size are kept as 10 and 0.0001, respectively.

The code to implement transfer learning is as follows:

```
import tensorflow as tf
from keras.layers import Dense, Flatten
from tensorflow.keras.applications import ResNet50
from tensorflow.keras.metrics import AUC, Accuracy,Recall, Precision
from tensorflow_addons.metrics import CohenKappa
// import any pretrained model here in the same way.
base = <model> //we will change this line keeping other things constant
for layer in base.layers:
    layer.trainable = False
layer = Flatten()(base.output)
x = Dense(512, activation='relu')(layer)
x = Dense(1, activation='sigmoid')(x)
model = tf.keras.models.Model(base.input, x)
model.compile(optimizer = tf.keras.optimizers.Adam(lr=0.0001), loss =
'binary_crossentropy',metrics = ['acc', CohenKappa(num_classes=2), AUC(),
Precision(), Recall()])
resnet = model.fit(train_generator, validation_data = valid_generator,
steps_per_epoch = 100, epochs = 10)
score = model.evaluate(test_generator, verbose=0)
d = dict(zip(model.metrics_names, score))
precision =d['precision']
recall = d['recall']
f1_score = (2*precision*recall/(precision+recall))
model.save(<model_name>.h5')
```

5.1 ResNet

ResNet, short for residual network, is a powerful DL architecture that was developed to address the challenges of training very deep neural networks that were previously prone to the problem of vanishing gradients. Vanishing gradients occur when the gradients during backpropagation become infinitesimally small as they propagate back through numerous layers, leading to slow convergence or even preventing the model from learning. The key innovation in ResNet is the use of residual blocks, also known as skip connections or shortcuts. These blocks facilitate the training of extremely deep networks by allowing information to flow directly across layers, bypassing one or more layers during the forward pass. This way, the gradient signal can also bypass those layers during backpropagation, mitigating the vanishing gradient problem and enabling the training of much deeper architectures (He et al., 2016).

5.1.1 Resnet101

ResNet-101 contains 101 layers, making it more capable of capturing intricate features and representations. It can be imported and used in the above pipeline as follows:

```
from tensorflow.keras.applications import ResNet101
base = ResNet101(input_shape = (300, 300, 3), include_top=False,
weights='imagenet')
```

5.1.2 Resnet50

It is a deep CNN that consists of 50 layers, making it a relatively deep model. It can be imported and used in the above pipeline as follows:

```
from tensorflow.keras.applications import ResNet50
base = ResNet50(input_shape = (300, 300, 3), include_top=False,
weights='imagenet')
```

5.2 DenseNet

DenseNet, short for densely connected convolutional networks, is a DL architecture that is designed to address some of the limitations of traditional CNNs and has gained popularity due to its efficient use of parameters and excellent performance in various computer vision tasks. The key idea behind DenseNet is the concept of dense connections between layers. In a standard CNN, information flows sequentially from one layer to the next, and each layer only receives input from the previous layer. However, in DenseNet, every layer receives direct inputs from all preceding layers within the same dense block. This dense connectivity allows for richer feature reuse and promotes gradient flow throughout the network, which can alleviate the vanishing gradient problem and improve training efficiency. In a DenseNet, the output of each layer is a concatenation of the feature maps from all preceding layers within the same dense block (Huang et al., 2017).

5.2.1 DenseNet121

DenseNet-121 is a specific variant of the DenseNet architecture with 121 layers. It can be imported and used in the above pipeline as follows:

```
from tensorflow.keras.applications import DenseNet121
base = DenseNet121(input_shape = (300, 300, 3), include_top=False,
weights='imagenet')
```

5.2.2 DenseNet 169

DenseNet-169 is a specific variant of the DenseNet architecture with 169 layers. It can be imported and used in the above pipeline as follows:

a balance between model size, speed, and accuracy. This makes MobileNet ideal for deploying DL models on mobile and embedded devices where computational resources are limited,

```
from           tensorflow.keras.applications        import        DenseNet169
base   =   DenseNet169(input_shape   =   (300,   300,   3),   include_top=False,
weights='imagenet')
```

5.3 MobileNet

MobileNet is a family of efficient DL architectures designed for running on mobile and embedded devices with limited computational resources. MobileNet's design choices, such as depthwise separable convolutions, width multiplier, and resolution multiplier, allow it to strike

enabling real-time and on-device inference for a wide range of computer vision tasks, including image classification, object detection, and semantic segmentation (Howard et al., 2017).

5.3.1 MobileNet

The code for MobileNet is given below.

```
from tensorflow.keras.applications import MobileNet
base = MobileNet(input_shape = (300, 300, 3), include_top=False,
weights='imagenet')
```

5.3.2 *MobileNetV2*

This variant consists of 53–87 convolution layers (Sandler et al., 2018, pp. 4510–4520), depending on the width multiplier and resolution multiplier used. The code for MobileNetV2 is given below.

```
from tensorflow.keras.applications import MobileNetV2
base = MobileNetV2(input_shape = (300, 300, 3), include_top=False,
      weights='imagenet')
```

5.4 VGG

VGG, short for visual geometry group, is a deep CNN architecture and is known for its simplicity and effectiveness, and it played a crucial role in advancing the field of computer vision by demonstrating the power of deep CNNs for image recognition tasks. The input image size of the model is 224 × 224 (Simonyan & Zisserman, 2014).

VGG has the following variants:

5.4.1 *VGG16*

It consists of 13 convolutional layers and three fully connected layers. The code for VGG16 is given below.

```
from tensorflow.keras.applications import VGG16
base = VGG16(input_shape = (224, 224, 3), include_top=False,
      weights='imagenet')
```

5.4.2 *VGG19*

VGG19 consists of 19 weight layers, which include 16 convolutional layers and three fully connected layers. The code for VGG19 is given below.

```
from tensorflow.keras.applications import VGG19
base = VGG19(input_shape = (224, 224, 3), include_top=False,
      weights='imagenet')
```

5.5 Inception/GoogLeNet

The inception model, also known as Google-Net (Szegedy et al., 2015, pp. 1–9), is a deep CNN architecture. The main innovation of the inception model is the use of "inception modules," which are carefully designed to capture multiscale features effectively while minimizing computational cost. These modules allow the network to learn and combine features at different spatial resolutions in parallel. The inception model consists of a series of inception modules, which are stacked together to create a deep network. Each inception module consists of multiple parallel convolutional operations of different filter sizes, followed by concatenation of their output feature maps. This allows the model to capture features at different scales. It has evolved into multiple versions namely, InceptionV2 (Szegedy et al., 2016, pp. 2818–2826), InceptionV3, and Inception-ResNet, which are introduced after further optimization.

InceptionV3 is an improved variant of the inception model architecture. It employs factorized convolutions, which are designed to reduce the number of parameters. Instead of using large-size convolutions directly, factorized convolutions break them into two smaller convolutions, such as 3 × 3 and 5 × 5, performed sequentially. This helps in capturing multiscale information while keeping the computational cost under control. The code for InceptionV3 is given below.

6. Detection of myocarditis with deep feature extraction

The next section discusses the deep feature extraction technique. Six different ML classifiers were tried on the extracted features from the proposed architecture. The images are preprocessed

```
from tensorflow.keras.applications import InceptionV3
base = InceptionV3(input_shape = (300, 300, 3), include_top=False,
weights='imagenet')
```

5.5.1 Inception ResNetV2

Inception-ResNet-v2 (Szegedy et al., 2017) replaces the traditional 7 × 7 convolutions used in the original inception model with factorized 7 × 1 and 1 × 7 convolutions. This factorization helps reduce the number of parameters and computations while retaining the ability to capture multiscale features. The code for Inception ResNetV2 is given below.

and sent through the pretrained CNN and the features are extracted from the penultimate fully connected layer (512 neurons), then a classifier head is attached to classify the images (Fig. 11.3). The code of the pipeline is presented below. We extract the features of the model from its penultimate layer with 512 neurons using Keras backend. We then attach a classifier head to produce the final results.

```
from tensorflow.keras.applications import InceptionResNetV2
base = InceptionResNetV2(input_shape = (300, 300, 3), include_top=False,
weights='imagenet')
```

```python
import numpy as np
from tensorflow.keras.preprocessing.image import ImageDataGenerator
from keras.layers import Flatten, Dense
import tensorflow as tf
from tensorflow.keras.metrics import AUC, Accuracy,Recall,Precision
from tensorflow_addons.metrics import F1Score, CohenKappa
from keras import backend as K
import os
import keras
path = './models/MobileNetV2.h5'
train_datagen = ImageDataGenerator()
val_datagen = ImageDataGenerator()
train_dir= './myocarditis/Train'
val_dir= './myocarditis/Validation'
test_dir= './myocarditis/Test'
train_generator = train_datagen.flow_from_directory(
    directory=train_dir,
    target_size=(300, 300),
    color_mode="rgb",
    batch_size=500,
    class_mode="binary",
    shuffle=True,
    seed=2023
)
valid_generator = val_datagen.flow_from_directory(
    directory=val_dir,
    target_size=(300, 300),
    color_mode="rgb",
    batch_size=200,
    class_mode="binary",
    shuffle=True,
    seed=2023
)
test_generator = val_datagen.flow_from_directory(
    directory=test_dir,
    target_size=(300, 300),
    color_mode="rgb",
    batch_size=100,
    class_mode="binary",
```

```
        shuffle=True,
        seed=2023
)

model = keras.models.load_model(path)
// extracting the features and storing it in layer
layer = K.function([model.layers[0].input], [model.layers[-2].output]
a,b = train_generator.next()
image = np.array(a)
label = np.array(b)
x_train = layer(image)
x_train = np.array(x_train)[0]
y_train = label

a,b = valid_generator.next()
image = np.array(a)
label = np.array(b)
layer = K.function([model.layers[0].input], [model.layers[-2].output])
x_val = layer(image)
x_val = np.array(x_val)[0]
y_val = label

a,b = test_generator.next()
image = np.array(a)
label = np.array(b)
layer = K.function([model.layers[0].input], [model.layers[-2].output])
x_test = layer(image)
x_test = np.array(x_test)[0]
y_test = label
```

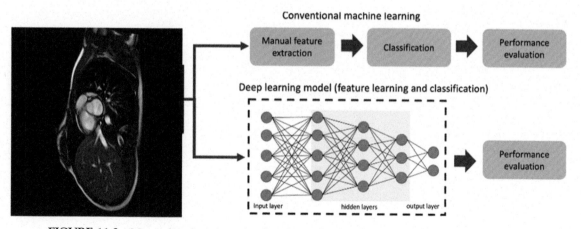

FIGURE 11.3 Myocarditis detection using deep feature extraction and conventional machine learning.

Now the extracted features from training, validation and testing images will be sent through a classifier to get classified.

6.1 k-nearest neighbors

In k-nearest neighbors (kNN), the algorithm makes predictions based on the nearest data points in the feature space to the input data point. The principal belief of kNN is that it supposes similar things to be in close proximity to each other. k essentially is the number of closest neighbors. The distance between the example points and the given query point is calculated and the first k distances in a sorted order are taken into consideration and the labels of these k entries are averaged in regression problems to get the correct answer, while in classification tasks, the mode of that data is taken (Peterson, 2009). The code for the KNN is as follows.

```
from sklearn.neighbors import KNeighborsClassifier
clf = KNeighborsClassifier()
clf.fit(x_train, y_train)
y_pred = clf.predict(x_train)
tr_acc = metrics.accuracy_score(y_train, y_pred)
y_pred = clf.predict(x_val)
val_acc = metrics.accuracy_score(y_val, y_pred)
y_pred = clf.predict(x_test)
t_acc = metrics.accuracy_score(y_test, y_pred)
re = metrics.recall_score(y_test, y_pred)
pre = metrics.precision_score(y_test, y_pred)
kap = metrics.cohen_kappa_score(y_test,y_pred)
```

6.2 Support vector machine

Support vector machine (SVM) is a powerful and widely used supervised ML algorithm. Its primary objective is to find an optimal hyperplane that best separates data points of different classes in a feature space. SVM achieves this by maximizing the margin between the classes, which is the distance between the hyperplane and the nearest data points of each class (support vectors). It can handle both linearly separable and nonlinearly separable data through the use of the kernel trick, which transforms the data into a higher-dimensional feature space. SVM's ability to handle high-dimensional data and find complex decision boundaries makes it effective for various classification tasks. By balancing the trade-off between maximizing the margin and minimizing classification errors, SVM provides a robust and generalized solution for binary classification and can be extended to multiclass classification as well as regression tasks (Hearst et al., 1998). The code for the SVM can be found below:

```
from sklearn import svm, metrics
clf = svm.SVC()
clf.fit(x_train, y_train)
y_pred = clf.predict(x_train)
tr_acc = metrics.accuracy_score(y_train, y_pred)
y_pred = clf.predict(x_val)
val_acc = metrics.accuracy_score(y_val, y_pred)
y_pred = clf.predict(x_test)
t_acc = metrics.accuracy_score(y_test, y_pred)
re = metrics.recall_score(y_test, y_pred)
pre = metrics.precision_score(y_test, y_pred)
kap = metrics.cohen_kappa_score(y_test,y_pred)
```

6.3 Random Forests

Random Forest is an ensemble learning method that excels in both classification and regression tasks. During its training phase, Random Forest builds multiple decision trees, each independently trained on random subsets of the data and features. This diversity reduces overfitting and ensures a robust model. When making predictions, each tree votes (for classification) or averages (for regression) its individual prediction, and the final output is determined by majority voting (in classification) or averaging (in regression). By combining the strengths of individual decision trees and mitigating their weaknesses, Random Forest provides highly accurate and stable predictions. It handles high-dimensional data, large datasets, and complex relationships between variables effectively. Its versatility, ease of implementation, and ability to handle nonlinear data make it a popular choice across various domains, from bioinformatics and finance to computer vision and natural language processing (Breiman, 2001).

```
from sklearn.ensemble import RandomForestClassifier
clf = RandomForestClassifier()
clf.fit(x_train, y_train)
y_pred = clf.predict(x_train)
tr_acc = metrics.accuracy_score(y_train, y_pred)
y_pred = clf.predict(x_val)
val_acc = metrics.accuracy_score(y_val, y_pred)
y_pred = clf.predict(x_test)
t_acc = metrics.accuracy_score(y_test, y_pred)
re = metrics.recall_score(y_test, y_pred)
pre = metrics.precision_score(y_test, y_pred)
kap = metrics.cohen_kappa_score(y_test,y_pred)
```

6.4 Bagging

Bootstrap aggregating (bagging) is an ensemble learning technique widely used in ML to improve model performance and reduce variance. The main idea behind bagging is to create multiple models by training them on different random subsets of the training data. During the training phase, bootstrap sampling is performed, where data points are randomly selected with replacements to form each subset. These subsets are used to train individual models independently. When making predictions, bagging combines the predictions of all models, usually through voting (for classification) or averaging (for regression), to arrive at a final prediction. Bagging reduces the risk of overfitting and increases the stability and accuracy of the overall model by reducing variance. It works well with high-variance algorithms like decision trees, creating a more robust and powerful ensemble model. Popular implementations of bagging include Random Forest for decision trees and Bagged Ensembles for other base models. Bagging has been proven effective in various domains, ranging from image recognition and sentiment analysis to financial forecasting and medical diagnostics (Breiman, 1996).

```
from sklearn.ensemble import BaggingClassifier
clf = BaggingClassifier()
clf.fit(x_train, y_train)
y_pred = clf.predict(x_train)
tr_acc = metrics.accuracy_score(y_train, y_pred)
y_pred = clf.predict(x_val)
val_acc = metrics.accuracy_score(y_val, y_pred)
y_pred = clf.predict(x_test)
t_acc = metrics.accuracy_score(y_test, y_pred)
re = metrics.recall_score(y_test, y_pred)
pre = metrics.precision_score(y_test, y_pred)
kap = metrics.cohen_kappa_score(y_test,y_pred)
```

6.5 AdaBoost

Adaptive boosting (AdaBoost) is a powerful ensemble learning technique used for binary classification tasks. It aims to improve the performance of weak learners by combining them into a strong classifier. In the training phase, AdaBoost assigns weights to individual data points, allowing it to focus on the samples that are difficult to classify correctly. It starts by training a weak learner on the data, and then iteratively updates the weights of misclassified samples to emphasize their importance in subsequent iterations. This "adaptive" weight updating process helps the algorithm prioritize challenging instances, leading to the creation of accurate classifiers. By combining the predictions of multiple

weak learners through weighted voting, AdaBoost produces a final prediction that leverages the strengths of individual models. AdaBoost has shown excellent generalization capabilities and robust performance in various domains, making it a popular and effective technique in the ML community (Freund & Schapire, 1997).

```
from sklearn.ensemble import AdaBoostClassifier
clf = AdaBoostClassifier()
clf.fit(x_train, y_train)
y_pred = clf.predict(x_train)
tr_acc = metrics.accuracy_score(y_train, y_pred)
y_pred = clf.predict(x_val)
val_acc = metrics.accuracy_score(y_val, y_pred)
y_pred = clf.predict(x_test)
t_acc = metrics.accuracy_score(y_test, y_pred)
re = metrics.recall_score(y_test, y_pred)
pre = metrics.precision_score(y_test, y_pred)
kap = metrics.cohen_kappa_score(y_test,y_pred)
```

6.6 XGBoost

eXtreme Gradient Boosting (XGBoost), is a state-of-the-art gradient-boosting algorithm that has gained significant popularity in the field of ML for both classification and regression tasks. It is an extension of the traditional gradient-boosting algorithm that optimizes performance and computational efficiency by incorporating several advanced techniques. XGBoost works by building an ensemble of weak learners, typically decision trees, in a sequential manner. In each iteration, it adds a new tree to the ensemble that corrects the errors made by the previous ones. The algorithm places a strong emphasis on regularization to prevent overfitting, using techniques such as max depth constraints, minimum child weight, and column subsampling. Additionally, it employs a customized loss function to optimize the learning process and handle specific types of problems effectively (Chen & Guestrin, 2016).

One of the key innovations of XGBoost lies in its gradient-based optimization approach, which utilizes the second-order derivative information (Hessian) to further refine the learning process. This technique significantly boosts the learning speed and makes XGBoost capable of handling large datasets and high-dimensional feature spaces efficiently. The code for the XGBoost can be found below.

```
import xgboost as xgb
clf = xgb.XGBClassifier()
clf.fit(x_train, y_train)
y_pred = clf.predict(x_train)
tr_acc.append(metrics.accuracy_score(y_train, y_pred))
y_pred = clf.predict(x_val)
tr_acc = metrics.accuracy_score(y_train, y_pred)
y_pred = clf.predict(x_val)
val_acc = metrics.accuracy_score(y_val, y_pred)
y_pred = clf.predict(x_test)
t_acc = metrics.accuracy_score(y_test, y_pred)
re = metrics.recall_score(y_test, y_pred)
pre = metrics.precision_score(y_test, y_pred)
kap = metrics.cohen_kappa_score(y_test,y_pred)
```

7. Results and discussions

7.1 Performance evaluation measures

Performance evaluation measures are used to assess the effectiveness and accuracy of ML models and algorithms. These measures provide valuable insights into how well the model is performing on the given task and help in comparing different models or tuning hyperparameters. The choice of performance evaluation measures depends on the type of task, such as classification, regression, or clustering. The dataset, in common practices, is divided into three unequal parts namely, training, validation, and testing set. The model is trained on the training data and it is evaluated and tested on the testing data. The validation set is utilized to fine-tune the hyperparameters involved in training the classifier. The metrics directly reflect the performance of the model and are extremely useful in quoting standard results.

True positive (TP) and true negative (TN) refer to the number of samples for which the model's classification is correct. A false positive (FP) occurs when the model incorrectly predicts a positive (yes) outcome when the actual outcome is negative (no). On the other hand, a false negative (FN) happens when the model incorrectly predicts a negative (no) outcome when the actual outcome is positive (yes). These metrics are essential in assessing a model's performance, especially in classification tasks, as they help identify the accuracy and reliability of the model's predictions.

Here are some commonly used performance evaluation measures for different types of tasks:

The proportion of correctly classified instances to the total number of instances in the dataset. It provides a general overview of the model's overall performance. The training, validation, and testing accuracy can be found using the following formula.

$$Accuracy = \frac{Number\ of\ correct\ predictions}{Total\ number\ of\ predictions}$$

Also known as positive predictive value, precision measures the percentage of true positive predictions out of all positive predictions made by the model. It is useful when the cost of false positives is high. It can be calculated as follows:

$$Precision = \frac{TP}{TP + FP}$$

Recall (sensitivity or true positive rate) measures the proportion of true positive predictions out of all actual positive instances in the dataset. It is essential when the cost of false negatives is high. It can be calculated as follows:

$$Recall = \frac{TP}{TP + FN}$$

The harmonic means of precision and recall, providing a balance between both metrics. It is useful when there is an imbalance between classes.

Then F1 measure can be formulated as:

$$F1 = \frac{2 * Precision * Recall}{Precision + Recall}$$

AUC represents the area under the receiver operating characteristic curve, which plots the true positive rate against the false positive rate. AUC measures the model's ability to distinguish between different classes and is insensitive to class imbalance. It plots the TPR against the FPR) at different threshold values. The TPR is also known as sensitivity or recall, and the FPR is the complement of the True Negative Rate (TNR), which is also known as specificity.

$$TP\ Rate = TP/(TP + FN)$$

$$FP\ Rate = FP/(FP + TN)$$

The ROC curve is created by plotting TPR (sensitivity) on the y-axis against FPR (1-specificity) on the x-axis. Each point on the curve corresponds to a specific threshold value used for classifying the data.

Kappa measures the agreement beyond chance between the two raters by taking into account the agreement that could be expected to occur by chance alone. It provides a more robust evaluation metric, especially when dealing with imbalanced datasets or when the accuracy alone might not be sufficient. It gives the idea of whether the model has guessed the answer or whether it is absolutely sure of the output.

The kappa score ranges from -1 to $+1$, where.

- Kappa $= 1$: Perfect agreement between the raters.
- Kappa $= 0$: Agreement equivalent to chance.
- Kappa < 0: Agreement is worse than chance.

It can be calculated using the following formula:

$$K = \frac{P_0 - P_e}{1 - P_e}$$

7.2 Experimental results

7.2.1 Results for transfer learning models

On training the model for 10 epochs all the models rest on high metrics the highest test accuracy being 99.8% by MobileNetV2, the results were as follows (Table 11.1):

7.2.2 Results for convolutional neural networks

Table 11.2 shows the experiment results for various custom CNNs having different numbers of convolutional layers. We were able to achieve a high-test accuracy of 99.6% with the convolutional network having four, six, and seven layers. F1 score, Cohen's kappa, and AUC were also found to be the highest with CNNs having four, six, and seven layers with values of 0.994, 0.991, and 1.00, respectively.

7.2.3 Feature extraction with pretrained models

The features extracted from the penultimate fully connected layer of the model have the

TABLE 11.1 Performance of transfer learning models.

Model	Train accuracy	Validation accuracy	Test accuracy	F1 score	Kappa	AUC
ResNet50	0.998	0.978	0.983	0.976	0.963	0.987
ResNet101	0.998	0.994	0.994	0.991	0.986	0.994
VGG16	0.998	0.976	0.985	0.979	0.968	0.990
VGG19	0.998	0.996	0.996	0.994	0.991	0.998
DenseNet121	0.961	0.916	0.899	0.829	0.760	0.854
DenseNet169	1.000	0.993	0.996	0.994	0.991	0.997
Inception_v3	0.997	0.992	0.996	0.994	0.991	0.995
InceptionResNetV2	0.961	0.990	0.994	0.991	0.986	0.992
MobileNetV2	0.996	0.990	0.998	0.997	0.995	0.997
MobileNet	0.995	0.988	0.996	0.994	0.991	0.997

TABLE 11.2 Performance of custom CNN models.

Model	Train accuracy	Validation accuracy	Test accuracy	F1 score	Kappa	AUC
CNN 2 layer	1.000	0.994	0.992	0.988	0.981	0.994
CNN 3 layer	1.000	0.996	0.994	0.991	0.986	0.998
CNN 4 layer	1.000	0.993	0.996	0.994	0.991	0.998
CNN 5 layer	1.000	0.994	0.572	0.177	0.072	0.469
CNN 6 layer	1.000	0.992	0.996	0.994	0.991	1.000
CNN 7 layer	1.000	0.998	0.996	0.994	0.991	1.000
CNN 8 layer	0.879	0.984	0.990	0.985	0.977	0.999

dimensions of 512 and they are sent to the classification models for classification.

7.2.3.1 k-Nearest neighbors

All the models give near-perfect results with the highest test accuracy of 1.00 and F1 score of 1, kappa score of 1 and precision, recall also as 1 (Table 11.3).

7.2.3.2 Support vector machine

We observed that the results for the SVM classifier were almost similar as compared to the K-NN classifier. With near-perfect results, the test accuracy, F1 score, kappa, precision, and recall all give near 1.00 values (Table 11.4).

7.2.3.3 Random Forest

In the case of Random Forest, the results are a little imperfect in InceptionResNetV2 and DenseNet121. Other models give near-perfect results and can be used for accurate predictions (Table 11.5).

7.2.3.4 AdaBoost and XGBoost

AdaBoost and XGBoost also give similar results, with InceptionResNetv2 and DensNet121

TABLE 11.3 Deep feature extraction along with K-NN classifier.

Model	Train accuracy	Validation accuracy	Test accuracy	F1 score	Kappa	Recall	Precision
ResNet50	1.00	1.00	1.00	1.00	1.00	1.00	1.00
ResNet101	1.00	0.99	1.00	1.00	1.00	1.00	1.00
MobileNetV2	1.00	0.98	1.00	1.00	1.00	1.00	1.00
MobileNet	1.00	0.99	1.00	1.00	1.00	1.00	1.00
InceptionV3	1.00	0.98	1.00	0.00	1.00	1.00	1.00
InceptionResNetV2	0.99	1.00	0.99	0.99	0.98	0.98	1.00
DenseNet169	1.00	0.99	1.00	1.00	1.00	1.00	1.00
DenseNet121	0.99	0.99	0.98	0.98	0.96	0.95	1.00

TABLE 11.4 Deep feature extraction along with SVM classifier.

Model	Train accuracy	Validation accuracy	Test accuracy	F1 score	Kappa	Recall	Precision
ResNet50	1.00	1.00	1.00	1.00	1.00	1.00	1.00
ResNet101	1.00	0.99	1.00	1.00	1.00	1.00	1.00
MobileNetV2	1.00	0.98	1.00	1.00	1.00	1.00	1.00
MobileNet	1.00	0.99	1.00	1.00	1.00	1.00	1.00
InceptionV3	1.00	0.98	1.00	1.00	1.00	1.00	1.00
InceptionResNetV2	1.00	1.00	1.00	1.00	1.00	1.00	1.00
DenseNet169	1.00	0.99	1.00	1.00	1.00	1.00	1.00
DenseNet121	0.99	0.98	0.98	0.98	0.96	0.96	1.00

TABLE 11.5 Deep feature extraction along with Random Forest classifier.

Model	Train accuracy	Validation accuracy	Test accuracy	F1 score	Kappa	Recall	Precision
ResNet50	1.00	0.99	1.00	1.00	1.00	1.00	1.00
ResNet101	1.00	0.99	1.00	1.00	1.00	1.00	1.00
MobileNetV2	1.00	0.98	1.00	1.00	1.00	1.00	1.00
MobileNet	1.00	0.99	1.00	1.00	1.00	1.00	1.00
InceptionV3	1.00	0.98	1.00	1.00	1.00	1.00	1.00
InceptionResNetV2	1.00	1.00	0.99	0.99	0.98	0.98	1.00
DenseNet169	1.00	0.99	1.00	1.00	1.00	1.00	1.00
DenseNet121	1.00	0.99	0.98	0.98	0.96	0.95	1.00

giving a little less perfect results, while the other pretrained models give the high-test accuracy and F1 score (Tables 11.6 and 11.7).

7.2.3.5 Bagging

Table 11.8 shows the results of the bagging classifier, which is slightly better than other techniques with the best test accuracy of 1.00, and 0.99 in InceptionResNetV2, and DenseNet121. The other models give perfect scores except for InceptionResNetV2 and DenseNet121.

7.2.2.6 Artificial neural network

Table 11.9 shows the results of the artificial neural network (ANN) classifier, which gives near-perfect results for most models except for InceptionResNetV2 and DenseNet121. With an accuracy of 0.98 and 0.99 in most cases, it provides the weakest results in comparison to the other classifiers.

For almost all the ML classifiers except for InceptionResNetV2 and DenseNet121 proved to give the highest results. The highest value obtained was 1.00 with almost all the classifiers.

TABLE 11.6   Deep feature extraction along with AdaBoost.

Model	Train accuracy	Validation accuracy	Test accuracy	F1 score	Kappa	Recall	Precision
ResNet50	1.00	1.00	1.00	1.00	1.00	1.00	1.00
ResNet101	1.00	0.99	1.00	1.00	1.00	1.00	1.00
MobileNetV2	1.00	0.98	1.00	1.00	1.00	1.00	1.00
MobileNet	1.00	0.99	1.00	1.00	1.00	1.00	1.00
InceptionV3	1.00	0.98	1.00	1.00	1.00	1.00	1.00
InceptionResNetV2	0.99	1.00	0.99	0.99	0.98	0.98	1.00
DenseNet169	1.00	0.99	1.00	1.00	1.00	1.00	1.00
DenseNet121	0.99	0.99	0.98	0.98	0.96	0.95	1.00

TABLE 11.7   Deep feature extraction along with XGBoost.

Model	Train accuracy	Validation accuracy	Test accuracy	F1 score	Kappa	Recall	Precision
ResNet50	1.00	0.99	1.00	1.00	1.00	1.00	1.00
ResNet101	1.00	0.99	1.00	1.00	1.00	1.00	1.00
MobileNetV2	1.00	0.98	1.00	1.00	1.00	1.00	1.00
MobileNet	1.00	0.99	1.00	1.00	1.00	1.00	1.00
InceptionV3	1.00	0.98	1.00	1.00	1.00	1.00	1.00
InceptionResNetV2	1.00	0.98	0.99	0.99	0.98	0.98	1.00
DenseNet169	1.00	0.99	1.00	1.00	1.00	1.00	1.00
DenseNet121	1.00	1.00	0.99	0.99	0.98	0.98	1.00

TABLE 11.8 Deep feature extraction along with bagging.

Model	Train accuracy	Validation accuracy	Test accuracy	F1 score	Kappa	Recall	Precision
ResNet50	1.00	0.99	1.00	1.00	1.00	1.00	1.00
ResNet101	1.00	0.99	1.00	1.00	1.00	1.00	1.00
MobileNetV2	1.00	0.98	1.00	1.00	1.00	1.00	1.00
MobileNet	1.00	0.99	1.00	1.00	1.00	1.00	1.00
InceptionV3	1.00	0.98	1.00	1.00	1.00	1.00	1.00
InceptionResNetV2	1.00	0.98	0.99	0.99	0.98	0.98	1.00
DenseNet169	1.00	0.99	1.00	1.00	1.00	1.00	1.00
DenseNet121	1.00	1.00	0.99	0.99	0.98	0.98	1.00

TABLE 11.9 Deep feature extraction along with ANN.

Model	Train accuracy	Validation accuracy	Test accuracy	F1 score	Kappa	Recall	Precision
ResNet50	1.00	1.00	0.99	0.99	0.98	1.00	0.98
ResNet101	1.00	0.99	1.00	1.00	1.00	1.00	1.00
MobileNetV2	1.00	0.98	1.00	1.00	1.00	1.00	1.00
MobileNet	1.00	0.98	1.00	1.00	1.00	1.00	1.00
InceptionV3	1.00	0.99	1.00	1.00	1.00	1.00	1.00
InceptionResNetV2	0.99	0.99	0.97	0.96	0.94	0.93	1.00
DenseNet169	1.00	0.98	1.00	1.00	1.00	1.00	1.00
DenseNet121	1.00	1.00	0.98	0.98	0.96	0.98	0.98

The F1 score, AUC, and Cohen's kappa were also the highest in this case. Among the ML classifiers, bagging seemed to work the best followed by KNN, SVM, XGBoost, Random Forest, AdaBoost, and ANN, respectively.

7.3 Discussion

DL techniques for detecting myocarditis, or inflammation of the heart muscles, have great promise for revolutionizing diagnostic treatments. DL, particularly CNNs, has shown promising results in a variety of medical contexts, including myocarditis identification. Its precision and efficiency allow for automatic interpretation of CMR images, assisting medical practitioners in making prompt and accurate diagnoses. Rapid processing of large amounts of data with great accuracy can help with early intervention and better patient outcomes.

A key advantage of employing DL for myocarditis detection is its potential to streamline CMR image interpretation, which is often time-consuming and relies on expertise. Healthcare personnel's workloads can be reduced by automated CMR image screening performed by computer-aided diagnostic tools, allowing them to concentrate on more difficult duties. DL algorithms also excel at identifying minor myocardial inflammation that might be missed by a human interpretation by collecting detailed patterns and nuances within CMR images. DL for myocarditis detection has a bright future ahead of it thanks to ongoing model improvements that incorporate transfer learning, attention processes, and reinforcement learning, which are anticipated to improve performance even more.

To improve the performance of a target task with little data, transfer learning uses pretrained models that have been trained on large datasets from similar activities. Utilizing pretrained models from multiple medical imaging datasets helps improve feature extraction and representation learning in the context of myocarditis identification. When there are few myocarditis-specific datasets available, this strategy is especially useful because transfer learning enables the model to profit from knowledge acquired in other medical fields.

The capacity of transfer learning to manage data scarcity is a significant benefit when using it to diagnose myocarditis. Training appropriate models from scratch might be challenging due to the scarcity of annotated myocarditis datasets. Transfer learning gets around this by improving pretrained models using a more constrained, niche dataset. By utilizing information from related jobs, this method not only saves time and resources but also enhances the generalization of the model.

When utilizing transfer learning techniques to detect myocarditis, there are a few obstacles to overcome. Making the proper pretrained model choice, fine-tuning the pertinent layers, and modifying the model to account for traits unique to myocarditis are crucial choices. While transfer learning techniques are useful, it is essential to check that the knowledge has been applied to the intended activity to prevent biases or errors.

Deep feature extraction techniques allow the automatic extraction of pertinent and discriminative characteristics from complicated medical pictures, which is frequently made possible by pretrained models. These techniques have the potential to detect subtle patterns and structural abnormalities in CMR images that may not be immediately visible to human viewers in the context of myocarditis identification. The ability of deep feature extraction techniques to handle vast and high-dimensional images is one of the key benefits of using them for myocarditis identification. The inadequacy of conventional manual feature extraction techniques to accurately detect complex patterns and variations within images can be a limitation. Deep feature extraction techniques, in contrast, are excellent at gathering both global and localized information, adding to a thorough analysis that helps with accurate detection and diagnosis.

The proposed methods and architecture beat various other DL models. The transfer learning approach presented a test accuracy of 99.6% and an AUC of 0.998. The feature extraction method produced near-perfect results in almost all the pretrained models except for DenseNet121 and InceptionResNetV2. The precision and recall in most of the algorithms produced a value of 1.00, which shows that the proposed model is good at extracting features.

In summary, using DL to diagnose myocarditis has the potential to drastically change the diagnosis. Although there are difficulties, the benefits of accuracy, efficiency, and the possibility of early intervention make it an appealing study direction. DL, cooperation, and data accessibility advancements are all pointing to AI's revolutionary effects on myocarditis detection, ultimately improving patient care and outcomes. Moreover, using transfer learning to identify myocarditis has the potential to advance diagnoses. This method can aid in a precise and quick diagnosis by modifying previously trained models to account for the particular traits of myocarditis. Transfer learning is likely to be essential in advancing myocarditis identification and patient management as it develops as domain-specific adaptations get better.

Furthermore, deep feature extraction methods are used to diagnose myocarditis, which is a significant development in medical picture analysis. These methods provide a sophisticated way to extract and use complex characteristics from CMR images, driven by CNNs and transfer learning methodologies.

8. Conclusion

A substantial advancement in medical imaging diagnostics has been made with the use of a DL approach for the identification of myocarditis. This method has the potential to revolutionize myocarditis detection, making it more precise, effective, and available by utilizing CNN capabilities. The future of myocarditis diagnosis is bright with concentrated efforts from the medical and ML communities, showcasing the potential of artificial intelligence to revolutionize healthcare procedures.

In this chapter, the various DL techniques and feature extraction techniques are discussed. Various experiments are conducted to find the best among the competitor models. It was found that the highest accuracy of 99.8% is achieved by the pretrained model MobileNetV2. The highest AUC, F1 score, and kappa score of 0.997, 0.995, and 0.997, respectively. The results were near perfect for all the pretrained models in the feature extraction method. In conclusion, this chapter serves as a valuable resource for researchers, clinicians, and data scientists interested in exploring the potential of DL techniques to revolutionize myocarditis detection and improve patient outcomes. By harnessing the power of artificial intelligence in medical diagnostics, the vision of enhanced precision medicine becomes ever more attainable.

References

Breiman, L. (1996). Bagging predictors. *Machine Learning, 24*, 123–140.

Breiman, L. (2001). Random forests. *Machine Learning, 45*, 5–32.

Caforio, A. L., Bottaro, S., & Iliceto, S. (2012). Dilated cardiomyopathy (DCM) and myocarditis: Classification, clinical and autoimmune features. *Applied Cardiopulmonary Pathophysiology, 16*(1), 82–95.

Chen, T., & Guestrin, C. (2016). *Xgboost: A scalable tree boosting system.* 785–794.

Fenoglio, J. J., Jr., Ursell, P. C., Kellogg, C. F., Drusin, R. E., & Weiss, M. B. (1983). Diagnosis and classification of myocarditis by endomyocardial biopsy. *New England Journal of Medicine, 308*(1), 12–18.

Freund, Y., & Schapire, R. E. (1997). A decision-theoretic generalization of on-line learning and an application to boosting. *Journal of Computer and System Sciences, 55*(1), 119–139.

Ghareeb, A.-N., Karim, S. A., Jani, V. P., Francis, W., Van den Eynde, J., Alkuwari, M., & Kutty, S. (2022). Patterns of cardiovascular magnetic resonance inflammation in acute myocarditis from South Asia and Middle East. *IJC Heart & Vasculature, 40*, 101029. https://doi.org/10.1016/j.ijcha.2022.101029

He, K., Zhang, X., Ren, S., & Sun, J. (2016). *Deep residual learning for image recognition, 770–778.*

Hearst, M. A., Dumais, S. T., Osuna, E., Platt, J., & Scholkopf, B. (1998). Support vector machines. *IEEE Intelligent Systems and Their Applications, 13*(4), 18–28.

Howard, A. G., Zhu, M., Chen, B., Kalenichenko, D., Wang, W., Weyand, T., Andreetto, M., & Adam, H. (2017). Mobilenets: Efficient convolutional neural networks for mobile vision applications. *ArXiv Preprint ArXiv:1704.04861.*

Huang, G., Liu, Z., Van Der Maaten, L., & Weinberger, K. Q. (2017). *Densely connected convolutional networks, 4700–4708.*

Kingma, D. P., & Ba, J. (2014). Adam: A method for stochastic optimization. *ArXiv Preprint ArXiv:1412.6980.*

Kvålseth, T. O. (1989). Note on Cohen's kappa. *Psychological Reports, 65*(1), 223–226.

LeCun, Y., & Bengio, Y. (1995). Convolutional networks for images, speech, and time series. *The handbook of brain theory and neural networks, 3361*(10), 1995.

Moravvej, S. V., Alizadehsani, R., Khanam, S., Sobhaninia, Z., Shoeibi, A., Khozeimeh, F., Sani, Z. A., Tan, R.-S.,

Khosravi, A., Nahavandi, S., Kadri, N. A., Azizan, M. M., Arunkumar, N., & Acharya, U. R. (2022). RLMD-PA: A reinforcement learning-based myocarditis diagnosis combined with a population-based algorithm for pretraining weights. *Contrast Media and Molecular Imaging, 2022,* 8733632. https://doi.org/10.1155/2022/8733632

Peterson, L. E. (2009). K-nearest neighbor. *Scholarpedia, 4*(2), 1883.

Sagar, S., Liu, P. P., & Cooper, L. T. (2012). Myocarditis. *The Lancet, 379*(9817), 738–747.

Sandler, M., Howard, A., Zhu, M., Zhmoginov, A., & Chen, L.-C. (2018). *Mobilenetv2: Inverted residuals and linear bottlenecks.* 4510–4520.

Sharifrazi, D., Alizadehsani, R., Joloudari, J. H., Shamshirband, S., Hussain, S., Sani, Z. A., Hasanzadeh, F., Shoaibi, A., Dehzangi, A., & Alinejad-Rokny, H. (2020). *CNN-KCL: Automatic myocarditis diagnosis using convolutional neural network combined with k-means clustering.*

Shoeibi, A., Ghassemi, N., Heras, J., Rezaei, M., & Gorriz, J. M. (2022). *Automatic diagnosis of myocarditis in cardiac magnetic images using CycleGAN and deep PreTrained models.* 145–155.

Simonyan, K., & Zisserman, A. (2014). Very deep convolutional networks for large-scale image recognition. *ArXiv Preprint ArXiv:1409.1556.*

Szegedy, C., Ioffe, S., Vanhoucke, V., & Alemi, A. (2017). Inception-v4, inception-resnet and the impact of residual connections on learning. *31*(1).

Szegedy, C., Liu, W., Jia, Y., Sermanet, P., Reed, S., Anguelov, D., Erhan, D., Vanhoucke, V., & Rabinovich, A. (2015). *Going deeper with convolutions.* 1–9.

Szegedy, C., Vanhoucke, V., Ioffe, S., Shlens, J., & Wojna, Z. (2016). *Rethinking the inception architecture for computer vision.* 2818–2826.

Woodruff, J. (1980). Viral myocarditis. A review. *The American Journal of Pathology, 101*(2), 425.

Artificial intelligence for 3D medical image analysis

Abdulhamit Subasi[1,2]

[1]Institute of Biomedicine, Faculty of Medicine, University of Turku, Turku, Finland; [2]Department of Computer Science, College of Engineering, Effat University, Jeddah, Saudi Arabia

1. Introduction

With the incorporation of artificial intelligence (AI) methods in recent years, the area of medical imaging has undergone tremendous upheaval. Among the many AI applications in healthcare, 3D medical image processing has emerged as a key field of study and development. The ability of AI systems to handle and interpret complicated volumetric data has created new opportunities for increasing diagnostic accuracy, treatment planning, and patient outcomes. This chapter explores the implications, problems, and probable future directions of artificial intelligence in 3D medical image processing.

The use of AI in 3D medical image analysis has the potential to completely transform the field of healthcare. Traditionally, radiologists and doctors depended primarily on manual analysis to interpret 3D medical images such as computed tomography (CT), magnetic resonance imaging (MRI), and ultrasound. This procedure is time-consuming, subjective, and sensitive to interobserver variation. Deep learning models, in particular, have emerged as

useful tools for automating and enhancing the image analysis workflow. Convolutional neural networks (CNNs), for example, may learn complicated patterns and features straight from raw imaging data, allowing for precise segmentation, classification, and detection of anomalies.

The accurate segmentation of anatomical components is a critical step in 3D medical image analysis. AI algorithms can learn to segment organs, tissues, and lesions from 3D images using large-scale annotated datasets. This permits quantitative and objective assessments of anatomical features, which can aid in the diagnosis, treatment planning, and monitoring of a variety of medical diseases. In neuroimaging, for example, AI-based segmentation can help with the identification and quantification of brain regions, leading to a better understanding and management of neurological disorders.

Additionally, AI algorithms have demonstrated promising capabilities in the detection and characterization of anomalies in 3D medical images. Deep learning models trained on vast datasets of normal and pathological images can learn to detect minute symptoms of pathology that human observers might miss. This early detection of anomalies can have a major impact on patient outcomes by allowing for timely intervention and therapy. In cancer, for example, AI-based algorithms can help with tumor detection and classification, allowing for early diagnosis and individualized therapy planning.

However, implementing AI in 3D medical image analysis presents its own set of obstacles. One major difficulty is the lack of high-quality, well-annotated datasets for training and validation. The diversity, quantity, and representativeness of the training data substantially influence the performance of AI algorithms. Collaboration between healthcare organizations, research groups, and regulatory agencies is critical for data exchange and the creation of standardized benchmark datasets that can drive innovation and improve transparency in this area.

Another difficulty is that AI models are difficult to interpret and explain. While deep learning algorithms achieve great results, they are frequently regarded as black-box models, making it difficult to grasp the reasoning behind their judgments. In the context of 3D medical image analysis, physicians must have faith in AI-based predictions and be able to evaluate the results. Efforts are being made to build explainable artificial intelligence systems that provide insights into the traits and regions of interest that contribute to the model's output. Explainable AI not only increases trust but also ensures that AI is used ethically and responsibly in healthcare.

As a result, incorporating AI into 3D medical picture analysis has the potential to revolutionize healthcare by enhancing diagnostic accuracy, treatment planning, and patient outcomes. AI can automate and augment the image analysis workflow by leveraging large-scale datasets and strong deep learning algorithms, enabling accurate segmentation, identification, and characterization of abnormalities. However, issues such as data availability, interpretability, and ethics must be addressed to enable the broad and responsible use of AI in the field. The entire potential of AI in 3D medical image processing can be realized through coordinated efforts among researchers, physicians, and regulatory agencies, leading to more precise and tailored healthcare solutions.

Sophisticated three-dimensional (3D) images are commonly generated using diverse scanning methods such as CT, MRI, positron emission tomography (PET), and ultrasound imaging (USI). Extracting meaningful information from these 3D images proves challenging due to concealed components. As a result, processing the 3D image data involves capturing the physical, morphological, and structural characteristics of these hidden elements in an abstract manner. For instance, an MRI is manipulated to uncover hidden visual intricacies. Achieving accurate insights into these concealed aspects necessitates dependable and resilient processing techniques. Lung cancer, a consequence of irregular cell growth in lung tissue often linked to smoking, can be effectively managed through early identification. Screening approaches can be utilized to detect nodules, referred to as white marks on lung tissue visible in X-ray and CT scan images (Raja et al., 2020).

2. Literature review

The discipline of medical image analysis has been transformed by AI, notably in the domain of 3D medical imaging. With the growing availability of high-resolution 3D imaging modalities such as CT, MRI, and PET, there is an increasing demand for advanced AI techniques to aid in the interpretation and analysis of these complex volumetric datasets. Accurate segmentation of anatomical features is critical in a variety of clinical applications. Deep learning-based approaches have demonstrated exceptional performance in automating segmentation. AI systems have demonstrated significant promise in detecting and classifying anomalies in 3D medical pictures. Deep learning models may learn complicated patterns and detect minor anomalies by training on enormous datasets.

Deep learning, a subset of Artificial Intelligence, has achieved remarkable success in the realm of image analysis and computer technology. Its potential can be harnessed to develop enhanced decision support systems in clinical radiology. An area where this can be particularly beneficial is in the detection and segmentation of brain tumor tissues. By employing deep learning and artificial intelligence, radiologists can obtain computer-based second opinions or decision support, aiding in the accurate and timely diagnosis of disease severity and patient survival. Gliomas, which are aggressive brain tumors with irregular shapes and unclear boundaries, pose significant challenges in detection. Their identification often necessitates a comprehensive analysis of diverse radiological scans. Chetty et al. (2022) introduced an entirely automated deep learning technique for segmenting brain tumors in multicontrast magnetic resonance image scans with various modalities. The proposed method employed a lightweight UNET architecture, comprising a computational model based on a multimodal CNN encoder-decoder framework. By leveraging the publicly available Brain Tumor Segmentation (BraTS) Challenge 2018 dataset from the Medical Image Computing and Computer Assisted Intervention (MICCAI) society, their novel approach utilizing the lightweight UNET model achieved improved performance compared to previous models in the challenge. Notably, their method does not require data augmentation or extensive computational resources, making it suitable for remote, resource-constrained healthcare settings.

The performance of deep learning is greatly influenced by the amount of training data available. Models that are pretrained on large datasets like ImageNet have proven to be effective in accelerating training convergence and enhancing accuracy. Similarly, in the domain of 3D medical imaging, models based on extensive datasets play a crucial role in advancing deep learning

techniques. However, constructing a sufficiently large dataset for 3D medical imaging is exceptionally challenging due to the difficulty of data acquisition and annotation. To address this issue, Chen et al. (2019) curated the 3DSeg-8 dataset by aggregating data from various medical challenges. This dataset encompasses diverse modalities, target organs, and pathologies. To extract general medical 3D features, they devised a heterogeneous 3D network called Med3D, which allows for the cotraining of the multidomain 3DSeg-8 dataset, resulting in a series of pretrained models. They applied the pretrained Med3D models to perform lung segmentation in the LIDC dataset, pulmonary nodule classification in the LIDC dataset, and liver segmentation in the LiTS challenge. They demonstrated that Med3D significantly accelerates the training convergence speed of the targeted 3D medical tasks by a factor of two compared to models pretrained on the Kinetics dataset and by a factor of 10 compared to training from scratch. Furthermore, Med3D improves accuracy by a range of 3%−20% across the different tasks. By transferring the Med3D model to the state-of-the-art DenseASPP segmentation network, they achieved a Dice coefficient of 94.6%, which closely approaches the results achieved by the most advanced algorithms in the LiTS challenge when using a single model.

Transfer learning has proven to be a highly effective approach in deep learning for medical image analysis, enabling the application of knowledge learned from natural images to medical images. However, this paradigm requires 3D imaging tasks, such as those involving CT and MRI, to be reformulated and addressed in 2D, leading to the loss of crucial 3D anatomical information and compromising performance. To address this limitation, Zhou et al. (2019) developed a set of models called Generic Autodidactic Models, also known as Models Genesis. These models are noteworthy as they are created from scratch, without manual labeling, and are self-taught through self-supervision.

Additionally, they are generic, serving as source models for generating application-specific target models. They demonstrated that Models Genesis surpasses the performance of learning from scratch in all five targeted 3D applications, encompassing both segmentation and classification tasks. Importantly, while learning a model from scratch in 3D may not necessarily outperform transfer learning from ImageNet in 2D, Models Genesis consistently outperforms any 2D approaches. This includes fine-tuning models pretrained on ImageNet as well as fine-tuning the 2D versions of Models Genesis. These results confirm the significance of 3D anatomical information and the importance of Models Genesis for 3D medical imaging. The exceptional performance of their models can be attributed to their unified self-supervised learning framework, built upon a straightforward yet powerful observation. They recognized that the intricate and recurrent anatomy present in medical images can serve as strong supervision signals, enabling deep models to autonomously learn common anatomical representations through self-supervision.

Currently, the application of deep learning in medical image processing is widespread. Nonetheless, deep neural networks typically require a substantial amount of labeled training data. Unfortunately, medical image segmentation tasks often encounter a scarcity of annotated data due to the resource-intensive and time-consuming process of labeling images. To tackle this challenge, Tang et al. (2019) introduced a new image augmentation technique. This method is founded on a statistical shape model and three-dimensional thin plate splines, enabling the creation of numerous synthetic images from a limited set of actual images. In this strategy, the shape details from labeled actual images are captured using a statistical shape model, which generates a variety of simulated shapes through sampling. These simulated shapes are then imbued with texture using three-dimensional thin plate splines, resulting in the generation of synthetic images.

The amalgamation of these synthetic images with genuine images is subsequently utilized for training deep neural networks. This proposed framework functions as a general data augmentation approach suitable for anatomical structure segmentation tasks using any deep neural network architecture. The effectiveness of this approach was assessed using two distinct datasets: a prostate MRI dataset and a liver CT dataset. Additionally, two diverse deep network structures, specifically multiscale 3D convolutional neural networks (multiscale 3D CNN) and U-net, were employed. Experimental outcomes revealed that the proposed data augmentation method enhances the precision of existing segmentation algorithms reliant on deep neural networks.

Accurate segmentation of medical images plays a crucial role in disease diagnosis and supporting medical decision systems. Alalwan et al. (2021) proposed a highly efficient 3D semantic segmentation deep learning model called "3D-DenseUNet-569" for liver and tumor segmentation. The key aspect of the proposed model is its fully 3D semantic segmentation approach, featuring a considerably deeper network and a reduced number of trainable parameters. To enhance efficiency, the proposed model incorporates depthwise separable convolution (DS-Conv) instead of traditional convolution. This DS-Conv significantly reduces GPU memory requirements and computational costs while maintaining high performance. Additionally, the 3D-DenseUNet-569 model leverages DensNet connections and UNet links, which effectively preserve low-level features and contribute to the production of accurate results. Through an experimental evaluation conducted on the widely used LiTS dataset, the results demonstrate the effectiveness and efficiency of the 3D-DenseNet-569 model compared to existing studies in the field. The model's performance in liver and tumor segmentation tasks showcases its potential for advancing medical image analysis.

With the advancement of deep learning, researchers are increasingly exploring the development of computer-aided diagnosis systems for 3D volumetric medical data. However, acquiring annotations for such 3D medical data is challenging, resulting in a scarcity of annotated images to effectively train deep learning networks. To overcome this limitation, self-supervised learning, which exploits the information within raw data, has emerged as a potential solution to alleviate the dependence on labeled training data. Zhuang et al. (2019) proposed a self-supervised learning framework for volumetric medical images. To pretrain 3D neural networks, the Rubik's cube recovery proxy task is introduced. The two procedures that make up this proxy task are cube rotation and cube rearrangement. The networks are encouraged by these processes to pick up translational and rotational invariant features from the unprocessed 3D data. The accuracy of many tasks, such as classifying brain hemorrhages and segmenting brain tumors, is increased when fine-tuning from a pretrained network as opposed to the conventional method of training networks from scratch. The findings show that, without the need for additional annotated data, the self-supervised learning strategy considerably improves the accuracy of 3D deep learning networks on volumetric medical datasets.

The creation of effective 3D CNNs for medical imaging encounters challenges due to limited available training data and difficulties in data acquisition and annotation. Prior attempts at 3D pretraining have often used self-supervised methods, employing predictive or contrastive learning on unlabeled data to generate invariant 3D representations. However, achieving semantically invariant and discriminative representations through these methods remains problematic due to the lack of comprehensive supervision information. To address this, Zhang et al. (2023) introduced a novel fully supervised 3D network pretraining framework that utilizes semantic supervision from large-scale 2D natural image datasets. This approach employs a redesigned 3D network architecture that uses transformed natural images to overcome the shortage of medical data, resulting in robust 3D representations. Their comprehensive experiments involving five benchmark datasets demonstrate the effectiveness of the proposed pretrained models. These models expedite convergence and enhance accuracy in various 3D medical imaging tasks such as classification, segmentation, and detection. Furthermore, their approach significantly reduces annotation efforts by up to 60% compared to training from scratch. Notably, on the NIH DeepLesion dataset, their method achieved state-of-the-art detection performance, outperforming previous self-supervised and fully supervised pretraining approaches, as well as methods relying on training from scratch.

The popularity of 3D CNNs for evaluating volumetric images is a result of the rapid advancement of deep learning and their efficiency in capturing rich 3D contextual information. Due to the dearth of training data for biomedical applications, the adoption of 3D convolutional kernels can result in a large increase in the number of trainable parameters. In their research, Qu et al. (2020) added a new module called the 3D dense separated convolution (3D-DSC) module to overcome this problem. The tightly connected 1D filters used in the construction of this module serve to replace the original 3D convolutional kernels. The 3D-DSC module decreases the risk of overfitting by reducing redundancy in a systematic manner and creates opportunities for deeper networks by breaking down the 3D kernel into 1D filters. Additionally, the module makes dense connections between the 1D filters and nonlinear layers to increase the network's representational capacity while retaining a small architecture. Experiments using volumetric medical image classification and segmentation tasks, which are known to be difficult in the field of biomedical images, were used to validate the effectiveness of the 3D-DSC module. The outcomes showed enhanced performance made possible by the 3D-

DSC module, highlighting its potential to advance the interpretation of medical images.

3. Artificial intelligence for 3D image classification

3.1 Convolutional neural networks

The CNN is a type of deep learning network widely used for image classification and recognition. CNN comprises an input layer and numerous layers responsible for detecting features. These feature detection layers perform three main actions: convolution, pooling, and Rectified Linear Unit (ReLU), which collectively make up the convolution process (Vasuki & Govindaraju, 2017). CNNs are inspired by biological systems and are applied in computer vision tasks for image classification and object recognition. The convolutional layers are designed to perform a convolution operation, where a filter is employed to transfer activations from one layer to the next. This filter, a weighted three-dimensional structure, has the same depth as the current layer but a smaller spatial area. By taking the dot product of the filter's weights and a spatial region's values in a layer, the hidden state in the subsequent layer is determined. This interaction is repeated across all available locations, generating the next layer with preserved spatial connections from the previous one. As each activation in a layer depends on a small spatial region in the prior layer, CNN connections tend to be sparse. With the exception of the last few levels, the layers maintain their spatial organization. This allows for a tangible visualization of how elements in an image influence activation in a layer. Initial layers capture basic shapes like lines, while higher-level layers identify more complex forms. Consequently, subsequent layers combine these

abstract features to generate meaningful patterns. Additionally, a subsampling layer reduces the size of the layers by a factor of 2 through averaging data in local 2×2 areas. CNNs have proven to be the most efficient among various neural network types. They are widely utilized for tasks like image recognition, object detection, localization, and even language processing (Aggarwal, 2018; Subasi, 2022).

Convolution involves running an image through convolution filters that activate specific features in the image. Pooling reduces data by nonlinear downsampling. ReLU retains positive values but zeros out negative ones. Just before the output layer, there's a classification layer. This fully connected layer has an N-dimensional output, with N being the number of categories for classification. This layer generates an N-dimensional vector, where each element signifies the likelihood of the input image belonging to one of the N classes. The final output layer, employing a SoftMax function, provides the categorized result. Data is processed in each layer, then passed to the next. CNNs are inspired by the visual cortex's biological structure. In CNN, prior layer's subregions are linked to neurons in a layer instead of being fully connected. Only specific subregions impact these neurons, and overlapping subregions allow for spatially related outcomes, setting CNN apart from traditional neural networks (Subasi, 2022; Vasuki & Govindaraju, 2017).

3.2 Implementation of 3D image classification from CT scans[1,2]

This example will demonstrate how to construct a 3D CNN to detect the presence of viral pneumonia in CT scans. 2D CNNs are often used to process RGB (3-channel) images. A 3D CNN is

[1] Adapted from https://keras.io/examples/vision/3D_image_classification/.

[2] https://github.com/hasibzunair/3D-image-classification-tutorial/blob/master/3D_image_classification.ipynb.

essentially the 3D equivalent: it takes as input a 3D volume or a sequence of 2D frames (e.g., slices in a CT scan). 3D CNNs are an effective model for learning volumetric data representations.

3.2.1 *Setup*

```
import os
import zipfile
import numpy as np
import tensorflow as tf
from tensorflow import keras
from tensorflow.keras import layers
```

3.2.2 *Downloading the MosMedData: chest CT scans with COVID-19 related findings*

In this example, we make use of the subset of the MosMedData: Chest CT Scans with COVID-19 Related Findings. This dataset includes lung CT scans both with and without COVID-19-related abnormalities. We will create a classifier to predict the existence of viral pneumonia using the accompanying radiological features from the CT images as labels. Consequently, the task is a problem of binary classification.

```
# Download url of normal CT scans.
url = "https://github.com/hasibzunair/3D-image-classification-
tutorial/releases/download/v0.2/CT-0.zip"
filename = os.path.join(os.getcwd(), "CT-0.zip")
keras.utils.get_file(filename, url)
# Download url of abnormal CT scans.
url = "https://github.com/hasibzunair/3D-image-classification-
tutorial/releases/download/v0.2/CT-23.zip"
filename = os.path.join(os.getcwd(), "CT-23.zip")
keras.utils.get_file(filename, url)
# Make a directory to store the data.
os.makedirs("MosMedData")
# Unzip data in the newly created directory.
with zipfile.ZipFile("CT-0.zip", "r") as z_fp:
    z_fp.extractall("./MosMedData/")
with zipfile.ZipFile("CT-23.zip", "r") as z_fp:
    z_fp.extractall("./MosMedData/")
```

3.2.3 Loading data and preprocessing

The data files are available in Nifti format, denoted by the extension.nii. To interpret the scans, we make use of the nibabel package, which can be installed through the command pip install nibabel. CT scans contain voxel intensity data presented as Hounsfield units (HU), ranging from -1024 to values exceeding 2000 in this dataset. Bones are represented by values above 400, each possessing distinct radiointensity, thus serving as an upper limit. To standardize CT scans, a common normalization range lies between -1000 and 400.

Our data processing pipeline involves the following steps.

- The volumes are initially rotated by 90 degrees to ensure a consistent orientation.
- The HU values are rescaled to fall within the range of $0-1$.
- Resizing is performed for width, height, and depth dimensions.

We establish several utility functions to facilitate data processing, which will be instrumental in constructing training and validation datasets.

```python
import nibabel as nib
from scipy import ndimage
def read_nifti_file(filepath):
    """Read and load volume"""
    # Read file
    scan = nib.load(filepath)
    # Get raw data
    scan = scan.get_fdata()
    return scan
def normalize(volume):
    """Normalize the volume"""
    min = -1000
    max = 400
    volume[volume < min] = min
    volume[volume > max] = max
    volume = (volume - min) / (max - min)
    volume = volume.astype("float32")
    return volume
def resize_volume(img):
    """Resize across z-axis"""
    # Set the desired depth
    desired_depth = 64
    desired_width = 128
    desired_height = 128
    # Get current depth
    current_depth = img.shape[-1]
    current_width = img.shape[0]
    current_height = img.shape[1]
    # Compute depth factor
    depth = current_depth / desired_depth
    width = current_width / desired_width
    height = current_height / desired_height
    depth_factor = 1 / depth
    width_factor = 1 / width
    height_factor = 1 / height
    # Rotate
    img = ndimage.rotate(img, 90, reshape=False)
    # Resize across z-axis
    img = ndimage.zoom(img, (width_factor, height_factor, depth_factor), order=1)
    return img
def process_scan(path):
    """Read and resize volume"""
    # Read scan
    volume = read_nifti_file(path)
```

```
# Normalize
volume = normalize(volume)
# Resize width, height and depth
volume = resize_volume(volume)
return volume
```

3.2.4 Build train and validation datasets

Assign labels after reading the scans from the class directories. Reduce the scans' resolution to 128 × 128 × 64. Rescale the unprocessed HU values to be between 0 and 1. Finally, divide the dataset into subgroups for train and validation.

Let's read the paths of the CT scans from the class directories.

```
# Folder "CT-0" consist of CT scans having normal lung tissue,
# no CT-signs of viral pneumonia.
normal_scan_paths = [
    os.path.join(os.getcwd(), "MosMedData/CT-0", x)
    for x in os.listdir("MosMedData/CT-0")
]
# Folder "CT-23" consist of CT scans having several ground-glass opacifications,
# involvement of lung parenchyma.
abnormal_scan_paths = [
    os.path.join(os.getcwd(), "MosMedData/CT-23", x)
    for x in os.listdir("MosMedData/CT-23")
]
print("CT scans with normal lung tissue: " + str(len(normal_scan_paths)))
print("CT scans with abnormal lung tissue: " + str(len(abnormal_scan_paths)))
```

```
# Read and process the scans.
# Each scan is resized across height, width, and depth and rescaled.
abnormal_scans = np.array([process_scan(path) for path in abnormal_scan_paths
])
normal_scans = np.array([process_scan(path) for path in normal_scan_paths])
# For the CT scans having presence of viral pneumonia
# assign 1, for the normal ones assign 0.
abnormal_labels = np.array([1 for _ in range(len(abnormal_scans))])
normal_labels = np.array([0 for _ in range(len(normal_scans))])
# Split data in the ratio 70-30 for training and validation.
x_train = np.concatenate((abnormal_scans[:70], normal_scans[:70]), axis=0)
y_train = np.concatenate((abnormal_labels[:70], normal_labels[:70]), axis=0)
x_val = np.concatenate((abnormal_scans[70:], normal_scans[70:]), axis=0)
y_val = np.concatenate((abnormal_labels[70:], normal_labels[70:]), axis=0)
print(
    "Number of samples in train and validation are %d and %d."
    % (x_train.shape[0], x_val.shape[0])
)
```

3.2.5 Data augmentation

The rotation of the CT scans at various angles during training is another enhancement. To execute 3D convolutions on the data, which is stored in rank-3 tensors of shape (samples, height, width, and depth), we add a dimension of size 1 at axis 4. Samples, height, breadth, and depth, along with one, make up the new shape. There are many alternative preprocessing and augmentation methods available; this example demonstrates a few straightforward ones.

```python
import random
from scipy import ndimage
@tf.function
def rotate(volume):
    """Rotate the volume by a few degrees"""
    def scipy_rotate(volume):
        # define some rotation angles
        angles = [-20, -10, -5, 5, 10, 20]
        # pick angles at random
        angle = random.choice(angles)
        # rotate volume
        volume = ndimage.rotate(volume, angle, reshape=False)
        volume[volume < 0] = 0
        volume[volume > 1] = 1
        return volume
    augmented_volume = tf.numpy_function(scipy_rotate, [volume]
, tf.float32)
    return augmented_volume
def train_preprocessing(volume, label):
    """Process training data by rotating and adding a channel."
""
    # Rotate volume
    volume = rotate(volume)
    volume = tf.expand_dims(volume, axis=3)
    return volume, label
def validation_preprocessing(volume, label):
    """Process validation data by only adding a channel."""
    volume = tf.expand_dims(volume, axis=3)
    return volume, label
```

The training data is fed through an augmentation function that randomly rotates volume at various angles while defining the train and validation data loaders. Be aware that the data for both training and validation have already been rescaled to have values between 0 and 1.

```python
# Define data loaders.
train_loader = tf.data.Dataset.from_tensor_slices((x_train, y_train))
validation_loader = tf.data.Dataset.from_tensor_slices((x_val, y_val))
batch_size = 2
# Augment the on the fly during training.
train_dataset = (
    train_loader.shuffle(len(x_train))
    .map(train_preprocessing)
    .batch(batch_size)
    .prefetch(2)
)
# Only rescale.
validation_dataset = (
    validation_loader.shuffle(len(x_val))
    .map(validation_preprocessing)
    .batch(batch_size)
    .prefetch(2)
)
```

Visualize an augmented CT scan.

```python
import matplotlib.pyplot as plt
data = train_dataset.take(1)
images, labels = list(data)[0]
images = images.numpy()
image = images[0]
print("Dimension of the CT scan is:", image.shape)
plt.imshow(np.squeeze(image[:, :, 30]), cmap="gray")
```

Since a CT scan has many slices, let's visualize
a montage of the slices.

```python
def plot_slices(num_rows, num_columns, width, height, data):
    """Plot a montage of 20 CT slices"""
    data = np.rot90(np.array(data))
    data = np.transpose(data)
    data = np.reshape(data, (num_rows, num_columns, width, height))
    rows_data, columns_data = data.shape[0], data.shape[1]
    heights = [slc[0].shape[0] for slc in data]
    widths = [slc.shape[1] for slc in data[0]]
    fig_width = 12.0
    fig_height = fig_width * sum(heights) / sum(widths)
    f, axarr = plt.subplots(
        rows_data,
        columns_data,
        figsize=(fig_width, fig_height),
        gridspec_kw={"height_ratios": heights},
    )
    for i in range(rows_data):
        for j in range(columns_data):
            axarr[i, j].imshow(data[i][j], cmap="gray")
            axarr[i, j].axis("off")
    plt.subplots_adjust(wspace=0, hspace=0, left=0, right=1, bottom=0, top=1
)
    plt.show()
# Visualize montage of slices.
# 4 rows and 10 columns for 100 slices of the CT scan.
plot_slices(4, 10, 128, 128, image[:, :, :40])
```

3.2.6 Define a 3D convolutional neural network

To make the model easier to understand, we structure it into blocks. The architecture of the 3D CNN used in this example is from Zunair et al. (2020).

```python
def get_model(width=128, height=128, depth=64):
    """Build a 3D convolutional neural network model."""
    inputs = keras.Input((width, height, depth, 1))
    x = layers.Conv3D(filters=64, kernel_size=3, activation="relu")(inputs)
    x = layers.MaxPool3D(pool_size=2)(x)
    x = layers.BatchNormalization()(x)
    x = layers.Conv3D(filters=64, kernel_size=3, activation="relu")(x)
    x = layers.MaxPool3D(pool_size=2)(x)
    x = layers.BatchNormalization()(x)
    x = layers.Conv3D(filters=128, kernel_size=3, activation="relu")(x)
    x = layers.MaxPool3D(pool_size=2)(x)
    x = layers.BatchNormalization()(x)
    x = layers.Conv3D(filters=256, kernel_size=3, activation="relu")(x)
    x = layers.MaxPool3D(pool_size=2)(x)
    x = layers.BatchNormalization()(x)
    x = layers.GlobalAveragePooling3D()(x)
    x = layers.Dense(units=512, activation="relu")(x)
    x = layers.Dropout(0.3)(x)
    outputs = layers.Dense(units=1, activation="sigmoid")(x)
    # Define the model.
    model = keras.Model(inputs, outputs, name="3dcnn")
    return model
# Build model.
model = get_model(width=128, height=128, depth=64)
model.summary()
```

3.2.7 Train model

```
# Compile model.
initial_learning_rate = 0.0001
lr_schedule = keras.optimizers.schedules.ExponentialDecay(
    initial_learning_rate, decay_steps=100000, decay_rate=0.96, staircase=Tr
ue
)
model.compile(
    loss="binary_crossentropy",
    optimizer=keras.optimizers.Adam(learning_rate=lr_schedule),
    metrics=["acc"],
)
# Define callbacks.
checkpoint_cb = keras.callbacks.ModelCheckpoint(
    "3d_image_classification.h5", save_best_only=True
)
early_stopping_cb = keras.callbacks.EarlyStopping(monitor="val_acc", patienc
e=15)
# Train the model, doing validation at the end of each epoch
epochs = 100
model.fit(
    train_dataset,
    validation_data=validation_dataset,
    epochs=epochs,
    shuffle=True,
    verbose=2,
    callbacks=[checkpoint_cb, early_stopping_cb],
)
```

The fact that we do not specify a random seed and that there are only 200 samples is significant. As a result, you can anticipate a wide range of results. You may access the complete dataset, which comprises more than 1000 CT scans, here.[3] An accuracy of 83% was attained using the entire dataset. In both circumstances, there is a 6%–7% range in the categorization performance.

3.2.8 Visualizing model performance

The accuracy and loss of the model for the training and validation sets are plotted here. Accuracy gives a fair assessment of the model's performance because the validation set is balanced across classes.

[3] https://www.medrxiv.org/content/10.1101/2020.05.20.20100362v1.

```
fig, ax = plt.subplots(1, 2, figsize=(20, 3))
ax = ax.ravel()


for i, metric in enumerate(["acc", "loss"]):
    ax[i].plot(model.history.history[metric])
    ax[i].plot(model.history.history["val_" + metric])
    ax[i].set_title("Model {}".format(metric))
    ax[i].set_xlabel("epochs")
    ax[i].set_ylabel(metric)
    ax[i].legend(["train", "val"])
```

3.2.9 Make predictions on a single CT scan

```
# Load best weights.
model.load_weights("3d_image_classification.h5")
prediction = model.predict(np.expand_dims(x_val[0], axis=0))[0]
scores = [1 - prediction[0], prediction[0]]

class_names = ["normal", "abnormal"]
for score, name in zip(scores, class_names):
    print(
        "This model is %.2f percent confident that CT scan is %s"
        % ((100 * score), name)
    )
```

4. Discussion

In the realm of 3D medical image analysis, AI has emerged as a strong tool, transforming how medical professionals interpret and analyze complex volumetric data. The use of artificial intelligence approaches in this domain has resulted in substantial advances in disease diagnosis, treatment planning, and patient management. This chapter will cover the important contributions of AI in 3D medical image processing, as well as the problems and future directions in this rapidly expanding field.

The automation of image segmentation tasks is one of the key contributions of AI in 3D medical image analysis. Segmentation is critical in extracting anatomical features or lesions from volumetric medical images, which provides vital information for diagnosis and therapy planning. CNNs and deep learning models, for example, have exhibited greater performance in accurately segmenting organs, tumors, and other structures of interest in 3D images. These algorithms take advantage of neural network's ability to understand complicated patterns and spatial correlations inside images, allowing for precise and efficient segmentation.

Disease detection and classification is another important contribution of AI in 3D medical image analysis. To discover patterns and traits associated with certain diseases, AI models have been trained on massive datasets of annotated 3D medical images. Deep learning algorithms, for example, have been effectively used to detect and classify lung lesions in 3D chest CT scans, assisting in the early diagnosis of lung cancer. Similarly, AI-based approaches in detecting brain tumors, breast cancer, and other pathologies in 3D medical imaging have shown promise. Furthermore, artificial intelligence techniques have aided in the development of predictive models for treatment planning and result prediction. AI models can provide insights into disease progression, therapy response, and patient prognosis by analyzing 3D medical images with clinical data.

However, several obstacles must be overcome before AI can be widely adopted and implemented in 3D medical image processing. One of the most significant issues is the lack of high-quality, annotated datasets for training and verifying AI models. Large-scale, diversified datasets are required for the development of strong, generalizable algorithms.

Another challenge is that AI models are difficult to interpret and explain. Healthcare practitioners must comprehend the underlying logic of AI forecasts and have trust in the decision-making process. AI algorithms that are transparent and interpretable, such as attention mechanisms and saliency maps, are being investigated to provide insights into the features and regions of interest that contribute to the model's output. Explainable AI techniques will not only increase clinician trust but will also ensure that AI is used ethically and responsibly in healthcare. In the integration of AI into 3D medical image analysis, ethical considerations are equally critical. Patient data privacy, security, and informed consent are all essential issues that must be addressed carefully. To ensure the proper use of AI technology in healthcare, regulatory frameworks must be implemented, protecting patient rights and upholding ethical standards.

In the future, AI will have enormous promise in 3D medical image analysis. AI algorithm advancements, such as multimodal data integration and the usage of generative models, will improve the accuracy and efficiency of 3D image analysis. Furthermore, the coupling of AI with other developing technologies, such as augmented reality and virtual reality, has the potential to improve surgical planning, image-guided treatments, and medical education.

5. Conclusion

AI has emerged as a strong tool in the field of 3D medical image processing, altering how medical professionals understand and use complex medical imaging data. This chapter discusses the use of AI approaches in 3D medical image processing, emphasizing its potential to improve diagnosis accuracy, treatment planning, and patient outcomes.

AI techniques, such as CNNs and deep learning models, have made tremendous progress in automating and expediting the examination of 3D medical images. These AI algorithms can extract subtle features and patterns from volumetric data, allowing for exact segmentation, classification, and abnormality detection in medical images. AI-based 3D medical image analysis supports reliable and reproducible outcomes across diverse medical practitioners and healthcare settings by decreasing the subjectivity and unpredictability associated with manual interpretation.

Furthermore, combining AI with 3D medical image analysis has yielded promising results in a variety of medical domains, including radiology, cancer, cardiology, and neurology. Radiologists can use AI algorithms to help them identify and characterize lesions, tumors, and anatomical features, resulting in more accurate diagnosis and treatment planning. AI-based image analysis in cancer treatment enables tumor segmentation, tracking treatment response, and forecasting patient outcomes. Furthermore, AI techniques make it easier to monitor cardiac function, brain disorders, and other complicated anatomical components, allowing for earlier identification and intervention.

In conclusion, the incorporation of AI approaches with 3D medical image processing has resulted in substantial advances in the field of healthcare. AI has the potential to improve diagnostic accuracy, treatment planning, and patient outcomes by automating and improving the interpretation of complicated 3D medical images. However, further research, collaboration, and standardization are required to solve data availability, interpretability, and clinical integration problems. AI-based 3D medical image analysis will play a transformative role in determining the future of healthcare with sustained innovation and responsible application.

References

Aggarwal, C. C. (2018). *Neural networks and deep learning.* Springer.

Alalwan, N., Abozeid, A., ElHabshy, A. A., & Alzahrani, A. (2021). Efficient 3D deep learning model for medical image semantic segmentation. *Alexandria Engineering Journal, 60*(1), 1231–1239.

Chen, S., Ma, K., & Zheng, Y. (2019). *Med3d: Transfer learning for 3d medical image analysis.* ArXiv Preprint ArXiv: 1904.00625.

Chetty, G., Yamin, M., & White, M. (2022). A low resource 3D U-Net based deep learning model for medical image analysis. *International Journal of Information Technology, 14*(1), 95–103.

Qu, L., Wu, C., & Zou, L. (2020). 3D dense separated convolution module for volumetric medical image analysis. *Applied Sciences, 10*(2), 485.

Raja, R., Kumar, S., Rani, S., & Laxmi, K. R. (2020). Lung segmentation and nodule detection in 3D medical images using convolution neural network. In *Artificial intelligence and machine learning in 2D/3D medical image processing* (pp. 179–188). CRC Press.

Subasi, A. (2022). *Applications of artificial intelligence in medical imaging.* Elsevier.

Tang, Z., Chen, K., Pan, M., Wang, M., & Song, Z. (2019). An augmentation strategy for medical image processing based on statistical shape model and 3D thin plate spline for deep learning. *IEEE Access, 7*, 133111–133121.

Vasuki, A., & Govindaraju, S. (2017). Deep neural networks for image classification. In *Deep learning for image processing applications, 31* p. 27). IOS Press.

Zhang, S., Li, Z., Zhou, H.-Y., Ma, J., & Yu, Y. (2023). Advancing 3D medical image analysis with variable dimension transform based supervised 3D pre-training. *Neurocomputing, 529*, 11–22.

Zhou, Z., Sodha, V., Rahman Siddiquee, M. M., Feng, R., Tajbakhsh, N., Gotway, M. B., & Liang, J. (2019). *Models genesis: Generic autodidactic models for 3d medical image analysis* (pp. 384–393).

Zhuang, X., Li, Y., Hu, Y., Ma, K., Yang, Y., & Zheng, Y. (2019). *Self-supervised feature learning for 3d medical images by playing a rubik's cube* (pp. 420–428).

Zunair, H., Rahman, A., Mohammed, N., & Cohen, J. P. (2020). *Uniformizing techniques to process CT scans with 3D CNNs for tuberculosis prediction* (pp. 156–168).

Medical image segmentation using artificial intelligence

Abdulhamit Subasi[1,2]

[1]Institute of Biomedicine, Faculty of Medicine, University of Turku, Turku, Finland; [2]Department of Computer Science, College of Engineering, Effat University, Jeddah, Saudi Arabia

1. Introduction

Human eyes have the capacity to transmit visual information to the brain, enabling us to recognize and comprehend a variety of objects. The human vision may fall short, nevertheless, when it comes to seeing objects inside the human body or really minute and unclear entities. This needs the application of specific techniques capable of image processing, analysis, and segmentation. To execute image segmentation, which is essential for object recognition and computer vision, the image must frequently first be denoised. Furthermore, processes are needed to improve the final image following the initial image segmentation. For instance, to lessen the influence of any residual noise, smoothing techniques like the Gaussian filter are used. In some cases, fragmented sections of the image are joined using region-merging and region-splitting procedures. The image segmentation algorithm, which is the essential element of the image segmentation system, determines how well images are segmented.

Medical diagnosis, treatment strategy, and biomedical research all heavily rely on biomedical image segmentation. It entails defining and identification of particular areas or objects of interest, such as organs, tumors, blood arteries, and cells, inside medical images. For the extraction of quantitative measurements, comprehension of disease patterns, and support of clinical decision-making, accurate and effective segmentation is crucial.

Artificial intelligence (AI) techniques, particularly deep learning-based algorithms, have transformed biomedical image segmentation in recent years. These powerful AI algorithms have excelled in analyzing complicated and heterogeneous medical image datasets. They can learn discriminative features automatically from large-scale training datasets and generalize well to new, previously unseen images. As a result, when compared to traditional manual or semiautomatic procedures, AI-based algorithms have greatly improved the accuracy, speed, and reliability of biomedical image segmentation.

One of the primary benefits of AI-based image segmentation is its capacity to handle images from many imaging modalities, such as magnetic resonance imaging (MRI), computed tomography (CT), positron emission tomography (PET), and microscopy. These modalities provide useful information about various aspects of human anatomy, physiology, and pathology. Healthcare practitioners and researchers may extract accurate and consistent information from multimodal images using AI algorithms, resulting in improved diagnostic accuracy and individualized treatment strategies.

We will look at recent improvements in biomedical image segmentation using artificial intelligence in this chapter. We go over the fundamentals of prominent AI approaches like convolutional neural networks (CNNs), and TransResUNet. Moreover, we implement a fully convolutional encoder–decoder model for lung segmentation named TransResUNet.

2. Background and literature review

The process of segmentation involves dividing an image into segments or groups based on shared characteristics, which may include color, intensity, or texture (Sharma & Aggarwal, 2010). These shared values are identified to define boundaries, locate objects, and differentiate between various elements. This procedure is typically accomplished through methods like region-based, boundary-based, or edge-based approaches. It is important to note that the initial outcome of segmentation may not always be the final one, often necessitating resegmentation due to potential overlapping, juxtaposition, or fragmentation of identified areas. Techniques employed for region segmentation usually differ from those used for the initial image segmentation and often cannot be directly applied to an unsegmented image.

Segmentation finds utility in diverse applications spanning fields from medicine to construction. For example, it plays a role in separating characters in computer engineering, identifying blood cells in microscopy, spotting defects or tumors in medical imaging, distinguishing between organs such as the heart, lungs, and ribs, and dividing satellite images into land-use categories in aerospace engineering (Rosenfeld, 1984). The choice of segmentation technique depends on the specific application and the nature of the image under analysis.

The exploration of image segmentation has been an area of research and experimentation for numerous decades, leading to the emergence of novel techniques over time. Nevertheless, a universal method that can be applied to all types of images does not exist. The underlying reason is straightforward: various methods fail to uphold output quality across different input scenarios, thus demonstrating specialization rather than universality. This intricacy remains a primary challenge in the domain of image processing and analysis.

Edge detection algorithms are commonly employed to identify abrupt changes within grayscale images. These alterations, referred to as discontinuities, manifest as sudden shifts in pixel intensity that demarcate object boundaries within a scene. An edge can be visualized as a group of connected pixels delineating the boundary between areas featuring significant differences, such as variations in gray levels, color distinctions, or texture changes. Such discrepancies in pixel concentration can be leveraged for image segmentation. Nonetheless, in images damaged by noise, the task of edge detection becomes intricate since both edges and noise comprise high-frequency components. Endeavors to mitigate noise often led to edges appearing blurred and distorted.

According to Muthukrishnan and Radha (2011), in images with a lot of noise, there are problems with edge identification, such as false edges being formed, it being difficult to detect the edges, some true edges disappearing, and there being a delay due to the time it takes to calculate the noise, among other things. Clearly, achieving a balance between the quantity of noise read and the precision of edge recognition is the key to successful segmentation. In brief, this method attempts to segment images by detecting edge pixels between different regions with quick intensity changes, which are then removed and linked to form closed object edges, yielding a binary image. The two basic edge-based segmentation approaches are the gray histogram and the gradient-based approach (Kang et al., 2009).

The extraction of local features, like edges and curves, which are defined by patterns of gray levels in specific image regions, is often a crucial task. Edge detection is frequently employed to segment images, especially when the pixel intensities exhibit considerable diversity, making binary outcomes challenging to attain. In numerous cases, even with such diverse pixel intensities, a discernible histogram can still be constructed, enabling differentiation between distinct regions. The prevalent approach for edge detection generally involves applying a local difference operator to individual pixels, resulting in significant magnitudes in the output where gray-level transitions are frequent, such as at edges. Typically, these operators calculate gray-level differences in two perpendicular directions, then determine the maximum absolute value to estimate magnitude, and utilize the arctangent to the quotient to estimate edge direction. It is important to note that while the differences are computed through convolution operations at each pixel, their combination to generate magnitude and direction involves a nonlinear process.

This form of identification is termed the gradient method, defining the gradient as the initial derivative of an image (Al-amri et al., 2010). According to Kang et al. (2009), this technique is effective when there is a sharp intensity shift at an edge coupled with minimal image noise. However, it is important to exercise caution as this approach can introduce

additional or omit edges in intricate and chaotic images. A variety of edge detection methods exist, each tailored to identify specific edge types. Some prominent techniques, as outlined by Kang et al. (2009), include the Sobel operator, Canny operator, Laplace operator, and Laplacian of Gaussian operator, among others. The Canny operator appears to outperform others, although it does entail a longer computation time compared to the Sobel operator. As mentioned by Gonzalez et al. (2009), Canny introduced three criteria for edge detection: attaining optimal detection outcomes, precise edge location, and minimal repetitive response. To fulfill these criteria, he devised an optimal linear filter that constitutes the first derivative of the Gaussian function. The subsequent sections will provide a brief examination of some of the prevalent operators employed in the gradient technique.

Threshold segmentation represents the most fundamental segmentation technique. It relies on the thresholding method, which uses a range of values applied to the pixel intensities in the image to distinguish regions. The outcome of this approach is a binary image, wherein pixel intensities determine their classification into either of the two groups. For instance, if a pixel's intensity falls below or outside the designated threshold range, it becomes a black pixel. Conversely, if the pixel's intensity lies within the threshold range, it becomes a white pixel. However, this algorithm introduces multiple challenges, with the primary concern being the subjective nature of threshold selection. There exists no gray area here; the pixel either falls within the range or it does not. This segmentation method can be categorized as either global thresholding or local thresholding. Additionally, Burnham et al. (1997) have included dynamic thresholding as a subcategory. Global thresholding involves applying a single threshold to the entire image under analysis, and it is particularly effective when a notable contrast exists between background and object pixel intensities. In this form of thresholding, the pixel attributes and grayscale levels solely determine the threshold value to be employed.

Local thresholding entails the application of distinct threshold values to specific image regions, and these values hinge on the pixel intensities within the vicinity. The procedure unfolds as follows: the image is divided into sections or regions, and a suitable threshold value is selected for each of these segments. Kang et al. (2009) mention that following the threshold application, it becomes necessary to employ a filter to equalize noncontinuous gray levels. Within this classification, the widely recognized techniques include those rooted in simple statistics, 2D entropy-based principles, and histogram transformation. In essence, local thresholding employs diverse threshold values for individual blocks within the image, whereas global thresholding utilizes a singular threshold value encompassing the entire image. Dynamic thresholding is simply the thresholding value that is applied to distinct items in a picture. It is called dynamic because the type of thresholding used is determined by the object to be segmented in the image, whether local or global thresholding. Kang et al. (2009) identify image thresholding, Watershed, and interpolatory thresholding as the primary dynamic thresholding techniques.

In comparison to edge-based segmentation, region-based segmentation is more intuitive but is also more susceptible to noise (Sarma & Gupta, 2021). These two segmentation approaches can be seen as contrasting methods. While edge detection seeks pronounced shifts in intensity, region detection searches for patterns encompassing color, texture, and intensities. This nuanced distinction contributes to longer computation times. Within this classification, two primary techniques emerge: region growing and splitting and merging regions. Region growing, as the name suggests, involves creating larger regions or clusters of pixels based on predefined criteria. The process involves establishing fundamental points in the original

image and then forming regions by attaching surrounding pixels with shared attributes, such as gray level or color, to each group of points. This territorial expansion adheres to a logical sequence: (a) selecting suitable foundational points to initiate the program; (b) defining the criteria for considering points as similar; and (c) determining a termination rule for program execution (Kang et al., 2009).

Region splitting and merging, on the other hand, divides an image into noninterconnected areas and then unites regions that fit the set requirements, or if necessary, separates them to meet those criteria. This strategy, in particular, is based on the image's uniformity. To begin, the given image is divided into four pieces, each of which is checked for homogeneity. Nonhomogeneous sections are broken into four portions again and assessed for homogeneity. The divisions will continue until the region is totally homogeneous, hence the name. The unification of homogenous components is referred to as region merging. Uncertainty, according to Bovik (2005), is a crucial component that contributes to poor results for fixed algorithms in image segmentation, analysis, recognition, and other phases of image processing. Furthermore, the results of earlier processing influence the performance of subsequent processing, demanding some degree of flexibility in image processing algorithms. According to Langote and Chaudhari (2012), clustering can be done using Fuzzy Set Theory, which allows for fuzzy borders between clustering. The key downside of this technique, they note, is that it is difficult to confirm the attributes of fuzzy members.

The process of segmentation using neural networks significantly deviates from traditional segmentation algorithms. Back in 1998, Qiaoping (1998) elucidated the procedure for applying neural network segmentation, outlining a series of steps: initially, an image is mapped onto a neural network, where each neuron corresponds to a pixel. Subsequently, employing dynamic equations, edge extraction takes place as the state of each neuron is steered toward the neural network's minimum energy state. Neural network-based segmentation is characterized by three key attributes: it boasts high levels of parallelism and rapid computational capacity, rendering it suitable for real-time applications; its unlimited nonlinear capability and strong interconnection among processing units allow modeling for diverse processes; and it displays robustness, rendering it resilient to noise interference. Nonetheless, employing neural networks for segmentation carries certain drawbacks. It necessitates familiarity with specific segmentation data types beforehand and is susceptible to image segmentation outcomes being influenced by initialization. Additionally, neural networks require pretraining through the learning process, which can be time-consuming and must be guarded against overtraining.

Traditional machine learning methods, such as clustering, thresholding, and region-based algorithms, were the focus of early attempts. While these approaches were successful in some situations, they frequently struggled with the complexity and variety of medical imaging data. The relative clarity, precision, specificity, and sensitivities of 16 global and 9 local thresholding approaches were estimated using the FIJI BioVoxxel tool plugin. For the first time, researchers analyzed, quantified, and confirmed the efficiency of various segmentation algorithms using z-stack images of ex vivo glia. Threshold algorithms differ greatly in terms of quality, specificity, accuracy, and sensitivity, with entropy-based thresholds scoring the highest for fluorescent staining (Healy, 2018). Bias is common during superpixel segmentation of MRI images due to nongray concentration and fuzzy edges. Researchers suggest a novel strategy by merging textural properties and improving basic linear repeating grouping (Wang et al., 2020).

Clustering is a critical issue in MRI brain imaging because it impacts the accuracy of disease detection, diagnosis, and treatment efficacy. Clustering techniques are used in the viewing and interpretation of brain images for various tasks (Mirzaei, 2018). Ji et al. (2014) introduced the MSFCM methodology for segmenting the brain MRI image, which combines the superpixel method with the FCM algorithm. The image was initially separated into many superpixels, and deep segmentation would focus on areas with the largest gray variance. It effectively mitigates the effects of noise and bias by increasing the granularity of grouping. In diffuse lower-grade gliomas with isocitrate dehydrogenase mutation and tumor aggressiveness predicted by diffusion- and perfusion-weighted MRI radiomics, a multiparametric MRI radiomics model with isocitrate dehydrogenase mutation increased diagnostic accuracy (Kim et al., 2020). The use of a sliding-window approach and symmetry analysis on 3D FSPGR brain MR data from ten patients resulted in automated brain tumor segmentation. The method was tested on 575 tumor-infected brain slices (Yanyun & Zhijian, 2014).

Biomedical image segmentation is a key task in medical imaging that is critical for assessing and comprehending complicated anatomical structures and pathological regions contained within medical images. Researchers and clinicians have observed a transformational shift in the field of biomedical image analysis as a result of significant breakthroughs in AI and deep learning. AI-based algorithms, particularly CNNs, have shown exceptional potential in effectively segmenting regions of interest in medical images, providing vital insights for diagnosis, therapy planning, and disease monitoring.

The introduction of AI and deep learning, particularly CNNs, resulted in a considerable improvement in segmentation performance. CNN architectures such as U-Net, SegNet, and DeepLab were quickly adopted by researchers to produce cutting-edge results in biomedical image segmentation. CNNs' capacity to learn hierarchical features and contextual information from massive datasets drastically transformed medical image analysis. Despite significant advances, problems in AI-driven biomedical image segmentation remain. One key barrier is the scarcity of annotated medical data, which impedes deep learning model training. To address data restrictions, researchers have investigated several methodologies such as transfer learning and data augmentation. Furthermore, the interpretability of AI models is becoming increasingly important in the medical field. To provide an understanding of model judgments and develop confidence with physicians, explainable AI methods such as attention processes and Grad-CAM have been presented. Furthermore, establishing the robustness and generalization of AI models across a wide range of patient groups and imaging modalities is an ongoing research topic.

The literature demonstrates the numerous uses of AI-driven image segmentation in different medical disciplines. CNNs have been used in radiology to segment organs, detect anomalies, and aid in disease classification. AI-based approaches for segmenting cellular components and recognizing malignant areas in histopathology slides have assisted pathologists. AI models in cardiology have enabled accurate segmentation of cardiac structures and improved heart disease diagnosis. AI has also made important advances in oncology, allowing for tumor segmentation and tracking therapy responses in radiographic images.

Akkus (2017) introduced deep learning for image segmentation. To begin, researchers examine current deep learning architectures for segmenting anatomical brain areas and disorders. The outcomes, speed, and characteristics of deep learning algorithms are then shown and discussed. Finally, the writer provides a critical evaluation of the current situation as well as projections for future events and trends. Deep learning has already been used extensively for brain image segmentation, with promising results. Lee (2020) provided a patch-wise U-net for automated brain segmentation in MRI. The proposed method divides MRI scan slices into nonoverlapping patches before feeding them into the network. The detection and classification of a brain tumor using modern imaging tools is a significant challenge. Bahadure et al. (2018) investigated the comparison method of different segmentation strategies and identified the best one by comparing their segmentation scores. The study achieved 91% accuracy, 91% specificity, 92% sensitivity, and a 94% dice similarity index coefficient.

TransResUNet is a new fully convolutional encoder–decoder model for lung segmentation. Reza et al. (2020) created this architecture by improving on the cutting-edge U-Net paradigm. As part of the suggested architecture's enhancements to the classical U-Net, they included a pretrained encoder, a specific skip connection, and a postprocessing module. As a consequence, the suggested model outperformed the baseline U-Net model in terms of accuracy and dice coefficient by 97.6% versus 94.9%. The proposed TransResUNet accomplished this feat while using approximately 24% fewer parameters than the basic U-Net.

Colorectal cancer (CRC) is a significant global health concern, necessitating timely screening for early detection. Colonoscopy, the primary diagnostic method, has a notable miss rate for identifying polyps and adenomas. Detecting polyps at a precancerous stage can reduce mortality and economic burden. Deep learning-based computer-aided diagnosis (CADx) systems offer potential assistance by improving polyp detection and being cost-effective for long-term colorectal cancer prevention. Tomar et al. (2022) introduced a novel deep learning architecture, Transformer ResU-Net (TransResU-Net), for automatic polyp segmentation. Built on ResNet-50 and incorporating transformer self-attention and dilated convolutions, TransResUNet demonstrates promising results on public polyp segmentation datasets. The model's high dice score and real-time speed suggest its suitability as a benchmark for constructing real-time polyp detection systems, contributing to early diagnosis, treatment, and prevention of colorectal cancer.

Chest X-ray (CXR) is extensively utilized for detecting and diagnosing lung diseases in children. Precisely segmenting the lung fields in digital CXR images is crucial for many computer-aided diagnosis systems. Chen et al. (2023) introduced a deep learning-based approach to enhance the accuracy and quality of lung segmentation in multicenter CXR images of children. The innovative aspect of this method lies in amalgamating the strengths of TransUNet and ResUNet. TransUNet incorporates a self-attention module to enhance feature learning, while ResUNet mitigates network degradation concerns. When applied to a test dataset comprising diverse center data, the proposed model achieves an impressive dice score of 0.9822.

As a conclusion, we can say that AI integration in biomedical image segmentation has a significant therapeutic impact. Improved diagnosis accuracy, lower interobserver variability, and better treatment planning have resulted from automated and precise segmentation. Furthermore, AI-driven segmentation has accelerated clinical operations, allowing physicians to focus more on patient care. AI-based biomedical image segmentation has a bright future. To make AI models more robust and useful in real-world clinical settings, ongoing research intends to solve existing obstacles such as data scarcity, interpretability, and generalization.

3. Deep learning methods for biomedical image segmentation

Deep learning systems, particularly CNNs, have gained attention in recent years as a result of their efficiency in different object recognition and medical image segmentation tasks (Krizhevsky et al., 2012). Based on prior developments, it is vital to stress that a single deep learning network can only incorporate images of many types and sources of information in the same way that people are prepared for a clinical or screening evaluation. Such complex and puzzling tasks can eventually be undertaken by a plethora of organizations, each specializing in a specific type of information index. Another option is to take action that requires using data from one company to achieve another activity.

The most advanced machine learning technology is deep learning. Deep learning surpasses standard methods for tumor diagnosis and tracking, according to preliminary studies. Despite the excitement surrounding this new generation of computer vision, there are significant barriers to developing and deploying CAD and AI technologies in medical care. Deep learning techniques are used to extract indicative morphometric features for better pathology identification. Researchers explore the limitations and possibilities of such technologies to improve cancer detection during diagnostic imaging. The expanding network neural net, in particular, has the potential to improve the performance and understandability of deep learning (He, 2022).

CNNs are capable of determining where the peaks of an area image are represented only when they are based on huge information indices and are adequate. CNNs are cutting-edge methodologies for segmenting brain tumors that produce cutting-edge outcomes. As inputs, image patches are used. The last layer is a softer max layer with six filters, one for each type of tissue to be categorized, allowing the results to be read as probabilities. CNNs possess a distinctive ability to establish intricate connections between information sources and outcomes that manual approximations struggle to capture. A seamless deep learning system enables self-contained computations. This facet is particularly crucial, as many existing image datasets lack a clear-cut output from the outset. In certain scenarios, CNNs can grasp new perspectives of images, similar to comprehending mathematics without reliance on a calculator. Even when applied to a limited dataset, CNNs can enhance calculation performance. An intriguing aspect of deep learning lies in its propensity to introduce unexpected inclusions, whether pertaining to collections of image data or otherwise. Addressing this challenge, ongoing research is dedicated to invigorating neural structures to externally manifest the original attributes or features of an image (Ciresan et al., 2012).

Another novel technique makes use of two distinct CNN architectures that are stacked. To determine the label of the central pixel, patches of 33 × 33 pixels for the local pathway and patches of 65 × 65 pixels for the global route are taken from each distinct MRI modality and centered in the same spot on the image. To eliminate class imbalances, two-phase training is used in conjunction with this innovative architectural method (Havaei et al., 2016). Furthermore, Maxout nonlinearity is used as a postprocessing step, as is the linked components technique. BRATS dice scores are 88% for the whole tumor region, 79% for the core tumor region, and 73% for the primary tumor location. It has also been suggested to employ two approaches with a single CNN. One of the most recent CNN techniques (Işın et al., 2016) investigated how successfully deeper CNN structures segmented brain tumors. Convolutional layers are composed of small 3 × 3 filters. This allows for the addition of more convolutional filters to the scheme while lowering the functional impression of larger filters.

Transfer learning, a popular machine learning technique, has received a lot of interest in the field of biomedical image segmentation. This method uses pretrained models on large and diverse datasets to improve segmentation performance on smaller, domain-specific datasets. Delineating and identifying regions of interest within medical pictures such as MRI scans, X-rays, and microscope images is what biomedical image segmentation is all about. Transfer learning has evolved as a viable technique to address these issues due to the scarcity of labeled data and the complexity of biomedical images. In transfer learning, a model is customized on a target domain made up of biomedical images after being pretrained on a source domain containing lots of data, such as natural images from the ImageNet dataset. The model is able to learn appropriate features for the target biomedical segmentation task with little labeled data, which takes advantage of the information learned from the source domain's feature extraction and pattern recognition skills. Pretrained models have the advantage of being able to recognize common elements like edges, textures, and forms, which are also necessary for biological images.

For biomedical image segmentation, several transfer learning approaches have been adapted. As a feature extractor, one typical way is to use the encoder of a pretrained CNN such as VGG, ResNet, or U-Net. This encoder collects hierarchical characteristics from input images, including low-level details as well as high-level abstract patterns. Using this encoder as a foundation, researchers can tune the subsequent layers to accomplish biomedical image segmentation tasks. This transfer learning method speeds up model convergence and improves segmentation accuracy, especially when the target dataset is small.

Furthermore, domain adaptation techniques are frequently used in transfer learning approaches to bridge the gap between the source and target domains. Domain adaptation approaches try to align the feature distributions of source and target data to mitigate domain shift-related issues. Techniques such as adversarial training, in which a domain discriminator leads feature learning, have been used to increase model generalization across different datasets in biomedical image segmentation tasks.

To summarize, transfer learning has evolved into an effective approach for overcoming the obstacles of biomedical image segmentation. Transfer learning approaches enable the building of accurate and stable segmentation models even in circumstances with minimal labeled data by relying on information gained from large-scale datasets. As research in this area progresses, the combination of transfer learning and domain adaptation shows promise for opening up new avenues in biomedical image processing and therapeutic applications.

4. Medical image segmentation with TransResUNet

By merging the strength of Transformers and U-Net, TransUNet is a cutting-edge deep learning architecture that revolutionizes medical image segmentation. The limits of conventional approaches in managing complicated medical images with variable structures and features are intended to be overcome by this novel fusion. The inclusion of Transformers, which have demonstrated remarkable performance in a variety of natural language processing tasks, is denoted by the "Trans" in the name TransUNet. The model's capacity to comprehend complex patterns within medical images is improved by the design, which makes use of the self-attention mechanism of Transformers to capture global contextual information while keeping local details.

The versatility of TransUNet in terms of image sizes is one of its defining features. TransUNet can handle images of various dimensions without image resizing or cropping, in contrast to standard approaches, making it extremely flexible for medical datasets that frequently include a range of image sizes. Due to its versatility, the architecture can continue to function effectively regardless of the size or quality of the input images. Utilizing the hierarchical feature extraction capabilities of U-Net, a well-liked architecture for semantic segmentation, TransUNet's localization accuracy is improved.

Together, Transformers' understanding of the global context and U-Net's accurate local information extraction produce a potent synergy.

Organ segmentation, lesion detection, and disease diagnosis are just a few of the medical imaging applications for which the TransUNet architecture has demonstrated exceptional performance. It presents a promising approach for improving accuracy and efficiency in medical image analysis due to its capability to efficiently collect both macroscopic and microscopic information inside medical pictures as well as its flexibility in handling various image dimensions. TransUNet stands out as a key development in the medical industry's continued use of cutting-edge technologies, opening the door for more precise and trustworthy diagnoses and treatments.

The U-Net model's architecture (Ronneberger et al., 2015) is composed of two core components: the encoder and the decoder. The encoder is responsible for extracting the spatial features from input images, while the decoder assembles the segmentation mapping utilizing the encoded features. The structure of the encoder module closely resembles that of a conventional convolutional network architecture. In this setup, the input image traverses through multiple blocks, each comprising two 3×3 convolution layers with ReLU activation and subsequent max-pooling. After each max-pooling step, the feature channels increase twofold, while the spatial dimensions are halved compared to their previous state. On the other hand, the decoder module works to up-sample the feature map, concatenating it with the corresponding decoder feature map, and then applying two 3×3 convolution operations with ReLU activation. During each up-sampling iteration, the decoder reduces the number of feature channels by half, while simultaneously doubling their spatial dimensions. This process preserves the original dimensions of the input image, allowing the network to generate the anticipated segmentation mask. The TransResUNet model was developed by modifying the classical U-Net architecture for lung segmentation. It incorporates a pretrained encoder from the VGG-16 model, employing only its initial seven layers to ensure compactness. The encoder includes three blocks, each comprising multiple convolutional layers and max-pooling. The U-Net's skip connections are optimized using a series of convolution blocks, like the Res path, to bridge the semantic gap between encoder and decoder features (Reza et al., 2020).

The decoder module is composed of three decoder blocks, each involving up-sampling and convolution layers with ReLU activation. Nearest neighbor up-sampling is used to mitigate artifacts at the predicted output's edges. The decoder's final convolution layer's output is concatenated with the corresponding encoded features through a residual connection. To address defects in predicted masks, a postprocessing step is introduced. It begins with a flood-fill algorithm to fill undesired holes in the mask. The algorithm starts at the top-left corner pixel of the mask, changing black pixels to white. Further refinements involve the use of Res path blocks in the TransResUNet model and a multistep postprocessing procedure to enhance the predicted mask quality, addressing issues like holes, overlaps, and extraneous objects. Second, some unwanted things can remain around the lung masks. To remove them, we identified all of the related components in the mask and deleted all components having an area less than a predetermined threshold value. Finally, we used an elliptical structuring element to execute a morphological opening on the predicated mask. Opening detaches the left and right masks on an overlapped mask sample (Reza et al., 2020).

4.1 Dataset

The Montgomery Country (MC) dataset (Candemir et al., 2013) is utilized for experimental purposes. The National Library of Medicine (NLM) and the Montgomery County, Maryland, Department of Health and Human Services work together to manage this dataset. It contains 138 posterior–anterior chest X-ray images, 80 of which are normal and 58 of which are pathological. The manual lung masks that correlate to these photos are also available, and they were annotated with the help of qualified radiologists. For our needs, we downsized all of the images to 512×512 pixels in size.

4.2 Code[1]

```
import os
import re
import numpy as np
import pandas as pd
import matplotlib.pyplot as plt

import cv2
from tqdm import tqdm
from glob import glob
from PIL import Image
from skimage.transform import resize
from sklearn.model_selection import train_test_split, KFold

import tensorflow as tf
import tensorflow.keras
from tensorflow.keras import backend as K
from tensorflow.keras.preprocessing.image import ImageDataGenerator
from tensorflow.keras.callbacks import EarlyStopping, ModelCheckpoint

K.set_image_data_format('channels_last')
```

Building the training dataset.
Let's look at the train image list

```
path = "../input/ultrasound-nerve-segmentation/train/"
file_list = os.listdir(path)
file_list[:20]
```

[1] GitHub — sakibreza/TransResUNet: Fully Convolutional Network for Lungs Segmentation.

Sort the file list in ascending order and separate it into images and masks

Each file has the form of either "subject_imageNum.tif" or "subject_imageNum_mask.tif", so we can extract subject and imageNum from each file name by using regular expression. "[0-9]+" means to find the first consecutive number.

Now, I try to load all image files and store them as variables X and y. After doing this, I recognize that it takes very much memory.

Please let me know if there are several efficient ways to store image files.

Let's modularize this work.

```python
train_image = []
train_mask = glob(path + '*_mask*')
for i in train_mask:
    train_image.append(i.replace('_mask', ''))


print(train_image[:10],"\n" ,train_mask[:10])
# Display the first image and mask of the first subject.
image1 = np.array(Image.open(path+"1_1.tif"))
image1_mask = np.array(Image.open(path+"1_1_mask.tif"))
image1_mask = np.ma.masked_where(image1_mask == 0, image1_mask)

fig, ax = plt.subplots(1,3,figsize = (16,12))
ax[0].imshow(image1, cmap = 'gray')
ax[1].imshow(image1_mask, cmap = 'gray')
ax[2].imshow(image1, cmap = 'gray', interpolation = 'none')
ax[2].imshow(image1_mask, cmap = 'jet', interpolation = 'none', alpha = 0.7)
```

```python
from tensorflow.keras.models import Model, load_model
from tensorflow.keras import Input
from tensorflow.keras.layers import Input, Activation, BatchNormalization, Dr
opout, Lambda, Conv2D, Conv2DTranspose, MaxPooling2D, concatenate,add
from tensorflow.keras.optimizers import Adam
from tensorflow.keras.callbacks import EarlyStopping, ModelCheckpoint
def dice_coef(y_true, y_pred):
    smooth = 0.0
    y_true_f = K.flatten(y_true)
    y_pred_f = K.flatten(y_pred)
    intersection = K.sum(y_true_f * y_pred_f)
    return (2. * intersection + smooth) / (K.sum(y_true_f) + K.sum(y_pred_f)
+ smooth)
def jacard(y_true, y_pred):
    y_true_f = K.flatten(y_true)
    y_pred_f = K.flatten(y_pred)
    intersection = K.sum ( y_true_f * y_pred_f)
    union = K.sum ( y_true_f + y_pred_f - y_true_f * y_pred_f)
    return intersection/union
def dice_coef_loss(y_true, y_pred):
    return -dice_coef(y_true, y_pred)
```

```
pos_mask = []
pos_img = []
neg_mask = []
neg_img = []


for mask_path, img_path in zip(train_mask, train_image):
    mask = cv2.imread(mask_path, cv2.IMREAD_GRAYSCALE)
    if np.sum(mask) == 0:
        neg_mask.append(mask_path)
        neg_img.append(img_path)
    else:
        pos_mask.append(mask_path)
        pos_img.append(img_path)
!mkdir generated
!mkdir generated/img
def flip_up_down(img):
    newImg = img.copy()
    return cv2.flip(newImg, 0)
def flip_right_left(img):
    newImg = img.copy()
    return cv2.flip(newImg, 1)
gen_img = []
gen_mask = []
for (img_path, mask_path) in tqdm(zip(pos_img, pos_mask)):
    image_name = img_path.split('/')[-1].split('.')[0]
    uf_img_path = 'generated/img/'+image_name+'_uf.jpg'
    uf_mask_path = 'generated/img/'+image_name+'_uf_mask.jpg'
    rf_img_path = 'generated/img/'+image_name+'_rf.jpg'
    rf_mask_path = 'generated/img/'+image_name+'_rf_mask.jpg'



        img = cv2.imread(img_path)
        mask = cv2.imread(mask_path)
        uf_img = flip_up_down(img)
        uf_mask = flip_up_down(mask)
        cv2.imwrite(uf_img_path, uf_img)
        cv2.imwrite(uf_mask_path, uf_mask)
        rf_img = flip_right_left(img)
        rf_mask = flip_right_left(mask)
        cv2.imwrite(rf_img_path, rf_img)
        cv2.imwrite(rf_mask_path, rf_mask)
        gen_img.append(uf_img_path)
        gen_mask.append(uf_mask_path)
        gen_img.append(rf_img_path)
        gen_mask.append(rf_mask_path)
    aug_img = gen_img + train_image
    aug_mask = gen_mask + train_mask
```

```python
df_ = pd.DataFrame(data={"filename": aug_img, 'mask' : aug_mask})
df = df_.sample(frac=1).reset_index(drop=True)
kf = KFold(n_splits = 5, shuffle=False)

from tensorflow.keras.applications.vgg16 import VGG16
from tensorflow.keras.layers import *
def res_block(inputs,filter_size):
    """
    res_block -- Residual block for building res path
    Arguments:
    inputs {<class 'tensorflow.python.framework.ops.Tensor'>} --
input for residual block
    filter_size {int} -- convolutional filter size
    Returns:
    add {<class 'tensorflow.python.framework.ops.Tensor'>} --
addition of two convolutional filter output
    """
    # First Conv2D layer
    cb1 = Conv2D(filter_size,(3,3),padding = 'same',activation="relu")(inputs
)
    # Second Conv2D layer parallel to the first one
    cb2 = Conv2D(filter_size,(1,1),padding = 'same',activation="relu")(inputs
)
    # Addition of cb1 and cb2
    add = Add()([cb1,cb2])
    return add



  def res_path(inputs,filter_size,path_number):
      """
      res_path -- residual path / modified skip connection
      Arguments:
      inputs {<class 'tensorflow.python.framework.ops.Tensor'>} --
  input for res path
      filter_size {int} -- convolutional filter size
      path_number {int} -- path identifier
      Returns:
      skip_connection {<class 'tensorflow.python.framework.ops.Tensor'>} --
  final res path
      """
      # Minimum one residual block for every res path
      skip_connection = res_block(inputs, filter_size)
      # Two serial residual blocks for res path 2
      if path_number == 2:
          skip_connection = res_block(skip_connection,filter_size)
```

```
        # Three serial residual blocks for res path 1
        elif path_number == 1:
            skip_connection = res_block(skip_connection,filter_size)
            skip_connection = res_block(skip_connection,filter_size)
        return skip_connection


    def decoder_block(inputs, res, out_channels, depth):

        """
        decoder_block -- decoder block formation
        Arguments:
        inputs {<class 'tensorflow.python.framework.ops.Tensor'>} --
    input for decoder block
        mid_channels {int} -- no. of mid channels
        out_channels {int} -- no. of out channels
        Returns:
        db {<class 'tensorflow.python.framework.ops.Tensor'>} --
    returning the decoder block
        """
        conv_kwargs = dict(
            activation='relu',
            padding='same',
            kernel_initializer='he_normal',
            data_format='channels_last'
        )



        # UpConvolutional layer
        db = Conv2DTranspose(out_channels, (2, 2), strides=(2, 2))(inputs)
        db = concatenate([db, res], axis=3)
        # First conv2D layer
        db = Conv2D(out_channels, 3, **conv_kwargs)(db)
        # Second conv2D layer
        db = Conv2D(out_channels, 3, **conv_kwargs)(db)

        if depth > 2:
            # Third conv2D layer
            db = Conv2D(out_channels, 3, **conv_kwargs)(db)
        return db

def TransResUNet(input_size=(512, 512, 1)):
    """
    TransResUNet -- main architecture of TransResUNet

    Arguments:
```

```
    input_size {tuple} -- size of input image


    Returns:
    model {<class 'tensorflow.python.keras.engine.training.Model'>} --
final model
    """
    # Input
    inputs = Input(input_size)
    # Handling input channels
    # input with 1 channel will be converted to 3 channels to be compatible w
ith VGG16 pretrained encoder
#    if input_size[-1] < 3:
#        inp = Conv2D(3, 1)(inputs)
#        input_shape = (input_size[0], input_size[0], 3)
#    else:
#        inp = inputs
#        input_shape = input_size
    # VGG16 with imagenet weights
    encoder = VGG16(include_top=False, weights='imagenet', input_shape=input_
size)
    # First encoder block
    enc1 = encoder.get_layer(name='block1_conv1')(inputs)
    enc1 = encoder.get_layer(name='block1_conv2')(enc1)
    enc2 = MaxPooling2D(pool_size=(2, 2))(enc1)
    # Second encoder block
    enc2 = encoder.get_layer(name='block2_conv1')(enc2)
    enc2 = encoder.get_layer(name='block2_conv2')(enc2)
    enc3 = MaxPooling2D(pool_size=(2, 2))(enc2)


    # Third encoder block
    enc3 = encoder.get_layer(name='block3_conv1')(enc3)
    enc3 = encoder.get_layer(name='block3_conv2')(enc3)
    enc3 = encoder.get_layer(name='block3_conv3')(enc3)
    enc4 = MaxPooling2D(pool_size=(2, 2))(enc3)
    # Fourth encoder block
    enc4 = encoder.get_layer(name='block4_conv1')(enc4)
    enc4 = encoder.get_layer(name='block4_conv2')(enc4)
    enc4 = encoder.get_layer(name='block4_conv3')(enc4)
    center = MaxPooling2D(pool_size=(2, 2))(enc4)
        # Center block
        center = Conv2D(1024, (3, 3), activation='relu', padding='same')(ce
    nter)
        center = Conv2D(1024, (3, 3), activation='relu', padding='same')(ce
    nter)    # classification branch
        cls = Conv2D(256, (3,3), activation='relu')(center)
    #   cls = Conv2D(128, (3,3), activation='relu')(cls)
```

```
cls = Conv2D(1, (1,1))(cls)
cls = GlobalAveragePooling2D()(cls)
cls = Activation('sigmoid', name='class')(cls)
clsr = Reshape((1, 1, 1))(cls)
# Decoder block corresponding to fourth encoder
res_path4 = res_path(enc4,256,4)
dec4 = decoder_block(center, res_path4, 512, 4)
# Decoder block corresponding to third encoder
res_path3 = res_path(enc3,128,3)
dec3 = decoder_block(dec4, res_path3, 256, 3)
# Decoder block corresponding to second encoder
res_path2 = res_path(enc2,64,2)
dec2 = decoder_block(dec3, res_path2, 128, 2)
# Final Block concatenation with first encoded feature
res_path1 = res_path(enc1,32,1)
dec1 = decoder_block(dec2, res_path1, 64, 1)
# Output
out = Conv2D(1, 1)(dec1)
out = Activation('sigmoid')(out)
out = multiply(inputs=[out,clsr], name='seg')
# Final model
model = Model(inputs=[inputs], outputs=[out, cls])
return model
# From: https://github.com/zhixuhao/unet/blob/master/data.py
from tensorflow.keras.applications.vgg16 import preprocess_input
def train_generator(data_frame, batch_size, train_path, aug_dict,


    image_color_mode="rgb",
    mask_color_mode="grayscale",
    image_save_prefix="image",
    mask_save_prefix="mask",
    save_to_dir=None,
    target_size=(256,256),
    seed=1):
'''
can generate image and mask at the same time use the same seed for
image_datagen and mask_datagen to ensure the transformation for image
and mask is the same if you want to visualize the results of generator,
set save_to_dir = "your path"
'''
image_datagen = ImageDataGenerator(**aug_dict)
mask_datagen = ImageDataGenerator(**aug_dict)
image_generator = image_datagen.flow_from_dataframe(
    data_frame,
    directory = train_path,
    x_col = "filename",
```

```
        class_mode = None,
        color_mode = image_color_mode,
        target_size = target_size,
        batch_size = batch_size,
        save_to_dir = save_to_dir,
        save_prefix  = image_save_prefix,
        seed = seed)

mask_generator = mask_datagen.flow_from_dataframe(
    data_frame,
    directory = train_path,
    x_col = "mask",
    class_mode = None,
    color_mode = mask_color_mode,
    target_size = target_size,
    batch_size = batch_size,
    save_to_dir = save_to_dir,
    save_prefix  = mask_save_prefix,
    seed = seed)

train_gen = zip(image_generator, mask_generator)
for (img, mask) in train_gen:
    img, mask, label = adjust_data(img, mask)
    yield (img,[mask,label])



    def adjust_data(img,mask):
        img = preprocess_input(img)
        mask = mask / 255
        mask[mask > 0.5] = 1
        mask[mask <= 0.5] = 0
        masks_sum = np.sum(mask, axis=(1,2,3)).reshape((-1, 1))
        class_lab = (masks_sum != 0) + 0.

        return (img, mask, class_lab)
    histories = []

    losses = []

    accuracies = []

    dicecoefs = []

    jacards = []

    train_generator_args = dict(rotation_range=0.2,
                                width_shift_range=0.05,
                                height_shift_range=0.05,
                                shear_range=0.05,
```

```
                              zoom_range=0.05,
                              horizontal_flip=True,
                              fill_mode='nearest')
EPOCHS = 40
BATCH_SIZE = 32
for k, (train_index, test_index) in enumerate(kf.split(df)):
    train_data_frame = df.iloc[train_index]
    test_data_frame = df.iloc[test_index]
    train_gen = train_generator(train_data_frame, BATCH_SIZE,
                                None,
                                train_generator_args,
                                target_size=(height, width))
    test_gener = train_generator(test_data_frame, BATCH_SIZE,
                                None,
                                dict(),
                                target_size=(height, width))
    model = TransResUNet(input_size=(height,width, 3))
    model.compile(optimizer=Adam(lr=5e-
6), loss={'seg':dice_coef_loss, 'class':'binary_crossentropy'}, \
                    loss_weights={'seg':50, 'class':1}, metrics=[jacard, di
ce_coef, 'binary_accuracy'])



    model.summary()
    model_checkpoint = ModelCheckpoint(str(k+1) + '_unet_ner_seg.hdf5',
                                        verbose=1,
                                        save_best_only=True)
    history = model.fit(train_gen,
                        steps_per_epoch=len(train_data_frame) // BATCH_SIZE
,
                        epochs=EPOCHS,
                        callbacks=[model_checkpoint],
                        validation_data = test_gener,
                        validation_steps=len(test_data_frame) // BATCH_SIZE
)


    model = load_model(str(k+1) + '_unet_ner_seg.hdf5', custom_objects={'dice
_coef_loss': dice_coef_loss, 'jacard': jacard, 'dice_coef': dice_coef})
```

```
    test_gen = train_generator(test_data_frame, BATCH_SIZE,
                                None,
                                dict(),
                                target_size=(height, width))
    results = model.evaluate(test_gen, steps=len(test_data_frame) // BATCH_SI
ZE)
    results = dict(zip(model.metrics_names,results))

    histories.append(history)
    accuracies.append(results['seg_binary_accuracy'])
    losses.append(results['seg_loss'])
    dicecoefs.append(results['seg_dice_coef'])
    jacards.append(results['seg_jacard'])


    break


import pickle

for h, history in enumerate(histories):

    keys = history.history.keys()
    fig, axs = plt.subplots(1, len(keys)//2, figsize = (25, 5))
    fig.suptitle('No. ' + str(h+1) + ' Fold Results', fontsize=30)

    for k, key in enumerate(list(keys)[:len(keys)//2]):
        training = history.history[key]
        validation = history.history['val_' + key]

        epoch_count = range(1, len(training) + 1)

        axs[k].plot(epoch_count, training, 'r--')
        axs[k].plot(epoch_count, validation, 'b-')
        axs[k].legend(['Training ' + key, 'Validation ' + key])

    with open(str(h+1) + '_lungs_trainHistoryDict', 'wb') as file_pi:
        pickle.dump(history.history, file_pi)
print('average accuracy : ', np.mean(np.array(accuracies)), '+-',
np.std(np.array(accuracies)))
print('average loss : ', np.mean(np.array(losses)), '+-',
np.std(np.array(losses)))
print('average jacard : ', np.mean(np.array(jacards)), '+-',
np.std(np.array(jacards)))
```

```
print('average dice_coe : ', np.mean(np.array(dicecoefs)), '+-',
np.std(np.array(dicecoefs)))


model = load_model('1_unet_ner_seg.hdf5', custom_objects={'dice_coef_loss':
dice_coef_loss, 'jacard': jacard, 'dice_coef': dice_coef})
for i in range(20):
    index=np.random.randint(0,len(test_data_frame.index))
    print(i+1, index)
    img = cv2.imread(test_data_frame['filename'].iloc[index])
    img = cv2.resize(img, (height, width))
    img = preprocess_input(img)
    img = img[np.newaxis, :, :, :]
    pred = model.predict(img)
    print(pred[1])

    plt.figure(figsize=(12,12))
    plt.subplot(1,3,1)

plt.imshow(cv2.resize(cv2.imread(test_data_frame['filename'].iloc[index]),
(height, width)))
    plt.title('Original Image')
    plt.subplot(1,3,2)

plt.imshow(np.squeeze(cv2.resize(cv2.imread(test_data_frame['mask'].iloc[inde
x]), (height, width))))
    plt.title('Original Mask')
    plt.subplot(1,3,3)
    plt.imshow(np.squeeze(pred[0]) > .5)
    plt.title('Prediction')
    plt.show()
```

5. Discussion

With the incorporation of AI techniques, notably deep learning algorithms, the field of biomedical image segmentation has seen tremendous breakthroughs. These breakthroughs have transformed medical imaging by enabling accurate and efficient segmentation of anatomical features and anomalies in a variety of modalities such as MRI, CT, and histopathology slides.

One of the key benefits of AI-based biomedical image segmentation is its ability to overcome manual segmentation restrictions. Manual segmentation is a time-consuming and labor-intensive method that significantly relies on radiologists' or pathologists' skills. It is also susceptible to inter- and intraobserver variability, which leads to inconsistency in results. Deep learning models, such as CNNs, have demonstrated exceptional performance in automating the segmentation process. These models can extract complex patterns and features from huge datasets, resulting in accurate and reproducible segmentation results. AI-based segmentation has the potential to improve clinical practice efficiency and consistency by minimizing human participation and unpredictability.

Across a range of medical disciplines, the impact of AI-based biomedical image segmentation is obvious. In oncology, precise tumor segmentation from medical images is essential for treatment planning, measuring patient response, and predicting outcomes. AI-driven segmentation can help with tumor delineation, volume estimation, and temporal tumor growth monitoring. Oncologists can use this information to develop individualized treatment plans, track treatment effectiveness, and forecast patient outcomes. Similarly, to diagnose and treat neurological disorders, precise segmentation of brain regions and lesions is crucial in neuroimaging. White matter hyperintensities, brain tumors, and stroke lesions are just a few examples of abnormalities that AI-based segmentation can help discover. This information is crucial for surgical planning, therapeutic monitoring, and disease progression analysis.

While AI-based biomedical image segmentation has many benefits, there are obstacles that must be overcome before it can be widely used and adopted. The need for big annotated datasets for deep learning model training is one of the main issues. Medical imaging data collection and annotation can be time-consuming, expensive, and fraught with ethical and privacy issues. Additionally, it could be difficult to gather diverse and representative datasets, which could result in biases and restricted generalization skills in AI models. To address these issues and enhance the effectiveness of AI segmentation models, researchers are investigating strategies like transfer learning, domain adaptation, and data augmentation.

The interpretability and explainability of AI models are another difficulty. Deep learning models are frequently regarded as black boxes, making it impossible to comprehend the reasoning behind their segmentation judgments. In the medical industry, where trust and transparency are critical, this lack of interpretability creates problems. Efforts are being made to build explainable AI techniques that provide insights into deep learning models' decision-making processes, allowing doctors to comprehend and trust the generated segmentations. Explainable AI strategies also help with regulatory compliance and ethical concerns when using AI-based segmentation solutions.

6. Conclusion

The primary aim of this chapter was to categorize and develop a simple algorithm for image segmentation. In addition to accomplishing this objective, an evaluation of image segmentation techniques was introduced to conduct a fundamental comparison of practical methods for segmentation. Image segmentation holds a promising future as a foundational technique in the realms of image processing and computer vision. This can be substantiated by observing the substantial focus on ongoing research dedicated to formulating a comprehensive segmentation approach.

While a plethora of segmentation algorithms emerge regularly, none can serve as a universal solution for all images due to their specific target purposes. Various factors, including image homogeneity, content, and texture, employ influence over the outcome of image segmentation. An effective image segmentation algorithm should be capable of considering all these factors.

Furthermore, although numerous options exist for image segmentation techniques, none can offer universal applicability, often constraining their effectiveness in comprehensively analyzing images. Thankfully, the integration of AI into image segmentation holds promise for addressing this limitation. Lastly, manual iterative segmentation is gaining prominence. Partitioning and discerning images come naturally to human perception, surpassing machine capabilities in image processing. The underlying reason lies in the human eye's ability to leverage synthesized knowledge when interpreting images. Given this reality, manual iterative segmentation appears poised for achieving more refined segmentation results.

AI-based biomedical image segmentation has significant promise for developing medical imaging and enhancing patient care. AI models have shown amazing performance in reliably and quickly segmenting biological images, such as MRI scans, CT scans, and histopathology slides, through the application of deep learning methods, notably CNNs. AI-based segmentation algorithms have the potential to improve disease monitoring, treatment planning, and patient outcomes, resulting in more individualized healthcare.

The ability of AI-based biomedical image segmentation to handle complicated and heterogeneous data is one of its main benefits. Frequently, complex architecture, minute changes, and noise in biomedical images make hand segmentation difficult and time-consuming. Deep learning algorithms have demonstrated the potential to automatically extract significant information and precisely designate regions of interest in biological images due to their capacity to learn hierarchical features from big datasets. To enable early detection and action, this can help in the identification and localization of anomalies, such as tumors, lesions, and anatomical structures.

Various medical specialties are where AI-based biomedical image segmentation is most noticeable. For instance, in cancer, precise tumor segmentation from medical imaging is essential for treatment planning, gauging patient response, and predicting prognosis. The use of AI-driven segmentation can help oncologists make informed choices and customize treatments for individual patients by assisting in the identification of tumor boundaries, quantifying tumor volume, and evaluating therapy response. The diagnosis and treatment of diseases are similarly aided by the accurate segmentation of brain structures and lesions.

In the future, there will be several fascinating research potentials and ideas for biomedical image segmentation utilizing AI. Segmentation models' precision and resilience can be improved by integrating multimodal data, which combines data from many imaging modalities. By utilizing the complementing data offered by several modalities, this integration enables thorough analysis. Additional active research areas include real-time segmentation, computational efficiency improvement, and integration with clinical decision support systems, which can improve the clinical value of AI-based segmentation in real-world healthcare settings.

The potential for AI-based biomedical image segmentation to advance medical imaging and enhance patient care is enormous, to say the least. The accurate and effective segmentation of biomedical images made possible by the combination of deep learning algorithms with AI approaches has many advantages for disease diagnosis, treatment planning, and disease monitoring. Unlocking the full potential of AI in biomedical image segmentation requires continual study and collaboration among researchers, physicians, and regulatory agencies. This is true despite obstacles relating to data accessibility, interpretability, and ethical issues. AI-driven segmentation has the potential to revolutionize healthcare delivery, improve clinical judgment, and ultimately better patient outcomes.

References

Akkus, Z. (2017). Deep learning for brain MRI segmentation: State of the Art and future directions. *Journal of Digital Imaging, 30*, 449–459.

Al-amri, S., Kalyankar, N. V., & Khatmitkar, S. D. (2010). Image segmentation by using edge detection. *International Journal on Computer Science and Engineering*, 804–807.

Bahadure, N. B., Ray, A. K., & Thethi, H. P. (2018). Comparative approach of MRI-based brain tumor segmentation and classification using Genetic algorithm. *Journal of Digital Imaging, 31*(4). https://doi.org/10.1007/s10278-018-0050-6. Article 4.

Bovik, A. (2005). *Handbook of image and video processing* (2nd ed.). In: https://www.elsevier.com/books/handbook-of-image-and-video-processing/bovik/978-0-12-119792-6.

Burnham, J., Hardy, J., Meadors, K., & Picone, J. (1997). Comparison of the roberts, sobel, robinson, canny, and hough image detection algorithms. In *Comparison of edge detection algorithms MS state DSP conference*. Starkville, MS, USA: Image Processing Group.

Candemir, S., Jaeger, S., Palaniappan, K., Musco, J. P., Singh, R. K., Xue, Z., Karargyris, A., Antani, S., Thoma, G., & McDonald, C. J. (2013). Lung segmentation in chest radiographs using anatomical atlases with nonrigid registration. *IEEE Transactions on Medical Imaging, 33*(2), 577–590.

Chen, L., Yu, Z., Huang, J., Shu, L., Kuosmanen, P., Shen, C., Ma, X., Li, J., Sun, C., & Li, Z. (2023). Development of lung segmentation method in x-ray images of children based on TransResUNet. *Frontiers in Radiology, 3*, 1190745.

Ciresan, D. C., Giusti, A., Gambardella, L. M., & Schmidhuber, J. (2012). Deep neural networks segment neuronal membranes in electron microscopy images. In *Nips* (pp. 2852–2860).

Gonzalez, R. C., Woods, R. E., & Masters, B. R. (2009). Digital image processing, third edition. *Journal of Biomedical Optics, 14*(2), 029901. https://doi.org/10.1117/1.3115362

Havaei, M., Davy, A., & Warde-Farley, D. (2016). *Brain tumor segmentation with deep neural networks*. Cornell University Library.

He, Y. (2022). Deep learning powers cancer diagnosis in digital pathology. *Computerized Medical Imaging and Graphics, 88*, 101820.

Healy, S. (2018). Threshold-based segmentation of fluorescent and chromogenic images of microglia, astrocytes and oligodendrocytes in FIJI. *Journal of Neuroscience Methods, 295*, 87–103.

Işın, A., Direkoğlu, C., & Şah, M. (2016). Review of MRI-based brain tumor image segmentation using deep learning methods. *Procedia Computer Science, 102*, 317–324.

Ji, S., Wei, B., Yu, Z., Yang, G., & Yin, Y. (2014). A new multistage medical segmentation method based on superpixel and fuzzy clustering. *Computational and Mathematical Methods in Medicine, 2014*, 747549. https://doi.org/10.1155/2014/747549

Kang, W.-X., Yang, Q.-Q., & Liang, R.-P. (2009). The comparative research on image segmentation algorithms. In *2009 first international workshop on education technology and computer science* (pp. 703–707). https://doi.org/10.1109/ETCS.2009.417

Kim, M., Jung, S. Y., Park, J. E., Jo, Y., Park, S. Y., Nam, S. J., Kim, J. H., & Kim, H. S. (2020). Diffusion- and perfusion-weighted MRI radiomics model may predict isocitrate dehydrogenase (IDH) mutation and tumor aggressiveness in diffuse lower grade glioma. *European Radiology, 30*(4). https://doi.org/10.1007/s00330-019-06548-3. Article 4.

Krizhevsky, A., Sutskever, I., & Hinton, G. E. (2012). ImageNet classification with deep convolutional neural networks. *Advances in Neural Information Processing Systems, 25*. https://proceedings.neurips.cc/paper/2012/hash/c399862d3b9d6b76c8436e924a68c45b-Abstract.html.

Langote, V. B., & Chaudhari, D. D. S. (2012). *Segmentation techniques for image analysis* (Vol. 4).

Lee, B. (2020). Automatic segmentation of brain MRI using a novel patch-wise U-net deep architecture. *Plos One, 15*, e0236493.

Mirzaei, G. (2018). Segmentation and clustering in brain MRI imaging. *Reviews in the Neurosciences, 30*, 31–44.

Muthukrishnan, R., & Radha, M. (2011). Edge detection techniques for image segmentation. *International Journal of Computer Science and Information Technology, 3*(6), 259–267. https://doi.org/10.5121/ijcsit.2011.3620

Qiaoping, W. (1998). One image segmentation technique based on wavelet analysis in the context of texture. *Data Collection and Processing, 13*, 12–16.

Reza, S., Amin, O. B., & Hashem, M. (2020). *Transresunet: Improving u-net architecture for robust lungs segmentation in chest x-rays* (pp. 1592–1595).

Ronneberger, O., Fischer, P., & Brox, T. (2015). *U-net: Convolutional networks for biomedical image segmentation* (pp. 234–241).

Rosenfeld, A. (1984). 7—Image analysis. In M. P. Ekstrom (Ed.), *Digital image processing techniques* (pp. 257–287). Academic Press. https://doi.org/10.1016/B978-0-12-236760-1.50012-2

Sarma, R., & Gupta, Y. K. (2021). A comparative study of new and existing segmentation techniques. *IOP Conference Series: Materials Science and Engineering, 1022*(1), 012027. https://doi.org/10.1088/1757-899X/1022/1/012027

Sharma, N., & Aggarwal, L. M. (2010). Automated medical image segmentation techniques. *Journal of Medical Physics/Association of Medical Physicists of India, 35*(1), 3–14. https://doi.org/10.4103/0971-6203.58777

Tomar, N. K., Shergill, A., Rieders, B., Bagci, U., & Jha, D. (2022). TransResU-Net: Transformer based ResU-Net for real-time colonoscopy polyp segmentation. *ArXiv Preprint ArXiv:2206.08985.*

Wang, Y., Qi, Q., & Shen, X. (2020). Image segmentation of brain MRI based on LTriDP and superpixels of improved SLIC. *Brain Sciences, 10*(2). https://doi.org/10.3390/brainsci10020116. Article 2.

Yanyun, L., & Zhijian, S. (2014). Automated brain tumor segmentation in magnetic resonance imaging based on sliding-window technique and symmetry analysis. *Chinese Medical Journal, 127*(3). https://doi.org/10.3760/cma.j.issn.0366-6999.20132554. Article 3.

DNA sequence classification using artificial intelligence

Abdulhamit Subasi[1,2]

[1]Institute of Biomedicine, Faculty of Medicine, University of Turku, Turku, Finland; [2]Department of Computer Science, College of Engineering, Effat University, Jeddah, Saudi Arabia

1. Introduction

A basic task in genomics is the classification of deoxyribonucleic acid (DNA) sequences, which has a wide range of applications, including the identification of diseases, the prediction of gene function, and evolutionary study. The integration of artificial intelligence (AI) tools into this field has been prompted by the intrinsic complexity of DNA sequences and the exponential expansion of genomic data. A new era of DNA sequence classification characterized by improved accuracy, efficiency, and scalability has begun as a result of this integration.

New methods have been made available for deciphering the complex patterns inherent in DNA sequences thanks to advances in AI, notably in machine learning (ML) and deep learning (DL). These patterns frequently elude typical computational methods, but they hold priceless knowledge regarding evolutionary links, regulatory components, and genetic features. Researchers have been able to find distinguishing patterns, recognize relevant features automatically, and then classify DNA sequences

with a level of accuracy previously unattainable by using AI. The goal of this chapter is to present a thorough analysis of the relationship between AI and DNA sequence classification. We explore the many AI approaches used in this sector, such as convolutional neural networks (CNNs), recurrent neural networks (RNNs), and support vector machines (SVMs), in light of the accessibility of large DNA sequence datasets. Researchers have advanced DNA sequence analysis beyond conventional sequence alignment approaches by utilizing these AI tools, allowing for a deeper investigation of genetic information (Gunasekaran et al., 2021).

This chapter examines the larger implications and uses of AI-driven DNA sequence classification in addition to its technical features. AI integration speeds up the classification process and reveals new information about the genetic underpinnings of certain traits and diseases. This study also emphasizes how important it is to use the right AI algorithms, model architectures, and feature representations to maximize classification performance. The combination of AI with DNA sequence classification holds enormous promise for the advancement of genomics. We anticipate discovering new genetic research frontiers as we begin this investigation into AI-driven DNA sequence classification. We also see the advent of cutting-edge applications in diagnostics, and our comprehension of evolutionary processes. When viewed through the lens of AI, DNA sequences go from being merely a collection of nucleotides to a dynamic and informational landscape that has the potential to completely alter how we understand genetics and how it affects the very nature of existence.

2. Literature review

To understand the intricate connections between genetic information, disease susceptibility, and evolutionary history, it is now essential to classify DNA sequences. The landscape of DNA sequence categorization has changed dramatically with the introduction of AI tools, particularly ML and DL algorithms. This has made it possible for researchers to meaningfully analyze enormous genomic datasets. Early attempts to classify DNA sequences mostly relied on labor-intensive manual sequence alignment techniques that were unable to capture complex genomic patterns. This discipline has undergone a revolution thanks to the incorporation of AI approaches, which has enabled academics to pursue new lines of inquiry. Numerous AI algorithms, including CNNs and RNNs, have shown an extraordinary ability to recognize minute sequence differences that are the basis of genetic features and disorders.

CNNs, which were originally created to recognize images, have been repurposed to interpret DNA sequences as one-dimensional data, efficiently capturing local patterns within nucleotide sequences. These networks have enabled accurate classification tasks, such as identifying disease-causing mutations or distinguishing functional sections inside genomes, by applying convolutional layers to detect

sequence motifs. Transfer learning techniques, pretrained on huge genomic datasets, were used by researchers to fine-tune CNNs for specific classification tasks, enhancing their performance even more. Classifying DNA sequences is a fundamental task within the broader scope of computational biomedical data analysis, and in recent years, several AI techniques have been employed to effectively tackle this challenge. However, the primary obstacle in this endeavor lies in the feature selection process. DNA sequences lack explicit features, and the conventional representations commonly used introduce a significant drawback by creating high-dimensional datasets. It is important to note that supervised ML methods, particularly those geared toward classification tasks, heavily rely on the feature extraction phase. To construct a meaningful representation, one must identify and quantify relevant characteristics of the items being classified. Notably, deep learning models (DLMs), specifically neural DL architectures, have demonstrated their ability to automatically extract valuable features from input patterns. Lo Bosco and Di Gangi (2017) introduced two distinct DL architectures designed for the classification of DNA sequences. They assessed their performance using a publicly available data set containing DNA sequences and evaluated their efficacy across five different classification tasks.

DL neural networks have the ability to extract significant patterns and features from raw data, which can then be utilized for various classification tasks. Rizzo et al. (2016) introduced a DL neural network specifically designed for classifying DNA sequences, leveraging a spectral sequence representation. They put this framework to the test using a dataset consisting of 16S genes, and they evaluated its performance by measuring accuracy and F1 score. To provide a comprehensive assessment, they compared the DLM's results to those of the general regression neural network, which has previously been employed for similar tasks, as well as to classifiers like Naïve Bayes, random forest, and support vector machine. Their findings indicate that the DL approach outperformed all other classifiers, particularly when classifying small sequence fragments that are 500 base pairs in length.

Targeted high-throughput DNA sequencing serves as a primary methodology in genomics, molecular diagnostics, and, more recently, as a tool for DNA information storage. However, the utilization of oligonucleotide probes to enrich specific gene loci introduces variations in hybridization kinetics, leading to nonuniform sequencing coverage. This nonuniformity not only increases the cost of sequencing but also diminishes its sensitivity. Zhang et al. (2021) introduced a DLM designed to predict the depth of next-generation sequencing (NGS) based on the sequences of DNA probes. They incorporated a bidirectional recurrent neural network, which takes into account both the identities of DNA nucleotides and the calculated probabilities of these nucleotides remaining unpaired. They applied DLM to three distinct NGS panels, namely a 39,145-plex panel for human single nucleotide polymorphisms (SNPs), a 2000-plex panel targeting human long noncoding RNA (lncRNA), and a 7373-plex panel aimed at nonhuman sequences for DNA information storage. Through cross-validation, DLM demonstrated an impressive ability to predict sequencing depth within a factor of 3, achieving 93% accuracy for the SNP panel and 99% accuracy for the nonhuman panel. When tested independently, DLM accurately predicted the sequencing depth of the lncRNA panel with an 89% accuracy rate, even when initially trained on the SNP panel. Additionally, the same model exhibited effectiveness in predicting the measured kinetic rate constants for single-plex DNA hybridization and strand displacement.

RNNs, on the other hand, are ideal for detecting sequential dependencies within DNA sequences. Their ability to retain memories over time steps makes it easier to identify long-term genetic patterns and regulatory elements. RNNs have been used by researchers to predict RNA secondary structures, infer DNA-binding sites, and classify sequences based on evolutionary conservation. In the realm of computational biomedical data analysis, the classification of DNA sequences represents a pivotal and challenging task. Over recent years, various ML techniques have been effectively utilized to tackle this challenge. This task gains significance in the context of identifying and classifying viruses, as it plays a crucial role in preventing outbreaks like the COVID-19 pandemic. However, one of the most formidable hurdles in this endeavor remains the process of feature selection. Conventional representations often exacerbate the problem, particularly when dealing with high-dimensional data, and DNA sequences inherently lack explicit features. Furthermore, this classification task contributes to the detection of virus effects and aids in drug design. In contemporary times, DLMs have emerged as powerful tools capable of automatically extracting relevant features from input data. Gunasekaran et al. (2021) employed CNN, CNN-LSTM, and CNN-bidirectional LSTM architectures, employing label and k-mer encoding methods for the classification of DNA sequences. These models underwent comprehensive evaluation based on various classification metrics. The experimental findings demonstrate that the CNN and CNN-Bidirectional LSTM models, when equipped with k-mer encoding, achieved remarkable accuracy, yielding 93.16% and 93.13%, respectively, on the testing dataset.

Eukaryotes' genetic information is primarily kept within their chromosomal DNA, which plays a vital role in their growth, development, and reproductive processes. It is well established that most chromosomal DNA sequences interact with histones, and being able to distinguish these specific DNA sequences from regular ones is critical for unraveling the genetic code governing life. The central challenge in addressing this problem is the feature selection process. DNA sequences inherently lack explicit features, and conventional representation methods, such as one-hot encoding, suffer from the significant drawback of generating high-dimensional data. Recent advancements in DLMs have demonstrated their capacity to automatically extract valuable features from input patterns. Du et al. (2020) explored the potential of various DL architectures to significantly enhance the field of DNA sequence classification, relying solely on sequence information. They introduced four distinct DL structures employing a combination of CNNs and long short-term memory networks for the classification of chromosomal DNA sequences. They employ natural language modeling techniques like Word2Vec to generate word embeddings from sequences, enabling feature extraction through DL. An extensive comparison of these four architectures is conducted using 10 chromosomal DNA datasets. The outcomes clearly indicate that the architecture incorporating both CNNs and long short-term memory networks outperforms other methods in terms of chromosomal DNA prediction accuracy.

Traditional ML algorithms, such as SVMs, have also found use in DNA sequence classification. SVMs use a kernel function to map sequences into a higher-dimensional space where they can be separated linearly. This method has been shown to be useful in tasks such as predicting protein-binding sites on DNA sequences and categorizing genetic variants linked to diseases. Hybrid models that integrate various algorithms have emerged as AI approaches continue to evolve. These models leverage the strengths of several AI approaches, resulting in improved accuracy and interpretability. Furthermore, attempts to democratize AI in genomics have resulted in the creation of user-friendly tools and frameworks that allow researchers from a variety of backgrounds to employ AI for DNA sequence classification.

The extraction of valuable insights from DNA plays a crucial role in bioinformatics research, with DNA sequence classification finding numerous applications in genomic and biomedical data analysis. Within the broader context of biomedical data processing, DNA sequence classification represents a significant challenge, and over recent years, various ML techniques have been employed to address this task effectively. ML is a data processing methodology that leverages training data to make informed decisions, predictions, classifications, and recognitions. Genomic researchers employ ML approaches to categorize DNA sequences into known groups to understand the functions of newly discovered proteins and genes. Therefore, the identification and characterization of these genes are of utmost importance. Sarkar et al. (2022) utilized several ML classifiers, including multinomial Naïve Bayes, SVM, KNN, and others, for DNA sequence classification using label and k-mer encoding. To assess the model's performance, various classification metrics were employed. Notably, the multinomial Naïve Bayes classifier and SVM, both utilizing k-mer encoding, demonstrated high accuracy, achieving 93.16% and 93.13%, respectively, on the testing dataset.

The recent COVID-19 pandemic, caused by the severe acute respiratory syndrome Coronavirus 2 (SARS-CoV-2), has rapidly and severely impacted the global population's health. With its swift transmission and extensive infection rates, it has become imperative to harness genome sequence analysis and advanced AI techniques to comprehend the genetic variations of this virus. Genome sequence analysis plays a pivotal role in unraveling the origin, behavior, and structure of COVID-19, offering valuable insights for the development of vaccines, antiviral drugs, and effective preventive measures. Ahmed and Jeon (2022) introduced an AI-based system designed for the genome sequence analysis of COVID-19 and related viruses like SARS, Middle East respiratory syndrome

(MERS), and Ebola. The system facilitates the extraction of crucial information from the genome sequences of these viruses, encompassing nucleotide composition, frequency, trinucleotide compositions, amino acid counts, sequence alignments, and DNA similarity data. Various visualization methods are employed to analyze these genome sequences, culminating in the application of a ML-based support vector machine classifier for the classification of distinct genome sequences. The dataset, comprising genome sequences of different viruses, is sourced from a publicly accessible online data center repository. Remarkably, the system achieves commendable classification results, boasting an accuracy rate of 97% for COVID-19, 96% for SARS, 95% for MERS, and 95% for Ebola genome sequences, respectively.

In the field of bioinformatics, the classification of unknown biological sequences plays a pivotal role in simplifying the organization, categorization, and study of organisms and their evolutionary relationships. Biological sequences can be conceptualized as data elements residing in high-dimensional spaces, with dimensions corresponding to sequence length. Computational techniques, such as one-hot encoding (OHE), have been commonly used for numerical representation in DNA sequence analysis. However, OHE suffers from two key limitations: (1) it fails to introduce additional informative variables, and (2) when dealing with sequences containing numerous classes, OHE substantially inflates the feature space. To address these shortcomings, Mahmoud and Guo (2021) introduced an efficient computational framework for classifying DNA sequences within the image domain. The proposed approach relies on a multilayer perceptron trained using a pseudoinverse learning autoencoder (PILAE) algorithm. Importantly, the PILAE training process eliminates the need to specify learning control parameters or the number of hidden layers, making the PILAE classifier outperform other deep neural network (DNN) methods such as VGG-16 and Xception

models in terms of performance. Experimental results have demonstrated that this novel approach achieves remarkable predictive accuracy while maintaining high computational efficiency across various datasets.

ML has played a pivotal role in enhancing decision-making processes by harnessing historical data effectively, a principle that extends to the domain of bioinformatics. Within bioinformatics, the classification of genes into natural and disease-affected categories, referred to as "valid" and "invalid" genes, represents a highly intricate challenge. This classification is crucial for assessing the applicability of new proteins discovered through genomic research, as it hinges on the categorization of DNA sequences. The identification of DNA sequence classes using a ML algorithm is primarily determined by the sequence of nucleotides. Even a slight mutation within the sequence results in a corresponding shift in the assigned class. Each numerical instance, representing a particular class, is associated with a gene family, encompassing categories, such as G protein-coupled receptors, tyrosine kinase, and synthase. Juneja et al. (2022) application of a classification algorithm to three distinct datasets, with the aim of determining the gene class to which they belong. To facilitate this analysis, the authors employed a technique involving the conversion of sequences into substrings of defined lengths, a process governed by the "k value," which serves as one approach for sequence analysis.

The process of taxonomic classification, which involves identifying and categorizing biological organisms based on their common origin and characteristics, is a fundamental task in the field of genetics. In contemporary genetics, taxonomic classification predominantly relies on comparing genome similarity with extensive genome databases. The quality of classification heavily hinges on the comprehensiveness of the database because it necessitates the presence of known representatives. Consequently, numerous genomic sequences either cannot be classified or exhibit a high rate of misclassification. Mock et al. (2021) introduced BERTax, a program that utilizes a DNN to precisely taxonomically classify DNA sequences into superkingdom, phylum, and genus without relying on the availability of known relatives in a database. BERTax accomplishes this by leveraging the natural language processing model BERT, which has been trained to represent DNA. They demonstrate that BERTax achieves performance levels at least on par with state-of-the-art methods when dealing with taxonomically similar species in the training data. However, in cases involving entirely new organisms, BERTax significantly outperforms all existing approaches. Furthermore, they illustrate how BERTax can be integrated with database-based methods to further enhance prediction accuracy. As BERTax does not rely on homologous entries in databases, it facilitates precise taxonomic classification for a broader spectrum of genomic sequences, resulting in a higher number of accurately classified sequences and thereby augmenting overall information acquisition.

Finally, the incorporation of AI approaches has redefined DNA sequence classification, allowing for remarkable precision in the discovery of complicated genomic patterns. AI methods have proven to be extremely effective at detecting genomic variants, regulatory elements, and disease-associated regions.

3. DNA sequencing with machine learning[1]

A genome constitutes the entirety of an organism's DNA, and it varies in size across different species. For instance, the human genome comprises 23 chromosomes, akin to organizing an encyclopedia into 23 volumes. If one were to count all the individual DNA "base pairs" in the human genome, the total would exceed six billion, making it an extensive compilation. The human genome, with approximately six billion characters represented by "A," "C," "G," and "T," can be likened to a book. Each person possesses a unique genome, yet scientists have observed that most segments of human genomes exhibit substantial similarity to one another.

Genomics, a data-driven scientific discipline, extensively relies on ML to discern patterns in data and formulate new biological hypotheses. However, the increasing volume of genomics data demands more powerful ML models to extract novel insights effectively. DL, known for its adeptness in harnessing extensive datasets, has made substantial contributions to various fields such as computer vision and natural language processing. In genomics, it has become the preferred approach for numerous modeling tasks, including predicting how genetic variations impact gene regulatory mechanisms like DNA accessibility and splicing. This example delves into the interpretation of DNA structure and the application of ML algorithms in constructing predictive models using DNA sequence data.

3.1 How is a DNA sequence represented?

Fig. 14.1 presents a segment of the DNA double helix structure. The double helix serves as the accurate chemical representation of DNA, but DNA possesses unique characteristics. It comprises nucleotides composed of four distinct types of nitrogen bases: Adenine (A), Thymine (T), Guanine (G), and Cytosine, often denoted as A, C, G, and T. These four chemical components connect through hydrogen bonds in various combinations, forming a chain that constitutes one strand of the DNA double helix. The complementary second strand of the double

---

[1] Adapted from GitHub—krishnaik06/DNA-Sequencing-Classifier.

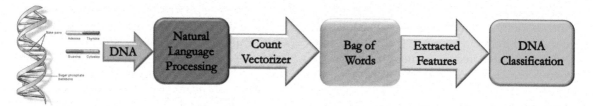

FIGURE 14.1 DNA classification framework.

helix complements the first strand; hence, if the first strand contains A, the second must contain T. Similarly, C always pairs with G, ensuring a balanced structure. Consequently, once you discern the sequence of one strand of the helix, you can deduce the sequence of the other. The arrangement or sequence of these nitrogen bases dictates the biological instructions encoded within a DNA strand. For instance, the sequence ATCGTT might encode instructions for blue eyes, whereas ATCGCT could correspond to instructions for brown eyes.

3.2 Python example

In this example, several classification models will be used to predict the function of a gene based solely on the DNA sequence of the coding sequence.

```
import numpy as np
import pandas as pd
import matplotlib.pyplot as plt
%matplotlib inline
human_data = pd.read_table('human_data.txt')
human_data.head()
```

We have some information and a class label for human DNA sequence coding regions. We also have data on Chimpanzees and the dog, a more divergent animal.

Treating DNA sequence as a "language"—k-mer counting

A persistent challenge in DNA sequence analysis is the lack of vectors with consistent lengths, a prerequisite for feeding data into classification or regression algorithms. Consequently, methods like sequence truncation or padding with characters such as "n" or "0" are often employed to achieve uniform-length vectors.

In a metaphorical sense, DNA and protein sequences can be likened to the language of life. This "biological language" encodes both instructions and functions for molecules present in all living organisms. Drawing parallels, the genome is akin to a book, subsequences (genes and gene families) represent sentences and chapters, while k-mers and peptides (motifs) function as words. Nucleotide bases and amino acids serve as the alphabet. Given the compelling nature of this analogy, the remarkable achievements in the field of natural language processing can potentially be adapted to analyze the natural language of DNA and protein sequences.

```
chimp_data = pd.read_table('chimp_data.txt')
dog_data = pd.read_table('dog_data.txt')
chimp_data.head()
dog_data.head()
```

The method described here is straightforward. Initially, a lengthy biological sequence is fragmented into overlapping "words" of a specified length, denoted as k-mers. For instance, when using hexamer "words" (word length of 6), the sequence "ATGCATGCA" is broken down into the following hexamers: "ATGCAT," "TGCATG," "GCATGC," and "CATGCA." Consequently, the example sequence is dissected into four hexamer words.

It is worth noting that the choice of hexamers is arbitrary, and the word length can be adjusted to suit specific applications. The optimal word length and degree of overlap must be determined empirically for each unique scenario. In genomics, this manipulation is referred to as "k-mer counting," where the goal is to calculate the occurrences of every possible k-mer sequence. While specialized tools exist for this purpose, Python's natural language processing tools provide a straightforward approach. Below, we present a function capable of converting any sequence string into overlapping k-mer words, allowing us to apply k-mer analysis to entire sequences.

```
# function to convert sequence strings into k-mer words, default size = 6
(hexamer words)
def getKmers(sequence, size=6):
    return [sequence[x:x+size].lower() for x in range(len(sequence) - size
+ 1)]
```

Our training data sequences can now be converted into short overlapping k-mers of length 6. Let us use our getKmers function to do this for each species of data we have.

```
human_data['words'] = human_data.apply(lambda x: getKmers(x['sequence']),
axis=1)
human_data = human_data.drop('sequence', axis=1)
chimp_data['words'] = chimp_data.apply(lambda x: getKmers(x['sequence']),
axis=1)
chimp_data = chimp_data.drop('sequence', axis=1)
dog_data['words']   =   dog_data.apply(lambda   x:   getKmers(x['sequence']),
axis=1)
dog_data = dog_data.drop('sequence', axis=1)
```

We need to convert the lists of k-mers for each gene into string sentences of words that the count vectorizer can utilize because we are going to use scikit-learn natural language processing tools to accomplish the k-mer counting. A "y" variable can also be created to hold the class labels.

```
human_texts = list(human_data['words'])
for item in range(len(human_texts)):
    human_texts[item] = ' '.join(human_texts[item])
y_data = human_data.iloc[:, 0].values
print(human_texts[2])
y_data
```

In a similar way, let us do for chimpanzee and dog.

```
chimp_texts = list(chimp_data['words'])
for item in range(len(chimp_texts)):
    chimp_texts[item] = ' '.join(chimp_texts[item])
y_chimp = chimp_data.iloc[:, 0].values          # y_c for
chimp

dog_texts = list(dog_data['words'])
for item in range(len(dog_texts)):
    dog_texts[item] = ' '.join(dog_texts[item])
y_dog = dog_data.iloc[:, 0].values
```

Now, we will utilize natural language processing (NLP) to apply the BAG of WORDS using CountVectorizer.

```python
# Creating the Bag of Words model using CountVectorizer()
# This is equivalent to k-mer counting
# The n-gram size of 4 was previously determined by testing
from sklearn.feature_extraction.text import CountVectorizer
cv = CountVectorizer(ngram_range=(4,4))
X = cv.fit_transform(human_texts)
X_chimp = cv.transform(chimp_texts)
X_dog = cv.transform(dog_texts)
print(X.shape)
print(X_chimp.shape)
print(X_dog.shape)
```

```python
human_data['class'].value_counts().sort_index().plot.bar()
# Splitting the human dataset into the training set and test set
from sklearn.model_selection import train_test_split
X_train, X_test, y_train, y_test = train_test_split(X,
                                        y_data,
                                        test_size = 0.20,
                                        random_state=42)
print(X_train.shape)
print(X_test.shape)

#Evaluate the Model and Print Performance Metrics
from sklearn import metrics
def print_metrics(ytest, ypred):
    print('Accuracy:', np.round(metrics.accuracy_score(ytest,ypred),4))
    print('Precision:', np.round(metrics.precision_score(ytest,
                        ypred,average='weighted'),4))
    print('Recall:', np.round(metrics.recall_score(ytest,ypred,
                                        average='weighted'),4))
    print('F1 Score:', np.round(metrics.f1_score(ytest,ypred,
                                        average='weighted'),4))
    print('Cohen Kappa Score:', np.round(metrics.cohen_kappa_score(ytest,
ypred),4))
    print('Matthews Corrcoef:', np.round(metrics.matthews_corrcoef(ytest,
ypred),4))
                        print('\t\tClassification                Report:\n',
metrics.classification_report(ypred, ytest))
    from sklearn.metrics import confusion_matrix
```

```
print("Confusion Matrix:\n",confusion_matrix(ytest, ypred))
#Plot Confusion Matrix
from sklearn.metrics import confusion_matrix
from io import BytesIO #neded for plot
import seaborn as sns; sns.set()
import matplotlib.pyplot as plt
mat = confusion_matrix(y_test, y_pred)
sns.heatmap(mat.T, square=True, annot=True, fmt='d', cbar=False)
plt.xlabel('true label')
plt.ylabel('predicted label');
plt.savefig("Confusion.jpg")
# Save SVG in a fake file object.
f = BytesIO()
plt.savefig(f, format="svg")


### Multinomial Naive Bayes Classifier ###
from sklearn.naive_bayes import MultinomialNB
model = MultinomialNB(alpha=0.1)
model.fit(X_train, y_train)
#Test the model with Testset
y_pred = model.predict(X_test)
#Print the performance metrics
print_metrics(y_test, y_pred)



### Random Forest Classifier
from sklearn.ensemble import RandomForestClassifier
model = RandomForestClassifier(n_estimators=200)
model.fit(X_train, y_train)
#Test the model with Testset
y_pred = model.predict(X_test)
#Print the performance metrics
print_metrics(y_test, y_pred)
```

```
### SVM Classifier
from sklearn import svm
""" The parameters and kernels of SVM classifier can be changed as follows
C = 10.0  # SVM regularization parameter
svm.SVC(kernel='linear', C=C)
svm.LinearSVC(C=C, max_iter=10000)
svm.SVC(kernel='rbf', gamma=0.7, C=C)
svm.SVC(kernel='poly', degree=3, gamma='auto', C=C))
"""
C = 50.0  # SVM regularization parameter
# fit model no training data
model  =  svm.SVC(kernel='linear', C=C)
model.fit(X_train, y_train)
#Test the model with Testset
y_pred = model.predict(X_test)
#Print the performance metrics
print_metrics(y_test, y_pred)



### k-NN Classifier
from sklearn.neighbors import KNeighborsClassifier
#Create the Model
model = KNeighborsClassifier(n_neighbors=5)
#Train the model with Training Dataset
model.fit(X_train, y_train)
#Test the model with Testset
y_pred = model.predict(X_test)
#Print the performance metrics
print_metrics(y_test, y_pred)



### ANN Classifier
from sklearn.neural_network import MLPClassifier
#Create the Model
model= MLPClassifier(hidden_layer_sizes=(100, ), learning_rate_init=0.001,
                     alpha=1, momentum=0.9,max_iter=1000)
#Train the model with Training Dataset
model.fit(X_train, y_train)
#Test the model with Testset
y_pred = model.predict(X_test)
#Print the performance metrics
print_metrics(y_test, y_pred)
```

```
### Decision Tree Classifier
from sklearn import tree
#Create the Model
model = tree.DecisionTreeClassifier()
#Train the model with Training Dataset
model.fit(X_train, y_train)
#Test the model with Testset
y_pred = model.predict(X_test)
#Print the perfromance metrics
print_metrics(y_test, y_pred)
```

4. Results

In Table 14.1, the classification performance of different ML models is given for our problem. Naïve Bayes achieved the best performance (95.536%) than other models. Artificial neural network (ANN) is the second one in the classification performance. Decision tree model achieved the worst performance.

5. Discussion

The use of AI approaches in DNA sequence classification has ushered in a new era of genomic analysis, allowing researchers to decipher complicated genetic links, forecast disease susceptibility, and gain insights into evolutionary processes. The ability of AI-driven DNA sequence classification to identify deep genomic patterns and correlations that were previously difficult to discern using standard computational methods is essential to its success. ML models have been shown to be effective at identifying local and long-range sequence motifs, resulting in accurate classification results. The use of AI has sped up the categorization process, allowing for the analysis of enormous genomic datasets in a fraction of the time that manual methods would take. AI-powered DNA sequence classification holds enormous promise for a variety of applications. Accurate classification of genomic variations can help in adapting treatment techniques to an individual's genetic makeup in personalized medicine. Other significant areas where AI may help are disease detection and risk prediction by discovering genetic markers associated with specific illnesses. Furthermore, the use of AI can help us better comprehend evolutionary links and genetic diversity among species.

While AI-powered DNA sequence classification provides incomparable possibilities, it also introduces new issues that must be addressed. The AI method and feature representation used can have a substantial impact on categorization accuracy. Based on the specific classification task and dataset features, researchers must carefully select algorithms and preprocess data.

TABLE 14.1 Comparison of AI models for DNA sequence classification.

Model	Accuracy	Precision	Recall	F1 score	Kappa	MCC
Naïve Bayes	98.4	98.43	98.4	98.4	98.05	98.05
Random forest	91.55	92.45	91.55	91.7	89.64	89.76
SVM	87.9	91.67	87.9	88.77	85.29	85.88
k-NN	82.42	87.41	82.42	82.71	77.84	79.4
ANN	96.12	96.19	96.12	96.12	95.25	95.27
Decision tree	81.05	82.71	81.05	81.5	77.01	77.18
XGBoost	89.84	91.24	89.84	89.96	87.38	87.77

Furthermore, the interpretability of AI models remains a challenge, particularly in genomics, where understanding the biological foundation of predictions is critical. The availability of high-quality, well-annotated genomic datasets is critical for training strong AI models. Data quality, biases, and imbalances can all have an impact on the generalizability of AI-driven classifiers. Efforts to gather broad and representative datasets are critical to ensuring that AI models are relevant across varied demographics and genetic origins.

Further advances in AI-driven DNA sequence classification are expected as AI techniques progress. Hybrid models that incorporate several AI approaches have the potential to improve classification accuracy and interpretability. Explainable AI (XAI) approaches are being investigated to shed light on the decision-making process of AI models, hence improving their transparency and trustworthiness. The integration of multiomics data, such as DNA sequences, gene expression, and epigenetic information, opens up new opportunities for complete biological insights. Synergistic analysis of various omics data using AI approaches can provide a more comprehensive understanding of gene regulation, disease mechanisms, and cellular processes.

Finally, the incorporation of AI into DNA sequence classification has transformed genomic research and analysis. CNNs, RNNs, and other AI algorithms have shown promise in finding genetic patterns and accurately classifying sequences. While obstacles including interpretability and dataset biases remain, current research and advances in AI approaches hold the possibility of revealing deeper insights into genetics, disease, and evolution. As AI advances, its impact on genetics will transform our knowledge of life's fundamental building components.

6. Conclusion

The combination of AI and DNA sequence classification constitutes a transformative milestone in genomics research, opening up new levels of understanding within the complex world of genetic information. The application of AI techniques has pushed the bounds of DNA sequence analysis, revealing hidden patterns, permitting accurate predictions, and expediting scientific discovery.

The importance of AI-driven DNA sequence classification is highlighted by its influence across multiple fields. AI-powered classifiers in personalized medicine have the potential to change healthcare by enabling individualized therapies based on an individual's genetic makeup. Accurate identification of genetic variants linked to diseases can aid in early diagnosis and intervention, thereby improving patient outcomes. Furthermore, using AI improves our grasp of evolutionary linkages, species adaptations, and the functional implications of genetic diversity. This paradigm change has been fraught with difficulties, demanding careful examination. The reliability and generalizability of classification models are influenced by the AI algorithms, feature representations, and dataset quality. Understanding the biological context of AI predictions is critical for researchers, doctors, and stakeholders, therefore interpretability remains a focus.

Looking ahead, the direction of AI-driven DNA sequence classification is promising. Hybrid models that combine various AI techniques may improve accuracy and interpretability. Explainable AI (XAI) approaches are positioned to decipher AI models' decision-making processes, boosting trust and enabling greater use in essential applications. The ability for AI to synergize with other "omics" data sources, such as transcriptomics and epigenomics, provides a comprehensive view of gene regulation, molecular connections, and complicated disease mechanisms. AI integration of these multimodal datasets promises to reveal unique insights and sophisticated biological networks.

Finally, AI-driven DNA sequence classification has reshaped genomics research, allowing scientists to understand genetic information with unparalleled precision and depth. The interaction of AI algorithms has revealed previously hidden genetic patterns, paving the way for new vistas in personalized medicine, disease understanding, and evolutionary biology. As AI approaches advance, their impact on genomics research has the potential to transform our understanding of the blueprint for life and its consequences for health, disease, and the natural environment.

References

Ahmed, I., & Jeon, G. (2022). Enabling artificial intelligence for genome sequence analysis of COVID-19 and alike viruses. *Interdisciplinary Sciences: Computational Life Sciences, 14*(2), 504–519. https://doi.org/10.1007/s125 39-021-00465-0

Du, Z., Xiao, X., & Uversky, V. N. (2020). Classification of chromosomal DNA sequences using hybrid deep learning architectures. *Current Bioinformatics, 15*(10), 1130–1136.

Gunasekaran, H., Ramalakshmi, K., Rex Macedo Arokiaraj, A., Deepa Kanmani, S., Venkatesan, C., & Suresh Gnana Dhas, C. (2021). Analysis of DNA sequence classification using CNN and hybrid models. *Computational and Mathematical Methods in Medicine, 2021*.

Juneja, S., Dhankhar, A., Juneja, A., & Bali, S. (2022). An approach to DNA sequence classification through machine learning: DNA sequencing, K Mer counting, thresholding, sequence analysis. *International Journal of Reliable and Quality E-Healthcare, 11*(2), 1–15.

Lo Bosco, G., & Di Gangi, M. A. (2017). Deep learning architectures for DNA sequence classification. In A. Petrosino, V. Loia, & W. Pedrycz (Eds.), *Fuzzy logic and soft computing applications* (pp. 162–171). Springer International Publishing.

Mahmoud, M. A. B., & Guo, P. (2021). DNA sequence classification based on MLP with PILAE algorithm. *Soft Computing, 25*(5), 4003–4014. https://doi.org/10.1007/s00500-020-05429-y

Mock, F., Kretschmer, F., Kriese, A., Böcker, S., & Marz, M. (2021). BERTax: Taxonomic classification of DNA sequences with deep neural networks. *BioRxiv*, Article 2021–07.

Rizzo, R., Fiannaca, A., La Rosa, M., & Urso, A. (2016). A deep learning approach to DNA sequence classification. In C. Angelini, P. M. Rancoita, & S. Rovetta (Eds.), *Computational intelligence methods for bioinformatics and biostatistics* (pp. 129–140). Springer International Publishing.

Sarkar, S., Mridha, K., Ghosh, A., & Shaw, R. N. (2022). Machine learning in bioinformatics: New technique for DNA sequencing classification. In R. N. Shaw, S. Das, V. Piuri, & M. Bianchini (Eds.), *Advanced computing and intelligent technologies* (pp. 335–355). Springer Nature Singapore.

Zhang, J. X., Yordanov, B., Gaunt, A., Wang, M. X., Dai, P., Chen, Y.-J., Zhang, K., Fang, J. Z., Dalchau, N., & Li, J. (2021). A deep learning model for predicting next-generation sequencing depth from DNA sequence. *Nature Communications, 12*(1), 4387.

15

Artificial intelligence in drug discovery and development

Abdulhamit Subasi[1,2]

[1]Institute of Biomedicine, Faculty of Medicine, University of Turku, Turku, Finland; [2]Department of Computer Science, College of Engineering, Effat University, Jeddah, Saudi Arabia

1. Introduction

Artificial intelligence (AI) has become a ground-breaking technology in several industries, revolutionizing how we approach challenging issues. The development of new drugs is one area where AI has made enormous improvements. Traditional approaches to drug development are frequently time-consuming, expensive, and fraught with problems. But by incorporating AI methods, scientists and researchers may now quicken the drug discovery process, improve the effectiveness of clinical trials, and ultimately transform the pharmaceutical sector. Recently, the combination of AI with drug discovery has created previously unimaginable possibilities for addressing some of the most urgent healthcare issues. The analysis of enormous volumes of data, including genomic and molecular structural data, by AI algorithms enables more accurate and rapid discovery of potential drug candidates. Researchers may examine through large datasets, anticipate chemical interactions, and model the behavior of pharmacological compounds in a fraction of

the time it would normally take by utilizing machine learning, deep learning, and other AI approaches.

The identification and development of new therapeutic agents to treat a variety of diseases and enhance patient outcomes depend heavily on the fields of drug discovery and development. The conventional drug discovery approach, however, is time consuming, expensive, and frequently has meager success rates. AI has become a promising technology in recent years with the potential to completely alter the landscape of drug discovery and development. Target identification, lead optimization, and clinical trials are just a few of the stages of the drug discovery process that can be sped up using AI techniques like machine learning, deep learning, and data mining. Target identification, which entails identifying certain molecular targets connected to a given disease, is an important phase in the drug discovery process. Large datasets, such as genetic and proteomic data, have been mined using AI algorithms to find prospective therapeutic targets. To examine gene expression data and find novel drug—

Applications of Artificial Intelligence in Healthcare and Biomedicine
https://doi.org/10.1016/B978-0-443-22308-2.00018-4

disease interactions, machine learning methods can be used. Similar to this, AI-based techniques have been applied to determine the druggability of target proteins and anticipate protein-protein interactions. These AI-driven methods could hasten the identification of new targets and make it easier to confirm their therapeutic utility. The application of AI to drug discovery and development offers a special chance to get around the drawbacks of conventional trial-and-error techniques. Researchers may create virtual tests, forecast drug efficacy and toxicity, and improve treatment candidates before they are tested in the lab by utilizing AI-powered models and algorithms. This not only lowers costs but also lessens the risks involved in drug development, allowing for a more focused and effective method of bringing life-saving drugs to market (Deng et al., 2022).

The improvement of clinical trials and the facilitation of medication repurposing initiatives have both benefited from the use of AI approaches. Clinical trials are necessary for determining the efficacy and safety of new medications, but they are frequently expensive and time-consuming. AI algorithms can help in trial design optimization, patient population identification, and clinical outcome prediction, resulting in more effective and productive trials. To find possible chances for drug repurposing, large-scale biomedical data have also been mined using AI-driven methodologies. AI algorithms can discover novel therapeutic uses for existing chemicals by examining existing medications and their recognized targets, potentially lowering the time and expense needed for drug development.

To improve their performance and safety characteristics, lead compounds are refined and optimized. By predicting the characteristics and actions of possible medication candidates, AI approaches have been used to speed up this procedure. For instance, lead optimization has benefited from the application of deep learning algorithms to estimate the binding affinity of small compounds and target proteins. Additionally, novel drug-like compounds with desirable features have been created using generative models like variational autoencoders (VAEs) and generative adversarial networks (GANs). These artificial intelligence-driven methods have the potential to expedite lead optimization and simplify the drug development process.

Furthermore, the impact of AI on precision therapies and personalized medicine cannot be emphasized. AI can find patterns and connections that help in treating patients individually by evaluating large databases containing patient information, genetic profiles, and clinical results. With the ability to administer the most efficacious medications at the best doses while reducing undesirable side effects, this tailored approach has the potential to change patient care.

Despite the impressive progress made thus far, obstacles still stand in the way of fully realizing AI's potential for drug discovery and development. The necessity for ongoing validation of AI models, ethical issues, and legal frameworks are all critical issues that must be addressed. The future, however, holds great promise for AI-driven drug discovery and development due to continuous research, industry and academic partnerships, and breakthroughs in AI technologies.

In this chapter, we will present the real world of applications of artificial intelligence in drug discovery and development, exploring its key applications, cutting-edge techniques, and the potential impact on healthcare. We will also discuss the challenges and ethical considerations surrounding AI in this domain, and shed light on the exciting possibilities that lie ahead.

2. Background and literature review

The pharmaceutical sector is encountering productivity declines, with clinical trial failures and high drug development costs. Recent strides in AI present opportunities to reverse this trend and encourage pharmaceutical research and development (R&D). While AI and deep learning concepts are not novel, progress has surged due to enhanced computing power, abundant annotated training data, and advanced frameworks for deep neural networks (DNNs). As of 2014, DNNs have surpassed human-level accuracy across tasks like image and voice recognition, autonomous driving, and more. These AI breakthroughs hold the potential to revolutionize pharmaceutical R&D by offering more precise and efficient data analysis, predictions, and innovation. In essence, the pharmaceutical industry faces productivity and cost challenges, but recent AI advancements, particularly in deep learning, show promise in mitigating these issues. Progress in computing, data availability, and DNN implementation can reshape pharmaceutical R&D, improving drug discovery and development processes (Zhavoronkov, 2018).

Over the past 20 years, the R&D productivity in the pharmaceutical industry has come under scrutiny due to high attrition rates and the substantial costs associated with drug development. There has been a debate within the research community regarding the merits of two main drug discovery approaches: phenotypic and target-based. Each approach offers distinct advantages in identifying new molecular entities that can successfully reach the market. However, both approaches have faced challenges in terms of translating their findings into clinical applications. The emergence of AI and the adoption of machine learning (ML) tools hold great promise for revolutionizing the field of drug development and overcoming obstacles in the drug discovery pipeline. In this context, Malandraki-Miller and Riley (2021) assessed the potential of both target-driven and phenotypic-based approaches. While the collaborations between AI/ML and pharmaceutical companies are still relatively new, they examine the potential and existing limitations, with a specific focus on phenotypic drug discovery. The potential of AI and ML tools in revolutionizing drug development was explored, with a focus on target-driven and phenotypic-based approaches.

AI is playing a transformative role in reshaping drug discovery and development within the pharmaceutical industry. Yang et al. (2019) presented an overview of AI's current status in this domain, examining its applications, challenges, and influence. Through an analysis of diverse studies and research articles, the review explores how AI methods are being harnessed to enhance multiple phases of drug discovery and development, spanning target identification, lead optimization, drug design, and clinical trials. The application of AI techniques shows great potential in expediting target identification and validation by efficiently analyzing extensive biological and genetic data. Additionally, AI offers valuable tools for optimizing lead compounds during the drug development process, a pivotal stage for enhancing candidate compound efficacy and safety.

In the realm of drug discovery, AI-powered computer-aided drug design (CADD) tools have become widely utilized by scientists, aiding in decision making throughout various stages of drug development programs. ML-based expert systems have gained significant popularity for their ability to predict biological activity, toxicity, physicochemical properties, formulation compositions, and drug—target interactions. Expert systems based on AI have been developed for both conventional and innovative drug delivery systems. The healthcare sector has also seen the application of AI in personalized treatments, medical diagnosis, and addressing epidemic outbreaks. In recent times, there has been a notable increase in collaborations between companies involved in pharmaceuticals, medical devices, life sciences, and AI technology, aiming to drive advancements in the healthcare field (Banerjee et al., 2022).

The utilization of AI in these sectors not only reduces the burden on human resources but also enables the achievement of targets within shorter timeframes. Paul et al. (2021) discussed the ongoing challenges associated with the implementation of AI and proposed strategies to overcome them. Additionally, they presented the future prospects of AI in the pharmaceutical industry, highlighting its potential and anticipated developments. They also highlighted the increasing use of AI in various sectors of the pharmaceutical industry, emphasizing its role in accelerating drug discovery, repurposing existing drugs, improving productivity, and enhancing the efficiency of clinical trials.

Molecular docking is a computational technique used to predict the binding affinity between small molecules and target proteins. AI-based approaches, including deep learning and reinforcement learning, have been utilized to enhance the accuracy and efficiency of molecular docking. Accurate prediction of drug pharmacokinetics and toxicity is essential for optimizing drug candidates and ensuring patient safety. AI techniques have been employed to develop models that can predict pharmacokinetic properties and assess drug toxicity. Moreover, machine learning approaches have been applied to predict drug toxicity and identify potential adverse effects. The application of AI tools in drug discovery is on the rise. While some proponents highlight the vast opportunities that these tools offer, others remain skeptical, awaiting clear evidence of their impact on drug discovery projects. The reality likely lies somewhere between these extremes. However, it is evident that AI is presenting new challenges not only for the scientists involved but also for the biopharmaceutical industry and its established processes for discovering and developing new drugs (Schneider et al., 2020).

Artificial neural networks (ANNs) are computational models inspired by human brain function, designed to mimic information processing. In contrast to conventional programming, ANNs acquire knowledge by discerning patterns and correlations within data, learning through experience. ANNs can effectively integrate experimental and literature-based data to address various challenges. Their applications encompass classification, pattern recognition, prediction, and modeling. ANNs hold significant potential in pharmaceutical sciences, with applications ranging from interpreting analytical data and drug/dosage form design to biopharmacy and clinical pharmacy (Agatonovic-Kustrin & Beresford, 2000).

With the continuous progress in the field of artificial intelligence and machine learning (AI/ML) applied to drug discovery, Zhavoronkov et al. (2020) aimed to explore the implications of recent AI/ML advancements in the realm of Clinical Pharmacology. In this context, they investigated the challenges faced and the developments made in AI/ML techniques for target identification. Additionally, they investigated the role of AI/ML in generative chemistry, specifically its application in the discovery of small molecule drugs. Furthermore, they examined the potential impact of AI/ML on evaluating clinical trial outcomes. Moreover, they provided a brief overview of current trends in the utilization of AI/ML in healthcare and its implications for the daily practices of clinical pharmacologists. By addressing these key areas, the transformative potential of AI/ML and its influence on the field of Clinical Pharmacology was explored.

Drug discovery and development are essential processes in addressing health issues, but they are time consuming, complex, and costly. The COVID-19 pandemic has underscored the urgency for rapid drug and vaccine development. AI has revolutionized drug development by utilizing bioinformatics tools, notably in computer-aided drug design. AI-driven techniques, especially machine learning, enable efficient handling of biological data and algorithm development. AI-based technologies offer effective solutions for drug discovery challenges, transforming the pharmaceutical industry's capabilities. The applications of AI extend beyond COVID-19, providing insights into mechanisms of action and aiding vaccine development. These approaches hold promise for addressing various health problems and advancing treatments for diverse diseases (Tripathi et al., 2023).

AI in conjunction with life science technologies is harnessed to leverage diverse datasets for predictive modeling and informed decision-making. The amalgamation of AI and ML transforms drug design and development by enhancing disease understanding, target identification, drug optimization, and clinical efficacy evaluation. Moingeon et al. (2022) exemplify the potential by bridging patient-specific characteristics and drug properties, paving the way for computational precision medicine. This approach tailors therapies to individual patients' unique attributes, encompassing physiology, disease features, and environmental influences. The integration of AI and life sciences opens the door to personalized medical interventions, offering treatments aligned with each patient's distinct needs and circumstances.

The integration of computer-aided drug design techniques into drug discovery and development processes has been facilitated by advancements in computing power and computational biology. AI, particularly through machine learning, has become a crucial tool for data mining due to its capabilities and the availability of pharmacological data. AI is utilized in multiple drug design aspects, including virtual screening, quantitative structure-activity relationship (QSAR) analysis, de novo drug design, and ADME/T evaluation. Despite the complexity of AI models, they have proven invaluable in driving drug discovery. Deep learning methods have gained prominence recently, enhancing AI's predictive and generative capabilities for molecular properties and compound generation. In conclusion, the synergy of computing power, computational advancements, and AI's data mining has enabled the incorporation of computer-aided drug design techniques into drug discovery. AI's role in various aspects of drug design, supported by deep learning, holds the potential for further enhancing drug development strategies (Zhong et al., 2018).

AI methods, including deep learning, have emerged as innovative solutions for handling complex and diverse datasets related to drug candidates. These approaches offer new ways to assess the effectiveness and safety of drug candidates by analyzing large datasets. AI-based models provide valuable insights throughout the drug development journey, from analyzing chemical structures to predicting in vitro, in vivo, and clinical outcomes. The integration of AI with big data has enabled rational drug development and optimization, promising significant advancements in drug discovery and public health. By leveraging extensive datasets and AI techniques, researchers can accelerate the development of safer and more effective drugs (Zhu, 2020).

Conventional methods such as wet laboratory testing, validations, and synthetic procedures are both time-consuming and costly in the context of drug discovery. Over the past few decades, a multitude of AI-based models have been developed to aid various stages of the drug discovery process. These models serve as valuable complements to traditional experiments, expediting the drug discovery timeline. Chen et al. (2023) introduced widely used data resources in drug discovery, including ChEMBL and DrugBank. Furthermore, they summarize the algorithms utilized in the development of AI-based models for drug discovery. Additionally, they investigated the application of AI-based models for de novo drug design, drug—target structure prediction, drug—target interactions, and binding affinity prediction. Moreover, they highlighted the applications of AI in areas such as drug synergism/antagonism prediction and nanomedicine design.

The integration of AI in drug discovery and development is still in its early stages, and there are numerous avenues for future research and development. Collaboration between AI experts, computational biologists, chemists, and pharmacologists is vital for leveraging the full potential of AI in this domain. Additionally, regulatory agencies and policymakers need to establish guidelines and frameworks to ensure the responsible and ethical implementation of AI in drug discovery and development. Deep learning holds great potential for drug discovery, offering advancements in areas such as image analysis, molecular structure and function prediction, and automated generation of novel chemical compounds with specific properties. Despite the increasing success of these prospective applications, the mathematical models underlying deep learning often lack interpretability for human understanding. There is a growing demand for "explainable" deep learning methods to provide insights into the machine language of molecular sciences. In essence, deep learning has promising implications for drug discovery, but there is a need for explainable models that can be understood by human experts (Jiménez-Luna et al., 2020). AI holds great promise in revolutionizing the field of drug discovery and development. The application of AI techniques, such as machine learning and deep learning, can accelerate the identification of potential drug targets, optimize drug design, predict pharmacokinetics and toxicity, and facilitate personalized medicine. However, challenges related to data availability, model interpretability, and ethical considerations need to be addressed. With ongoing advancements and collaborative efforts, AI has the potential to transform the way new drugs are discovered, developed, and delivered to patients.

3. Artificial intelligence for DDD

3.1 Implementation of AI methods for DDD

This example adapted from[1]

1. Download Bioactivity Data

ChEMBL Database

The *ChEMBL Database* is a database that contains curated bioactivity data of more than two million compounds. It is compiled from more than 2.4M compounds, 88,000 documents, 1.6 million assays, 14K drugs, the data spans with 15K targets, 2K cells and 45K indications. [Data as of December 14, 2023; ChEMBL version 33].

Installing libraries

Install the ChEMBL web service package so that we can retrieve bioactivity data from the ChEMBL Database.

```
! pip install chembl_webresource_client
```

Importing libraries

```
# Import necessary libraries
import pandas as pd
from chembl_webresource_client.new_client import new_client
```

Search for Target protein
Target search for aromatase

```
# Target search for Aromatase
target = new_client.target
target_query = target.search('aromatase')
targets = pd.DataFrame.from_dict(target_query)
targets
```

Select and retrieve bioactivity data for Cytochrome P450 19A1 (first entry)

---

[1] https://github.com/dataprofessor/code/blob/master/python/CDD_ML_Part_4_Acetylcholinesterase_Regression_Random_Forest.ipynb.

We will assign the first entry (which corresponds to the target protein, single protein) to the *selected_target* variable.

```
selected_target = targets.target_chembl_id[0]
selected_target
```

Here, we will retrieve only bioactivity data for (CHEMBL1978) that are reported as IC50 values in nM (nanomolar) unit.

```
activity = new_client.activity
res = activity.filter(target_chembl_id=selected_target).filter(standard_type="IC50")
df = pd.DataFrame.from_dict(res)
df.head(3)
df.info()
df.standard_type.unique()
array(['IC50'], dtype=object)
```

Finally, we will save the resulting bioactivity data to a CSV file **bioactivity_data.csv**.

```
df.to_csv('bioactivity_data.csv', index=False)
```

Handling missing data

If any compound has a missing value for the **standard_value** column, then drop it.

```
df2 = df[df.standard_value.notna()]
df2
df2["standard_value"].value_counts()
df2[df2["standard_value"]==0]
```

Apparently, for this dataset, there is no missing data. But we can use the above code cell for bioactivity data of other target proteins.

Data pre-processing of the bioactivity data
Labeling compounds as either being active, inactive, or intermediate

The bioactivity data is in the IC50 unit. Compounds having values of less than 1000 nM will be considered to be **active** while those greater than 10,000 nM will be considered to be **inactive**.

As for those values between 1000 and 10,000 nM will be referred to as **intermediate**.

```
bioactivity_class = []
for i in df2.standard_value:
  if float(i) >= 10000:
    bioactivity_class.append("inactive")
  elif float(i) <= 1000:
    bioactivity_class.append("active")
  else:
    bioactivity_class.append("intermediate")
```

Iterate the *molecule_chembl_id* to a list

```
mol_cid = []
for i in df2.molecule_chembl_id:
  mol_cid.append(i)
```

Iterate *canonical_smiles* to a list

```
canonical_smiles = []
for i in df2.canonical_smiles:
  canonical_smiles.append(i)
```

Iterate *standard_value* to a list

```
standard_value = []
for i in df2.standard_value:
  standard_value.append(i)
```

Combine the 4 lists into a dataframe

```
data_tuples = list(zip(mol_cid, canonical_smiles, bioactivity_class, standard_value))
df3 = pd.DataFrame( data_tuples, columns=['molecule_chembl_id', 'canonical_smiles', 'bioactivity_class', 'standard_value'])
df3
df3[(df3['bioactivity_class']=='active')]
```

Saves dataframe to CSV file.

```
df3.to_csv('bioactivity_preprocessed_data.csv', index=False)
```

2. Exploratory Data Analysis

Install conda and rdkit

```
! wget https://repo.anaconda.com/miniconda/Miniconda3-py37_4.8.2-Linux-x86_64.sh
! chmod +x Miniconda3-py37_4.8.2-Linux-x86_64.sh
! bash ./Miniconda3-py37_4.8.2-Linux-x86_64.sh -b -f -p /usr/local
! conda install -c rdkit rdkit -y
import sys
```

Load bioactivity data

```
import pandas as pd
df = pd.read_csv('bioactivity_preprocessed_data.csv')
df
```

Calculate Lipinski descriptors

Christopher Lipinski, a scientist at Pfizer, came up with a set of rule-of-thumb for evaluating the **druglikeness** of compounds. Such druglikeness is based on the Absorption, Distribution, Metabolism, and Excretion (ADME) that is also known as the pharmacokinetic profile. Lipinski analyzed all orally active FDA-approved drugs in the formulation of what is to be known as the **Rule-of-Five** or **Lipinski's Rule**.

The Lipinski's Rule stated the following:

- Molecular weight < 500 Da
- Octanol-water partition coefficient (LogP) < 5
- Hydrogen bond donors < 5
- Hydrogen bond acceptors < 10

Import libraries

```
import numpy as np
from rdkit import Chem
from rdkit.Chem import Descriptors, Lipinski
```

Calculate descriptors

```
# Inspired by: https://codeocean.com/explore/capsules?query=tag:data-curation
def lipinski(smiles, verbose=False):
  moldata= []
  for elem in smiles:
    mol=Chem.MolFromSmiles(elem)
    moldata.append(mol)

  baseData= np.arange(1,1)
  i=0
  for mol in moldata:

    desc_MolWt = Descriptors.MolWt(mol)
    desc_MolLogP = Descriptors.MolLogP(mol)
    desc_NumHDonors = Lipinski.NumHDonors(mol)
    desc_NumHAcceptors = Lipinski.NumHAcceptors(mol)

    row = np.array([desc_MolWt,
            desc_MolLogP,
            desc_NumHDonors,
            desc_NumHAcceptors])


    if(i==0):
      baseData=row
    else:
      baseData=np.vstack([baseData, row])
    i=i+1

  columnNames=["MW","LogP","NumHDonors","NumHAcceptors"]
  descriptors = pd.DataFrame(data=baseData,columns=columnNames)

  return descriptors
df_lipinski = lipinski(df.canonical_smiles)
```

Combine DataFrames

Let's take a look at the 2 DataFrames that will be combined.

df_lipinski

df

Now, let's combine the 2 DataFrame.

```
df_combined = pd.concat([df,df_lipinski], axis=1)
df_combined
df_combined[df_combined["standard_value"]==0]
df_combined.drop([2149], axis=0, inplace=True)
df_combined[df_combined["standard_value"]==0]
```

Convert IC50 to pIC50

To allow **IC50** data to be more uniformly distributed, we will convert **IC50** to the negative logarithmic scale, which is essentially $-\log10(IC50)$.

This custom function pIC50() will accept a DataFrame as input and will:

- Take the IC50 values from the standard_value column and convert them from nM to M by multiplying the value by 10−9
- Take the molar value and apply −log10
- Delete the standard_value column and create a new pIC50 column

```
import numpy as np
def pIC50(input):
  pIC50 = []
  for i in input['standard_value_norm']:
    molar = i*(10**-9) # Converts nM to M
    pIC50.append(-np.log10(molar))
  input['pIC50'] = pIC50
  x = input.drop('standard_value_norm', 1)
  return x
```

Point to note: Values greater than 100,000,000 will be fixed at 100,000,000 otherwise the negative logarithmic value will become negative.

```
df_combined.standard_value.describe()
-np.log10( (10**-9)* 100000000 )
-np.log10( (10**-9)* 10000000000 )
def norm_value(input):
  norm = []
  for i in input['standard_value']:
    if i > 100000000:
    i = 100000000
    norm.append(i)
  input['standard_value_norm'] = norm
  x = input.drop('standard_value', 1)
  return x
```

We will first apply the norm_value() function so that the values in the standard_value column are normalized.

```
df_norm = norm_value(df_combined)
df_norm
df_norm.standard_value_norm.describe()

df_final = pIC50(df_norm)
df_final
df_final.pIC50.describe()
```

Removing the "intermediate" bioactivity class

Here, we will be removing the intermediate class from our data set.

```
df_2class = df_final[df_final.bioactivity_class != 'intermediate']
df_2class
```

Exploratory Data Analysis (Chemical Space Analysis) via Lipinski descriptors
Import library

```
import seaborn as sns
sns.set(style='ticks')
import matplotlib.pyplot as plt
```

Frequency plot of the 2 bioactivity classes

```
plt.figure(figsize=(5.5, 5.5))
sns.countplot(x='bioactivity_class', data=df_2class, edgecolor='black')
plt.xlabel('Bioactivity class', fontsize=14, fontweight='bold')
plt.ylabel('Frequency', fontsize=14, fontweight='bold')
plt.savefig('plot_bioactivity_class.pdf')
```

Scatter plot of MW versus LogP

It can be seen that the 2 bioactivity classes are spanning similar chemical spaces as evidenced by the scatter plot of MW versus LogP.

```
plt.figure(figsize=(5.5, 5.5))
sns.scatterplot(x='MW', y='LogP', data=df_2class, hue='bioactivity_class', size='pIC50', edgecolor='black', alpha=0.7)
plt.xlabel('MW', fontsize=14, fontweight='bold')
plt.ylabel('LogP', fontsize=14, fontweight='bold')
plt.legend(bbox_to_anchor=(1.05, 1), loc=2, borderaxespad=0)
#plt.savefig('plot_MW_vs_LogP.pdf')
```

Box plots
pIC50 value

```
plt.figure(figsize=(5.5, 5.5))
sns.boxplot(x = 'bioactivity_class', y = 'pIC50', data = df_2class)
plt.xlabel('Bioactivity class', fontsize=14, fontweight='bold')
plt.ylabel('pIC50 value', fontsize=14, fontweight='bold')
plt.savefig('plot_ic50.pdf')
```

Statistical analysis | Mann–Whitney U Test

```python
def mannwhitney(descriptor, verbose=False):
    # https://machinelearningmastery.com/nonparametric-statistical-significance-tests-in-python/
    from numpy.random import seed
    from numpy.random import randn
    from scipy.stats import mannwhitneyu
    # seed the random number generator
    seed(1)
    # actives and inactives
    selection = [descriptor, 'bioactivity_class']
    df = df_2class[selection]
    active = df[df.bioactivity_class == 'active']
    active = active[descriptor]
    selection = [descriptor, 'bioactivity_class']
    df = df_2class[selection]
    inactive = df[df.bioactivity_class == 'inactive']
    inactive = inactive[descriptor]
    # compare samples
    stat, p = mannwhitneyu(active, inactive)
    #print('Statistics=%.3f, p=%.3f' % (stat, p))
    # interpret
    alpha = 0.05
```

```python
    if p > alpha:
        interpretation = 'Same distribution (fail to reject H0)'
    else:
        interpretation = 'Different distribution (reject H0)'

    results = pd.DataFrame({'Descriptor':descriptor,
                            'Statistics':stat,
                            'p':p,
                            'alpha':alpha,
                            'Interpretation':interpretation}, index=[0])
    filename = 'mannwhitneyu_' + descriptor + '.csv'
    results.to_csv(filename)

    return results
mannwhitney('pIC50')
```

MW

```python
plt.figure(figsize=(5.5, 5.5))
sns.boxplot(x = 'bioactivity_class', y = 'MW', data = df_2class)
plt.xlabel('Bioactivity class', fontsize=14, fontweight='bold')
plt.ylabel('MW', fontsize=14, fontweight='bold')
plt.savefig('plot_MW.pdf')
mannwhitney('MW')
```

LogP

```python
plt.figure(figsize=(5.5, 5.5))
sns.boxplot(x = 'bioactivity_class', y = 'LogP', data = df_2class)
plt.xlabel('Bioactivity class', fontsize=14, fontweight='bold')
plt.ylabel('LogP', fontsize=14, fontweight='bold')

plt.savefig('plot_LogP.pdf')
mannwhitney('LogP')
```

NumHDonors

```
plt.figure(figsize=(5.5, 5.5))
sns.boxplot(x = 'bioactivity_class', y = 'NumHDonors', data = df_2class)
plt.xlabel('Bioactivity class', fontsize=14, fontweight='bold')
plt.ylabel('NumHDonors', fontsize=14, fontweight='bold')
plt.savefig('plot_NumHDonors.pdf')
mannwhitney('NumHDonors')
```

NumHAcceptors

```
plt.figure(figsize=(5.5, 5.5))
sns.boxplot(x = 'bioactivity_class', y = 'NumHAcceptors', data = df_2class)
plt.xlabel('Bioactivity class', fontsize=14, fontweight='bold')
plt.ylabel('NumHAcceptors', fontsize=14, fontweight='bold')
plt.savefig('plot_NumHAcceptors.pdf')
mannwhitney('NumHAcceptors')
```

Interpretation of Statistical Results
Box Plots
pIC50 values

Taking a look at pIC50 values, the **actives** and **inactives** displayed *statistically significant difference*, which is to be expected since threshold values (IC50 < 1000 nM = Actives while IC50 > 10,000 nM = Inactives, corresponding to pIC50 > 6 = Actives and pIC50 < 5 = Inactives) were used to define actives and inactives.

Lipinski's descriptors

Of the 4 Lipinski's descriptors (MW, LogP, NumHDonors, and NumHAcceptors), only LogP exhibited *no difference* between the **actives** and **inactives** while the other 3 descriptors (MW, NumHDonors, and NumHAcceptors) shows *statistically significant difference* between **actives** and **inactives**.

Zip files

```
! zip -r results.zip . -i *.csv *.pdf
```

3. Descriptor Calculation and Dataset Preparation

Download PaDEL-Descriptor

```
! wget https://github.com/dataprofessor/bioinformatics/raw/master/padel.zip
! wget https://github.com/dataprofessor/bioinformatics/raw/master/padel.sh
! unzip padel.zip
df_2class
df_2class.to_csv('df_2class.csv', index=False)

selection = ['canonical_smiles','molecule_chembl_id']
df_selection = df_2class[selection]
df_selection.to_csv('molecule.smi', sep='\t', index=False, header=False)
! cat molecule.smi | head -5
! cat molecule.smi | wc -l
```

Calculate fingerprint descriptors
Calculate PaDEL descriptors

```
! cat padel.sh
! bash padel.sh
```

Preparing the X and Y Data Matrices
X data matrix

```
import pandas as pd
import numpy as np
df_X = pd.read_csv('descriptors_output.csv')
df_2class = pd.read_csv('df_2class.csv')
df_X
X = df_X.drop(columns=['Name'])
df_2class.head()
df_2class['bioactivity_class']=df_2class['bioactivity_class'].map(lambda x:1 if x== "active" else 0)

Y = df_2class['bioactivity_class']
```

Input features

The *Aromatase* dataset contains 881 input features and 1 output variable (pIC50 values).

```
X
```

Output features

```
Y
Y.value_counts()
```

Let's examine the data dimension

```
X.shape
Y.shape
from sklearn.model_selection import train_test_split
X_train, X_test, y_train, y_test = train_test_split( X, Y, test_size=0.20, random_state=42)
```

Feature Selection

1. VarianceThreshold

```
from sklearn.feature_selection import VarianceThreshold
selection = VarianceThreshold(threshold=(.8 * (1 - .8)))
X = selection.fit_transform(X)
models = GetBasedModel()
names,results = BasedLine2(X_train, y_train, models)
PlotBoxR().PlotResult(names,results)
```

2. RandomForest

```
from sklearn.ensemble import RandomForestClassifier
from sklearn.feature_selection import RFECV
rf_model= RFECV(estimator=RandomForestClassifier())
model_rf_model = RandomForestClassifier(n_estimators=100, random_state=92116)
rf = RandomForestClassifier(random_state=0)
 rf.fit(X_train,y_train)
RandomForestClassifier(random_state=0)
 rfe = RFECV(rf,cv=5,scoring="accuracy")
 rfe.fit(X_train,y_train)
RFECV(cv=5, estimator=RandomForestClassifier(random_state=0),
    scoring='accuracy')
selected_features = np.array(X.columns)[rfe.get_support()]
selected_features.shape
X_info=X[(selected_features)]
X_info
X_train, X_test, y_train, y_test = train_test_split(X_info, Y, test_size=0.2, random_state = 42)

models = GetBasedModel()
names,results = BasedLine2(X_train, y_train, models)
PlotBoxR().PlotResult(names,results)
```

3. Information Gain — Classification Problems

```
from sklearn.feature_selection import SelectKBest, mutual_info_regression,mutual_info_classif
infogain_classif = SelectKBest(score_func=mutual_info_classif, k=6)
infogain_classif.fit(X_train, y_train)
d1 = X.columns.tolist()
# what are scores for the features
for i in range(len(infogain_classif.scores_)):
    print(f'Feature {i} : {round(infogain_classif.scores_[i],3)}')
print()
# plot the scores
plt.bar([d1[i] for i in range(len(infogain_classif.scores_))], infogain_classif.scores_)
plt.xticks(rotation=90)
plt.rcParams["figure.figsize"] = (8,6)
plt.show()
infogain_classif.scores_
infogain_classif_scores_df = pd.DataFrame(infogain_classif.scores_,index=X.columns,columns=['Rank']).sort_
values(by='Rank',ascending=False)
infogain_classif_scores_df.head()
X_infogain_selected=infogain_classif_scores_df[infogain_classif_scores_df["Rank"]>=0.01]
X_infogain_selected
X_infogain=list(X_infogain_selected.index)
X_infogain
X_info=X[(X_infogain)]
X_info
X_train, X_test, Y_train, Y_test = train_test_split(X_info, Y, test_size=0.2, random_state = 42)

models = GetBasedModel()
names,results = BasedLine2(X_train, y_train, models)
PlotBoxR().PlotResult(names,results)
```

4. Select KBest for Classification Problems

```python
from sklearn.feature_selection import SelectKBest,SelectPercentile,f_classif,f_regression
Kbest_classif = SelectKBest(score_func=f_classif, k=6)
Kbest_classif.fit(X_train, y_train)
d1 = X.columns.tolist()
# what are scores for the features
for i in range(len(Kbest_classif.scores_)):
    print(f'Feature {i} : {round(Kbest_classif.scores_[i],3)}')
print()
plt.bar([d1[ i] for i in range(len(Kbest_classif.scores_))], Kbest_classif.scores_)
plt.xticks(rotation=90)
plt.rcParams["figure.figsize"] = (12,16)
plt.show()
Kbest_classif_scores_df = pd.DataFrame(Kbest_classif.scores_,index=X.columns,columns=['Rank']).sort_value
s(by='Rank',ascending=False)
Kbest_classif_scores_df.head(20)
Kbest_classif_scores_df.describe()
X_kbest_selected=Kbest_classif_scores_df[Kbest_classif_scores_df["Rank"]>=10]
X_kbest_selected
X_kbest=list(X_kbest_selected.index)
X_kbest
X_kbest=X[(X_kbest)]
X_kbest
X_train, X_test, y_train, y_test = train_test_split(X_kbest, Y, test_size=0.2, random_state = 42)

models = GetBasedModel()
names,results = BasedLine2(X_train, y_train, models)
PlotBoxR().PlotResult(names,results)
```

5. Select Percentile

```
percentile = SelectPercentile(percentile=50)
percentile.fit(X_train, y_train)
percentile.get_support()
Select_Percentile_=pd.DataFrame(percentile.get_support(), index=X.columns,columns=['Rank'])
Select_Percentile_df = pd.DataFrame(Select_Percentile_,index=X.columns,columns=['Rank']).sort_values(by=
'Rank',ascending=False)
Select_Percentile_df
X_Select_Percentile=Select_Percentile_df[Select_Percentile_df["Rank"]==True]
X_Select_Percentile
X_Select=list(X_Select_Percentile.index)
X_Select=X[(X_Select)]
X_Select
X_train, X_test, y_train, y_test = train_test_split(X_Select, Y, test_size=0.2, random_state = 42)

models = GetBasedModel()
names,results = BasedLine2(X_train, y_train, models)
PlotBoxR().PlotResult(names,results)
```

sampling with feature selection
BEST RESULT!!!

```
# resample by adding samples to minority class using SMOTE
from collections import Counter
from imblearn.combine import SMOTETomek
from imblearn.under_sampling import TomekLinks
# 'majority': resample only the majority class;
resample = SMOTETomek(tomek=TomekLinks(sampling_strategy='majority'))
X_oversamp, Y_over_samp = resample.fit_resample(X_Select, Y)
#split data train and test
from sklearn.model_selection import train_test_split
X_train, X_test, y_train, y_test = train_test_split(X_oversamp,
                    Y_over_samp,
                    test_size=0.3,
                    random_state = 10)
print(f"Class counts after resampling {Counter( Y_over_samp)}")
Class counts after resampling Counter({1: 1386, 0: 1361})
models = GetBasedModel()
names,results = BasedLine2(X_train, y_train, models)
PlotBoxR().PlotResult(names,results)
ex=ExtraTreesClassifier()
extra_model = ex.fit(X_train, y_train)
y_pred=extra_model.predict(X_test)
mse = mean_squared_error(y_test, extra_model.predict(X_test))
plot_matrics(y_pred)
```

Base Models

```
!pip install chart_studio
# Load libraries
# plot
import warnings
warnings.filterwarnings("ignore")

from sklearn.metrics import mean_squared_error
from sklearn.metrics import mean_absolute_error
from sklearn.metrics import mean_squared_error, r2_score
import numpy as np
import matplotlib.pyplot as plt
#from chart_studio.plotly import plot, iplot
import plotly.offline as py
import plotly.graph_objs as go
import plotly.tools as tls
py.init_notebook_mode(connected=True)
import matplotlib.font_manager
#models
from pandas import set_option
from pandas.plotting import scatter_matrix
from sklearn.preprocessing import StandardScaler
from sklearn.model_selection import KFold
from sklearn.model_selection import StratifiedKFold
from sklearn.model_selection import cross_val_score
from sklearn.model_selection import GridSearchCV
from sklearn.metrics import classification_report
from sklearn.metrics import confusion_matrix
from sklearn.metrics import accuracy_score
```

```python
from sklearn.pipeline import Pipeline
from sklearn.linear_model import LogisticRegression
from sklearn.tree import DecisionTreeClassifier
from sklearn.neighbors import KNeighborsClassifier
from sklearn.discriminant_analysis import LinearDiscriminantAnalysis
from sklearn.naive_bayes import GaussianNB
from sklearn.svm import SVC
from sklearn.ensemble import AdaBoostClassifier
from sklearn.ensemble import GradientBoostingClassifier
from sklearn.ensemble import RandomForestClassifier
from sklearn.ensemble import ExtraTreesClassifier
from sklearn.ensemble import BaggingClassifier
from xgboost import XGBClassifier
# Spot-Check Algorithms
def GetBasedModel():
    basedModels = []
    basedModels.append(('LR'  , LogisticRegression(max_iter= 1000)))
    basedModels.append(('LDA' , LinearDiscriminantAnalysis()))
    basedModels.append(('KNN' , KNeighborsClassifier()))
    basedModels.append(('CART', DecisionTreeClassifier()))
    basedModels.append(('NB'  , GaussianNB()))
    basedModels.append(('SVM' , SVC(probability=True)))
    basedModels.append(('ADA' , AdaBoostClassifier()))
    basedModels.append(('GBM' , GradientBoostingClassifier()))
    basedModels.append(('RF'  , RandomForestClassifier()))
    basedModels.append(('ET'  , ExtraTreesClassifier()))
    basedModels.append(('Bagging'  , BaggingClassifier()))
```

```python
    basedModels.append(('XGBoost'   , XGBClassifier(eval_metric='mlogloss')))
    return basedModels


def BasedLine2(X_train, y_train, models):
    # Test options and evaluation metric
    num_folds = 10
    scoring = 'accuracy'
    results = []
    names = []
    for name, model in models:
        kfold = StratifiedKFold(n_splits=num_folds, random_state=42, shuffle=True)
        cv_results = cross_val_score(model, X_train, y_train, cv=kfold, scoring=scoring)
        results.append(cv_results)
        names.append(name)
        msg = "%s: %f (%f)" % (name, cv_results.mean(), cv_results.std())
        print(msg)
    return names, results
class PlotBoxR(object):
    def _Trace(self,nameOfFeature,value):
        trace = go.Box(
            y=value,
            name = nameOfFeature,
            marker = dict(
                color = 'rgb(0, 128, 128)',
            )
        )
        return trace
```

```
def PlotResult(self,names,results):
    data = []
    for i in range(len(names)):
        data.append(self._Trace(names[i],results[i]))
    py.iplot(data)
models = GetBasedModel()
names,results = BasedLine2(X_train, y_train, models)
PlotBoxR().PlotResult(names,results)
def ScoreDataFrame(names,results):
    def floatingDecimals(f_val, dec=3):
        prc = "{:."+str(dec)+"f}"

        return float(prc.format(f_val))
    scores = []
    for r in results:
        scores.append(floatingDecimals(r.mean(),4))
    scoreDataFrame = pd.DataFrame({'Model':names, 'Score': scores})
    return scoreDataFrame
basedLineScore = ScoreDataFrame(names,results)
basedLineScore
```

SCALER

```python
from sklearn.preprocessing import StandardScaler
from sklearn.preprocessing import MinMaxScaler
from sklearn.preprocessing import RobustScaler
def GetScaledModel(nameOfScaler):

  if nameOfScaler == 'standard':
    scaler = StandardScaler()
  elif nameOfScaler =='minmax':
    scaler = MinMaxScaler()
  elif nameOfScaler =='robust':
    scaler = RobustScaler()

  pipelines = []
  pipelines.append((nameOfScaler+'LR' , Pipeline([('Scaler', scaler),('LR' , LogisticRegression())])))
  pipelines.append((nameOfScaler+'LDA' , Pipeline([('Scaler', scaler),('LDA' , LinearDiscriminantAnalysis())]
)))
  pipelines.append((nameOfScaler+'KNN' , Pipeline([('Scaler', scaler),('KNN' , KNeighborsClassifier())])))
  pipelines.append((nameOfScaler+'CART', Pipeline([('Scaler', scaler),('CART', DecisionTreeClassifier())])))
  pipelines.append((nameOfScaler+'NB' , Pipeline([('Scaler', scaler),('NB' , GaussianNB())])))
  pipelines.append((nameOfScaler+'SVM' , Pipeline([('Scaler', scaler),('SVM' , SVC())])))
  pipelines.append((nameOfScaler+'AB' , Pipeline([('Scaler', scaler),('AB' , AdaBoostClassifier())]) ))
  pipelines.append((nameOfScaler+'GBM' , Pipeline([('Scaler', scaler),('GMB' , GradientBoostingClassifier())]
) ))
  pipelines.append((nameOfScaler+'RF' , Pipeline([('Scaler', scaler),('RF' , RandomForestClassifier())]) ))
  pipelines.append((nameOfScaler+'ET' , Pipeline([('Scaler', scaler),('ET' , ExtraTreesClassifier())]) ))
  pipelines.append((nameOfScaler+'Bagging', Pipeline([('Scaler', scaler),('Bagging', BaggingClassifier())])))
  pipelines.append((nameOfScaler+'XGBoost', Pipeline([('Scaler', scaler),('Bagging', XGBClassifier(eval_metr
ic='mlogloss'))])))

  return pipelines
```

```
models = GetScaledModel('standard')
names,results = BasedLine2(X_train, y_train, models)
PlotBoxR().PlotResult(names,results)
scaledScoreStandard = ScoreDataFrame(names,results)
compareModels = pd.concat([basedLineScore,
            scaledScoreStandard], axis=1)
compareModels
models = GetScaledModel('minmax')
names,results = BasedLine2(X_train, y_train,models)
PlotBoxR().PlotResult(names,results)
scaledScoreMinMax = ScoreDataFrame(names,results)
compareModels = pd.concat([basedLineScore,
            scaledScoreStandard,
            scaledScoreMinMax], axis=1)
compareModels
models = GetScaledModel('robust')
names,results = BasedLine2(X_train, y_train,models)
PlotBoxR().PlotResult(names,results)
scaledScoreMinMax = ScoreDataFrame(names,results)
compareModels = pd.concat([basedLineScore,
            scaledScoreStandard,
            scaledScoreMinMax], axis=1)
compareModels
```

Metrics with best model

```python
def plot_matrics(y_pred):
    #y_pred = pipeline.predict(X_test)
    from sklearn import metrics
    from sklearn.metrics import roc_auc_score,precision_recall_curve,roc_curve
    from sklearn.metrics import confusion_matrix
    from sklearn.metrics import accuracy_score
    print('Accuracy:', np.round(metrics.accuracy_score(y_test, y_pred),4))
    print('Precision:', np.round(metrics.precision_score(y_test, y_pred,average='weighted'),4))
    print('Recall:', np.round(metrics.recall_score(y_test, y_pred,
                       average='weighted'),4))
    print('F1 Score:', np.round(metrics.f1_score(y_test, y_pred,
                       average='weighted'),4))
    print('Cohen Kappa Score:', np.round(metrics.cohen_kappa_score(y_test, y_pred),4))
    print('Matthews Corrcoef:', np.round(metrics.matthews_corrcoef(y_test, y_pred),4))
    from sklearn.metrics import roc_auc_score
    print("roc_auc_score:", roc_auc_score(y_test, y_pred, average=None))
    print('\t\tClassification Report:\n', metrics.classification_report(y_pred,y_test))
    from sklearn.metrics import confusion_matrix
    print("Confusion Matrix:\n",confusion_matrix(y_test, y_pred))
    #Plot Confusion Matrix
    from sklearn.metrics import confusion_matrix
    from io import BytesIO #neded for plot
    import seaborn as sns; sns.set()
    import matplotlib.pyplot as plt
    mat = confusion_matrix(y_test, y_pred)
    sns.heatmap(mat.T, square=True, annot=True, fmt='d', cbar=False)
    plt.xlabel('true label')
    plt.ylabel('predicted label');



plt.savefig("Confusion.jpg")
# Save SVG in a fake file object.
f = BytesIO()
plt.savefig(f, format="svg")
ex=ExtraTreesClassifier()
#scaler = StandardScaler()
extra_model = ex.fit(X_train, y_train)
y_pred=extra_model.predict(X_test)
mse = mean_squared_error(y_test, extra_model.predict(X_test))
plot_matrics(y_pred)
```

Automated machine learning

```
!pip3 install auto-sklearn
!pip install scikit-learn==0.24.0


from autosklearn.experimental.askl2 import AutoSklearn2Classifier
import autosklearn.classification
cls = autosklearn.classification.AutoSklearnClassifier()
cls.fit(X_train, y_train)
predictions = cls.predict(X_test)
import sklearn.metrics
print("Accuracy score:", sklearn.metrics.accuracy_score(y_test, predictions))
```

Model 1: sampling_strategy = "majority"

```
# resample by adding samples to minority class using SMOTE
from collections import Counter
from imblearn.combine import SMOTETomek
from imblearn.under_sampling import TomekLinks
# 'majority': resample only the majority class;
resample = SMOTETomek(tomek=TomekLinks(sampling_strategy='majority'))
X_oversamp, Y_over_samp = resample.fit_resample(X, Y)
#split data train and test
from sklearn.model_selection import train_test_split
X_train, X_test, y_train, y_test = train_test_split(X_oversamp,
                          Y_over_samp,
                          test_size=0.3,
                          random_state = 10)
print(f"Class counts after resampling {Counter( Y_over_samp)}")
models = GetBasedModel()
names,results = BasedLine2(X_train, y_train, models)
PlotBoxR().PlotResult(names,results)
ex=ExtraTreesClassifier()
extra_model = ex.fit(X_train, y_train)
y_pred=extra_model.predict(X_test)
mse = mean_squared_error(y_test, extra_model.predict(X_test))
plot_matrics(y_pred)
from sklearn import svm
C=200
""" The parameters and kernels of SVM classifierr can be changed as follows
C = 10.0  # SVM regularization parameter
svm.SVC(kernel='linear', C=C)
svm.LinearSVC(C=C, max_iter=10000)
```

```
svm.SVC(kernel='rbf', gamma=0.7, C=C)
svm.SVC(kernel='poly', degree=3, gamma='auto', C=C))
"""
clf = svm.SVC(kernel='rbf', gamma=0.7, C=C)
#clf = svm.SVC(kernel='poly', degree=3, gamma='auto', C=C)
model=clf.fit(X_train, y_train)
y_pred=model.predict(X_test)
plot_matrics(y_pred)
```

RobustScaler

```
from sklearn.preprocessing import RobustScaler
robust = RobustScaler()
X_train_std = robust.fit_transform(X_train)
X_test_std = robust.transform(X_test)
ex=ExtraTreesClassifier()
extra_model = ex.fit(X_train_std, y_train)
y_pred=extra_model.predict(X_test)
mse = mean_squared_error(y_test, extra_model.predict(X_test))
plot_matrics(y_pred)
```

Model 2: sampling_strategy: not minority

```
# resample by adding samples to minority class using SMOTE
from collections import Counter
from imblearn.combine import SMOTETomek
from imblearn.under_sampling import TomekLinks
# 'not minority': resample all classes but the minority class;
resample = SMOTETomek(tomek=TomekLinks(sampling_strategy='not minority'))
X_oversamp, Y_over_samp = resample.fit_resample(X, Y)
#split data train and test
from sklearn.model_selection import train_test_split
X_train, X_test, y_train, y_test = train_test_split(X_oversamp,
                    Y_over_samp,
                    test_size=0.3,
                    random_state = 10)
print(f"Class counts after resampling {Counter( Y_over_samp)}")
models = GetBasedModel()
names,results = BasedLine2(X_train, y_train, models)
PlotBoxR().PlotResult(names,results)
```

Model 3: sampling_strategy: not majority

```python
# resample by adding samples to minority class using SMOTE
from collections import Counter
from imblearn.combine import SMOTETomek
from imblearn.under_sampling import TomekLinks
# 'not majority': resample all classes but the majority class;
resample = SMOTETomek(tomek=TomekLinks(sampling_strategy='not majority'))
X_oversamp, Y_over_samp = resample.fit_resample(X, Y)
#split data train and test
from sklearn.model_selection import train_test_split
X_train, X_test, y_train, y_test = train_test_split(X_oversamp,
                        Y_over_samp,
                        test_size=0.3,
                        random_state = 10)
print(f"Class counts after resampling {Counter( Y_over_samp)}")
models = GetBasedModel()
names,results = BasedLine2(X_train, y_train, models)
PlotBoxR().PlotResult(names,results)
```

DEEP LEARNING CONV1D

```python
from keras.models import Sequential
from keras.layers import Dense, Conv1D, Flatten, MaxPooling1D
from sklearn.model_selection import train_test_split
from sklearn.metrics import confusion_matrix
def plot_matrics(pred_y):
    #y_pred = pipeline.predict(X_test)
    from sklearn import metrics
    from sklearn.metrics import roc_auc_score,precision_recall_curve,roc_curve
    from sklearn.metrics import confusion_matrix
    from sklearn.metrics import accuracy_score
    print('Accuracy:', np.round(metrics.accuracy_score(y_test, pred_y),4))
    print('Precision:', np.round(metrics.precision_score(y_test, pred_y,average='weighted'),4))
    print('Recall:', np.round(metrics.recall_score(y_test, pred_y,
                        average='weighted'),4))
    print('F1 Score:', np.round(metrics.f1_score(y_test,pred_y,
                        average='weighted'),4))
    print('Cohen Kappa Score:', np.round(metrics.cohen_kappa_score(y_test, pred_y),4))
    print('Matthews Corrcoef:', np.round(metrics.matthews_corrcoef(y_test, pred_y),4))
    from sklearn.metrics import roc_auc_score
    print("roc_auc_score:", roc_auc_score(y_test, pred_y, average=None))
    print('\t\tClassification Report:\n', metrics.classification_report(pred_y, y_test))
    from sklearn.metrics import confusion_matrix
    print("Confusion Matrix:\n",confusion_matrix(y_test, pred_y))
    #Plot Confusion Matrix
    from sklearn.metrics import confusion_matrix
    from io import BytesIO #neded for plot
    import seaborn as sns; sns.set()
    import matplotlib.pyplot as plt
```

```
mat = confusion_matrix(y_test, pred_y)
sns.heatmap(mat.T, square=True, annot=True, fmt='d', cbar=False)
plt.xlabel('true label')
plt.ylabel('predicted label');
plt.savefig("Confusion.jpg")
# Save SVG in a fake file object.
f = BytesIO()
plt.savefig(f, format="svg")
model = Sequential()
model.add(Conv1D(128, 2, activation="relu", input_shape=(881,1)))
model.add(Dense(64, activation="relu"))
model.add(MaxPooling1D())
model.add(Dense(32, activation="relu"))
model.add(MaxPooling1D())
model.add(Dense(16, activation="relu"))
model.add(MaxPooling1D())
model.add(Flatten())
model.add(Dense(2, activation = 'softmax'))
model.compile(loss = 'sparse_categorical_crossentropy',
    optimizer = "adam",



        metrics = ['accuracy'])
model.summary()
model.fit(X_train, y_train, batch_size=16,epochs=100, verbose=0)
acc = model.evaluate(X_train, y_train)
print("Loss:", acc[0], " Accuracy:", acc[1])


pred = model.predict(X_test)
pred_y = pred.argmax(axis=-1)
cm = confusion_matrix(y_test, pred_y)
print(cm)
plot_matrics(pred_y)
```

sampling_strategy = "majority"

```
# resample by adding samples to minority class using SMOTE
from collections import Counter
from imblearn.combine import SMOTETomek
from imblearn.under_sampling import TomekLinks
# 'majority': resample only the majority class;
resample = SMOTETomek(tomek=TomekLinks(sampling_strategy='majority'))
X_oversamp, Y_over_samp = resample.fit_resample(X, Y)
#split data train and test
from sklearn.model_selection import train_test_split
X_train, X_test, y_train, y_test = train_test_split(X_oversamp,
                        Y_over_samp,
                        test_size=0.3,
                        random_state = 10)
print(f"Class counts after resampling {Counter( Y_over_samp)}")
model = Sequential()
model.add(Conv1D(128, 2, activation="relu", input_shape=(881,1)))
model.add(Dense(64, activation="relu"))
model.add(MaxPooling1D())
model.add(Dense(32, activation="relu"))
model.add(MaxPooling1D())
model.add(Dense(16, activation="relu"))
```

```
model.add(MaxPooling1D())
model.add(Flatten())
model.add(Dense(2, activation = 'softmax'))
model.compile(loss = 'sparse_categorical_crossentropy',
    optimizer = "adam",
        metrics = ['accuracy'])
model.summary()
model.fit(X_train, y_train, batch_size=16,epochs=100, verbose=0)
acc = model.evaluate(X_train, y_train)
print("Loss:", acc[0], " Accuracy:", acc[1])
pred = model.predict(X_test)
pred_y = pred.argmax(axis=-1)
cm = confusion_matrix(y_test, pred_y)
print(cm)
plot_matrics(pred_y)
```

LSTM

```python
import numpy
from keras.models import Sequential
from keras.layers import Dense
from keras.layers import LSTM
from keras.layers.embeddings import Embedding
from keras.preprocessing import sequence
# fix random seed for reproducibility
numpy.random.seed(7)
from tensorflow.keras.layers import Dense, SimpleRNN, LSTM, Dropout, GRU, Bidirectional, Activation
from keras.regularizers import l1,l2,l1_l2
history = model.fit(X_train, y_train,epochs=10, batch_size=32,validation_data=(X_test,y_test),shuffle=True)
model.evaluate(X_test, y_test)
y_pred = model.predict(X_test)
from sklearn.metrics import accuracy_score
def split_sequence(sequence, n_steps):
    X, y = list(), list()
    for i in range(len(sequence)):
        # find the end of this pattern
        end_ix = i + n_steps
        # check if we are beyond the sequence
        if end_ix > len(sequence)-1:
            break
        # gather input and output parts of the pattern
        seq_x, seq_y = sequence[i:end_ix], sequence[end_ix]
        X.append(seq_x)
        y.append(seq_y)
    return array(X), array(y)
n_steps = 5
# split into samples
X_train, y_train = split_sequence(X_train, n_steps)
X_test, y_test = split_sequence(X_test, n_steps)
print('Training input shape: {}'.format(X_train.shape))
print('Training output shape: {}'.format(y_train.shape))
print('Test input shape: {}'.format(X_test.shape))
print('Test output shape: {}'.format(y_test.shape))
n_features = 1
X_train = X_train.reshape((X_train.shape[0], X_train.shape[1], n_features))
# define model
model = Sequential()
model.add(LSTM(50, activation='relu', input_shape=(n_steps, n_features)))
model.add(Dense(1))
```

```
model.compile(optimizer='adam', loss='mse')
# fit model
model.fit(X_train, y_train, epochs=30, verbose=0)
#%%
# demonstrate prediction
X_test = X_test.reshape((X_test.shape[0], X_test.shape[1], n_features))
#x_input = Xtest.reshape((1, n_steps, n_features))
y_pred = model.predict(X_test, verbose=2)
ex=ExtraTreesClassifier()
extra_model = ex.fit(X_train, y_train)
y_pred=extra_model.predict(X_test)
mse = mean_squared_error(y_test, extra_model.predict(X_test))
plot_matrics(y_pred)
# ============================================================================
# ExtraTreesClassifier with Cross Validation
# ============================================================================
from sklearn.ensemble import ExtraTreesClassifier
from sklearn.model_selection import cross_val_score
#In order to change to accuracy increase n_estimators
"""RandomForestClassifier(n_estimators='warn', criterion='gini', max_depth=None,




min_samples_split=2, min_samples_leaf=1, min_weight_fraction_leaf=0.0,
max_features='auto', max_leaf_nodes=None, min_impurity_decrease=0.0,
min_impurity_split=None, bootstrap=True, oob_score=False, n_jobs=None,
random_state=None, verbose=0, warm_start=False, class_weight=None)"""
# fit model no training data
model = ExtraTreesClassifier()
CV=10 #10-Fold Cross Validation
#Evaluate Model Using 10-Fold Cross Validation and Print Performance Metrics
Acc_scores = cross_val_score(model, X, Y, cv=CV)
print("Accuracy: %0.3f (+/- %0.3f)" % (Acc_scores.mean(), Acc_scores.std() * 2))
f1_scores = cross_val_score(model, X, Y, cv=CV,scoring='f1_macro')
print("F1 score: %0.3f (+/- %0.3f)" % (f1_scores.mean(), f1_scores.std() * 2))
Precision_scores = cross_val_score(model, X, Y, cv=CV,scoring='precision_macro')
print("Precision score: %0.3f (+/- %0.3f)" % (Precision_scores.mean(), Precision_scores.std() * 2))
Recall_scores = cross_val_score(model, X, Y, cv=CV,scoring='recall_macro')
print("Recall score: %0.3f (+/- %0.3f)" % (Recall_scores.mean(), Recall_scores.std() * 2))
from sklearn.metrics import cohen_kappa_score, make_scorer
kappa_scorer = make_scorer(cohen_kappa_score)
Kappa_scores = cross_val_score(model, X, Y, cv=5,scoring=kappa_scorer)
print("Kappa score: %0.3f (+/- %0.3f)" % (Kappa_scores.mean(), Kappa_scores.std() * 2))
```

```
# ==================================================================================
# ExtraTreesClassifier with Cross Validation
# ==================================================================================
from sklearn.ensemble import ExtraTreesClassifier
from sklearn.model_selection import cross_val_score
# fit model no training data
model = ExtraTreesClassifier(n_estimators=20)


CV=10 #10-Fold Cross Validation
#Evaluate Model Using 10-Fold Cross Validation and Print Performance Metrics
Acc_scores = cross_val_score(model, X, Y, cv=CV)
print("Accuracy: %0.3f (+/- %0.3f)" % (Acc_scores.mean(), Acc_scores.std() * 2))
rf=RandomForestClassifier()
#scaler = StandardScaler()
random_model = rf.fit(X_train, y_train)
y_pred=random_model.predict(X_test)
mse = mean_squared_error(y_test, random_model.predict(X_test))
plot_matrics(y_pred)
```

```
# ==================================================================================
# RandomForestClassifier with Cross Validation
# ==================================================================================
from sklearn.ensemble import RandomForestClassifier
from sklearn.model_selection import cross_val_score
#In order to change to accuracy increase  n_estimators
"""RandomForestClassifier(n_estimators=’warn’, criterion=’gini’, max_depth=None,
min_samples_split=2, min_samples_leaf=1, min_weight_fraction_leaf=0.0,
max_features=’auto’, max_leaf_nodes=None, min_impurity_decrease=0.0,
min_impurity_split=None, bootstrap=True, oob_score=False, n_jobs=None,
random_state=None, verbose=0, warm_start=False, class_weight=None)"""
# fit model no training data
model = RandomForestClassifier()
CV=10 #10-Fold Cross Validation
#Evaluate Model Using 10-Fold Cross Validation and Print Performance Metrics
Acc_scores = cross_val_score(model, X, Y, cv=CV)
print("Accuracy: %0.3f (+/- %0.3f)" % (Acc_scores.mean(), Acc_scores.std() * 2))
f1_scores = cross_val_score(model, X, Y, cv=CV,scoring='f1_macro')
print("F1 score: %0.3f (+/- %0.3f)" % (f1_scores.mean(), f1_scores.std() * 2))
Precision_scores = cross_val_score(model, X, Y, cv=CV,scoring='precision_macro')
print("Precision score: %0.3f (+/- %0.3f)" % (Precision_scores.mean(), Precision_scores.std() * 2))
```

```
Recall_scores = cross_val_score(model, X, Y, cv=CV,scoring='recall_macro')
print("Recall score: %0.3f (+/- %0.3f)" % (Recall_scores.mean(), Recall_scores.std() * 2))
from sklearn.metrics import cohen_kappa_score, make_scorer
kappa_scorer = make_scorer(cohen_kappa_score)
Kappa_scores = cross_val_score(model, X, Y, cv=5,scoring=kappa_scorer)
print("Kappa score: %0.3f (+/- %0.3f)" % (Kappa_scores.mean(), Kappa_scores.std() * 2))
# ====================================================================
# RandomForestClassifier with Cross Validation
# ====================================================================
from sklearn.ensemble import RandomForestClassifier
from sklearn.model_selection import cross_val_score
# fit model no training data
model = RandomForestClassifier(n_estimators=10)
CV=10 #10-Fold Cross Validation
#Evaluate Model Using 10-
Fold Cross Validation and Print Performance Metrics
Acc_scores = cross_val_score(model, X, Y, cv=CV)
print("Accuracy: %0.3f (+/-
%0.3f)" % (Acc_scores.mean(), Acc_scores.std() * 2))
```

4. Discussion

Accelerating the process of identifying prospective drug candidates is one of AI's most important benefits in drug research. Traditional approaches to drug discovery entail considerable research and experimentation that might take years or even decades. Contrarily, AI algorithms can scan enormous volumes of data and produce predictions and insights that help identify new medication candidates. AI can drastically shorten the time and money needed to bring novel medications to market by speeding up this early phase.

Predictive modeling and drug compound optimization rely on AI techniques like machine learning and deep learning. Large datasets comprising information on molecular structures, genetic information, and clinical outcomes can be analyzed by AI algorithms to find patterns and connections that might not be obvious to human researchers. With the use of this predictive modeling, drug candidates can be virtually tested, allowing researchers to improve their characteristics and forecast their efficacy and safety profiles without having to invest time and money in costly laboratory tests.

Identification and verification of therapeutic targets are both greatly aided by AI. Massive volumes of genomic and proteomic data can be analyzed by AI algorithms to find possible disease targets that might have gone unnoticed by more conventional approaches. This broadens the range of potential medication targets and enhances our understanding of the fundamental pathophysiology of illnesses. Insights about the possible efficacy and safety of targeting particular disease-related proteins or pathways can be provided by AI, which can also help with target validation.

Drug repurposing is the process of identifying and assessing existing medications for potential new therapeutic applications. AI algorithms can help with this process. AI can find prospective drug candidates that may be useful against many diseases by evaluating extensive datasets and molecular structures. By using this method, the time and expense involved in creating new medications from scratch can be greatly reduced. AI can also help in the selection of the best combination therapies, in which a number of medications are taken simultaneously to boost effectiveness and combat drug resistance.

As a result, artificial intelligence has become a game-changer in the research and development of new drugs. AI speeds up the drug development process, improves medicinal molecules, and enables personalized therapy by utilizing its predictive modeling skills. It has the potential to transform patient care, improve recovery after treatments, and meet unmet medical requirements. To fully utilize AI's promise in this field, however, obstacles like ethical dilemmas and legal restrictions must be overcome. AI has the potential to revolutionize the pharmaceutical sector and open the door to more precise and effective treatments with ongoing research and collaboration.

5. Conclusion

AI has transformed the pharmaceutical sector by becoming a potent tool for drug discovery and development. Target discovery, lead generation, molecular docking, drug design, pharmacokinetic prediction, and toxicity evaluation are a few areas of the drug development process where the application of AI approaches has shown substantial promise. Large-scale data analysis, pattern recognition, and prediction-making capabilities of AI systems are remarkable. This could shorten the time it takes to identify new drugs, lower expenses, and increase the likelihood that innovative therapies will be successful when they are introduced to the market.

There are various advantages of using AI in drug research and development. It lets researchers examine big datasets and find hidden insights that would not be seen using conventional techniques. AI programs are able to find possible pharmacological targets, create new substances, predict their pharmacokinetic characteristics, and evaluate their toxicity. These abilities not only speed up the drug development process but also improve the identification of strong candidates, increasing the likelihood that clinical trials will be successful.

To fully utilize AI in drug discovery and development in the future, more research, collaboration, and regulatory guidance are required. Even more potential exists when AI is combined with other cutting-edge technologies like high-throughput screening, genomics, and proteomics. The pharmaceutical industry may hasten the development of new drugs, enhance patient outcomes, and meet unmet medical needs by utilizing AI.

In conclusion, artificial intelligence has the potential to revolutionize the research and development of pharmaceuticals. We may anticipate more effective, precise, and tailored methods for drug discovery as AI techniques progress. This will eventually result in the creation of safer and more potent treatments for a variety of ailments. It is a big step forward for the pharmaceutical business to incorporate AI, and it has a lot of potential for the future of medicine.

References

Agatonovic-Kustrin, S., & Beresford, R. (2000). Basic concepts of artificial neural network (ANN) modeling and its application in pharmaceutical research. *Journal of Pharmaceutical and Biomedical Analysis, 22*(5), 717−727.

Banerjee, D., Rajput, D., Banerjee, S., & Saharan, V. A. (2022). Artificial intelligence and its applications in drug discovery, formulation development, and healthcare. In *Computer aided pharmaceutics and drug delivery: An application guide for students and researchers of pharmaceutical sciences* (pp. 309−380). Springer.

Chen, W., Liu, X., Zhang, S., & Chen, S. (2023). Artificial intelligence for drug discovery: Resources, methods, and applications. *Molecular Therapy-Nucleic Acids, 31*, 691–702. https://doi.org/10.1016/j.omtn.2023.02.019

Deng, J., Yang, Z., Ojima, I., Samaras, D., & Wang, F. (2022). Artificial intelligence in drug discovery: Applications and techniques. *Briefings in Bioinformatics, 23*(1), bbab430.

Jiménez-Luna, J., Grisoni, F., & Schneider, G. (2020). Drug discovery with explainable artificial intelligence. *Nature Machine Intelligence, 2*(10), 573–584. https://doi.org/10.1038/s42256-020-00236-4

Malandraki-Miller, S., & Riley, P. R. (2021). Use of artificial intelligence to enhance phenotypic drug discovery. *Drug Discovery Today, 26*(4), 887–901.

Moingeon, P., Kuenemann, M., & Guedj, M. (2022). Artificial intelligence-enhanced drug design and development: Toward a computational precision medicine. *Drug Discovery Today, 27*(1), 215–222.

Paul, D., Sanap, G., Shenoy, S., Kalyane, D., Kalia, K., & Tekade, R. K. (2021). Artificial intelligence in drug discovery and development. *Drug Discovery Today, 26*(1), 80.

Schneider, P., Walters, W. P., Plowright, A. T., Sieroka, N., Listgarten, J., Goodnow, R. A., Jr., Fisher, J., Jansen, J. M., Duca, J. S., & Rush, T. S. (2020). Rethinking drug design in the artificial intelligence era. *Nature Reviews Drug Discovery, 19*(5), 353–364.

Tripathi, A., Misra, K., Dhanuka, R., & Singh, J. P. (2023). Artificial intelligence in accelerating drug discovery and development. *Recent Patents on Biotechnology, 17*(1), 9–23.

Yang, X., Wang, Y., Byrne, R., Schneider, G., & Yang, S. (2019). Concepts of artificial intelligence for computer-assisted drug discovery. *Chemical Reviews, 119*(18), 10520–10594.

Zhavoronkov, A. (2018). Artificial intelligence for drug discovery, biomarker development, and generation of novel chemistry. *Molecular Pharmaceutics, 15*(10), 4311–4313.

Zhavoronkov, A., Vanhaelen, Q., & Oprea, T. I. (2020). Will artificial intelligence for drug discovery impact clinical pharmacology? *Clinical Pharmacology & Therapeutics, 107*(4), 780–785.

Zhong, F., Xing, J., Li, X., Liu, X., Fu, Z., Xiong, Z., Lu, D., Wu, X., Zhao, J., & Tan, X. (2018). Artificial intelligence in drug design. *Science China Life Sciences, 61*, 1191–1204.

Zhu, H. (2020). Big data and artificial intelligence modeling for drug discovery. *Annual Review of Pharmacology and Toxicology, 60*, 573–589.

Hospital readmission forecasting using artificial intelligence

Abdulhamit Subasi[1,2]

[1]Institute of Biomedicine, Faculty of Medicine, University of Turku, Turku, Finland; [2]Department of
Computer Science, College of Engineering, Effat University, Jeddah, Saudi Arabia

1. Introduction

Hospital readmissions have become a significant challenge in healthcare systems worldwide. High rates of hospital readmissions not only impose a burden on healthcare resources but also indicate gaps in the quality of patient care and the effectiveness of care transitions. Therefore, there is a growing need for accurate prediction models that can identify patients at risk of readmission and enable timely interventions to prevent unnecessary hospitalizations. Artificial intelligence (AI) techniques, such as machine learning and deep learning, have emerged as powerful tools in healthcare analytics, offering the potential to improve readmission forecasting.

Hospital readmissions have significant implications for patients, healthcare providers, and the healthcare system as a whole. They often indicate unresolved health issues, complications, or inadequate care during the initial hospitalization. High readmission rates not only affect patient well-being but also impose a financial burden on healthcare systems. By accurately predicting readmissions, healthcare providers can intervene in a timely manner, implement preventive strategies, and improve patient outcomes. AI techniques offer the potential to enhance the accuracy and effectiveness of readmission forecasting models, enabling healthcare providers to allocate resources efficiently and improve patient care (Huang et al., 2021).

AI techniques, such as machine learning and deep learning, have shown promise in hospital readmission forecasting. These techniques utilize historical patient data, including demographic information, clinical variables, laboratory results, and previous healthcare utilization, to develop predictive models. Machine learning algorithms with electronic health records (EHRs) can be utilized to predict readmissions. AI techniques offer several advantages in hospital readmission forecasting. First, they can handle large and diverse datasets, allowing for the integration of various data sources to capture a comprehensive view of patient health. Second, AI models can identify complex relationships and patterns in the data that may not be apparent through traditional statistical approaches. Third, AI techniques can continuously learn and adapt from new data, improving the accuracy and robustness of readmission forecasting models over time. Finally, AI models can provide interpretable insights into the factors driving readmissions, enabling healthcare providers to target interventions and improve care processes (Chen et al., 2022). Hence, using AI approaches to forecast hospital readmissions has enormous potential to enhance patient care and healthcare outcomes. In comparison to conventional methods, AI models can predict readmissions with improved accuracy by utilizing huge and diverse datasets. Healthcare practitioners can adopt targeted interventions, enhance care coordination, and efficiently allocate resources by enabling the early identification of patients who are more likely to require readmission.

2. Background and literature review

AI-based hospital readmission forecasting has become a potential method for enhancing patient care and making the most use of available medical resources. For predicting hospital readmissions, supervised learning, unsupervised learning, and deep learning techniques have all been used. Predictive models based on patient data, clinical characteristics, and demographic factors have been created using supervised learning methods such as logistic regression, decision trees, and support vector machines. Recurrent neural networks and convolutional neural networks are two examples of deep learning models that have shown promise in capturing temporal and geographical relationships in patient data, allowing precise readmission predictions.

Implementing AI-based readmission-predicting models presents difficulties and ethical quandaries. One significant problem is incorporating AI algorithms into existing healthcare systems, notably electronic health record systems, which necessitate robust data preprocessing and interoperability solutions. It is critical to ensure the openness and interpretability of AI models to earn clinician trust and facilitate clinical decision-making. Furthermore, mitigating biases in AI systems, as well as ensuring patient privacy and data security, are critical ethical considerations that must be addressed.

The field of artificial intelligence-based hospital readmission predictions is rapidly evolving. Future research should concentrate on constructing understandable AI models that provide insights into the underlying elements that contribute to readmission risks. The incorporation of real-time data sources, such as wearable devices and remote monitoring systems, can improve forecast accuracy and allow for timely interventions. To establish uniform criteria for the deployment and evaluation of AI models in clinical practice, researchers, doctors, and policymakers must work together.

While recent data mining efforts have focused on predicting healthcare costs or the risk of hospital readmission, Sushmita et al. (2016) introduced a dual predictive modeling approach that leverages healthcare data to predict both the risk and cost of any hospital readmission (referred to as "all-cause" readmission). To achieve this, they utilize machine learning algorithms to make accurate predictions regarding healthcare costs and the likelihood of readmission within 30 days. The results of risk prediction for "all-cause" readmission, when compared to the standardized readmission tool (LACE), show promise, and proposed techniques for cost prediction consistently outperform baseline models, demonstrating significantly lower mean absolute error (MAE).

Avoidable hospital readmissions not only contribute to high healthcare costs but also harm patient care quality. EHR adoption provides an opportunity to identify high-risk patients and implement interventions to prevent readmissions. Previous machine learning models have predicted 30-day readmissions, but a real-time and accurate model for hospitals is needed. Jamei et al. (2017) used EHR data from over 300,000 hospital stays in California, utilizing Google's TensorFlow library to develop an artificial neural network (NN) model. Compared to conventional and nonconventional models, they showed that NNs effectively capture complex relationships within EHR data. The standard model, LACE, had a precision (PPV) of 0.20 for identifying high-risk patients, while the proposed NN model achieved a PPV of 0.24, a 20% improvement. They also explored Social Determinants of Health (SDoH) data's predictive capabilities and presented a cost analysis to aid hospitalists in postdischarge interventions.

Unplanned patient readmission (UPRA) is a frequent and costly occurrence in healthcare settings. Currently, there are no identified indicators during hospitalization that clinicians can utilize to effectively identify patients at a high risk of UPRA. Tey et al. (2021) aimed to develop a prediction model for the early detection of 14-day UPRA in patients with pneumonia. Data from patients diagnosed with pneumonia at three hospitals in Taiwan between 2016 and 2018 were collected, resulting in a total of 21,892 cases, out of which 1208 (6%) experienced UPRA. Two models, namely the artificial neural network (ANN) and the convolutional neural network (CNN), were employed and compared using a training set (n = 15,324; approximately 70%) and a test set (n = 6568; approximately 30%) to assess the accuracy of the models. An application was developed to facilitate the prediction and classification of UPRA. The results indicated that (i) the 17 feature variables utilized in the proposed study achieved a high area under the receiver operating characteristic curve (AUC) of 0.75 when using the ANN model and (ii) the ANN model outperformed the CNN model, as evidenced by a higher AUC (0.73) compared to the CNN (0.50).

Reducing unplanned hospital readmissions is a key objective in improving healthcare quality. To accurately evaluate hospital performance, prediction models using healthcare claims data are employed. Regression-based models are commonly used but lack precision. In a study by Liu et al. (2020), four prediction models for unplanned readmissions in patients with acute myocardial infarction (AMI), congestive heart failure (HF), and pneumonia (PNA) were compared. They assessed hierarchical logistic regression, gradient boosting, and two artificial neural network models. By incorporating unsupervised Global Vector for Word Representations embedding representations of administrative claims data with artificial neural networks, the prediction of 30-day readmissions was significantly improved. The best models enhanced the area under the curve (AUC) for 30-day readmission predictions, increasing from 0.68−0.72 for AMI, 0.60−0.64 for HF, and 0.63−0.68 for PNA, when compared to hierarchical logistic regression. Moreover, when risk-standardized hospital readmission rates were calculated using the artificial neural network model with embeddings, about 10% of hospitals experienced reclassification across performance categories.

Healthcare systems, payers, and regulators have prioritized efforts to reduce the length of hospital stays (LOS) and early readmissions, despite the uncertain benefits. Transparent and interpretable machine learning (ML) techniques can aid in identifying the risk of important outcomes. Hilton et al. (2020) trained various medical, sociodemographic, and institutional variables to predict readmission, LOS, and death within 48−72 h. Prediction performance was evaluated using metrics such as the area under the receiver operator characteristic curve (AUC) and the Brier score loss (BSL), which quantifies the match between predicted and observed probabilities. Multiple feature extraction algorithms were employed to generate interpretations of the predictions.

Predicting hospital readmission rates for chronic diseases like diabetes is a crucial challenge in healthcare, as it helps in allocating necessary resources such as beds, rooms, specialists, and medical staff to ensure adequate quality of service. However, there is a scarcity of research studies addressing this specific problem, as most studies focus on predicting the diseases themselves rather than readmission likelihood. While numerous machine learning techniques can be employed for prediction, there is a lack of comprehensive comparative studies that identify the most suitable techniques for this task. To address this gap, Alajmani and Elazhary (2019) presented a comparative study of five commonly used techniques for predicting hospital readmission likelihood in diabetic patients. These techniques include logistic regression (LR) analysis, multilayer perceptron (MLP), Naïve Bayesian (NB) classifier, decision tree, and support vector machine (SVM). Realistic data obtained from multiple hospitals in the United States is used for the comparative analysis. The study reveals that SVM performs the best among the evaluated techniques, while the NB classifier and LR analysis exhibit poorer performance.

The study by Romero-Brufau et al. (2020) focused on reducing unplanned hospital readmissions using an AI-based clinical decision support system. The system, implemented in a regional hospital, assessed the risk of readmission for patients in general care units and provided targeted interventions. The AI tool demonstrated a sensitivity of 65% and specificity of 89% in identifying high-risk patients. After implementation, the hospital's readmission rates dropped from 11.4% to 8.1% over a 6-month period, resulting in a 25% relative reduction when accounting for control hospitals. The AI intervention proved effective in decreasing hospital readmissions.

Hospital readmissions are considered a significant economic factor in healthcare systems and are often used as an indicator of service quality provided by healthcare institutions. Being able to predict readmissions in advance enables

timely interventions and improved postdischarge strategies, leading to the prevention of life-threatening events and the reduction of medical costs for both patients and healthcare systems. Michailidis et al. (2022) used four machine learning models, namely support vector machines with a linear kernel, support vector machines with an radial basis function (RBF) kernel, balanced random forests, and weighted random forests, for predicting readmissions. These records encompass administrative, medical-clinical, and operational variables. The experimental findings demonstrate that the balanced random forest model outperforms the other models, achieving a sensitivity of 0.70 and an AUC value of 0.78.

AI is transforming healthcare, but transparency for medical experts is essential. Previous studies lacked explanations for disease prediction models, impacting trust. To address this, explainable AI (XAI) methods create interpretable "white box" models. Unplanned healthcare readmissions are costly. Lu and Uddin (2022) introduced a stacked model predicting 30-day readmissions in diabetic patients. They balanced classes using Random Under-Sampling, performed feature selection via SelectFromModel, and built a stacking model. Their model surpasses existing ones in prediction and remains interpretable. Key predictors include inpatient count, primary diagnosis, home discharge with service, and emergencies. Individual-level explainability is provided using local interpretable model-agnostic explanations.

While hospital readmission prediction has been extensively researched in medical patients, its application to surgical patients has received less attention. Mišić et al. (2020) aimed to predict readmissions in postoperative patients, focusing on those from the emergency department, and doing so earlier than discharge. They collected data from a tertiary care academic medical center, including surgical, demographic, lab, medication, and procedural information from electronic health records. Their goal was to predict emergency-department-originating hospital readmissions within 30 days of surgery. They evaluated different machine learning models and feature combinations, using the area under the receiver–operator characteristic curve (AUC) as the performance measure. The analysis covered 34,532 surgical admissions from April 2013 to December 2016. Incorporating surgical and demographic features led to moderate prediction ability (AUC: 0.74–0.76). Adding lab features improved prediction significantly, with gradient-boosted trees performing best (AUC: 0.866). This performance was validated in 2017–18 data (AUC: 0.85–0.88). Impressively, predictions made at 36 h postsurgery (AUC: 0.88–0.89) closely matched those at discharge. This study showcases a machine learning approach's potential to accurately predict 30-day readmissions via the emergency department for postoperative patients, allowing early predictions independent of discharge data.

Chronic obstructive pulmonary disease (COPD) affects numerous people globally and often leads to hospital readmissions within 30 days of discharge due to multiple factors. Min et al. (2019) explored machine learning and deep learning techniques to predict COPD patient readmission risks. They tested these models on a real-world database with medical claims data from 111,992 patients, spanning from January 2004 to September 2015. The models incorporated both knowledge-driven features from clinical understanding potentially linked to COPD readmission and data-driven features extracted from patient data. Combining both types of features, specifically using a 1-year claims history before discharge, enhanced prediction performance, increasing the Area Under the Receiver Operating Characteristic (ROC) Curve (AUC) from about 0.60 to 0.653. Interestingly, employing complex deep learning models did not significantly improve prediction performance; the highest AUC achieved was around 0.65.

Diabetes is a chronic condition characterized by the body's inability to metabolize blood glucose properly. EHRs have become essential in tracking disease trends for individuals and populations. Machine and deep learning techniques have shown superior predictive capabilities compared to traditional assessments. Phu and Wang (2021) applied machine and deep learning models using EHRs, harnessing the power of these models with various features and hyperparameter optimization. Hyperparameter optimization involves the utilization of random search optimization to minimize a predetermined loss function using independent data. Through a comparative evaluation of different methods, including logistic regression, artificial neural network, Naïve Bayesian classifier, support vector machine, and XGBoost, they observed improved results in terms of metrics such as accuracy, recall, F1 score, and AUC score when compared to previous models on the same reprocessed public dataset.

Hospital readmissions are commonly seen as a sign of substandard care, often resulting from insufficient discharge planning and coordination. Furthermore, experts widely believe that a significant number of readmissions are unnecessary and could have been prevented. Chopra et al. (2017) developed a recurrent neural network (RNN) model to forecast whether a patient would be readmitted to the hospital. They compared the accuracy of this model with basic classifiers such as SVM, random forest, and simple neural networks. The RNN model exhibited the highest predictive power among all the models utilized, indicating its potential usefulness for hospitals in identifying high-risk patients and taking proactive measures to prevent recurrent admissions.

Amritphale et al. (2021) examined the rates and causes of unplanned readmissions within 30 days after carotid artery stenting (CAS) and to developed a prediction model using artificial intelligence. Predicting unplanned readmissions following CAS remains challenging, and there is a need to leverage deep learning algorithms to create robust tools for early readmission prediction. They analyzed data from the US Nationwide Readmission Database (NRD) for patients who underwent inpatient CAS in 2017. The rates, predictors, and costs of unplanned 30-day readmissions were evaluated. Logistic regression, SVM, deep neural network (DNN), random forest, and decision tree models were assessed to generate a reliable prediction model. A total of 16,745 CAS patients were identified, with a readmission rate of 7.4% within 30 days. Depression, heart failure, cancer, in-hospital bleeding, and coagulation disorders were identified as the strongest predictors of readmission. The DNN model based on machine learning demonstrated a C-statistic value of 0.79 (0.73 in validation) in predicting patients at risk of all-cause unplanned readmission within 30 days after the index CAS discharge.

Predicting readmission within 30 days of hospital discharge is a crucial research task in personalized healthcare. However, the challenges of imbalanced class distribution and the heterogeneity of electronic health records pose significant obstacles in developing an effective machine learning model for this task. To address these issues, Du et al. (2021) proposed a novel algorithm for predicting 30-day readmission that aimed to improve performance. They tackled the problem of class imbalance in readmission prediction by learning sample weights using the hypothesis margin loss. Additionally, they devised an optimization framework that involves two variables: sample weights and source weights. Through iterative optimization, they derived the prediction results for readmission. To validate the effectiveness of our proposed method, they conducted experiments using three real-world readmission datasets. The experimental results demonstrated that the proposed algorithm offers distinct advantages in addressing the challenge of predicting readmission within 30 days of hospital discharge.

Unplanned readmissions pose a substantial financial burden and are indicative of healthcare quality and hospital performance. However, challenges persist in creating accurate predictive models due to limited predictor understanding and shortcomings of traditional statistical methods. Jiang et al. (2018) introduced a comprehensive approach for hospital readmission risk prediction, integrating feature selection algorithms and machine learning models to overcome these issues. They developed an enhanced version of multi-objective bare-bones particle swarm optimization (EMOBPSO) for effective feature selection, incorporating mutual information-based assessment and a greedy local search strategy (GLS) for optimal feature subset size. For modeling, manifold machine learning models like SVM, random forest, and deep neural network were employed with preprocessed data sets. In a real-world case study at a Chinese hospital, their methodology was tested on hospital data, showcasing the effectiveness of EMOBPSO and EMOBPSO-GLS. Combining EMOBPSO with deep neural networks demonstrated robust predictive capabilities across various datasets. The approach also yielded insights into high-risk patient identification and resource allocation for interventions.

Intensive care unit (ICU) readmissions are associated with high mortality rates and prolonged hospital stays. Current prediction models focus on physiological variables using snapshot measurements but fail to capture the predictive information in temporal trends of physiological and medication variables. Xue et al. (2019) identified strong predictors by analyzing temporal trends and building accurate prediction models for 30-day ICU readmission risk. They used data from the MIMIC-II clinical dataset, transforming variables into trend graphs and extracting important trends through subgraph mining. These trends were grouped to represent patients' pathophysiological states and medication profiles. A logistic regression model was trained using snapshot measurements and grouped trends. The model outperformed other models, achieving nearly 4% improvement in the area under the receiver operating characteristic curve (AUC).

Reddy et al. (2020) proposed the use of a deep learning algorithm called Deep Belief Network to effectively predict early readmission of diabetes patients to a hospital. The Pima Indian Diabetes Dataset was utilized for experimentation. The proposed algorithm was compared with existing algorithms to evaluate its effectiveness, using measures such as precision, accuracy, specificity, negative predictive value (NPV), and F1-score. The findings demonstrated that the Deep Belief Network algorithm can be employed to predict the likelihood of early readmission for diabetes patients.

During the COVID-19 pandemic, hospitals have faced resource shortages, and reducing readmissions related to the virus is crucial for maintaining capacity. Afrash et al. (2022) identified influential features associated with COVID-19 readmission and compared the predictive capabilities of various ML algorithms using these features. The analysis included data from 5791 hospitalized COVID-19 patients, with the LASSO feature selection algorithm used to identify key predictors. Multiple ML classifiers, such as Hist-GradientBoosting, Bagging, MLP, SVM, and XGBoost, were employed for prediction. Performance evaluation was conducted using a 10-fold cross-validation method with six metrics. Out of the 42 features considered, 14 were identified as the most relevant predictors for COVID-19 readmission. The XGBoost classifier exhibited the highest performance, with an average accuracy of 91.7%, specificity of 91.3%, sensitivity of 91.6%, F-measure of 91.8%, and an AUC of 0.91%. These findings demonstrate the effectiveness of ML models in predicting COVID-19 readmissions.

Accurately predicting the hospitalization needs of diabetic patients is crucial for effective management of their condition. However, conventional machine learning approaches may introduce biases related to social determinants,

leading to health disparities. To tackle this issue, Raza (2022) introduced a machine learning pipeline that not only makes predictions but also detects and mitigates biases in the data and model predictions. The pipeline thoroughly analyzed clinical data, identified biases, and applied techniques to remove them before making predictions. The method's performance was evaluated on a clinical dataset using measures of accuracy and fairness, showing that early bias mitigation leads to fairer and less discriminatory predictions. By integrating bias detection and mitigation, the proposed approach improved the equity and accuracy of machine learning predictions for diabetic patient hospitalization needs, enhancing healthcare decision-making.

Hammoudeh et al. (2018) proposed a novel approach that combines data engineering techniques with the representation learning capabilities of CNNs for predicting hospital readmissions in diabetic patients. The integration of CNNs and data engineering was found to yield superior results when compared to other machine learning algorithms.

The use of electronic medical records (EMRs) presents a valuable opportunity to improve hospital care and patient survival rates through data mining techniques. Early prediction of hospital readmissions is crucial for timely interventions and cost reduction. However, existing methods often overlook class label imbalances and varying costs of misclassification errors. Wang et al. (2018) utilized CNNs to automatically learn features from vital sign time series data, and categorical feature embedding to encode diverse clinical features. These features are then used to predict readmissions using a MLP model. To address class label skewness, they incorporated a cost-sensitive formulation during MLP training. Experiments on real medical datasets showed that the proposed approach outperforms existing methods, with an AUC of 0.70 for 30-day readmission prediction in general hospital ward data.

Predicting hospital readmissions, which involve patients being admitted again within a specific period, is complex due to imbalanced data. Turgeman and May (2016) developed a predictive model that balances transparency and accuracy while considering database characteristics. Their model combined a boosted C5.0 tree and a SVM as base and secondary classifiers. They used anonymized administrative records from VHA hospitals for congestive heart failure (CHF) patients. SVM predictions showed higher sensitivity than C5.0 predictions at various ROC curve cutoffs. The ensemble model achieved an accuracy of 81%–85%. Different predictors, such as comorbidities, lab values, and vital signs, had distinct roles. This mixed-ensemble model efficiently discovers knowledge and controls classification errors for positive readmission cases. By combining classifiers, it surpasses individual models and traditional ensembling methods, enhancing accuracy in predicting positive readmission instances, particularly when strong predictors are limited.

Jovanovic et al. (2016) developed a model for accurate and interpretable prediction of unplanned readmissions in a general pediatric patient population. They addressed the challenges related to high dimensionality, sparsity, and class imbalance in electronic health data. By combining a data-driven model (sparse logistic regression) with domain knowledge based on the ICD-9-CM hierarchy of diseases, the approach allows for model interpretation. The analysis involves over 66,000 pediatric hospital discharge records from California State Inpatient Databases. The proposed Tree-Lasso regularized logistic regression model, integrated with the ICD-9-CM hierarchy, achieves comparable prediction accuracy while offering more interpretable models with high-level diagnoses. The results align with the existing medical understanding of pediatric readmissions. The best-performing models reach AUC values of 0.783 and 0.779 for traditional Lasso and Tree-Lasso, respectively.

3. Artificial intelligence for hospital readmission forecasting

3.1 Implementation of AI methods for hospital readmission forecasting[1]

3.1.1 General overview

Dataset Resources: UC Irvine's Machine Learning Repository.

Kaggle Link.

The dataset covers 10 years of clinical care at 130 US hospitals and integrated delivery networks (1999–2008). It has about 50 attributes that indicate patient and hospital outcomes. Patient number, race, gender, age, admission type, time in hospital, medical specialty of admitting physician, number of lab tests performed, HbA1c test result, diagnosis, number of medications, diabetic medications, number of outpatient, inpatient, and emergency visits in the year preceding the hospitalization, and so on are all included in the data.

Import libraries

```
import pandas as pd
import numpy as np
import matplotlib.pyplot as plt
import seaborn as sns
from scipy import stats
import re

# to avoid warnings
import warnings
warnings.filterwarnings('ignore')
warnings.warn("this will not show")

sns.set(style='darkgrid')
%matplotlib inline
```

---

[1] This part is adapted from https://www.kaggle.com/code/kirshoff/diabetic-patient-readmission-prediction/notebook.

Import dataset

```
pd.read_csv("../input/uci-diabetes-
dataset/dataset_diabetes/diabetic_data.csv").head()
# csv contains "?" for missing values. We replace it with NaN
data = pd.read_csv("../input/uci-diabetes-
dataset/dataset_diabetes/diabetic_data.csv", na_values=["?"])
df= data.copy()
df.head()
```

Import features dataset

```
features = pd.read_csv('features.csv',index_col='Unnamed: 0')
info = lambda attribute:print(f"{attribute.upper()} : {features[features['Fea
ture']==attribute]['Description'].values[0]}\n")
features.head()
info('encounter_id')
```

Check duplicates

```
df.duplicated().value_counts()
# df = df.drop_duplicates()
```

Descriptive analysis

```python
def summary(df, pred=None):
    obs = df.shape[0]
    Types = df.dtypes
    Counts = df.apply(lambda x: x.count())
    Min = df.min()
    Max = df.max()
    Uniques = df.apply(lambda x: x.unique().shape[0])
    Nulls = df.apply(lambda x: x.isnull().sum())
    print('Data shape:', df.shape)

    if pred is None:
        cols = ['Types', 'Counts', 'Uniques', 'Nulls', 'Min', 'Max']
        str = pd.concat([Types, Counts, Uniques, Nulls, Min, Max], axis = 1,
sort=True)

    str.columns = cols
    print('_____\nData Types:')
    print(str.Types.value_counts())
    print('_____')
    return str

display(summary(df).sort_values(by='Nulls', ascending=False))
```

- "citoglipton' and 'examide' features that the number of uniques is 1 are dropped.
- all values of ' encounter_id' column are unique. It has to be dropped.

```python
df = df.drop(['citoglipton','examide','encounter_id'],axis=1)
```

FOCUS ON 'gender'

```
df.gender.value_counts(dropna=False)
```

We regard the observations of 'Unknown/ Invalid' gender as null values and drop them.

```
gender_index = df[df.gender == 'Unknown/Invalid'].index
df = df.drop(gender_index, axis=0)
# confirm removal
df.gender.value_counts(dropna=False)
```

FOCUS ON 'readmitted'

```
df.readmitted.value_counts(dropna=False)
```

Patients readmitted to the hospital within and after 30 days will be combined into one column because these patients ultimately returned.

```
df = df.replace(['<30', '>30'], 'YES')
```

FOCUS ON 'patient_nbr'

```
info('patient_nbr')
df['patient_nbr'].duplicated().value_counts(dropna=False)
```

- We can think of 'patient_nbr' as the id number of each patient.
- It turned out that the dataset is the data of 71,515 unique patients.
- Some patients visited the hospital multiple times for treatment so to avoid overrepresenting any particular individual, only the first encounter with a patient will be used/kept in this dataset.

```
# total unique patients
len(df.patient_nbr), df.patient_nbr.nunique()
# locate number of patient visits using patient_id
df.patient_nbr.value_counts()
# keep only one record for each patient, the first visit
df = df.drop_duplicates(['patient_nbr'], keep='first')
df.shape
df.patient_nbr.nunique()
```

- As patient_nbr is unique, it is no longer
  needed.

```
df = df.drop('patient_nbr', axis=1)
```

Dropping irrelevant columns

```
def null_values(df):
    """a function to show null values with percentage"""
    nv=pd.concat([df.isnull().sum(), 100 * df.isnull().sum()/df.shape[0]],axi
s=1).rename(columns={0:'Missing_Records', 1:'Percentage (%)'})
    return nv[nv.Missing_Records>0].sort_values('Missing_Records', ascending=
False)
# columns with missing values
null_values(df)
for i in ['weight','medical_specialty','payer_code']: info(i)
```

- The majority of patients do not have a weight
  listed so this column can be dropped.
- Medical specialty and payer code are also
  missing for about half of the patients.
- We do not need to know how the patients
  paid for their treatments.
- we do not have enough information to figure
  out which medical unit they went to.

```
df = df.drop(['weight','medical_specialty','payer_code'], axis=1)
summary(df).sort_values(by='Uniques', ascending=False)[:20]
for i in ['admission_type_id', 'discharge_disposition_id', 'admission_source_
id']: info(i)
```

- We do not need 'admission_type_id,' 'discharge_disposition_id', 'admission_source_id' columns

- You can reach the extensive diagnosis description on this website by querying with the ICD9 code: http://icd9.chrisendres.com/

```
# drop columns
drop_cols = ['admission_type_id', 'discharge_disposition_id', 'admission_sour
ce_id']
df = df.drop(drop_cols, axis=1)
```

Handling missing values

```
null_values(df)
df.race.value_counts(dropna=False)
```

- As there is no way to know the race of the patient using existing information, the best option is to remove those rows.

```
df = df.dropna(axis=0, subset=['race'])
null_values(df)
for i in ['diag_1', 'diag_2', 'diag_3']: info(i)
```

Now, we are down to three columns with missing information: diagnoses 1, 2, and 3.

- Diagnosis 1 is described as the primary diagnosis made during the patient's visit while diagnosis 2 is the second and 3 is any additional diagnoses made after that.
- Looking at the patients' rows that are missing a primary diagnosis, most of them have a second diagnosis or even a third.
- As it does not make sense to have a second (or third) but not a primary diagnosis, we will remove these columns from the dataset.

```
info('number_diagnoses')
df[['diag_1', 'diag_2', 'diag_3','number_diagnoses']][df.diag_1.isnull() & df
.diag_2.notnull() & df.diag_3.notnull() & df.number_diagnoses.notnull()]
```

The number of diagnoses column shows the total number of conditions a patient is diagnosed with. Only the first three are recorded, so those that are missing the first diagnosis but still a second or third are in error.

```
# remove rows where diagnosis 1 is missing
df = df.dropna(axis=0, subset=['diag_1'])
```

There are two remaining diagnosis columns with missing values. Each number correlates to a specific condition so if there is a missing value, then it is likely that the patient only has one diagnosed condition. The number of diagnoses column lists the total number of diagnosed conditions. When looking at all three diagnosis columns, if the number is one, then 2 and 3 can be filled in with a 0 to show that there is no additional diagnosis. If diagnosis 2 or 3 is missing a value and the number of diagnoses is greater than 1, then some diagnoses were not recorded, and the rows should be removed.

```
null_values(df)
df[['diag_1','diag_2', 'diag_3','number_diagnoses']][df.diag_2.isnull() & (df
.diag_3.notnull()|(df.number_diagnoses > 1))]
# remove rows where diagnosis 2 is missing and number of diagnoses is greater
  than 1
diag_2_indexes = df[df.diag_2.isnull() & (df.diag_3.notnull()|(df.number_diag
noses > 1))].index
df = df.drop(index = diag_2_indexes, axis=0)
null_values(df)
```

Diagnosis 3 is the last column left with unaccounted missing values. As some patients have 1 or 2 diagnosed conditions, the diagnosis 3 column is left intentionally blank. The goal here is to remove the rows that have a diagnosis number greater than two.

- 'diag_1', 'diag_2' and 'diag_3' columns contain codes for the types of conditions patients are diagnosed with.
- There are too many unique codes throughout this dataset.
- We can group the related icd9 diagnosis codes among themselves. In this way, we use

```
# list of affected rows
df[['diag_1','diag_2', 'diag_3', 'number_diagnoses']][df.diag_3.isnull() & (d
f.number_diagnoses > 2)]
# remove rows with missing diagnosis 3 and number of diagnoses is greater tha
n 2
diag_3_indexes = df[(df.diag_3.isnull()) & (df.number_diagnoses > 2)].index
df = df.drop(index=diag_3_indexes, axis=0)
null_values(df)
sns.heatmap(df[['diag_1','diag_2', 'diag_3','number_diagnoses']].isnull(),yti
cklabels=False,cbar=False,cmap='viridis');
# replace NaN with None in diagnosis 2 and 3 to show there is no additional d
iagnosis
df.fillna('None', inplace=True)
# confirm there are no more NaN values
null_values(df)
```

Grouping diagnosis codes

```
summary(df[['diag_1','diag_2', 'diag_3']])
```

categorical group names instead of numerical codes.

- The grouping is based on the research paper table (https://www.hindawi.com/journals/bmri/2014/781670/tab2/).

Group names

```
1-Circulatory
2-Respiratory
3-Digestive
4-Diabetes
5-Injury
6-Musculoskeletal
7-Genitourinary
8-Neoplasms
9-Other
```

```python
# Circulatory
codes =[str(i) for i in list(range(390,460)) + [785]]
df = df.replace(codes, 'Circulatory')
# Respiratory
codes =[str(i) for i in list(range(460,520)) + [786]]
df = df.replace(codes, 'Respiratory')
# Digestive
codes =[str(i) for i in list(range(520,580)) + [787]]
df = df.replace(codes, 'Digestive')
# Diabetes
df = df.replace(regex=r'^250.*', value='Diabetes')
# Injury
codes =[str(i) for i in range(800,1000)]
df = df.replace(codes, 'Injury')
# Musculoskeletal
codes =[str(i) for i in range(710,740)]
df = df.replace(codes, 'Musculoskeletal')
# Genitourinary
codes =[str(i) for i in list(range(580,630)) + [788]]
df = df.replace(codes, 'Genitourinary')
# Neoplasms
codes =[str(i) for i in range(140,240)]
df = df.replace(codes, 'Neoplasms')
# Other
df = df.replace(regex=r'^[E,V].*', value='Other')

codes =[str(i) for i in range(0,1000)]
df = df.replace(codes, 'Other')
df[['diag_1', 'diag_2', 'diag_3']].head()
# Unique Values of Each Features:
for i in df[['diag_1', 'diag_2', 'diag_3']]:
    print(f'{i}:\n{sorted(df[i].unique())}\n')
# need to add 365.44 to Other
df = df.replace('365.44', 'Other')
```

Analysis of diagnosis

```
plt.figure(figsize=(20, 8))
for diag in ['diag_1','diag_2','diag_3']:
    sns.lineplot(x=df[diag].value_counts().sort_index().index, y= df[diag].va
lue_counts().sort_index().values, marker='o')
plt.legend(['diag_1','diag_2','diag_3'])
plt.show()
```

- Looking at the graph above, we can say that there is a high correlation between the diagnoses. So we drop diag_2 and diag_3.
- Also as the most common diagnoses are prevalent in all three diagnoses listed, we are only using the primary diagnosis variable to build the machine learning model

```
# drop diagnoses 2 and 3
df = df.drop(columns=['diag_2', 'diag_3'])
```

Outlier detection

Based on the basic statistics describing the dataset, it looks there are outliers that influence skewness in the data. To represent the majority of samples and build clean models, we are going to remove outliers that have z-scores greater than 3.0 or less than −3.0. This means that we are removing samples that are more (or less) than 3 times the standard deviation from the mean.

FOCUS ON "number_diagnoses"

```
plt.figure(figsize=(8,5))
ax = df.number_diagnoses.value_counts().sort_index().plot.bar()
def labels(ax, df=df):
    for p in ax.patches:
            ax.annotate('{:.0f}'.format(p.get_height()),
                        (p.get_x(), p.get_height()+100),size=10)
labels(ax)
```

- For a small number of observations with number of diagnoses greater than 9, let's change the number of diagnoses to 9.

```
df.number_diagnoses = df.number_diagnoses.replace([10,11,12,13,14,15,16],9)
```

```
df.describe().T
features = df.describe().columns
def col_plot(df,col_name):
    plt.figure(figsize=(15,6))
    plt.subplot(141)
    plt.hist(df[col_name], bins = 20)
    f=lambda x:(np.sqrt(x) if x>=0 else -np.sqrt(-x))
    plt.axvline(x=df[col_name].mean() + 3*df[col_name].std(),color='red')
    plt.axvline(x=df[col_name].mean() - 3*df[col_name].std(),color='red')
    plt.xlabel(col_name)
    plt.tight_layout
    plt.xlabel("Histogram ±3z")
    plt.ylabel(col_name)
    plt.subplot(142)
    plt.boxplot(df[col_name])
    plt.xlabel("IQR=1.5")
    plt.subplot(143)
    plt.boxplot(df[col_name].apply(f), whis = 2.5)
    plt.xlabel("ROOT SQUARE - IQR=2.5")
    plt.subplot(144)
    plt.boxplot(np.log(df[col_name]+0.1), whis = 2.5)
    plt.xlabel("LOGARITMIC - IQR=2.5")
    plt.show()
for i in features:
    col_plot(df,i)
from scipy.stats.mstats import winsorize

def plot_winsorize(df,col_name,up=0.1,down=0):
    plt.figure(figsize = (15, 6))
    winsor=winsorize(df[col_name], (down,up))
    logr=np.log(df[col_name]+0.1)
    plt.subplot(141)
    plt.hist(winsor, bins = 22)
    plt.axvline(x=winsor.mean()+3*winsor.std(),color='red')
    plt.axvline(x=winsor.mean()-3*winsor.std(),color='red')
    plt.xlabel('Winsorize_Histogram')
    plt.ylabel(col_name)
    plt.tight_layout
    plt.subplot(142)
    plt.boxplot(winsor, whis = 1.5)
    plt.xlabel('Winsorize - IQR:1.5')
```

```
    plt.subplot(143)
    plt.hist(logr, bins=22)
    plt.axvline(x=logr.mean()+3*logr.std(),color='red')
    plt.axvline(x=logr.mean()-3*logr.std(),color='red')
    plt.xlabel('Logr_col_name')
    plt.subplot(144)
    plt.boxplot(logr, whis = 1.5)
    plt.xlabel("Logaritmic - IQR=1.5")
    plt.show()

for i in features:
    plot_winsorize(df,i)


df_winsorised=df.copy()
for i in features:
    df_winsorised[i]=winsorize(df_winsorised[i], (0,0.1))
df_log=df.copy()
for i in features:
    df_log[i]=np.log(df_log[i])
df_root=df.copy()
f=lambda x:(np.sqrt(x) if x>=0 else -np.sqrt(-x))
for i in features:
    df_root[i]=df_root[i].apply(f)
from numpy import percentile
from scipy.stats import zscore
from scipy import stats
def outlier_zscore(df, col, min_z=1, max_z = 5, step = 0.1, print_list = Fals
e):
    z_scores = zscore(df[col].dropna())
    threshold_list = []
    for threshold in np.arange(min_z, max_z, step):
        threshold_list.append((threshold, len(np.where(z_scores > threshold)[
0])))

        df_outlier = pd.DataFrame(threshold_list, columns = ['threshold', 'ou
tlier_count'])
        df_outlier['pct'] = (df_outlier.outlier_count -
 df_outlier.outlier_count.shift(-1))/df_outlier.outlier_count*100
    plt.plot(df_outlier.threshold, df_outlier.outlier_count)
    best_treshold = round(df_outlier.iloc[df_outlier.pct.argmax(), 0],2)
    outlier_limit = int(df[col].dropna().mean() + (df[col].dropna().std()) *
df_outlier.iloc[df_outlier.pct.argmax(), 0])
    percentile_threshold = stats.percentileofscore(df[col].dropna(), outlier_
limit)
    plt.vlines(best_treshold, 0, df_outlier.outlier_count.max(),
               colors="r", ls = ":")
```

```
                    )
        plt.annotate("Zscore : {}\nValue : {}\nPercentile : {}".format(best_tresh
old, outlier_limit, (np.round(percentile_threshold, 3), np.round(100-
percentile_threshold, 3))), (best_treshold, df_outlier.outlier_count.max()/2)
)
        #plt.show()
        if print_list:
            print(df_outlier)
        return (plt, df_outlier, best_treshold, outlier_limit, percentile)
from scipy.stats import zscore
from scipy import stats

def outlier_inspect(df, col, min_z=1, max_z = 5, step = 0.5, max_hist = None,
 bins = 50):
    fig = plt.figure(figsize=(20, 6))
    fig.suptitle(col, fontsize=16)
    plt.subplot(1,3,1)
    if max_hist == None:
        sns.distplot(df[col], kde=False, bins = 50)
    else :
        sns.distplot(df[df[col]<=max_hist][col], kde=False, bins = 50)
    plt.subplot(1,3,2)
    sns.boxplot(df[col])
    plt.subplot(1,3,3)
    z_score_inspect = outlier_zscore(df, col, min_z=min_z, max_z = max_z, ste
p = step)
    plt.subplot(1,3,1)
    plt.axvline(x=df[col].mean() + z_score_inspect[2]*df[col].std(),color='re
d',linewidth=1,linestyle ="--")
    plt.axvline(x=df[col].mean() -
 z_score_inspect[2]*df[col].std(),color='red',linewidth=1,linestyle ="--")
    plt.show()

    return z_score_inspect
def detect_outliers(df:pd.DataFrame, col_name:str, p=1.5) ->int:
    '''
    this function detects outliers based on 3 time IQR and
    returns the number of lower and uper limit and number of outliers respect
ively
    '''
    first_quartile = np.percentile(np.array(df[col_name].tolist()), 25)
    third_quartile = np.percentile(np.array(df[col_name].tolist()), 75)
    IQR = third_quartile - first_quartile

    upper_limit = third_quartile+(p*IQR)
```

```
        lower_limit = first_quartile-(p*IQR)
        outlier_count = 0

        for value in df[col_name].tolist():
            if (value < lower_limit) | (value > upper_limit):
                outlier_count +=1
        return lower_limit, upper_limit, outlier_count
k=3
print(f"Number of Outliers for {k}*IQR\n")
total=0
for col in features:
    if detect_outliers(df, col)[2] > 0:
        outliers=detect_outliers(df, col, k)[2]
        total+=outliers
        print("{} outliers in '{}'".format(outliers,col))
print("\n{} OUTLIERS TOTALLY".format(total))
k=3
print(f"Number of Outliers for {k}*IQR after Root Square\n")
total=0
for col in features:
    if detect_outliers(df_root, col)[2] > 0:
        outliers=detect_outliers(df_root, col, k)[2]
        total+=outliers
        print("{} outliers in '{}'".format(outliers,col))
print("\n{} OUTLIERS TOTALLY".format(total))
k=3
print(f"Number of Outliers for {k}*IQR after Winsorised\n")
total=0
for col in features:
    if detect_outliers(df_winsorised, col)[2] > 0:
        outliers=detect_outliers(df_winsorised, col, k)[2]
        total+=outliers
        print("{} outliers in '{}'".format(outliers,col))
print("\n{} OUTLIERS TOTALLY".format(total))
k=3
print(f"Number of Outliers for {k}*IQR after Logarithmed\n")
total=0
for col in features:
    if detect_outliers(df_log, col)[2] > 0:
        outliers=detect_outliers(df_log, col, k)[2]
        total+=outliers
        print("{} outliers in '{}'".format(outliers,col))
print("\n{} OUTLIERS TOTALLY".format(total))
```

```
z_scores=[]
for i in features:
    z_scores.append(outlier_inspect(df,i)[2])
z_scores
features
# create columns for z scores, new column with z score
df_3z=df.copy()
for x in features:
    df_3z[x + '_z'] = stats.zscore(df_3z[x])

for x in df_3z.columns[-len(features):]:
    df_3z = df_3z[(df_3z[x] < 3) & (df_3z[x] > -3)]
# drop _z columns
df_3z = df_3z.drop(columns=df_3z.columns[-8:])
print('Number of Outliers:',len(df)-len(df_3z))
df_3z.describe().T.round(2)
df.describe().T.round(2)
```

Check unique values

Investigate the unique values of each column and look for error entries.

```
summary(df_3z)
```

Drop the columns where the number of uniques is 1.

```
df_3z = df_3z.drop(['acetohexamide','glimepiride-pioglitazone','metformin-
rosiglitazone'],axis=1)
```

Export cleaned dataset

```
df_3z = df_3z.reset_index(drop=True)
df_3z.to_csv('diabetic_data_cleaned.csv')
```

3.1.2 *Visualization*

- We are looking for correlations between the independent variables and the target variable, the likelihood of being readmitted to the hospital, using graphs and plots.
- This is also a good time to get a better understanding of patient demographics, their experiences at the hospital, medications being used/not used, and any diagnosed conditions.

Import libraries

```
import pandas as pd
import numpy as np
import matplotlib.pyplot as plt
import seaborn as sns
from scipy import stats
from pylab import rcParams
rcParams['figure.figsize'] = 12,6
# to avoid warnings
import warnings
warnings.filterwarnings('ignore')
warnings.warn("this will not show")
sns.set(style='darkgrid')
%matplotlib inline
```

Import dataset

```python
data = pd.read_csv('diabetic_data_cleaned.csv', index_col=0)
df = data.copy()
df.head()
features = pd.read_csv('features.csv',index_col='Unnamed: 0')
info = lambda attribute:print(f"{attribute.upper()} : {features[features['Fea
ture']==attribute]['Description'].values[0]}\n")
features.head()
def summary(df, pred=None):
    obs = df.shape[0]
    Types = df.dtypes
    Counts = df.apply(lambda x: x.count())
    Min = df.min()
    Max = df.max()
    Uniques = df.apply(lambda x: x.unique().shape[0])
    Nulls = df.apply(lambda x: x.isnull().sum())
    print('Data shape:', df.shape)

    if pred is None:
        cols = ['Types', 'Counts', 'Uniques', 'Nulls', 'Min', 'Max']
        str = pd.concat([Types, Counts, Uniques, Nulls, Min, Max], axis = 1,
sort=True)

    str.columns = cols
    print('_____\nData Types:')
    print(str.Types.value_counts())
    print('_____')
    return str

summary(df)
round(df.describe(), 2)
df.shape
sns.pairplot(df, hue='readmitted');
plt.figure(figsize=(20,10))
sns.heatmap(df.corr(), annot=True, cmap="coolwarm");
```

FOCUS ON 'readmitted' patients overall

```python
info('readmitted')
def labels(ax):
    for p in ax.patches:
            ax.annotate('%{:.1f}\n{:.0f}'.format(100*p.get_height()/len(df),p
.get_height()),
                            (p.get_x()+0.3, p.get_height()-1900),size=11)
ax = sns.countplot(x='readmitted', palette='husl', data=df)
labels(ax)
# sns.catplot(x='readmitted', kind='count', palette='husl', data=df)   # alter
native
plt.title('Readmit Rates')
plt.show()
```

FOCUS ON 'race'

```python
def labels(ax):
    for bar in ax.patches:
            ax.annotate('%{:.1f}\n{:.0f}'.format(100*bar.get_height()/len(df),bar
.get_height()), (bar.get_x() + bar.get_width() / 2,
                            bar.get_height()), ha='center', va='center',
                        size=10, xytext=(0, 8),
                        textcoords='offset points')

rcParams['figure.figsize'] = 12,6
ax = sns.countplot(x='race', hue='readmitted', palette='husl', data=df)
labels(ax)
# sns.catplot(x='race', hue='readmitted', kind='count', palette='husl', data=
df, aspect=2, legend_out=False)
plt.title('Patient Demographic Readmissions')
plt.show()
pd.crosstab(df.race, df.readmitted, margins=True, margins_name='Total')
```

FOCUS ON 'gender'

```python
rcParams['figure.figsize'] = 12,6
ax = sns.countplot(x='gender', hue='readmitted', palette='husl', data=df)
labels(ax)
plt.title('Readmissions by Gender')
plt.show()
pd.crosstab(df.gender, df.readmitted, margins=True, margins_name='Total')
```

FOCUS ON 'age' groups

```
ax = sns.countplot(x='age', palette='husl', data=df.sort_values('age'))
labels(ax)
plt.title('Patient Demographics')
plt.show()
```

It looks like most patients are older, 50+ years old, though there are not many patients over 90.

```
ax = sns.countplot(x='age', hue='readmitted', palette='husl', data=df.sort_va
lues('age'))
labels(ax)
plt.title('Readmits By Age Group')
plt.show()
pd.crosstab(df.age, df.readmitted, margins=True, margins_name='Total').T
```

In every age group, more patients are not readmitted. The 70–80 age group account has the highest number of readmitted and not readmitted patients.

FOCUS ON 'time_in_hospital'

```
sns.countplot(x='time_in_hospital', palette='muted', data=df)
mean, median = np.mean(df.time_in_hospital), np.median(df.time_in_hospital)
plt.axvline(mean-
df.time_in_hospital.min(), color='blue', label=f'mean:{round(mean,2)}')
plt.axvline(median-
df.time_in_hospital.min(), color='red', label=f'median:{round(median,2)}')
plt.title('Duration of Hospital Visit in Days')
plt.legend()
plt.show()
```

Does the amount of time spent in the hospital impact a patient's chances of readmission?

```
sns.catplot(x='time_in_hospital', hue='readmitted', kind='count', palette='hu
sl', aspect=3, data=df, legend_out=False)
plt.title('Readmission Based on Time in Hospital')
plt.show()
sns.displot(x='time_in_hospital', hue='readmitted', data=df, height=7, aspect
=3)
plt.title('Readmission Based on Time in Hospital')
plt.show()
```

Based on the graph, the longer a patient spends in the hospital, the likelier their chances are of being readmitted. Patients who spend more than a week in the hospital usually have a serious illness or complication that may reoccur depending on their ability to recover, which is why they may need to revisit the hospital.

Which age group is spending the most time in hospitals during visits?

```python
def box_labels(ax, df,col1,col2):
    medians = df.groupby([col1])[col2].median()
    vertical_offset = df[col2].median() * 0.05 # offset from median for displ
ay

    for xtick in ax.get_xticks():
        ax.text(xtick,medians[xtick] + vertical_offset,medians[xtick],
                horizontalalignment='center',size='x-
small',color='w',weight='semibold')

ax = sns.boxplot(x='age', y='time_in_hospital', data=df.sort_values('age'))
box_labels(ax, df.sort_values('age'),'age','time_in_hospital')
plt.title('Length of Hospital Stay Based on Age')
plt.show()
```

What is the comparison of time in hospital for readmitted patients?

```python
ax = sns.boxplot(x='readmitted', y='time_in_hospital', data=df.sort_values('r
eadmitted'))
box_labels(ax, df.sort_values('readmitted'),'readmitted','time_in_hospital')
plt.title('Length of Hospital Stay for Readmitted Patients')
plt.show()
```

Readmitted patients stay longer in the hospital on average compared to those who are not readmitted.

FOCUS ON 'number of lab procedures'

```
info("num_lab_procedures")
rcParams['figure.figsize'] = 25,10
sns.countplot(x='num_lab_procedures', data=df)
mean, median = np.mean(df.num_lab_procedures), np.median(df.num_lab_procedure
s)
plt.axvline(mean-
df.num_lab_procedures.min(), color='blue', label=f'mean:{round(mean,2)}')
plt.axvline(median-
df.num_lab_procedures.min(), color='black', label=f'median:{round(median,2)}'
)
plt.title('Number of Lab Procedures Performed During Visit')
plt.legend()
plt.show()
df.groupby('readmitted')['num_lab_procedures'].describe().round(2)
```

Do the patients with longer hospital stays have more lab tests?

```
def box_labels(ax, df,col1,col2):
    medians = df.groupby([col1])[col2].median()
    vertical_offset = df[col2].median() * 0.05 # offset from median for displ
ay
    for xtick in ax.get_xticks():
        ax.text(xtick,medians[xtick] + vertical_offset,medians[xtick],
                horizontalalignment='center',size=12,color='w',weight='semibo
ld')
ax = sns.boxplot(x='time_in_hospital', y='num_lab_procedures', data=df.sort_v
alues('time_in_hospital'))
# box_labels(ax, df.sort_values('time_in_hospital'),'time_in_hospital','num_l
ab_procedures')
plt.title('Lab Procedures Based on Length of Hospital Visit')
plt.show()
```

- There is a positive correlation between time spent in the hospital and number of lab tests completed.

- This makes sense because patients with longer stays had more tests completed to properly diagnose their conditions.

Do readmitted patients have more lab tests?

```
plt.figure(figsize=(10, 8))
ax = sns.boxplot(x='readmitted', y='num_lab_procedures', data=df.sort_values(
'readmitted'))
box_labels(ax, df.sort_values('readmitted'),'readmitted','num_lab_procedures'
)
plt.title('Lab Procedures for Readmitted Patients')
plt.show()
```

- The average number of lab procedures is about equal for readmitted and not readmitted patients.
- Not readmitted patients have a slightly lower number of lab procedures done during their visit.

FOCUS ON "number of procedures" (other than lab)

```
info('num_procedures')
sns.catplot(x='num_procedures', kind='count', palette='muted', data=df)
mean, median = np.mean(df.num_procedures), np.median(df.num_procedures)
plt.axvline(mean, color='blue', label=f'mean:{round(mean,2)}')
plt.axvline(median, color='black', label=f'median:{round(median,2)}')
plt.title('Number of Procedures Performed (Except Lab)')
plt.legend()
plt.show()
```

Do the number of tests performed indicate whether a patient will be readmitted?

```
def labels(ax):
    for bar in ax.patches:
        ax.annotate('%{:.1f}\n{:.0f}'.format(100*bar.get_height()/len(df),bar
.get_height()), (bar.get_x() + bar.get_width() / 2,
                      bar.get_height()-400), ha='center', va='center',
                    size=14, xytext=(0, 8),
                    textcoords='offset points')

ax = sns.countplot(x='num_procedures', hue='readmitted', palette='husl', data
=df)
labels(ax)
plt.title('Readmits Based on Procedures (Sans Lab)')
plt.show()
```

FOCUS ON 'number of medications'

```
info('num_medications')
rcParams['figure.figsize'] = 25,10
sns.countplot(x='num_medications', data=df)
mean, median = np.mean(df.num_medications), np.median(df.num_medications)
plt.axvline(mean-
df.num_medications.min(), color='blue', label=f'mean:{round(mean,2)}')
plt.axvline(median-
df.num_medications.min(), color='black', label=f'median:{round(median,2)}')
plt.title('Number of Distinct Generic Medications Administered During Visit')
plt.legend()
plt.show()


df.groupby('readmitted')['num_medications'].describe()
```

How many medications are patients receiving during their visit?

```
ax = sns.boxplot(x='time_in_hospital', y='num_medications', data=df)
# box_labels(ax, df.sort_values('time_in_hospital'),'time_in_hospital','num_m
edications')
plt.title('Medications Administered Based on Length of Hospital Visit')
plt.show()
```

Patients who spend more time in the hospital receive more medications, but there are a few that receive over 60 different kinds of medications.

How many medications are patients receiving during their visit?

```
ax = sns.boxplot(x='readmitted', y='num_medications', data=df.sort_values('re
admitted'))
box_labels(ax, df.sort_values('readmitted'),'readmitted','num_medications')
plt.title('Medications Administered')
plt.show()
```

The distribution is almost equal for readmitted and not readmitted patients, with readmits being slightly higher on average.

FOCUS ON "number of outpatient" visits

```
info('number_outpatient')
def labels(ax):
    for bar in ax.patches:
        ax.annotate('%{:.1f}\n{:.0f}'.format(100*bar.get_height()/len(df),bar
.get_height()), (bar.get_x() + bar.get_width() / 2,
                        bar.get_height()+750), ha='center', va='center',
                        size=16, xytext=(0, 8),
                        textcoords='offset points')

ax = sns.countplot(x='number_outpatient',data=df)
labels(ax)
plt.title('Number of Outpatient Visits Prior to Encounter')
plt.show()
# outpatient visit stats
df.groupby('readmitted')['number_outpatient'].describe()
# outpatient vists and readmissions
ax = sns.countplot(x='number_outpatient',data=df, hue='readmitted')
labels(ax)
plt.title('Outpatient Vists and Readmissions')
plt.show()
pd.crosstab(df.readmitted, df.number_outpatient, margins=True, margins_name='
Total')
```

Most patients did not have any outpatient visits before the recorded one.

FOCUS ON "number of emergency" visits

```
info('number_emergency')
# plt.figure(figsize=(20,5))
ax = sns.countplot(x='number_emergency', data=df)
labels(ax)
plt.title('Number of Emergency Visits Prior to Encounter')
plt.show()
# emergency vists and readmissions
ax = sns.countplot(x='number_emergency', hue='readmitted', data=df)
labels(ax)
plt.title('Emergency Vists and Readmissions')
plt.show()
```

Most patients did not visit the emergency room before their recorded visit.

```
pd.crosstab(df.readmitted, df.number_emergency, margins=True, margins_name='T
otal')
```

How many emergency visits did patients have before this visit?

```
plt.figure(figsize=(5, 5))
sns.boxplot(x='readmitted', y='number_emergency', data=df)
plt.title('Readmits for Emergency Vists')
plt.show()
```

FOCUS ON "number of inpatient" visits

```
info('number_inpatient') # onceki yildaki yatarak tedavi sayisi
ax = sns.countplot(x='number_inpatient',data=df)
labels(ax)
plt.title('Number of Inpatient Visits Prior to Encounter')
plt.show()
# inpatient visits and readmissions
ax = sns.countplot(x='number_inpatient', hue='readmitted',data=df)
labels(ax)
plt.title('Inpatient Visits and Readmissions')
plt.show()
```

Inpatient visits are not common for most patients before this visit.

```
pd.crosstab(df.readmitted, df.number_inpatient, margins=True, margins_name='T
otal')
```

FOCUS ON "number of diagnoses"

```
info('number_diagnoses')
ax = sns.countplot(x='number_diagnoses',data=df)
mean, median = np.mean(df.number_diagnoses), np.median(df.number_diagnoses)
plt.axvline(mean-
df.number_diagnoses.min(), color='blue', label=f'mean:{round(mean,2)}')
plt.axvline(median-
df.number_diagnoses.min(), color='red', label=f'median:{round(median,2)}')
plt.title('Number of Diagnoses')
plt.legend()
plt.show()
# number of diagnoses and readmit rate
ax = sns.countplot(x='number_diagnoses', hue='readmitted', palette='Accent',
data=df)
# labels(ax)
plt.title('Readmits By Number of Diagnoses')
plt.show()
pd.DataFrame(df.number_diagnoses.describe()).T.round(2)
df.groupby('readmitted')['number_diagnoses'].describe().round(2)
# number of diagnoses
pd.crosstab(df.readmitted, df.number_diagnoses, margins=True, margins_name='T
otal')
```

- Most patients have up to nine diagnosed conditions during their visit, after that, only a handful have more than nine in one visit.
- Readmitted patients tend to have more diagnosed conditions but their average is only slightly higher than those not readmitted.

How many diagnoses do readmitted patients have?

```
plt.figure(figsize=(8, 6))
ax = sns.boxplot(x='readmitted', y='number_diagnoses', data=df.sort_values('r
eadmitted'))
box_labels(ax, df.sort_values('readmitted'),'readmitted','number_diagnoses')
plt.title('Number of Diagnoses for Re/admitted Patients')
plt.show()
```

FOCUS ON "glucose serum test results"

```
info('max_glu_serum')
ax = sns.countplot(x='max_glu_serum', data=df)
labels(ax)
plt.title('Glucose Serum Test Results')
plt.show()
```

As the majority of patients do not have a glucose reading, they will be excluded from the next graph to show the readmit rates for patients who do have a reading.

```
def labels(ax, df=df):
    for p in ax.patches:
            ax.annotate('%{:.1f}\n{:.0f}'.format(100*p.get_height()/len(df),p
.get_height()),
                        (p.get_x()+0.2, p.get_height()-27),size=16)
    # exclude patients without a glucose reading
    glucose_none = df[df.max_glu_serum != 'None']
    # glucose serum results and readmit impact
    ax = sns.countplot(x='max_glu_serum', hue='readmitted', palette='Accent', dat
    a=glucose_none)
    labels(ax,glucose_none)
    plt.title('Readmits By Glucose Serum Levels')
    plt.show()
```

Patients with a glucose serum reading of over 300 have a 50–50 chance of being readmitted. High blood sugar levels are often dangerous for older patients due to the medical complications involved, so it is understandable that more patients return to the hospital for additional care.

```
pd.crosstab(df.readmitted, df.max_glu_serum, margins=True, margins_name='Tota
l')
```

Unsupported Cell Type. Double-Click to inspect/edit the content.

FOCUS ON "A1C results"

```
info('A1Cresult')
ax = sns.countplot(x='A1Cresult', palette='husl', data=df)
labels(ax)
plt.title('A1c Test Results')
plt.show()
```

- Similar to the glucose reading, the majority of patients also do not have a HbA1c test reading.

- To understand the impact of A1c tests on readmit rates, patients without a reading will be excluded in the graph below.

```
# exclude patients without an A1C reading
a1c_none = df[df.A1Cresult != 'None']
# A1C results and readmit impact
ax = sns.countplot(x='A1Cresult', hue='readmitted', palette='Accent', data=a1
c_none)
labels(ax, a1c_none)
plt.title('Readmits By A1C Test Results')
plt.show()
pd.crosstab(df.readmitted, df.A1Cresult, margins=True, margins_name='Total')
```

FOCUS ON "change" column

```
info('change')
```

Change in medications, dosage, or brand

```
# change in medications
ax = sns.countplot(x='change', hue='readmitted', data=df)
labels(ax)
plt.title('Change in Diabetic Medications')
plt.show()
pd.crosstab(df.change, df.readmitted, margins=True, margins_name='Total')
```

Who is likely to have a change in medication?

```
ax = sns.countplot(x='gender', hue='change', palette='Set2', data=df)
labels(ax)
plt.title('Change in Medication Based on Gender')
plt.show()
pd.crosstab(df.gender, df.change, margins=True, margins_name='Total')
```

FOCUS ON "diabetesMed"

```
info('diabetesMed')
ax = sns.countplot(x='diabetesMed', hue='readmitted', data=df)
labels(ax)
plt.title('Prescribed Diabetic Medications During Visit')
plt.show()
pd.crosstab(df.diabetesMed, df.readmitted, margins=True, margins_name='Total'
)
```

Who is likely or not likely to have a change in medication?

```
sns.catplot(x='diabetesMed', hue='readmitted', col='gender', palette='Accent'
, data=df, kind='count', height=4, aspect=1)
plt.show()
```

Medications used by patients

```
columns=['metformin', 'repaglinide', 'nateglinide',
        'chlorpropamide', 'glimepiride', 'glipizide', 'glyburide',
        'tolbutamide', 'pioglitazone', 'rosiglitazone', 'acarbose', 'miglitol'
,
        'troglitazone', 'tolazamide', 'insulin', 'glyburide-metformin',
        'glipizide-metformin', 'metformin-pioglitazone']

plt.figure(figsize=(26, 26))
for i,col in enumerate(columns):
    plt.subplot(6,3,i+1)
    sns.countplot(x=df[col])
```

Dosages for insulin show the most activity out of all diabetic medications, most of which are not prescribed to patients.

```
info('insulin')
sns.countplot(x='insulin', hue='readmitted', data=df)
plt.title('Readmit Rates by Medication: Insulin')
plt.show()
```

3.1.3 General overview—statistical analysis

- We want to analyze the variables in this dataset to understand any relationships between them and their overall effects.
- To do this,

convert all categorical variables to numeric variables using dummy variables. This function turns the string values in a variable to columns labeled 0 or 1 relative to the string. In addition, we will standardize the original numerical variables with a mean of 0 and a standard deviation of 1.

```
    * `Chi-square test` for categorical variables relationship
    * We have to analyze numerical variables using `analysis of var
iance` or `ANOVA test`.
```

- The goal of these tests is to see if there is a statistically significant link between the objective variable, readmissions, and the independent variable. Our P-value is 0.01, and anything higher than that prevents us from rejecting the null hypothesis.
- Because a machine learning model can comprehend integers as well as texts, we must

- Finally, we examine the correlation coefficients between the independent variables to ensure that they do not have a significant influence on one another. We chose a threshold of $-0.7 < x < 0.7$.

Import libraries

```python
import pandas as pd
import numpy as np
import matplotlib.pyplot as plt
import seaborn as sns
from scipy import stats
from pylab import rcParams
rcParams['figure.figsize'] = 12,6
# to avoid warnings
import warnings
warnings.filterwarnings('ignore')
warnings.warn("this will not show")
sns.set(style='darkgrid')
%matplotlib inline
import statsmodels.api as sm
from statsmodels.formula.api import ols
from scipy.stats import chi2_contingency
from numpy.random import seed
from sklearn.preprocessing import LabelEncoder, MinMaxScaler
data = pd.read_csv('diabetic_data_cleaned.csv', index_col=0) # import data
df = data.copy() # save a copy of data as diabetes
features = pd.read_csv('features.csv',index_col='Unnamed: 0')
info = lambda attribute:print(f"{attribute.upper()} : {features[features['Fea
ture']==attribute]['Description'].values[0]}\n")
def summary(df, pred=None):
    obs = df.shape[0]
    Types = df.dtypes
    Counts = df.apply(lambda x: x.count())
    Min = df.min()
    Max = df.max()
    Uniques = df.apply(lambda x: x.unique().shape[0])
    Nulls = df.apply(lambda x: x.isnull().sum())
    print('Data shape:', df.shape)
    if pred is None:
        cols = ['Types', 'Counts', 'Uniques', 'Nulls', 'Min', 'Max']
        str = pd.concat([Types, Counts, Uniques, Nulls, Min, Max], axis = 1,
sort=True)
    str.columns = cols
    print('_____\nData Types:')
    print(str.Types.value_counts())
    print('_____')
    return str
summary(df)
df.describe().round(2).T
plt.figure(figsize=(6,6))
explode = [0,0.1]
plt.pie(df['readmitted'].value_counts(),explode=explode,autopct='%1.1f%%',sha
dow=True,startangle=60)
plt.legend(labels=df.readmitted.value_counts().index)
plt.title('Readmitted Patients')
plt.axis('off')
plt.show()
```

3.1.4 Feature selection
Categorical variables

```
print('Unique Values of Each Features:\n')
for i in df:
    print(f'{i}:\n{sorted(df[i].unique())}\n')
categorical=df.select_dtypes(include='object').columns.tolist()
print(categorical)
```

- The categorical variables are as follows:

  -
        ['race', 'gender', 'age', 'diag_1', 'max_glu_serum', 'A1Cresult',
        'metformin', 'repaglinide', 'nateglinide', 'chlorpropamide',
        'glimepiride', 'glipizide', 'glyburide', 'tolbutamide',
        'pioglitazone', 'rosiglitazone', 'acarbose', 'miglitol',
        'troglitazone', 'tolazamide', 'insulin', 'glyburide-metformin',
        'glipizide-metformin', 'metformin-pioglitazone', 'change',
        'diabetesMed', 'readmitted']

- To reject the null hypothesis, we use the chi-square test for association with a P-value of 0.01.

Chi-square test for association

```
# define a function that returns a table, a chi-square value, and a p value
def chisquare_test(df, var_list, target, null_list=[]):
    for var in var_list:
        print(var.upper())
        chi_test = pd.crosstab(df[var], df[target])
        display(chi_test)
            chisq_value, pvalue, dataframe, expected = chi2_contingency(chi_t
est)
            print(f"""Chi-square value: {chisq_value:.2f}
p-value\t\t: {pvalue:.3f}\n""")
        if pvalue > 0.01: # adds variables that fail to reject the null hypot
hesis
            null_list.append(var)
    print(f'Fail to reject null hypothesis: {null_list}')
cols_cat = ['race','gender', 'age', 'diag_1', 'max_glu_serum', 'A1Cresult', '
change', 'diabetesMed']
null_list=[]
chisquare_test(df, cols_cat,'readmitted',null_list)
```

We may fairly conclude that there is no relationship between the independent variables and the target variable based on the chi-square value and *P*-value.

Medications

```
medications = ['metformin', 'repaglinide', 'nateglinide', 'chlorpropamide', '
glimepiride',
            'glipizide', 'glyburide', 'tolbutamide', 'pioglitazone',
              'rosiglitazone', 'acarbose', 'miglitol', 'troglitazone', 'tola
zamide',
              'insulin', 'glyburide-metformin', 'glipizide-
metformin', 'metformin-pioglitazone']
chisquare_test(df, medications,'readmitted', null_list)
```

- Nateglinide, chlorpropamide, glimepiride, acetohexamide, glyburide, tolbutamide, miglitol, troglitazone, tolazamide, glipizide-metformin, and metformin-pioglitazone all failed the test due to P-values larger than 0.01.
- We are deleting these variables from the dataset because they are not independent of the target variable.

```
print(null_list)
# drop columns that do not pass the p-value test
df = df.drop(columns=null_list)
```

3.1.5 Numerical variables

Statistical testing—analysis of variance

```
# The numerical variables
numerical=df.select_dtypes(include=['int64','float']).columns.tolist()
print(numerical)
```

- We want to see if there is a statistically significant link between a numerical variable and a category target variable using the analysis of variance (ANOVA) test. Our P-value cutoff is 0.01.

We can reduce the number of emergency visits based on the ANOVA test because we cannot reject the null hypothesis that the averages for each class are equal, and the P-value is bigger than our threshold of 0.01.

```
# drop number_emergency column
df = df.drop(columns=['number_emergency'])
```

3.1.6 One hot encoding

Binary columns will be replaced with 0 for No and 1 for Yes. In the gender column, Male and Female will be replaced with 0 and 1, respectively.

```
# Unique Values of Each Features
for i in df:
    print(f'{i}:\n{sorted(df[i].unique())}\n')
df_dummy = pd.get_dummies(df,drop_first=True)
df_dummy.head()
```

```
df.describe().T.round(2)
# define a function that performs the ANOVA test and returns a table
def anova_table(var_list, null_list=[]):
    for var in var_list:
        print(var.upper())

        anova = ols('time_in_hospital ~ {}'.format(var), data=df).fit()
        table = sm.stats.anova_lm(anova, typ=2)
        pvalue=table['PR(>F)'][0]
        if pvalue > 0.01: # adds variables that fail to reject the null hypot
hesis
            null_list.append(var)
        display(table)
    print(f'Fail to reject null hypothesis: {null_list}')
anova_vars = ['readmitted']+numerical
anova_table(anova_vars)
```

3.1.7 Are the features that affect readmissions correlated with each other?

If the correlation value is larger than 0.7 or less than −0.7, one of the two columns must be removed.

For this notebook, the correlation map is fairly huge. Instead, we will look at each correlation coefficient individually and identify those with coefficients more than 0.7 and less than −0.7.

```python
plt.figure(figsize=(20,5))
df_dummy.corr()["readmitted_YES"].sort_values()[:-1].plot.bar();
plt.figure(figsize=(20,20))
sns.heatmap(df_dummy.corr(), cmap="coolwarm");
def corrank(X, threshold=0):
    import itertools
    df = pd.DataFrame([[i,j,X.corr().abs().loc[i,j]] for i,j in list(itertool
s.combinations(X.corr().abs(), 2))],columns=['Feature1','Feature2','corr'])

    df = df.sort_values(by='corr',ascending=False).reset_index(drop=True)
    return df[df['corr']>threshold]

# prints a descending list of correlation pair (Max on top)
corrank(df_dummy, 0.7)
# Remove the highly collinear features from data
def remove_collinear_features(x, threshold):
    # Calculate the correlation matrix
    corr_matrix = x.corr()
    iters = range(len(corr_matrix.columns) - 1)
    drop_cols = []

    # Iterate through the correlation matrix and compare correlations
    for i in iters:
        for j in range(i+1):
            item = corr_matrix.iloc[j:(j+1), (i+1):(i+2)]
            col = item.columns
            row = item.index
            val = abs(item.values)

            # If correlation exceeds the threshold
            if val >= threshold:
                # Print the correlated features and the correlation value
                print(col.values[0], "|", row.values[0], "|", round(val[0][0]
, 2))
                drop_cols.append(col.values[0])
    # Drop one of each pair of correlated columns
    drops = set(drop_cols)
    x = x.drop(columns=drops)
    return x
#Remove columns having more than 70% correlation
#Both positive and negative correlations are considered here
df_dummy = remove_collinear_features(df_dummy,0.70)
df_dummy.shape
```

Saving machine learning dataset

```
# save dataset to new file for machine learning
df_dummy.to_csv('diabetic_data_cleaned_dummy.csv')
```

3.1.8 General overview—machine learning
Import and load

```
# for basic operations
import numpy as np
import pandas as pd
# for visualizations
import matplotlib.pyplot as plt
import seaborn as sns
from pylab import rcParams
# rcParams['figure.figsize'] = 4,4
# plt.style.use('fivethirtyeight')
from collections import Counter
# for modeling
import sklearn
from xgboost.sklearn import XGBClassifier
from sklearn.metrics import confusion_matrix, classification_report, plot_pre
cision_recall_curve, precision_recall_curve
from sklearn.metrics import roc_curve, roc_auc_score
from sklearn.model_selection import GridSearchCV, RandomizedSearchCV, cross_v
al_score, train_test_split, KFold
from sklearn.preprocessing import StandardScaler
from sklearn.ensemble import RandomForestClassifier
from sklearn.linear_model import LogisticRegression
from sklearn import datasets, metrics
from sklearn.tree import DecisionTreeClassifier
from sklearn.naive_bayes import GaussianNB
from sklearn.svm import SVC
from sklearn.decomposition import PCA
import imblearn
from imblearn.under_sampling import RandomUnderSampler
from imblearn.over_sampling import SMOTE
# to avoid warnings
import warnings
warnings.filterwarnings('ignore')
warnings.warn("this will not show")
data = pd.read_csv('diabetic_data_cleaned_dummy.csv', index_col=0)
df = data.copy()
df.head()
```

Lazy predict with 5000 samples

```
from lazypredict.Supervised import LazyClassifier
df_5000 = df.sample(5000,random_state=42)
y = df_5000['readmitted_YES']
X = df_5000.drop('readmitted_YES', axis=1)
X_train, X_test, y_train, y_test = train_test_split(X, y, stratify=y, random_
state =42)
sc = StandardScaler()
X_train = sc.fit_transform(X_train)
X_test = sc.transform(X_test)
clf = LazyClassifier(verbose=0,ignore_warnings=True, custom_metric=None)
models,predictions = clf.fit(X_train, X_test, y_train, y_test)
models
```

Split data

```
ax = df['readmitted_YES'].value_counts(normalize=True).plot.bar()
def labels(ax):
    for p in ax.patches:
        ax.annotate(f"%{p.get_height()*100:.2f}", (p.get_x() + 0.15, p.get_he
ight() * 1.005),size=11)
labels(ax)
# separating the dependent and independent data
X = df.drop('readmitted_YES', axis=1)
y = df['readmitted_YES']
# the function train_test_split creates random data samples (default: 75-25%)
X_train, X_test, y_train, y_test = train_test_split(X, y, stratify=y, random_
state=42)
# getting the shapes
print(f"""shape of X_train: {X_train.shape}
shape of X_test\t: {X_test.shape}
shape of y_train: {y_train.shape}
shape of y_test\t: {y_test.shape}""")
```

Data scaling

```
# creating a standard scaler
sc = StandardScaler()
# fitting independent data to the model
X_train = sc.fit_transform(X_train)
X_test = sc.transform(X_test)
```

Iteration 1: (unbalanced data)

```
cv_acc_train = {}
cv_acc_test = {}
cv_TPR = {}
cv_FPR = {}
cv_AUC = {}
def plot_result(model, name:str):
    model.fit(X_train, y_train)
    y_pred = model.predict(X_test)
    # Evaluation based on a 10-fold cross-validation
    scoring = ['balanced_accuracy', 'recall_macro']
    scores_train = cross_val_score(model, X_train, y_train, cv=10, scoring =
'balanced_accuracy')
    scores_test = cross_val_score(model, X_test, y_test, cv=10, scoring = 'ba
lanced_accuracy')
    cv_acc_train[name] = round(scores_train.mean(), 4)*100  # balanced accura
cy
    cv_acc_test[name] = round(scores_test.mean(), 4)*100  # balanced accuracy
    cv_TPR[name] = (confusion_matrix(y_test, y_pred)[1][1]/confusion_matrix(y
_test, y_pred)[1].sum())*100  # recall (Max)
    cv_FPR[name] = (confusion_matrix(y_test, y_pred)[0][1]/confusion_matrix(y
_test, y_pred)[0].sum())*100  # fallout (Min)
    # accuracy scores
    print('Average Balanced Accuracy (CV=10), Test Set:', scores_test.mean())

    print('Average Balanced Accuracy (CV=10), Training Set: ', scores_train.m
ean())
    # print classification report
    print(classification_report(y_test, y_pred, zero_division=0))
    # Plot Confusion Matrix
    plot_confusion_matrix(model, X_test, y_test)
    plt.show()
```

1-Decision tree

```
from sklearn.tree import DecisionTreeClassifier, plot_tree
from sklearn.model_selection import cross_val_score
from sklearn.metrics import plot_confusion_matrix, classification_report, con
fusion_matrix
dtc = DecisionTreeClassifier()
plot_result(dtc, "dtc")
# plot tree
# plt.figure(figsize=(16,6))
# plot_tree(dtc, filled = True, class_names=["-
1", "1"], feature_names=X.columns, fontsize=11);
cv_acc_train, cv_acc_test, cv_TPR, cv_FPR
```

2-Logistic regression

```
from sklearn.linear_model import LogisticRegression
lr = LogisticRegression()
plot_result(lr, "lr")
```

3-SVC

```
# svc = SVC(probability=True)  # default values
# plot_result(svc, "svc")
```

4-NearestCentroid

```
from sklearn.neighbors import NearestCentroid
from sklearn.metrics import plot_confusion_matrix, classification_report, con
fusion_matrix
nc = NearestCentroid()
plot_result(nc, "nc")
```

5-RandomForest

```python
from sklearn.ensemble import RandomForestClassifier
rfc = RandomForestClassifier()
plot_result(rfc, "rfc")
def plot_feature_importances(model):
    feature_imp = pd.Series(model.feature_importances_,index=X.columns).sort_
values(ascending=False)[:10]
    sns.barplot(x=feature_imp, y=feature_imp.index)
    plt.title("Feature Importance")
    plt.show()
    print(f"Top 10 Feature Importance for {str(model).split('(')[0]}\n\n",fea
ture_imp[:10],sep='')
plot_feature_importances(rfc)
```

6-GradientBoosting

```python
from sklearn.ensemble import GradientBoostingClassifier
gbc = GradientBoostingClassifier(random_state=42)
plot_result(gbc, "gbc")
plot_feature_importances(gbc)
```

7-Naive bayes

```python
from sklearn.naive_bayes import GaussianNB
nb = GaussianNB()
plot_result(nb, "nb")
```

8-kNN

```python
from sklearn.neighbors import KNeighborsClassifier
knn = KNeighborsClassifier()
plot_result(knn, "knn")
```

9-XGBOOST

```
from xgboost import XGBClassifier
from sklearn.model_selection import GridSearchCV, RandomizedSearchCV, cross_v
al_score
xgb = XGBClassifier(eval_metric = "logloss")
plot_result(xgb, "xgb")
plot_feature_importances(xgb)
from xgboost import plot_importance
plot_importance(xgb,max_num_features=10)
plt.xlabel('The F-Score for each features')
plt.ylabel('Importances')
plt.show()
```

Evaluation (iteration 1)

```python
def AUC(cv_AUC, X_test=X_test):
    dtc_auc= roc_auc_score(y_test,dtc.predict(X_test)) #Decision Tree Classif
ier
    lr_auc= roc_auc_score(y_test, lr.decision_function(X_test))#logistic regr
ession
#    svc_auc= roc_auc_score(y_test, svc.decision_function(X_test))#Support V
ector Classifier
    nc_auc= roc_auc_score(y_test, nc.predict(X_test))#Nearest Centroid Classi
fier
    rfc_auc= roc_auc_score(y_test, rfc.predict_proba(X_test)[:,1])#Randomfore
st Classifier
    gbc_auc= roc_auc_score(y_test, gbc.predict_proba(X_test)[:,1])#GradientBo
osting Classifier
    nb_auc= roc_auc_score(y_test, nb.predict_proba(X_test)[:,1])#Naive Bayes
Classifier
    knn_auc= roc_auc_score(y_test, knn.predict(X_test))#KNeighbors Classifier
    xgb_auc= roc_auc_score(y_test, xgb.predict_proba(X_test)[:,1])#XGBoost Cl
assifier

    cv_AUC={'dtc': dtc_auc,
            'lr': lr_auc,
#           'svc':svc_auc,
            'nc':nc_auc,
            'rfc':rfc_auc,
            'gbc':gbc_auc,
            'nb':nb_auc,
            'knn':knn_auc,
            'xgb':xgb_auc}
    return cv_AUC
cv_AUC = AUC(cv_AUC)
df_eval = pd.DataFrame(data={'model': list(cv_acc_test.keys()),
                        'bal_acc_train':list(cv_acc_train.values()),
                        'bal_acc_test': list(cv_acc_test.values()),
                        'recall': list(cv_TPR.values()),
                        'fallout':list(cv_FPR.values()),
                         'AUC': list(cv_AUC.values())}).round(2)
df_eval
def plot_ROC(X_test=X_test, y_test=y_test):
    fpr_dtc, tpr_dtc, thresholds = roc_curve(y_test,dtc.predict(X_test)) #Dec
ision Tree Classifier
    fpr_lr, tpr_lr, thresholds = roc_curve(y_test, lr.decision_function(X_tes
t))#logistic regression
```

```
#    fpr_svc, tpr_svc, thresholds = roc_curve(y_test, svc.decision_function(
X_test))#Support Vector Classifier
    fpr_nc, tpr_nc, thresholds = roc_curve(y_test, nc.predict(X_test))#Neares
t Centroid Classifier
    fpr_rfc, tpr_rfc, thresholds = roc_curve(y_test, rfc.predict_proba(X_test
)[:,1])#Randomforest Classifier
    fpr_gbc, tpr_gbc, thresholds = roc_curve(y_test, gbc.predict_proba(X_test
)[:,1])#GradientBoosting Classifier
    fpr_nb, tpr_nb, thresholds = roc_curve(y_test, nb.predict_proba(X_test)[:
,1])#Naive Bayes Classifier
    fpr_knn, tpr_knn, thresholds = roc_curve(y_test, knn.predict(X_test))#KNe
ighbors Classifier
    fpr_xgb, tpr_xgb, thresholds = roc_curve(y_test, xgb.predict_proba(X_test
)[:,1])#XGBoost Classifier
    #compare the ROC curve between different models
    plt.figure(figsize=(10,10))
    plt.plot(fpr_dtc, tpr_dtc, label='Decision Tree Classifier')
    plt.plot(fpr_lr, tpr_lr, label='Logistic Regression')
#    plt.plot(fpr_svc, tpr_svc, label='Support Vector Classifier')
    plt.plot(fpr_nc, tpr_nc, label='Nearest Centroid Classifier')
    plt.plot(fpr_rfc, tpr_rfc, label='Randomforest Classifier')
    plt.plot(fpr_gbc, tpr_gbc, label='GradientBoosting Classifier')
    plt.plot(fpr_nb, tpr_nb, label='Naive Bayes Classifier')
    plt.plot(fpr_knn, tpr_knn, label='KNeighbors Classifier')
    plt.plot(fpr_xgb, tpr_xgb, label='XGBoost Classifier')
    plt.plot([0, 1], [0, 1], linestyle='--', lw=2, color='r',
            label='random', alpha=.8)
    plt.xlim([0,1])
    plt.ylim([0,1])
    plt.xticks(np.arange(0,1.1,0.1))
    plt.yticks(np.arange(0,1.1,0.1))
    plt.grid()
    plt.legend()
    plt.axes().set_aspect('equal')
    plt.xlabel('False Positive Rate')
    plt.ylabel('True Positive Rate')
plot_ROC()
fig, ax = plt.subplots(1,4, figsize=(20, 4))
sns.barplot(x="bal_acc_train", y="model", data=df_eval.sort_values(by="recall
"), ax=ax[0])
ax[0].set_title("Unbalanced Train Acc")
sns.barplot(x="bal_acc_test", y="model", data=df_eval.sort_values(by="recall"
), ax=ax[1])
ax[1].set_title("Unbalanced Test Acc")
sns.barplot(x="recall", y="model", data=df_eval.sort_values(by="recall"), ax=
ax[2])
ax[2].set_title("Unbalanced Test TPR")
sns.barplot(x="fallout", y="model", data=df_eval.sort_values(by="recall"), ax
=ax[3])
ax[3].set_title("Unbalanced Test FPR")
plt.show()
```

NaiveBayes gave high BalanceAccuracy and TPR_Score (Recall), but it gave a poor FPR_Score (Fallout) in this unbalanced dataset.

Iteration 2: (oversampling with SMOTE)
Balancing data

```python
y_test.value_counts(normalize=True)
y_train.value_counts(normalize=True)
# pip install imblearn
from imblearn import under_sampling, over_sampling
from imblearn.over_sampling import SMOTE
oversmote = SMOTE()
X_train_os, y_train_os= oversmote.fit_resample(X_train, y_train)
ax = y_train_os.value_counts().plot.bar(color=["blue", "red"])
def labels(ax):
    for p in ax.patches:
        ax.annotate(f"{p.get_height()}", (p.get_x() + 0.15, p.get_height()+20
0),size=8)
labels(ax)
plt.show()
X_train_os.shape
```

Use algorithms

```python
cv_acc_balance_train = {}
cv_acc_balance_test = {}
cv_TPR_balance = {}
cv_FPR_balance = {}
cv_AUC_balance = {}
def plot_result_smote(model, name:str):
    model.fit(X_train_os, y_train_os)
    y_pred = model.predict(X_test)
    # Evaluation based on a 10-fold cross-validation
    scoring = ['balanced_accuracy', 'recall_macro']
    scores_train = cross_val_score(model, X_train, y_train, cv=10, scoring =
'balanced_accuracy')
    scores_test = cross_val_score(model, X_test, y_test, cv=10, scoring = 'ba
lanced_accuracy')
    cv_acc_balance_train[name] = round(scores_train.mean(), 4)*100  # balance
d accuracy
    cv_acc_balance_test[name] = round(scores_test.mean(), 4)*100   # balanced
accuracy
    cv_TPR_balance[name] = (confusion_matrix(y_test, y_pred)[1][1]/confusion_
matrix(y_test, y_pred)[1].sum())*100  # recall (max)
    cv_FPR_balance[name] = (confusion_matrix(y_test, y_pred)[0][1]/confusion_
matrix(y_test, y_pred)[0].sum())*100  # fallout (min)
    # accuracy scores
    print('Average Balanced Accuracy (CV=10), Test Set:', scores_test.mean())

    print('Average Balanced Accuracy (CV=10), Training Set: ', scores_train.m
ean())
    # print classification report
    print(classification_report(y_test, y_pred, zero_division=0))
    # Plot Confusion Matrix
    plot_confusion_matrix(model, X_test, y_test)
    plt.show()
# Decision tree
dtc = DecisionTreeClassifier()
plot_result_smote(dtc, "dtc")
# Logistic Regression
lr = LogisticRegression()
plot_result_smote(lr, "lr")
# NearestCentroid
nc = NearestCentroid()
plot_result_smote(nc, "nc")
```

```
# # SVC
# svc = SVC()
# plot_result_smote(svc, "svc")
# Random Forest
rfc = RandomForestClassifier()
plot_result_smote(rfc, "rfc")
# Gradient Boost
gbc = GradientBoostingClassifier(random_state=42)
plot_result_smote(gbc, "gbc")
# Naive Bayes
nb = GaussianNB()
plot_result_smote(nb, "nb")
# kNN
knn = KNeighborsClassifier()
plot_result_smote(knn, "knn")
# XGBOOST
xgb = XGBClassifier(eval_metric = "logloss", random_state=42)
plot_result_smote(xgb, "xgb")
cv_AUC_balance = AUC(cv_AUC_balance)
df_eval_smote = pd.DataFrame(data={'model': list(cv_acc_balance_test.keys()),

                                   'bal_acc_train':list(cv_acc_balance_train.
values()),

                                   'bal_acc_test': list(cv_acc_balance_test.v
alues()),

                                   'recall': list(cv_TPR_balance.values()),
                                   'fallout':list(cv_FPR_balance.values()),
                                   'AUC': list(cv_AUC_balance.values())}).rou
nd(2)
df_eval_smote
fig, ax = plt.subplots(2,4, figsize=(20, 8))
sns.barplot(x="bal_acc_train", y="model", data=df_eval.sort_values(by="recall
"), ax=ax[0,0])
ax[0,0].set_title("Unbalanced Train Acc")
sns.barplot(x="bal_acc_test", y="model", data=df_eval.sort_values(by="recall"
), ax=ax[0,1])
ax[0,1].set_title("Unbalanced Test Acc")
sns.barplot(x="recall", y="model", data=df_eval.sort_values(by="recall"), ax=
ax[0,2])
ax[0,2].set_title("Unbalanced Test TPR")
sns.barplot(x="fallout", y="model", data=df_eval.sort_values(by="recall"), ax
=ax[0,3])
ax[0,3].set_title("Unbalanced Test FPR")
sns.barplot(x="bal_acc_train", y="model", data=df_eval_smote.sort_values(by="
recall"), ax=ax[1,0])
```

```
ax[1,0].set_title("Smote Model Train Acc")
sns.barplot(x="bal_acc_test", y="model", data=df_eval_smote.sort_values(by="r
ecall"), ax=ax[1,1])
ax[1,1].set_title("Smote Model Test Acc")
sns.barplot(x="recall", y="model", data=df_eval_smote.sort_values(by="recall"
), ax=ax[1,2])
ax[1,2].set_title("Smote Model Test TPR")
sns.barplot(x="fallout", y="model", data=df_eval_smote.sort_values(by="recall
"), ax=ax[1,3])
ax[1,3].set_title("Smote Model Test FPR")
plt.tight_layout()
plt.show()
plot_ROC()
```

GradientBoosting yielded the optimized result as better FPR and relative mean strong recall scores. The balance accuracy is also relatively good.

Iteration 3: (with RandomUnderSampling)

```
import imblearn
from imblearn.under_sampling import RandomUnderSampler, EditedNearestNeighbou
rs, NearMiss
under_sampler = RandomUnderSampler(random_state=42)
X_train_rus, y_train_rus = under_sampler.fit_sample(X_train, y_train)
ax = y_train_rus.value_counts().plot.bar(color=["blue", "red"])
labels(ax)
plt.show()
```

Use algorithm

```python
cv_acc_rus_train = {}
cv_acc_rus_test = {}
cv_TPR_rus = {}
cv_FPR_rus = {}
cv_AUC_rus = {}
def plot_result_rus(model, name:str):
    model.fit(X_train_rus, y_train_rus)
    y_pred = model.predict(X_test)
    # Evaluation based on a 10-fold cross-validation
    scoring = ['balanced_accuracy', 'recall_macro']
    scores_train = cross_val_score(model, X_train, y_train, cv=10, scoring = 'balanced_accuracy')
    scores_test = cross_val_score(model, X_test, y_test, cv=10, scoring = 'balanced_accuracy')
    cv_acc_rus_train[name] = round(scores_train.mean(), 4)*100  # balanced accuracy
    cv_acc_rus_test[name] = round(scores_test.mean(), 4)*100  # balanced accuracy
    cv_TPR_rus[name] = (confusion_matrix(y_test, y_pred)[1][1]/confusion_matrix(y_test, y_pred)[1].sum())*100  # recall (max)
    cv_FPR_rus[name] = (confusion_matrix(y_test, y_pred)[0][1]/confusion_matrix(y_test, y_pred)[0].sum())*100  # fallout (min)
    # accuracy scores
    print('Average Balanced Accuracy (CV=10), Test Set:', scores_test.mean())

    print('Average Balanced Accuracy (CV=10), Training Set: ', scores_train.mean())
    # print classification report
    print(classification_report(y_test, y_pred, zero_division=0))
    # Plot Confusion Matrix
    plot_confusion_matrix(model, X_test, y_test)
    plt.show()
# Decision tree
dtc = DecisionTreeClassifier()
plot_result_rus(dtc, "dtc")
# Logistic Regression
lr = LogisticRegression()
plot_result_rus(lr, "lr")
# NearestCentroid
nc = NearestCentroid()
plot_result_rus(nc, "nc")
```

```python
# # SVC
# svc = SVC()
# plot_result_rus(svc, "svc")
# Random Forest
rfc = RandomForestClassifier()
plot_result_rus(rfc, "rfc")
# Gradient Boost
gbc = GradientBoostingClassifier(random_state=42)
plot_result_rus(gbc, "gbc")
# Naive Bayes
nb = GaussianNB()
plot_result_rus(nb, "nb")
# kNN
knn = KNeighborsClassifier()
plot_result_rus(knn, "knn")
# XGBOOST
xgb = XGBClassifier(eval_metric = "logloss",random_state=42)
plot_result_rus(xgb, "xgb");
cv_AUC_rus = AUC(cv_AUC_rus)
df_eval_rus = pd.DataFrame(data={'model': list(cv_acc_rus_train.keys()),
                                 'bal_acc_train':list(cv_acc_rus_train.values()),
                                 'bal_acc_test': list(cv_acc_rus_test.values()),
                                 'recall': list(cv_TPR_rus.values()),
                                 'fallout':list(cv_FPR_rus.values()),
                                 'AUC': list(cv_AUC_rus.values())}).round(2)
df_eval_rus
fig, ax = plt.subplots(3,4, figsize=(20, 12))
sns.barplot(x="bal_acc_train", y="model", data=df_eval.sort_values(by="recall
"), ax=ax[0,0])
ax[0,0].set_title("Unbalanced Train Acc")
sns.barplot(x="bal_acc_test", y="model", data=df_eval.sort_values(by="recall"
), ax=ax[0,1])
ax[0,1].set_title("Unbalanced Test Acc")
sns.barplot(x="recall", y="model", data=df_eval.sort_values(by="recall"), ax=
ax[0,2])
ax[0,2].set_title("Unbalanced Test TPR")
sns.barplot(x="fallout", y="model", data=df_eval.sort_values(by="recall"), ax
=ax[0,3])
ax[0,3].set_title("Unbalanced Test FPR")
sns.barplot(x="bal_acc_train", y="model", data=df_eval_smote.sort_values(by="
recall"), ax=ax[1,0])
ax[1,0].set_title("Smote Model Train Acc")
sns.barplot(x="bal_acc_test", y="model", data=df_eval_smote.sort_values(by="r
ecall"), ax=ax[1,1])
ax[1,1].set_title("Smote Model Test Acc")
```

```
sns.barplot(x="recall", y="model", data=df_eval_smote.sort_values(by="recall"
), ax=ax[1,2])
ax[1,2].set_title("Smote Model Test TPR")
sns.barplot(x="fallout", y="model", data=df_eval_smote.sort_values(by="recall
"), ax=ax[1,3])
ax[1,3].set_title("Smote Model Test FPR")
sns.barplot(x="bal_acc_train", y="model", data=df_eval_rus.sort_values(by="re
call"), ax=ax[2,0])
ax[2,0].set_title("RUS_Featured Model Test Acc")
sns.barplot(x="bal_acc_test", y="model", data=df_eval_rus.sort_values(by="rec
all"), ax=ax[2,1])
ax[2,1].set_title("RUS_Featured Model Test Acc")
sns.barplot(x="recall", y="model", data=df_eval_rus.sort_values(by="recall"),
 ax=ax[2,2])
ax[2,2].set_title("RUS_Featured Model Test TPR")
sns.barplot(x="fallout", y="model", data=df_eval_rus.sort_values(by="recall")
, ax=ax[2,3])
ax[2,3].set_title("RUS_Featured Model Test FPR")
plt.tight_layout()
plt.show()
plot_ROC()
```

Iteration 4: (with SMOTE and PCA)

```python
from sklearn.decomposition import PCA
pca = PCA().fit(X_train_os)
fig, ax = plt.subplots(figsize=(20,8))
xi = np.arange(0, 54, step=1)
y = np.cumsum(pca.explained_variance_ratio_[0:160:1])
plt.ylim(0.0,1.1)
plt.plot(xi, y, marker='.', linestyle='--', color='b')
plt.xlabel('Number of Components')
plt.xticks(np.arange(0, 54, step=2), rotation=90) #change from 0-
based array index to 1-based human-readable label
plt.ylabel('Cumulative variance (%)')
plt.title('The number of components needed to explain variance')
plt.axhline(y=0.95, color='r', linestyle='-')
plt.text(0.5, 0.85, '95% cut-off threshold', color = 'red', fontsize=16)
ax.grid(axis='x')
plt.show()
```

It looks like n_components = 43 is suitable for% 95 total explained variance.

```python
pca = PCA(n_components=43)
pca.fit(X_train_os)
per_var = np.round(pca.explained_variance_ratio_ * 100, 1)
labels = ['PC' + str(x) for x in range(1,len(per_var)+1)]
plt.figure(figsize=(20,6))
plt.bar(x=range(len(per_var)), height=per_var, tick_label=labels)
plt.title('Total explained variance {}'.format(np.round(sum(per_var),2)))
plt.ylabel('Explained variance in percent')
plt.xticks(rotation=90)
plt.show()
X_train_os_pca = pca.transform(X_train_os)
pd.DataFrame(X_train_os_pca)
```

The loads (loading scores) indicate "how high a variable X loads on a factor Y."

(The i-th principal components can be selected via i in pca.components_ [0]).

```python
# Top 20 columns that have the greatest impact
loading_scores = pd.Series(pca.components_[0], index=X.columns)
loading_scores.abs().sort_values(ascending=False)[:20]
```

Use algorithm

```
X_test_pca = pca.transform(X_test)
cv_acc_balance_train_pca = {}
cv_acc_balance_test_pca = {}
cv_TPR_balance_pca = {}
cv_FPR_balance_pca = {}
cv_AUC_balance_pca = {}
def plot_result_smoted_pca(model, name:str):
    model.fit(X_train_os_pca, y_train_os)
    y_pred = model.predict(X_test_pca)
    # Evaluation based on a 10-fold cross-validation
    scoring = ['balanced_accuracy', 'recall_macro']
    scores_train = cross_val_score(model, X_train_os_pca, y_train_os, cv=10,
scoring = 'balanced_accuracy')
    scores_test = cross_val_score(model, X_test_pca, y_test, cv=10, scoring =
 'balanced_accuracy')
    cv_acc_balance_train_pca[name] = round(scores_train.mean(), 4)*100  # bal
anced accuracy
    cv_acc_balance_test_pca[name] = round(scores_test.mean(), 4)*100  # balan
ced accuracy
    cv_TPR_balance_pca[name] = (confusion_matrix(y_test, y_pred)[1][1]/confus
ion_matrix(y_test, y_pred)[1].sum())*100  # recall (max)
    cv_FPR_balance_pca[name] = (confusion_matrix(y_test, y_pred)[0][1]/confus
ion_matrix(y_test, y_pred)[0].sum())*100  # fallout (min)
    # accuracy scores
    print('Average Balanced Accuracy (CV=10), Test Set:', scores_test.mean())

    print('Average Balanced Accuracy (CV=10), Training Set: ', scores_train.m
ean())
    # print classification report
    print(classification_report(y_test, y_pred, zero_division=0))
    # Plot confusion matrix
    plt.figure(figsize=(3,3))
    plot_confusion_matrix(model, X_test_pca, y_test)
    plt.show()
# Decision tree
dtc = DecisionTreeClassifier()
plot_result_smoted_pca(dtc, "dtc")
# Logistic Regression
lr = LogisticRegression()
plot_result_smoted_pca(lr, "lr")
```

```
# NearestCentroid
nc = NearestCentroid()
plot_result_smoted_pca(nc, "nc")
# # SVC
# svc = SVC()
# plot_result_smoted_pca(svc, "svc")
# Random Forest
rfc = RandomForestClassifier()
plot_result_smoted_pca(rfc, "rfc")
# Gradient Boost
gbc = GradientBoostingClassifier()
plot_result_smoted_pca(gbc, "gbc")
# Naive Bayes
nb = GaussianNB()
plot_result_smoted_pca(nb, "nb")
# kNN
knn = KNeighborsClassifier()
plot_result_smoted_pca(knn, "knn")
# XGBOOST
xgb = XGBClassifier(eval_metric = "logloss")
plot_result_smoted_pca(xgb, "xgb");
cv_AUC_balance_pca = AUC(cv_AUC_balance_pca, X_test_pca)
cv_AUC_balance_pca
df_eval_smote_pca = pd.DataFrame(data={'model': list(cv_acc_balance_train_pca
.keys()),
                                        'bal_acc_train':list(cv_acc_balance_tr
ain_pca.values()),
                                        'bal_acc_test': list(cv_acc_balance_te
st_pca.values()),
                                        'recall': list(cv_TPR_balance_pca.valu
es()),
                                        'fallout':list(cv_FPR_balance_pca.valu
es()),
                                        'AUC': list(cv_AUC_rus.values())}).rou
nd(2)
df_eval_smote_pca
fig, ax = plt.subplots(4,4, figsize=(20, 16))
sns.barplot(x="bal_acc_train", y="model", data=df_eval.sort_values(by="recall
"), ax=ax[0,0])
ax[0,0].set_title("Unbalanced Train Acc")
sns.barplot(x="bal_acc_test", y="model", data=df_eval.sort_values(by="recall"
), ax=ax[0,1])
ax[0,1].set_title("Unbalanced Test Acc")
sns.barplot(x="recall", y="model", data=df_eval.sort_values(by="recall"), ax=
ax[0,2])
```

```
ax[0,2].set_title("Unbalanced Test TPR")
sns.barplot(x="fallout", y="model", data=df_eval.sort_values(by="recall"), ax
=ax[0,3])
ax[0,3].set_title("Unbalanced Test FPR")
sns.barplot(x="bal_acc_train", y="model", data=df_eval_smote.sort_values(by="
recall"), ax=ax[1,0])
ax[1,0].set_title("Smote Model Train Acc")
sns.barplot(x="bal_acc_test", y="model", data=df_eval_smote.sort_values(by="r
ecall"), ax=ax[1,1])
ax[1,1].set_title("Smote Model Test Acc")
sns.barplot(x="recall", y="model", data=df_eval_smote.sort_values(by="recall"
), ax=ax[1,2])
ax[1,2].set_title("Smote Model Test TPR")
sns.barplot(x="fallout", y="model", data=df_eval_smote.sort_values(by="recall
"), ax=ax[1,3])
ax[1,3].set_title("Smote Model Test FPR")
sns.barplot(x="bal_acc_train", y="model", data=df_eval_rus.sort_values(by="re
call"), ax=ax[2,0])
ax[2,0].set_title("RUS_Featured Model Test Acc")
sns.barplot(x="bal_acc_test", y="model", data=df_eval_rus.sort_values(by="rec
all"), ax=ax[2,1])
ax[2,1].set_title("RUS_Featured Model Test Acc")
sns.barplot(x="recall", y="model", data=df_eval_rus.sort_values(by="recall"),
 ax=ax[2,2])
ax[2,2].set_title("RUS_Featured Model Test TPR")
sns.barplot(x="fallout", y="model", data=df_eval_rus.sort_values(by="recall")
, ax=ax[2,3])
ax[2,3].set_title("RUS_Featured Model Test FPR")
sns.barplot(x="bal_acc_train", y="model", data=df_eval_smote_pca.sort_values(
by="recall"), ax=ax[3,0])
ax[3,0].set_title("Smoted_PCA Model Train Acc")
sns.barplot(x="bal_acc_test", y="model", data=df_eval_smote_pca.sort_values(b
y="recall"), ax=ax[3,1])
ax[3,1].set_title("Smoted_PCA Model Test Acc")
sns.barplot(x="recall", y="model", data=df_eval_smote_pca.sort_values(by="rec
all"), ax=ax[3,2])
ax[3,2].set_title("Smoted_PCA Model Test TPR")
sns.barplot(x="fallout", y="model", data=df_eval_smote_pca.sort_values(by="re
call"), ax=ax[3,3])
ax[3,3].set_title("Smoted_PCA Model Test FPR")
plt.tight_layout()
plt.show()
plot_ROC(X_test_pca)
```

According to SMOTE and PCA, none of the models really gave relatively good results.

```
df_eval["type"] = "Unbalanced"
df_eval_smote["type"] = "Smote"
df_eval_rus["type"] = "RUS"
df_eval_smote_pca["type"] = "Smote_PCA"
frames = [df_eval, df_eval_smote, df_eval_rus, df_eval_smote_pca]
df_result = pd.concat(frames, ignore_index=True)
df_result['model'] = df_result['model'].str.upper()
df_result[["recall", "fallout", "bal_acc_train", "bal_acc_test",'AUC']] = df_
result[["recall", "fallout",  "bal_acc_train", "bal_acc_test",'AUC']].apply(l
ambda x: np.round(x, 2))
df_result
sns.relplot(x="recall", y="AUC", hue="model", size="bal_acc_test",
            sizes=(40, 400), col="type", alpha=1, palette="bright", height=4,
 legend='full', data=df_result)
```

- In this plot, it looks like GradientBoosting in Smote_PCA has the best scores. But There is an overfitting there, GradientBoosting in RUS (Random Under Sampling) is better. There is no overfitting. Recall:57.27, AUC:0.62, F1:58
- In the last iteration, we will make hyperparameter optimization with GradientBoosting in RUS. We try to reach a better score.

Iteration 5: (with RUS and hyperparameter optimization)

At the end of four iterations, GradientBoost with only undersampled and scaled dataset gave better results. In this iteration, we try to improve the GradientBoost Model with hyperparameter optimization.

```
# Gradient Boosting Classifier
from sklearn.ensemble import GradientBoostingClassifier
under_sampler = RandomUnderSampler(random_state=42)
X_train_rus, y_train_rus = under_sampler.fit_sample(X_train, y_train)
params={"learning_rate": [1],
      "min_samples_split": [50, 10, 2],
       "min_samples_leaf": [1, 5, 10],
       "max_depth":[3,4,5],
       "subsample":[0.5, 1.0],
       "n_estimators":[10, 50, 100],
       "random_state":[42]}
gbc_tunned = GridSearchCV(GradientBoostingClassifier(),
                            params,
                            n_jobs=-1,
                            verbose=2,
                            ).fit(X_train_rus, y_train_rus)
from sklearn.metrics import plot_confusion_matrix, classification_report, con
fusion_matrix
print(gbc_tunned.best_estimator_)
y_pred = gbc_tunned.predict(X_test)
# AUC Score
print('AUC:', roc_auc_score(y_test, gbc_tunned.predict_proba(X_test)[:,1]))

# print classification report
print(classification_report(y_test, y_pred, zero_division=0))
# Plot confusion matrix
plt.figure(figsize=(3,3))
plot_confusion_matrix(gbc_tunned, X_test, y_test)
plt.show()
```

The tunned GradientBoost Model did not give a better result.

3.1.9 Implementation summary

- In this project, the diabetic_data.csv dataset was classified using machine learning algorithms with five iterations. With each iteration, one attempted to improve the model outcome incrementally.
- Eight different algorithms were utilized (DecisionTree, Logistic Regression, Random Forest, Gradient Boost, NaiveBayes, Nearest Centroid, XGBOOST, and kNearestNeigbour).
- Because there were numerous large and small features, the dataset was scaled with StandartScaler after the data cleaning and EDA procedure. Then, in each cycle, something unusual (oversampling, feature selection, feature extraction, hyperparameter tuning) was applied.
- In the end, GradientBoost performed better with only undersampled and scaled datasets.

4. Discussion

AI-based hospital readmission forecasting is a developing area of interest and research in healthcare. The capacity to effectively identify which patients are most likely to require readmission can have a significant impact on healthcare outcomes and resource allocation. The ability of AI to evaluate and interpret vast amounts of patient data is one of the primary benefits of employing AI for hospital readmission predictions. Data types that AI algorithms can process include EHRs, patient demographics, medical history, lab results, and even societal determinants of health. AI models can detect patterns, risk factors, and indications that contribute to hospital readmission by integrating and analyzing various disparate data sources.

In hospital readmission predictions, AI models can use a variety of approaches such as machine learning, deep learning, and natural language processing. These algorithms can discover significant traits and make reliable predictions using historical patient data. They can identify patients who are at increased risk of readmission by capturing complicated interactions between patient features, clinical variables, and healthcare interventions. Such insights can assist healthcare practitioners in intervening early, implementing preventive measures, and effectively allocating resources to lower readmission rates.

Furthermore, AI can help with personalized risk prediction by taking specific patient features and medical histories into account. Healthcare professionals can build tailored interventions and care plans by adapting forecasts to particular patient profiles, improving patient outcomes, and lowering the likelihood of readmission.

5. Conclusion

AI-powered hospital readmission forecasting has emerged as a viable strategy for improving patient care, lowering healthcare costs, and optimizing resource allocation. AI approaches, such as machine learning and deep learning algorithms, have been used to forecast hospital readmissions in areas such as risk prediction, feature selection, and decision support. These methods use large-scale and heterogeneous datasets to uncover hidden patterns and relationships in patient data, allowing for reliable forecasts of readmission risk. AI models highlight key elements leading to readmission risk and enable focused interventions for improved patient outcomes by offering decision support to healthcare practitioners. We looked at the applications, benefits, problems, and future directions of applying AI in hospital readmission forecasting in this chapter.

There are various advantages of using AI in hospital readmission prediction. AI algorithms are capable of handling complicated and diverse data sources, including clinical data, claims data, genetic information, and social determinants of health. This integration enhances the accuracy of readmission-predicting models while also providing a thorough grasp of patient risk profiles. Furthermore, AI models can continuously learn and adapt to new data, allowing for real-time updates and improved forecasting ability.

Efforts should be directed in the future toward the establishment of standardized data protocols, the implementation of explainable AI models, and the introduction of AI into clinical workflows and decision support systems. Collaboration among healthcare organizations, data scientists, and policymakers is required to address the hurdles and fully realize the potential of AI in hospital readmission predictions.

Finally, by employing advanced algorithms and data analysis, AI has shown considerable potential in hospital readmission prediction. Healthcare systems may proactively identify patients at risk of readmission, execute tailored interventions, and ultimately improve patient outcomes while optimizing budget allocation by leveraging the power of AI.

References

Afrash, M. R., Kazemi-Arpanahi, H., Shanbehzadeh, M., Nopour, R., & Mirbagheri, E. (2022). Predicting hospital readmission risk in patients with COVID-19: A machine learning approach. *Informatics in Medicine Unlocked, 30*, 100908. https://doi.org/10.1016/j.imu.2022.100908

Alajmani, S., & Elazhary, H. (2019). Hospital readmission prediction using machine learning techniques. *International Journal of Advanced Computer Science and Applications, 10*(4).

Amritphale, A., Chatterjee, R., Chatterjee, S., Amritphale, N., Rahnavard, A., Awan, G. M., Omar, B., & Fonarow, G. C. (2021). Predictors of 30-day unplanned readmission after carotid artery stenting using artificial intelligence. *Advances in Therapy, 38*, 2954—2972.

Chen, T., Madanian, S., Airehrour, D., & Cherrington, M. (2022). Machine learning methods for hospital readmission prediction: Systematic analysis of literature. *Journal of Reliable Intelligent Environments, 8*(1), 49—66.

Chopra, C., Sinha, S., Jaroli, S., Shukla, A., & Maheshwari, S. (2017). *Recurrent Neural Networks with Non-sequential Data to Predict Hospital Readmission of Diabetic Patients*, 18—23.

Du, G., Zhang, J., Li, S., & Li, C. (2021). Learning from class-imbalance and heterogeneous data for 30-day hospital readmission. *Neurocomputing, 420*, 27—35.

Hammoudeh, A., Al-Naymat, G., Ghannam, I., & Obied, N. (2018). Predicting hospital readmission among diabetics using deep learning. In *The 9th international conference on emerging ubiquitous systems and pervasive networks (EUSPN-2018)/the 8th international conference on current and future trends of information and communication technologies in healthcare (ICTH-2018)/affiliated workshops* (pp. 484—489). https://doi.org/10.1016/j.procs.2018.10.138, 141.

Hilton, C. B., Milinovich, A., Felix, C., Vakharia, N., Crone, T., Donovan, C., Proctor, A., & Nazha, A. (2020). Personalized predictions of patient outcomes during and after hospitalization using artificial intelligence. *NPJ Digital Medicine, 3*(1), 51.

Huang, Y., Talwar, A., Chatterjee, S., & Aparasu, R. R. (2021). Application of machine learning in predicting hospital readmissions: A scoping review of the literature. *BMC Medical Research Methodology, 21*(1), 1—14.

Wang, H., Cui, Z., Chen, Y., Avidan, M., Abdallah, A. B., & Kronzer, A. (2018). Predicting hospital readmission via cost-sensitive deep learning. *IEEE/ACM Transactions on Computational Biology and Bioinformatics, 15*(6), 1968—1978. https://doi.org/10.1109/TCBB.2018.2827029

Jamei, M., Nisnevich, A., Wetchler, E., Sudat, S., & Liu, E. (2017). Predicting all-cause risk of 30-day hospital readmission using artificial neural networks. *PloS One, 12*(7), e0181173.

Jiang, S., Chin, K.-S., Qu, G., & Tsui, K. L. (2018). An integrated machine learning framework for hospital readmission prediction. *Knowledge-Based Systems, 146*, 73—90.

Jovanovic, M., Radovanovic, S., Vukicevic, M., Van Poucke, S., & Delibasic, B. (2016). Building interpretable predictive models for pediatric hospital readmission using Tree-Lasso logistic regression. *Artificial Intelligence in Medicine, 72*, 12—21. https://doi.org/10.1016/j.artmed.2016.07.003

Liu, W., Stansbury, C., Singh, K., Ryan, A. M., Sukul, D., Mahmoudi, E., Waljee, A., Zhu, J., & Nallamothu, B. K. (2020). Predicting 30-day hospital readmissions using artificial neural networks with medical code embedding. *PloS One, 15*(4), e0221606.

Lu, H., & Uddin, S. (2022). Explainable stacking-based model for predicting hospital readmission for diabetic patients. *Information, 13*(9), 436.

Michailidis, P., Dimitriadou, A., Papadimitriou, T., & Gogas, P. (2022). Forecasting hospital readmissions with machine learning. *Pubmed, 10*(6), 981.

Min, X., Yu, B., & Wang, F. (2019). Predictive modeling of the hospital readmission risk from patients' claims data using machine learning: A case study on COPD. *Scientific Reports, 9*(1), 2362.

Mišić, V. V., Gabel, E., Hofer, I., Rajaram, K., & Mahajan, A. (2020). Machine learning prediction of postoperative emergency department hospital readmission. *Anesthesiology, 132*(5), 968—980.

Phu, C. L. D., & Wang, D. (2021). A comparison of machine learning methods to predict hospital readmission of diabetic patient. *Asia Proceedings of Social Sciences, 7*(2), 164—168.

Raza, S. (2022). A machine learning model for predicting, diagnosing, and mitigating health disparities in hospital readmission. *Healthcare Analytics, 2*, 100100. https://doi.org/10.1016/j.health.2022.100100

Romero-Brufau, S., Wyatt, K. D., Boyum, P., Mickelson, M., Moore, M., & Cognetta-Rieke, C. (2020). Implementation of artificial intelligence-based clinical decision support to reduce hospital readmissions at a regional hospital. *Applied Clinical Informatics, 11*(04), 570—577.

Reddy, S. S., Sethi, N., & Rajender, R. (2020). Evaluation of deep Belief network to predict hospital readmission of diabetic patients. In *2020 second international conference on inventive research in computing applications (ICIRCA)* (pp. 5—9). https://doi.org/10.1109/ICIRCA48905.2020.9182800

Sushmita, S., Khulbe, G., Hasan, A., Newman, S., Ravindra, P., Roy, S. B., De Cock, M., & Teredesai, A. (2016). Predicting 30-day risk and cost of" all-cause" hospital readmissions. In *Workshops at the thirtieth AAAI conference on artificial intelligence*.

Tey, S.-F., Liu, C.-F., Chien, T.-W., Hsu, C.-W., Chan, K.-C., Chen, C.-J., Cheng, T.-J., & Wu, W.-S. (2021). Predicting the 14-day hospital readmission of patients with pneumonia using artificial neural networks (ANN). *International Journal of Environmental Research and Public Health*, *18*(10), 5110.

Turgeman, L., & May, J. H. (2016). A mixed-ensemble model for hospital readmission. *Artificial Intelligence in Medicine*, *72*, 72–82. https://doi.org/10.1016/j.artmed.2016.08.005

Xue, Y., Klabjan, D., & Luo, Y. (2019). Predicting ICU readmission using grouped physiological and medication trends. *Artificial Intelligence in Medicine*, *95*, 27–37. https://doi.org/10.1016/j.artmed.2018.08.004

Further reading

Guo, A., Pasque, M., Loh, F., Mann, D. L., & Payne, P. R. O. (2020). Heart failure diagnosis, readmission, and mortality prediction using machine learning and artificial intelligence models. *Current Epidemiology Reports*, *7*(4), 212–219. https://doi.org/10.1007/s40471-020-00259-w

Index

Note: 'Page numbers followed by *f* indicate figures and *t* indicate tables.'

Printed and bound by CPI Group (UK) Ltd, Croydon, CR0 4YY

21/06/2024

01013844-0018